PRINCIPLES AND PRACTICE

Springer

Berlin
Heidelberg
New York
Hong Kong
London
Milan
Paris
Tokyo

Roza Maria Kamp
Juan J. Calvete
Theodora Choli-Papadopoulou (Eds.)

Methods in Proteome and Protein Analysis

With 144 Figures, 19 in Color

Springer

Professor Dr. Roza Maria Kamp
Technische Fachhochschule Berlin
Seestraße 65
13347 Berlin
Germany
email: kamp@tfh-berlin.de

Dr. Juan J. Calvete
Instituto de Biomedicina de Valencia, C.S.I.C.
Jaime Roig 11
46010 Valencia
Spain
email: jcalvete@ibv.csic.es

Ass. Prof. Dr. Theodora Choli-Papadopoulou
Aristotle University of Thessaloniki
School of Chemistry, Laboratory of Biochemistry
Thessaloniki 54006
Greece
email: TCHOLI@CHEM.AUTH.GR

Selected papers presented at the 14th meeting on Methods in Protein Structural Analysis (MPSA), September 2002, Valencia, Spain

ISBN 3-540-20222-6 Springer-Verlag Berlin Heidelberg New York

Library of Congress Cataloging-in-Publication Data.

Methods in proteome and protein analysis / Roza Maria Kamp, Juan J. Calvete, Theodora Choli-Papadopoulou (Eds).
p. cm. – (Principles and practice)
Includes bibliographical references and index.
ISBN 3-540-20222-6 (alk. paper)
1. Proteins–Analysis–Congresses. 2. Proteomics–Congresses. I. Kamp, R. M. (Roza Maria), 1951- II. Calvete, Juan J. III. Choli-Papadopoulou, T. (Theodora), 1956- IV. Series.

QP551.M399 2004
572'.6–dc22 2003066408

Springer-Verlag Berlin Heidelberg New York
Springer-Verlag is a part of Springer Science+Business Media

springeronline.com

Printed in Germany

Production and typesetting: Friedmut Kröner, 69115 Heidelberg, Germany
Cover design: *design & production* GmbH, 69126 Heidelberg, Germany

31/3150 YK – 5 4 3 2 1 0 – Printed on acid free paper

Preface

The 14th meeting on Methods in Protein Structural Analysis (MPSA) was held in Valencia (Spain) September 8–12, 2002. Approximately 200 researchers from more than two dozen countries, representing both the academic and the industrial worlds, attended the conference. The MPSA meetings began in 1974 as a small workshop organized with the aim of exchanging information on newly developed instruments and chemistry for N-terminal sequencing of polypeptides. Since then, MPSA conferences are held every two years, usually alternating between both sides of the Atlantic Ocean. Starting with the 13th conference (Charlottesville, USA, 2000), the biennial MPSA meetings are now sponsored by the International Association of Protein Structure Analysis and Proteomics (IAPSAP), a non-profit organization established in 1999 to promote the discovery and exchange of new methods and techniques for the analysis of protein structures [1]. With the "omics" revolutions in the so-called post-genomic era, the scope of the MPSA conferences has also expanded from protein sequence analysis to proteome (and protein) structure analyses. Thus, major topics of MPSA2002 included different experimental approaches (X-ray crystallography, mass spectrometry, cryo-electron microscopy, tomography) for studying very large multi-subunit molecular nanomaschines; international initiatives committed to developing high-throughput methods for large-scale protein expression and purification, sample preparation, and automatic data acquisition for structure determination by both X-ray diffraction and NMR spectroscopy; mechanisms of protein folding and misfolding in vitro and in vivo; protein–protein interactions; analysis of post-translational modifications; and the classification, structure prediction, and evolution of protein folds and functions [2]. MPSA2002 program and abstracts are available on the meeting website (http://www.mpsa2002.ibv.csic.es). This book, which contains contributions presented by speakers and papers selected from poster presentations, is published with the aim to inspire new ideas for the advancement of the rapid growing field of protein science. I am glad to note the significant contribution of Spanish (senior and young) researchers, which reflects the scientific level reached in a country where less than 1 % of its annual budget is invested in research.

As President of the Organizing Committee of MPSA2002, I wish to express special thanks to my colleagues of the Scientific Program Committee, Ettore Appella, Carl W. Anderson, and Jay W. Fox, who devoted much time to the organization of the scientific program that made MPSA2002 a very exciting meeting. The many corporate sponsors and institutions that contributed both economically and scientifically to the success of the conference are also gratefully acknowledged.

JUAN J. CALVETE Valencia, March 2003

[1] Apella, E., Fox, J.W. & Anderson, C.W. (2001) Meeting Report. Protein Sci. 10, 459–461
[2] Anderson, C.W., Calvete, J.J., Fox, J.W. & Appella, E. (2003) Meeting Report. Protein Sci. 12, 398–400

Contents

Contributors

ALBAN, ANDREW

Amersham Biosciences UK Limited, The Grove Centre, White Lion Road, Amersham, Buckinghamshire, HP7 9LL, UK

ALGOTSSON, MATTIAS

Amersham Biosciences AB, Björkgatan 30, 751 84, Uppsala, Sweden

AMARANTOS, IOANNIS

University of Patras, School of Medicine, Laboratory of Biochemistry, 26500 Patras, Greece

ANTOKU, SUSUMU

Department of Genetics and Developmental Biology, University of Connecticut Health Center, 263 Farmington Avenue, Farmington, Connecticut 06030-3301, USA

APWEILER, ROLF

The EMBL Outstation, European Bioinformatics Institute, Wellcome Trust Genome Campus, Hinxton, Cambridge, CB10 1SD, UK

AVILÉS, FRANCESC X.

(e-mail: fxaviles@einstein.uab.es, Tel.: +34-93-5811315,
Fax: +34-93-5812011)
Institut de Biotecnologia i de Biomedicina, Departament de Bioquímica, Universitat Autonoma de Barcelona, 08193 Bellaterra, Barcelona, Spain

BAUDOT, ANAÏS

Laboratoire de Génétique et Physiologie du Développement, Institut de Biologie du Développement de Marseille, Marseille, France

Bench, Graham

Center for Accelerator Mass Spectrometry, Lawrence Livermore National Laboratory; Livermore, California 94551, USA

Bhikhabhai, Rama

(e-mail: rama.bhikhabhai@amersham.com, Tel.: +46-18-6120000, Fax: +46-18-6121844)
Amersham Biosciences AB, Björkgatan 30, 751 84, Uppsala, Sweden

Brun, Christine

Laboratoire de Génétique et Physiologie du Développement, Institut de Biologie du Développement de Marseille, Marseille, France

Burgess, Antony W.

The Ludwig Institute for Cancer Research, Melbourne Tumour Biology Branch, P.O. Royal Melbourne Hospital, Parkville, Victoria 3050, Australia, and The CRC for Cellular Growth Factors, Parkville, Victoria, Australia

Calvete, Juan J.

(e-mail: jcalvete@ibv.csic.es); Instituto de Biomedicina de Valencia, CSIC, Jaime Roig 11, 46010 Valencia, Spain

Carazo, José María

Centro Nacional de Biotecnología, Campus Universidad Autónoma, Cantoblanco, 28049 Madrid, Spain

Carlsson, Ulrika

Amersham Biosciences AB, Björkgatan 30, 751 84, Uppsala, Sweden

Catimel, Bruno

The Ludwig Institute for Cancer Research, Melbourne Tumour Biology Branch, P.O. Royal Melbourne Hospital, Parkville, Victoria 3050, Australia

Cavada, Benildo S.

BioMol-Lab, Departamento de Bioquímica e Biologia Molecular, Universida de Federal de Ceará, Fortaleza, Brasil

Ceccatto, Vania M.

BioMol-Lab, Departamento de Bioquímica e Biologia Molecular, Universida de Federal de Ceará, Fortaleza, Brasil

Chagoyen, Mónica

Centro Nacional de Biotecnología, Campus Universidad Autónoma, Cantoblanco, 28049 Madrid, Spain

Choli-Papadopoulou, Theodora

(e-mail: tcholi@chem.auth.gr, Tel.: +30-23-10997806, Fax: +-30-23-10997689)
Laboratory of Biochemistry, School of Chemistry, Aristotle University of Thessaloniki, 54124 Thessaloniki, Greece

Chow, Nicole L.

Genencor International Inc., 925 Page Mill Road, Palo Alto, California 94304, USA

de-Alarcón, Pedro Antonio

Centro Nacional de Biotecnología, Campus Universidad Autónoma, Cantoblanco, 28049 Madrid, Spain

Gallego Del Sol, Francisca

Instituto de Biomedicina de Valencia, CSIC, Jaime Roig 11, 46010 Valencia, Spain

DiGruccio, Micheal

The Salk Institute, 10010 N. Torrey Pines Road, La Jolla, California 92037, USA

Donaire, Antonio

(e-mail: adonaire@umh.es, Tel.: +34-96-6658942, Fax: +34-96-6658758)
Instituto de Biología Molecular y Celular, Universidad Miguel Hernández, Edificio Torregaitán, Avda. Ferrocarril s/n, 03202-Elche, Alicante, Spain

Elortza, Felix

Protein Research Group, Department of Biochemistry and Molecular Biology, University of Southern Denmark, 5230 Odense M, Denmark

Fischer, Wolfgang H.

(e-mail: Fischer@salk.edu, Tel.: +85-84-53 4100)
The Salk Institute, 10010 N. Torrey Pines Road, La Jolla, California 92037, USA

Flensburg, John

Amersham Biosciences AB, Björkgatan 30, 751 84, Uppsala, Sweden

FOSTER, LEONARD J.

Protein Research Group, Department of Biochemistry and Molecular Biology, University of Southern Denmark, 5230 Odense M, Denmark

FRYKSDALE, BETH G.

Genencor International, Inc., 925 Page Mill Road, Palo Alto, California 94304, USA

GAERTNER, ALFRED L.

(e-mail: agaertner@genecor.com)
Genencor International, Inc., 925 Page Mill Road, Palo Alto, California 94304, USA

GARRETT, THOMAS P.J.

The CRC for Cellular Growth Factors, Parkville, Victoria, Australia, and The Walter and Eliza Hall Institute of Medical Research, Parkville, Victoria, Australia

GÓMEZ, MANUEL J.

(e-mail: mjgommo@cnb.uam.es)
Centro Nacional de Biotecnología, CSIC, Campus UAM, Cantoblanco 28049 Madrid, Spain

GÓMEZ-PUERTAS, PAULINO

(e-mail: pagomez@cnb.uam.es)
Centro de Astrobiología, CSIC-INTA. Ctra. Torrejón – Ajalvir, Km 4. Torrejón de Ardoz, 28850 Madrid, Spain

GRANGEIRO, THALLES B.

Departamento de Biologia, Universidade Federal de Ceará, Fortaleza, Brasil

GRANT, PATRICK G.

Center for Accelerator Mass Spectrometry, Lawrence Livermore National Laboratory; Livermore, California 94551, USA

GUÉNOCHE, ALAIN

Institut de Mathématiques de Luminy, Parc Scientifique de Luminy, Case 907, 13288 Marseille Cedex 9, France

HALL, NATHAN

The Ludwig Institute for Cancer Research, Melbourne Tumour Biology Branch, P.O. Royal Melbourne Hospital, Parkville, Victoria 3050 Australia, and The CRC for Cellular Growth Factors, Parkville, Victoria, Australia

HAWKINS, EDWARD

(e-mail: edward.hawkins@amersham.com)
Amersham Biosciences UK Limited, The Grove Centre, White Lion Road, Amersham, Buckinghamshire, HP7 9LL, UK

HILLEGONDS, DARREN J.

Center for Accelerator Mass Spectrometry, Lawrence Livermore National Laboratory; Livermore, California 94551, USA

HIRANO, HISASHI

(e-mail: hirano@yokohama-cu.ac.jp, Tel./Fax: +81-45-8201901)
Yokohama City University, Maioka-cho 641–12, Totsuka-ku, Yokohama, 2440813, Japan

HJERNØ, KARIN

Department of Biochemistry and Molecular Biology, University of Southern Denmark, Campusvej 55, 5230 Odense M, Denmark

HØJRUP, PETER

(e-mail: php@bmb.sdu.dk, Tel.: +45-65-502371, Fax: +45-65-502467)
Department of Biochemistry and Molecular Biology, University of Southern Denmark, Campusvej 55, 5230 Odense M, Denmark

HÖRNSTEN, LENA

Amersham Biosciences AB, Björkgatan 30, 751 84, Uppsala, Sweden

HUGHES, BARRY

Amersham Biosciences UK Limited, The Grove Centre, White Lion Road, Amersham, Buckinghamshire, HP7 9LL, UK

ISLAM, NAZRUL

Yokohama City University, Maioka-cho 641–12, Totsuka-ku, Yokohama, 2440813, Japan

JACQ, BERNARD

(e-mail: jacq@lgpd.univ-mrs.fr)
Laboratoire de Génétique et Physiologie du Développement, Institut de Biologie du Développement de Marseille, Marseille, France

JEDRZEJEWSKI, PAUL

Genencor International, Inc., 925 Page Mill Road ,Palo Alto, California, 94304, USA

JENSEN, OLE N.

(e-mail: jenseno@bmb.sdu.dk, URL: www.protein.sdu.dk, Tel.: +45 -65-502368, Fax: +45-65-502467)
Protein Research Group, Department of Biochemistry and Molecular Biology, University of Southern Denmark, 5230 Odense M, Denmark

JIMÉNEZ, BEATRIZ

Departamento de Química Inorgánica, Universitat de València. C/Dr. Moliner, 50, 46100-Burjassot, Valencia, Spain

JIMÉNEZ-LOZANO, NATALIA

Centro Nacional de Biotecnología, Campus Universidad Autómona, Cantoblanco, 28049 Madrid, Spain

JIN, HONG

Department of Biotechnology, University of Applied Sciences, Seestrasse 64, 13347 Berlin, Germany

KALPAXIS, DIMITRIOS L.

(e-mail: Dimkal@med.upatras.gr, Tel.: +30-26-10996124, Fax: +30-26-10997690)
University of Patras, School of Medicine, Laboratory of Biochemistry, 26500 Patras, Greece

KAMP, ROZA MARIA

(e-mail: kamp@tfh-berlin.de, Tel.: +49-30-45043923, Fax: +49-30-45043959)
Department of Biotechnology, University of Applied Sciences, Seestrasse 64, 13347 Berlin, Germany

KANG, JAMES

The Salk Institute, 10010 N. Torrey Pines Road, La Jolla, California 92037, USA

KARAS, MICHAEL

ConSequence GmbH, Potsdamer Strasse 18a, 14513 Teltow, Germany and Institute of Pharmaceutical Chemistry, Frankfurt, Germany

KIRBY, DEAN

The Salk Institute, 10010 N.Torrey Pines Road, La Jolla, California 92037, USA

KOTTAKIS, FILIPPOS

Laboratory of Biochemistry, School of Chemistry, Aristotle University of Thessaloniki, 54124 Thessaloniki, Greece

KOUIDOU, SOFIA

Laboratory of Biological Chemistry, Department of Medicine, Aristotle University of Thessaloniki, 54124 Thessaloniki, Greece

KRELL, TINO

(e-mail: tino.krell@aventis.com, Tel.: +33-43-7379012, Fax: +33-43-7373180)
Aventis Pasteur, 1541 avenue Marcel Mérieux, 69280 Marcy l'Etoile, France

KRÜGER, RALF

Institute of Pharmaceutical Chemistry, Frankfurt, Germany

LARSSON, CAMILLA

Amersham Biosciences AB, Björkgatan 30, 751 84, Uppsala, Sweden

LEONTIADOU, FOTINI

Laboratory of Biochemistry, School of Chemistry, University of Thessaloniki, 54006 Thessaloniki, Greece

LILLO, MARIA PILAR

(e-mail: pilar.lillo@iqfr.csic.es,Tel.: +34-91-5619400. Fax: +34-91-5642431)
Department of Biophysics. Instituto Química Física Rocasolano, IQFR, CSIC, Serrano 119, 28006 Madrid, Spain

LIMINGA, MARIA

Amersham Biosciences AB, Björkgatan 30, 751 84, Uppsala, Sweden

LÓPEZ-ROMERO, PEDRO

(e-mail: plromero@cnb.uam.es)
Centro Nacional de Biotecnología, CSIC, Campus UAM, Cantoblanco 28049 Madrid, Spain

LOW, WILLIAM

The Salk Institute, 10010 N. Torrey Pines Road, La Jolla, California 92037, USA

Maloisel, Jean-Luc

Amersham Biosciences AB, Björkgatan 30, 751 84, Uppsala, Sweden

Matragkou, Christina

Laboratory of Biochemistry, School of Chemistry, Aristotle University of Thessaloniki, 54124 Thessaloniki, Greece

Mayer, Bruce J.

(e-mail: bmayern@neuron.uchc.edu, Tel.: +86-06-791836, Fax: t+86-06-798345)
Department of Genetics and Developmental Biology, University of Connecticut Health Center, 263 Farmington Avenue, Farmington, Connecticut 06030–3301, USA

McKern, Neil M.

The CRC for Cellular Growth Factors, Parkville, Victoria, Australia, and CSIRO Health Sciences and Nutrition, Parkville, Victoria, Australia

Miller, Brian S.

Genencor International, Inc., 925 Page Mill Road, Palo Alto, California 94304, USA

Mingarro, Ismael

(e-mail: Ismael.Mingarro@uv.es, Tel.: +34-96-3543796, Fax: +34-96-3544635)
Departament de Bioquímica i Biologia Molecular, Universitat de València, 46 100 Burjassot, Spain

Moratal, José María

Departamento de Química Inorgánica. Universitat de València. C/Dr. Moliner, 50. 46100-Burjassot, Valencia, Spain

Moreno, Frederico B.M.B.

BioMol-Lab, Departamento de Bioquímica e Biologia Molecular, Universida de Federal de Ceará, Fortaleza, Brasil

Mulder, Nicola Jane

(e-mail: mulder@ebi.ac.uk, Tel.: +44-12-23494602, Fax: +44-12-23494468)
The EMBL Outstation, European Bioinformatics Institute, Wellcome Trust Genome Campus, Hinxton, Cambridge, CB10 1SD, UK

NAGANO, CELSO S.

Departamento de Engenharia de Pesca, Universidade Federal de Ceará, Fortaleza, Brasil

NICE, EDOUARD C.

(e-mail: Ed.Nice@ludwig.edu.au, Tel.: + 61-39-3413135, Fax: +61-39-3413104)
The Ludwig Institute for Cancer Research, Melbourne Tumour Biology Branch, P.O. Royal Melbourne Hospital, Parkville, Victoria 3050, Australia and The CRC for Cellular Growth Factors, Parkville, Victoria, Australia

ORZÁEZ, MAR

Departament de Bioquímica i Biologia Molecular, Universitat de València, 46 100 Burjassot, Spain

PAECH, SIGRID

Genencor International, Inc., 925 Page Mill Road, Palo Alto, California 94304, USA

PALMBLAD, MAGNUS

Center for Accelerator Mass Spectrometry, Lawrence Livermore National Laboratory, Livermore, California 94551, USA

PALMGREN, RONNIE

Amersham Biosciences AB, Björkgatan 30, 751 84, Uppsala, Sweden

PÉREZ-PAYÁ, ENRIQUE

Departament de Bioquímica i Biologia Molecular, Universitat de València, 46 100 Burjassot, Spain

PERRIN, MARILYN

The Salk Institute, 10010 N. Torrey Pines Road, La Jolla, California 92037, USA

PESULA, MARI-ANN

Amersham Biosciences AB, Björkgatan 30, 751 84, Uppsala, Sweden

PICCIOLI, MARIO

Department of Chemistry and CERM, University of Florence, Via L. Sacconi, 6–50019 Sesto, Fiorentino, Italy

PRIME, JOHN

Amersham Biosciences UK Limited, The Grove Centre, White Lion Road, Amersham, Buckinghamshire, HP7 9LL, UK

PRUESS, MANUELA

The EMBL Outstation, European Bioinformatics Institute, Wellcome Trust Genome Campus, Hinxton, Cambridge, CB10 1SD, UK

QUEROL, ENRIQUE

Institut de Biotecnologia i de Biomedicina, and Departament de Bioquímica, Universitat Autonoma de Barcelona, 08193 Bellaterra, Barcelona, Spain

RADCKE, CHRISTOPH

WITA GmbH, Teltow, Germany

RENAULD-MONGÉNIE, GENEVIÈVE

Aventis Pasteur, 1541 avenue Marcel Mérieux, 69280 Marcy l'Etoile, France

RIVAS, GERMAN

(e-mail: grivas@cib.csic.es)
Department of Structure and Function of Proteins, Centro de Investigaciones Biológicas, CIB, SCIC, Ramiro de Maeztu 9, 28040 Madrid, Spain

ROTHACKER, JULIE A.

The Ludwig Institute for Cancer Research, Melbourne Tumour Biology Branch, P.O. Royal Melbourne Hospital, Parkville, Victoria 3050, Australia and The CRC for Cellular Growth Factors, Parkville, Victoria, Australia

SAMPAIO, ALEXANDRE H.

Departamento de Engenharia de Pesca, Universidade Federal de Ceará, Fortaleza, Brasil

SERRANO, LUIS

European Molecular Biology Laboratory (EMBL), Meyerhofstrasse 1, 69117, Heidelberg, Germany

SHARMA, ALOK

Department of Genetics and Developmental Biology, University of Connecticut Health Center, 263 Farmington Avenue, Farmington, Connecticut 06030-3301, USA

Stensballe, Allan

Protein Research Group, Department of Biochemistry and Molecular Biology, University of Southern Denmark, 5230 Odense M, Denmark

Taberner, Francisco J.

Departament de Bioquímica i Biologia Molecular, Universitat de València, 46 100 Burjassot, Spain

Tsiftsoglou, Asterios S.

Laboratory of Pharmacology, Department of Pharmaceutical Sciences, Aristotle University of Thessaloniki, 54124, Thessaloniki, Greece

Valencia, Alfonso

(e-mail: valencia@cnb.uam.es, Tel.: +34-915854669, Fax: +34-915854506)
Centro Nacional de Biotecnología, CSIC, Campus UAM, Cantoblanco
28049 Madrid, Spain

Van Regenmortel, Marc H.V.

(e-mail: vanregen@esbs.u-strasbg.fr, Tel.: +33-39-0244812,
Fax: +33-39-0244811)
Ecole Supérieure de Biotechnologie de Strasbourg, UMR 7100, CNRS,
Boulevard Sébastien Brandt, Illkirch 67400, France

Villanueva, Josep

Institut de Biotecnologia i de Biomedicina, and Departament de Bioquímica, Universitat Autonoma de Barcelona, 08193 Bellaterra, Barcelona, Spain

Vizirianakis, Ioannis S.

Laboratory of Pharmacology, Department of Pharmaceutical Sciences, Aristotle University of Thessaloniki, 54124, Thessaloniki, Greece

Vogel, John S.

(e-mail: jsvogel@llnl.gov, Tel.: +92-5-4-234232, Fax: +92-5-4237884)
Center for AMS, Lawrence Livermore National Laboratory,
7000 East Avenue, L-397, Livermore, California 94551, USA

Ward, Colin W.

The CRC for Cellular Growth Factors, Parkville, Victoria, Australia, and CSIRO Health Sciences and Nutrition, Parkville, Victoria, Australia

Wittmann-Liebold, Brigitte

ConSequence GmbH, Potsdamer Strasse 18a, 14513 Teltow, Germany and WITA GmbH, Teltow, Germany

Wong, David L.

Genencor International, Inc., 925 Page Mill Road, Palo Alto, California 94304, USA

Wurzel, Christian

(e-mail: Wurzel@snafu.de)
ConSequence GmbH, Potsdamer Strasse 18a, 14513 Teltow, Germany

Xaplanteri, Maria A.

University of Patras, School of Medicine, Laboratory of Biochemistry, 26500 Patras, Greece

Yanes, Oscar

Institut de Biotecnologia i de Biomedicina, and Departament de Bioquímica, Universitat Autonoma de Barcelona, 08193 Bellaterra, Barcelona, Spain

Zorrilla, Silvia

(e- mail: szorrilla@pop3.iib.uam.es,Tel.: +34-91-5619400.
Fax: +34-91-5642431)
Department of Biophysics, Instituto Química Física Rocasolano, IQFR, CSIC, Serrano 119, 28006 Madrid, Spain.

zu Lynar, Barbara

ConSequence GmbH, Potsdamer Strasse 18a, 14513 Teltow, Germany

1 Helix-Helix Packing Between Transmembrane Fragments

MAR ORZÁEZ, FRANCISCO J. TABERNER, ENRIQUE PÉREZ-PAYÁ and ISMAEL MINGARRO

Abstract

The rules that govern the folding of membrane proteins are still not completely understood when compared with the well-detailed set of principles described for the folding of soluble proteins. Although the molecular determinants of the folding mechanism should be basically the same for both types of proteins, the main difference, which in turn could also represent the main difficulty, is the media that surrounds the protein. In this sense, it would be useful to develop new molecular techniques that allow for the study of the folding mechanism of membrane proteins in their natural media, the membrane. In the present work we have collected a series of studies devoted to understanding the principles that govern the molecular mechanism of folding and packing of membrane proteins, an important group of the proteome. In particular, we have focused our attention on Glycophorin A (GpA), a single span membrane protein, which has been extensively used as a model system to study helix-helix transmembrane (TM) packing. Here, we report on the importance of the molecular distance between the critical oligomerisation motif and polar residues in TM fragments. We have also given an account of the influence of the length of the TM fragment as well as the effect of proline residues on the packing of TM alpha helices.

1.1 Introduction

It is estimated that membrane proteins constitute about 30% of fully sequenced genomes– a major part of biological life (Wallin and von Heijne, 1998). These types of protein play an important role in cell function, acting as cell receptors, transporters, channels and as essential components of respiratory and photosynthetic complexes. Furthermore, membrane proteins are

Principles and Practice
Methods in Proteome and Protein Analysis
R.M. Kamp, J. J. Calvete, T. Choli-Papadopoulou (Eds.)

also implicated in essential processes inside the cell. They have attracted the interest not only of basic, but also of more applied fields like medical chemistry, due to their potential as targets for therapeutic intervention. However, due to their hydrophobic character, the fine characterisation of different aspects of the folding, membrane insertion and structure-function relationships of membrane proteins has been delayed more than expected. For example, very few membrane protein structures are known at atomic resolution (~20 out of over ~20,000 known structures), which is a major stumbling block in our understanding of how they function, how to correct their malfunctions and how to exploit them. In soluble proteins, the knowledge of protein structures paralleled with the definition of the molecular mechanisms that govern protein folding has permitted a better understanding of protein function. This has facilitated the rational design of new pharmaceuticals directed to selected target proteins. It is reasonable to believe that the same approach could be extended to membrane proteins. Thus, there is increasing interest in the definition of the molecular mechanism that drives the folding of proteins in the membrane environment, and in particular in the rules that permit TM segments to be energetically stabilised through packing interactions.

When studying membrane proteins one has to clearly realise that these proteins "*live*" in an environment completely different from the aqueous media, the membrane. The cell membrane is a highly heterogeneous media, composed mainly of phospholipids that are self-organised in two leaflets giving rise to the formation of a bilayer. The hydrocarbon core (HC) is the hydrophobic part of the membrane that is approximately 30 Å thick. The polar heads of the phospholipids define the lipid/water interphases (IF) and add 15 Å to the thickness of each leaflet. It is this complex environment, with physical and chemical properties different from aqueous media, into which membrane proteins are integrated.

All membrane protein structures solved to date show that TM domains fold as either α-helices or β-strands, due to the physical and chemical constraints imposed by the hydrophobic environment (White and Wimley, 1999). The α-helical-type proteins are most abundant and can be made up of a single helix or of multiple helices packed together in bundles. The folding of constitutive α-helical membrane proteins has been conceptualised, in its simplest form, as a two-stage process (Fig. 1.1), in which the helices are first independently formed across the membrane and then laterally assembled to form the native protein (Popot and Engelman, 1990). The formation of the individual helices is a consequence of main-chain hydrogen bonding, as the hydrophobic effect of the lipid bilayer has an influence, restraining other unordered peptide structures that would expose polar peptide bonds (reviewed in (White *et al.*, 1998)). The side-to-side helix association or protein assembly, the second stage, is driven by different interactions, like van der Waals forces, electrostatic effects, steric clashes, or differential effect of asymmetrically distributed lipids (Popot and Engelman, 2000; White and Wimley, 1999).

a)

b)

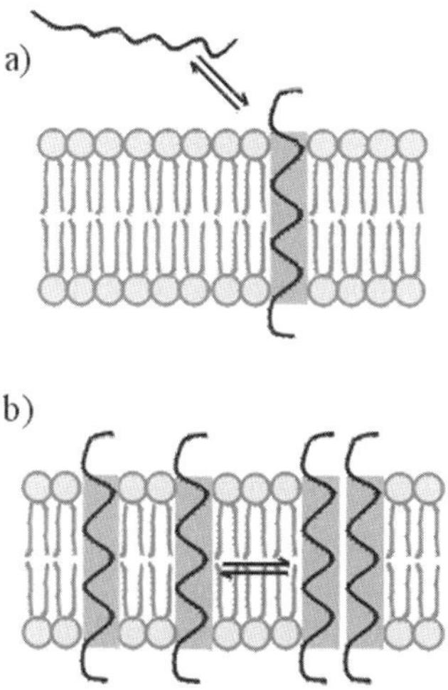

Fig. 1.1. The two-stage model. **a** The first stage, insertion of the prefolded helix in the lipid bilayer. **b** The second stage, lateral association between inserted *TM* helices to produce the folded structure

These theoretical principles have been, to some extent, demonstrated by means of membrane spanning synthetic peptides that become folded in the membrane media and by experimental systems that allow the study of membrane multispanning helices. However, the main problem is to obtain detailed information on protein structures. This is due, in part, to the loss of structural stability of the proteins outside their natural media, and on the extreme complexity that the membrane introduces when classical methods of protein structure analysis are contemplated (DeGrado et al., 2003). Thus, the development of new, or the improvement of "classical", strategies to study membrane proteins at the atomic level is one of the key areas in structural biology. In this sense, the use of well-characterised model proteins could allow the design of experimental techniques directed to the study of complex questions still to be solved regarding membrane proteins. One of the best-suited models of a membrane protein that oligomerises (more specifically, dimerises) through interactions of its TM α-helices is undoubtedly Glycophorin A (Arkin, 2002; Lemmon and Engelman, 1994).

1.2 Glycophorin A as a Model System

Glycophorin A (GpA) was the first membrane protein whose sequence was determined. A long stretch of hydrophobic amino acid residues was identified from these studies (Tomita and Marchesi, 1975). Furthermore, GpA has provided the first clear example of non-covalent membrane protein oligomerisation due to specific interactions of its TM α-helices (see Bormann and Engelman, 1992 for a review). The free energy decrease associated with the dimerisation process is enough to confer dimerisation of an unrelated water-soluble protein when fused to the TM domain of GpA to form a chimeric protein (Lemmon et al., 1992a). Moreover, the addition of a His-tag at the extreme C-terminus of this fusion for purification purposes did not perturb TM association (Mingarro et al., 1996).

The wide use of this protein as a model membrane protein is also based on its intrinsic simplicity, since its single TM fragment drives a detergent resistant homodimerisation of the protein. Thus, the dimerisation process and those factors that could affect or modify it can be analysed using SDS-PAGE. The GpA homodimer, defines a dimerisation interface that has been extensively studied by diverse techniques such as saturation mutagenesis (Lemmon et al., 1992b), alanine-insertion scanning (Mingarro et al., 1996), computational modelling (Adams et al., 1996), solution NMR in dodecylphosphocholine micelles (MacKenzie et al., 1997) and solid-state NMR in lipid membranes (Smith *et al.*, 2001). The output of these studies describes a dimerisation motif in the TM fragment composed of seven residues, $L^{75}IxxGVxxGVxxxT^{87}$, which is responsible for the dimerisation process (Fig. 1.2).

In the present study, we have focused on three major factors that have an influence on the molecular mechanism of helix-helix packing in a membrane-like environment. First, we analysed how important it is to keep a minimum distance between the dimerisation motif and the flanking charged residues on the cytoplasmic side of the protein. Secondly, we described the minimisation of the GpA dimerisation motif that allows dimer formation in homogeneous poly-leucine stretches of different length. Finally, we would like to achieve an experimental rationalisation for the observed and unexpected over-representation of proline residues in TM fragments as well as their role in membrane protein assembly.

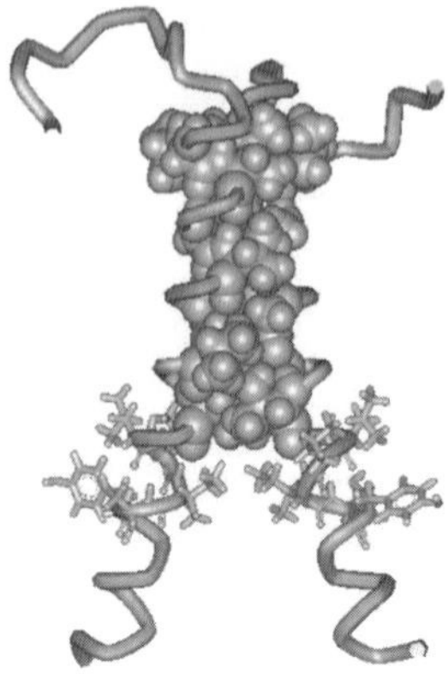

Fig. 1.2. Ribbon drawing derived from the *GpA TM* helix of the experimental *NMR* structure (PDB file 1AFO). The seven critical interface residues are shown in space-filling mode, residues Leu89-Tyr93 in ball-and-sticks mode. The drawing was generated using the program WebLab Viewer Pro 3.7 (Molecular Simulations, San Diego, California)

1.3 Influence of the Distance Between the Dimerisation Motif and the Flanking Charged Residues on the Packing Process Between TM Helices

The presence of positive charged residues at the cytoplasmic side of TM fragments is a feature of many membrane proteins. In most organisms the orientation of TM fragments in the membrane seems to be influenced by the charge distribution flanking the hydrophobic core of the TM segment. Positively charged amino acid residues often direct a charged domain to remain on the cytosolic side of the membrane (the side of protein synthesis). This experimental observation is formally known as the "*positive-inside* rule" (von Heijne, 1992).

Glycophorin A has at its C-terminus, adjacent to the TM fragment, arginine and lysine residues that corroborate the general rule. In particular, the TM fragment of GpA has a stretch of basic residues located eight residues downstream from the last amino acid of the dimerisation motif (Thr87). Although the residues included in the hydrophobic stretch (Ile88-Ile95) are not directly implicated in the net of van der Waals interactions that maintain the dimeric structure of this protein, deletions in this region (mutants Δ89/93, Δ89/91and Δ91/93, for example) abrogate dimer formation, (see Table 1.1). These last three constructs all contain a deletion of the amino acid residue Ile91, located at the contact interface of the two TM helices (Fig. 1.2). However, a mutant protein with a point deletion of this residue (Δ91) dimerised as efficiently as the wild type (Table 1.1), indicating that Ile91 is probably not specifically

Table 1. Dimerisation capacity of GpA C-terminal mutants

Mutant	Sequence	Dimer (%)
Wt	^{72}EITLIIFGVMAGVIGTILLISYGIRRLIKK^{101}	84
Δ89/93	^{72}EITLIIFGVMAGVIGTI---GIRRLIKK^{101}	0
Δ89/91	^{72}EITLIIFGVMAGVIGTI--SYGIRRLIKK^{101}	0
Δ91/93	^{72}EITLIIFGVMAGVIGTILL--GIRRLIKK^{101}	9
Δ91	^{72}EITLIIFGVMAGVIGTILL-SYGIRRLIKK^{101}	86
91–93L	^{72}EITLIIFGVMAGVIGTILLLLLGIRRLIKK^{101}	78
91–95L	^{72}EITLIIFGVMAGVIGTILLLLLLLRRLIKK^{101}	77
Δ89/91,Δ96/97	^{72}EITLIIFGVMAGVIGTI--SYGI-LIKK^{101}	2
Δ89/91,96/97L	^{72}EITLIIFGVMAGVIGTI--SYGILLLIKK^{101}	2
Δ89/91,96/97L,100/101L	^{72}EITLIIFGVMAGVIGTI--SYGILLLILL^{101}	73
Δ91,Δ96/97	^{72}EITLIIFGVMAGVIGTILL-SYGI-LIKK^{101}	0
Δ91,96/97L	^{72}EITLIIFGVMAGVIGTILL-SYGILLLIKK^{101}	45
Δ96/97	^{72}EITLIIFGVMAGVIGTILLISYGI-LIKK^{101}	69
96/97L	^{72}EITLIIFGVMAGVIGTILLISYGILLLIKK^{101}	74
96/97L,100/101L	^{72}EITLIIFGVMAGVIGTILLISYGILLLILL^{101}	55

required for dimerisation raising the possibility of an important role of the helix length in the dimerisation process.

Next, we evaluated whether an amino acid-dependent specificity event is by itself responsible for the lack of protein dimerisation in these regions. The GpA TM fragment is entirely α-helical (MacKenzie et al., 1997). In order to minimise any putative local secondary structure perturbation due to point mutation, we selected leucine as a replacement (Table 1.1) for wild-type residues in the GpA TM fragment. Leucine has a high α-helical propensity in membrane environments (Li and Deber, 1994) and it seemed unlikely that leucine mutations would cause gross secondary structure perturbation. Moreover, this residue is one of the more abundant residues in membrane environments (Arkin and Brunger, 1998; Bywater et al., 2001; Ulmschneider and Sansom, 2001). Mutants 91–93L and indeed 91–95L have an oligomerisation capacity indistinguishable from the wild type sequence, suggesting that the specific amino acid sequence is not involved in key residue interactions when the dimerisation of the TM fragment takes place. This is probably done by keeping the appropriate distance between the flanking charged residues and the dimerisation motif.

In order to probe this idea a series of new mutants were constructed. Thus, mutants Δ89/91-Δ96/97 and Δ89/91–96/97L show that deletion or substitution of R96/R97 is not enough to recuperate dimeric structure, this is probably due to the presence of the neighbouring lysines (K100/K101) in positions that could interfere with dimer formation. In fact, the mutant Δ89/91,96/97L,100/101L, in which these two last residues are substituted by leucines the capacity to dimerise was recovered. Altogether these results highlight the importance of the C-terminal fragment of the TM segment of GpA in maintaining the charged residues at an adequate distance and orientation in order to keep the interacting interface intact.

1.4 Length of the Hydrophobic Fragment and Oligomerisation Processes

The hydrocarbon core (HC) of a membrane has an average thickness of 30 Å. A polypeptide segment of 20 amino acid residues in length, when folded in an α-helical conformation, would span the length of the HC and could be defined as a classical TM fragment. However, in cell membranes there is a wide range of amino acid lengths that defines specific TM segments. It has been postulated that a TM with a shorter amino acid length would force local lipid rearrangements which could have biological relevance (Dumas et al., 1999; Killian, 1998). Furthermore, those TM fragments with an amino acid length greater than the ideal would tilt the helical axis in the membrane (Killian and von Heijne, 2000).

In the *two stage* model for folding of membrane proteins, as introduced above, the *second stage* consists of the packing between TM helices. In oligomeric TM proteins this process occurs as a consequence of the energetic balance between the loss of lipid-protein interactions and the increase in the number of lipid-lipid and protein-protein interactions (White and Wimley, 1999). The equilibrium established between the monomeric and the oligomeric forms could be displaced towards oligomer formation, depending on several factors that could modify the energetic balance. Among the factors that affect the oligomerisation processes, the mismatch between the length of the TM fragment and the membrane width is probably, although of great importance, one of the less studied. This is mainly due to the absence of direct experimental evidence that the concept of hydrophobic mismatch operates in biological membranes (Dumas et al., 1999). In addition, the wide diversity in acyl chain length and structure that lipids display in biological membranes makes it difficult to perform systematic studies.

In this scenario, we investigated the role that the length of the hydrophobic TM fragment could play on the capacity to drive dimerisation of the GpA oligomerisation motif in a membrane-mimetic environment (SDS micelles). A set of chimeric polyleucine fragments with different hydrophobic lengths were used to provide a homogeneous, but non-dimerising scaffold in SDS (Zhou et al., 2000). The capacity to induce dimerisation of the poly-leucine segments (see Table 1.2), is dependent on the length of the hydrophobic region, indicating that, in fact, the energetic balance that drives the monomer-dimer equilibrium can be displaced as a function of the hydrophobic mismatch. Insertion of the GpA dimerisation motif in hydrophobic regions of 15 leucines in length (15L) is not enough to induce dimer formation between these artificial TM helices, and also a decrease in dimerisation levels is observed for hydrophobic segments longer than 24L, showing that in both too short and too long TM fragments, the oligomerisation equilibrium is displaced towards the monomeric form of the protein. In order to check if the position of the dimerisation motif inside the hydrophobic region could perform a role in the process, we designed mutant 18Lb

Table 2. Influence of hydrophobic fragment length on the dimerisation capacity induced by the GVxxGVxxT motif

Mutant	Dimerisation percentage
$L_3GVL_2GVL_2TL_3$(15L)	0
$L_6GVL_2GVL_2TL_3$(18L)	68±4
$L_3GVL_2GVL_2TL_6$(18Lb)	46±12
$L_6GVL_2GVL_2TL_6$(21L)	68±3
$L_9GVL_2GVL_2TL_6$(24L)	64±18
$L_9GVL_2GVL_2TL_9$(27L)	31±13

(Table 1.2), in which this motif was moved approximately one helical turn upstream. We found that the18Lb construct is capable of inducing dimerisation, suggesting that the number of leucine residues at the N-terminus of the hydrophobic region is not the main reason for the observed differences in dimerisation between mutants 15L and 18L (Table 1.2), thus indicating that the length of the hydrophobic fragment is mainly responsible for losing dimerisation capacity.

These results can be explained by considering the energetics of the system. Introduction of a TM fragment within the hydrophobic media when there is a hydrophobic mismatch produces distortion on the surrounding acyl chains, this is done in order to avoid unfavourable exposure of hydrophobic protein surfaces to a hydrophilic environment. As a consequence, an increase of the entropy of the system may be produced when a positive as well as a negative mismatch occurs. In this context, oligomerisation between TM fragments produces a decrease on the surface of protein-acyl chains interactions causing a reorganisation of the media that renders a more ordered environment, that is, a decrease in entropy in the system. When there is a wide mismatch, gains in protein-protein interactions as well as improvement in protein-acyl chains interactions are not enough to compensate for the loss of entropy produced by the oligomerisation processes resulting in the observed bell-shape profile (Table 1.2). Interestingly, the same profile was found in a statistical analysis of length distribution of predicted TM α-helices, where the average length was roughly 21 hydrophobic residues for multispannig proteins, and one to two residues longer in single spanning membrane proteins (Arkin and Brunger, 1998). These lengths coincide both with the minimum length of an α-helix required to traverse a 30-Å-thick lipid bilayer (as stated above), and with the maximum dimerisation efficiency observed in the present study.

All in all, mismatching could influence behaviour of hydrophobic TM fragments on the membrane by modifying not only insertion, orientation and sorting of membrane proteins but also influencing helix-helix association and/or disassociation. This is an interesting issue in view of the increasing body of evidence for coexisting membrane domains with different lipids compositions, and (apparently) different widths.

1.5 Prolines in Transmembrane Helix Packing

Proline is among the twenty different residues that form part of proteins and is unique in some of its properties. Its lateral side chain forms a pyrrolidine ring by bonding with the amine group of its backbone. The main consequence of having this structure is that the capacity to form structures in proteins such as α-helices or β-sheets is restricted. In the case of β-sheets the phi angle is about -120° to -140°, and proline is limited to -60° to +25° (Li et al., 1996). In contrast, proline angles are not incompatible with Ramachandran diagram for α-helical structures. Nevertheless, this residue lacks the amide proton and this therefore produces a disruption of the hydrogen bonding network required to maintain helical structure. At the same time the bulkiness of the pyrrolidine ring produces steric folding problems, inducing kink formation on the helix backbone. For these reasons proline residues in soluble proteins usually form part of turns, unstructured regions or in the case of α-helices is located preferentially on the first turn, where these problems seem to be less crucial (Kim and Kang, 1999).

The fact that the presence of a proline in an α-helical structure leaves a carbonyl group which without hydrogen bonding could lead to the idea that in membrane environments the presence of this residue should be avoided due to its polar character. Notwithstanding this idea, the proline residue is found to be widely distributed in the putative TM helices of many integral membrane proteins (Li et al., 1996; Sansom, 1992), and it has been implicated in relevant cellular processes such as channel gating (Tieleman et al., 2001), G protein coupled receptors (GPCRs) function (Sansom and Weinstein, 2000), or in membrane protein folding precluding β-structure formation (Wigley et al., 2002). Moreover, the existence of two hydrogen bonds between the Cδ protons of the proline side chain and the carbonyl groups in the preceding turn of the helix has been described in some cases (Chakrabarti and Chakrabarti, 1998). This scenario alleviates the problems of a free carbonyl group in a hydrophobic membrane environment (Chakrabarti and Chakrabarti, 1998).

In an attempt to understand the role that proline residues perform in TM helices, or more precisely in helix packing, we have used the GpA system as a scaffold, where we have replaced every amino acid of the *seven residue motif* with proline, allowing us to study its effect on the helix-helix association process.

As expected (in all but one case; Fig. 1.3), the presence of proline on the dimerisation interface of GpA precludes dimer formation. The interpretation is somewhat more complicated for the dimerising Leu75Pro mutant, since as it was shown in a previous exhaustive saturation mutagenesis study (Lemmon et al., 1992b), a very subtle alteration in the side-chain structure at Leu75 position has a profound effect upon the propensity of the helices to dimerise.

Although the presence of proline residues normally compromise regular secondary structure formation, the dimerisation degree observed in Leu75Pro mutant should be compatible with α-helical structures. In order to test this hypotheses TM peptides containing wild-type and proline 75 GpA sequences were chemically synthesised. The secondary structure adopted by both synthetic TM peptides was evaluated by using circular dichroism (CD) spectroscopy in the presence of dodecylphosphocholine (DPC) micelles, the media previously used to obtain the solution NMR structure (MacKenzie et al., 1997). As seen in Fig. 1.4, only small differences were found for both sequences in this membrane mimetic environment. In addition, in CD spectroscopy, the ratio $\theta_{220\ nm}/\theta_{208\ nm}$ has been used to differentiate between

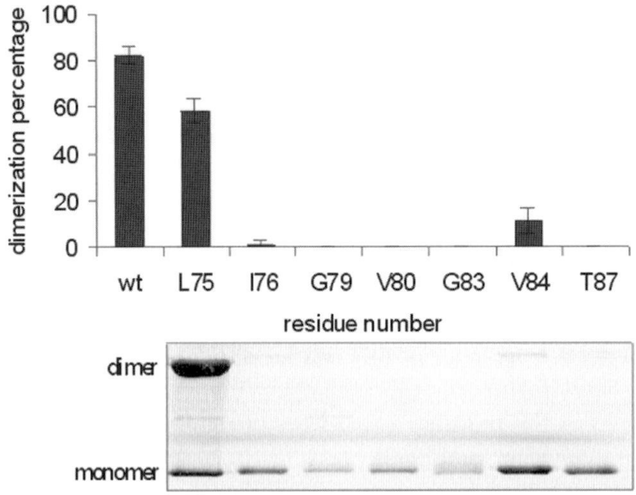

Fig. 1.3. Substitution of the amino acids of the seven residue motif of *GpA* by proline. Samples were run on a 12% SDS-PAGE. The effect that these replacements have on the dimerisation capacity of the protein are quantified in the graphic shown at the *top*

monomeric or coiled α-helices. From the CD spectra of the two peptides we obtained a $\theta_{220\,nm}/\theta_{208\,nm}$ ratio close to 1.0, which has been proposed for a two stranded α-helical coiled-coil (Lau et al., 1984), suggesting dimer formation in both peptides.

These results point to a similar folding of both peptides, probably by maintaining intact the core of the motif formed mainly by the most C-terminal residues, i.e. ^{79}GVxxGVxxT^{87}. It is important to note that proline is unique in that it is an imino (rather than amino) acid lacking the amide hydrogen atom, and is thus unable to act as a hydrogen bond donor. A necessary function of this, is that it is incapable of forming (at location *i*) the hydrogen bond with the carbonyl group of a residue (at $i - 4$) in the preceding turn of a canonical α-helix, without major distortion of the C-terminus of the helix which includes the rest of the dimerisation motif.

To summarise, although in soluble proteins the presence of proline residues in α-helices usually produces disruption of this structure; unfolded proteins have the possibility of forming H-bonds with the aqueous medium. However, in membrane environments, this local unfolded situation is only possible at the ends of TM fragments at the interface of the membrane where the environment is more polar. In fact, that would explain the results obtained for the Leu75Pro mutant. Statistical studies about abundance of proline residues in TM fragments (Ulmschneider and Sansom, 2001) show also that in natural membrane proteins, prolines are usually located at the ends of TM helices. Experiments studying the influence of proline residues replacing the

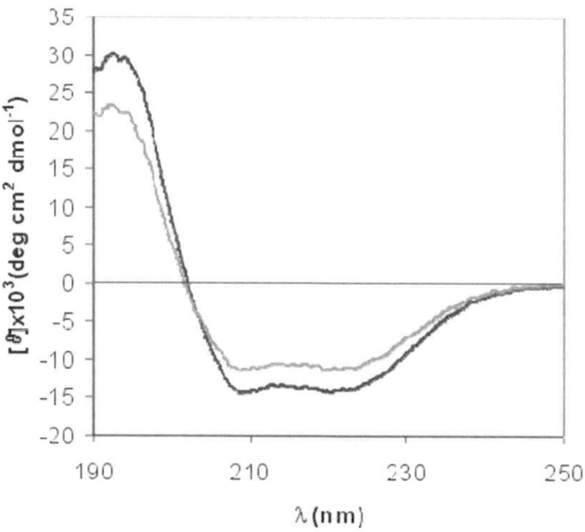

Fig. 1.4. Circular dichroism spectra of the synthetic peptides corresponding to the *TM* fragment of *GpA* in the presence of 10 mM DPC. The *black line* shows the results obtained for the wild-type sequence and the *grey line* for the Leu75Pro mutant. Peptide concentration was 30 μM in both cases

rest of hydrophobic residues of the GpA TM fragment are currently in progress in our laboratory.

1.6 Future Prospects for Membrane Protein Analysis

Wide distribution of membrane proteins and their implication in many essential cellular processes has increased the necessity of understanding more about how they work. The association, folding, and misfolding of TM domains play important roles in physiological as well as pathophysiological processes (Partridge et al., 2002; Wigley et al., 2002). Furthermore, transient associations of TM domains are also believed to be important for the regulation of a variety of proteins (Giancotti, 2003).

A variety of theoretical and experimental complications must be considered when attempting to characterise a system involving integral membrane protein folding or oligomerisation. It is therefore necessary to explore simple model systems in order to wade through this difficult problem. In our approach, it is possible that for some constructs we do not detect dimerisation in SDS (merely due to the relatively harsh conditions of SDS-PAGE, as pointed out previously in this system (Schneider and Engelman 2003), but they could associate it in an actual membrane. Despite this caveat, and taking into account paucity of structural information on membrane proteins, the results obtained with the GpA system contribute to a better understanding of the factors that take part in the process of packing between TM helices. The results presented here highlight the importance of flanking polar residues positioning and mismatching, in the oligomerisation processes. We have also introduced results regarding the consequences of the presence of proline residues in a dimerisation interface. Until techniques evolve to give us the opportunity of using more accurate systems, the type of strategy described here must be used in order to obtain valuable information about how the membrane proteins world work.

Acknowledgments. We would like to thank Dr. P. Whitley (University of Bath) for critical reading of the manuscript. Work performed in our laboratory was supported by the Spanish MCyT project no. BMC2000–1448 and Generalitat Valenciana grant GV00–040–5 to I.M. M.O. was supported as a recipient of a predoctoral fellowship from the Generalitat Valenciana (Spain).

References

Adams, P., Engelman, D. & Brünger, A. (1996). Improved prediction for the structure of a dimeric transmembrane domain of glycophorin A obtained through global searching. *PROTEINS – Struct Funct Genet* 26, 257–261

Arkin, I. T. & Brunger, A. T. (1998). Statistical analysis of predicted transmembrane alpha-helices. *Biochim Biophys Acta* 8(1), 113–28

Arkin, I. T. (2002). Structural aspects of oligomerization taking place between the transmembrane α-helices of bitopic membrane proteins. *Biochimica et Biophysica Acta (BBA) - Biomembranes* 1565(2), 347–363

Bormann, B. J. & Engelman, D. M. (1992). Intramembrane Helix-Helix Association in Oligomerization and Transmembrane Signaling. *Annu Rev Biophys Biomol Struc* 21, 223–242

Bywater, R. P., Thomas, D. & Vriend, G. (2001). A sequence and structural study of transmembrane helices. *J Comput Aided Mol Des* 15(6), 533–52

Chakrabarti, P. & Chakrabarti, S. (1998). C–H...O hydrogen bond involving proline residues in alpha-helices. *J Mol Biol* 284(4), 867–73

DeGrado, W. F., Gratkowski, H. & Lear, J. D. (2003). How do helix-helix interactions help determine the folds of membrane proteins? Perspectives from the study of homo-oligomeric helical bundles. *Protein Sci* 12(4), 647–665

Dumas, F., Lebrun, M. C. & Tocanne, J. F. (1999). Is the protein/lipid hydrophobic matching principle relevant to membrane organization and functions? *FEBS Lett* 458(3), 271–7

Giancotti, F. G. (2003). A structural view of integrin activation and signaling. *Dev Cell* 4(2), 149–51

Killian, J. A. (1998). Hydrophobic mismatch between proteins and lipids in membranes. *Biochim Biophys Acta* 10(3), 401–15

Killian, J. A. & von Heijne, G. (2000). How proteins adapt to a membrane-water interface. *Trends-in-Biochemical-Sciences* 25(9), 429–434

Kim, M. K. & Kang, Y. K. (1999). Positional preference of proline in alpha-helices. *Protein Sci* 8(7), 1492–9

Lau, S. Y., Taneja, A. K. & Hodges, R. S. (1984). Synthesis of a model protein of defined secondary and quaternary structure. Effect of chain length on the stabilization and formation of two-stranded alpha-helical coiled-coils. *J Biol Chem* 259(21), 13253–61

Lemmon, M. A. & Engelman, D. M. (1994). Specificity and promiscuity in membrane helix interactions. *Q Rev Biophys* 27(2), 157–218

Lemmon, M. A., Flanagan, J. M., Hunt, J. F., Adair, B. D., Bormann, B.-J., Dempsey, C. E. & Engelman, D. M. (1992a). Glycophorin A dimerization is driven by specific interactions between transmembrane α-helices. *J Biol Chem* 267, 7683–7689

Lemmon, M. A., Flanagan, J. M., Treutlein, H. R., Zhang, J. & Engelman, D. M. (1992b). Sequence specificity in the dimerization of transmembrane α-helices. *Biochemistry* 31(51), 12719–12725

Li, S.-C. & Deber, C. M. (1994). A measure of helical propensity for amino acids in membrane environments. *Nature Struct Biol* 1, 368–373

Li, S.-C., Goto, N. K., Williams, K. A. & Deber, C. M. (1996). α-helical but not β-sheet, propensity of proline is determined by peptide environment. *Proc Natl Acad Sci USA* 93, 6676–6681

MacKenzie, K. R., Prestegard, J. H. & Engelman, D. M. (1997). A transmembrane helix dimer: Structure and implications. *Science* 276, 131–133

Mingarro, I., Whitley, P., Lemmon, M. A. & von Heijne, G. (1996). Alα–insertion scanning mutagenesis of the glycophorin A transmembrane helix. A rapid way to map helix-helix interactions in integral membrane proteins. *Protein Sci.* 5, 1339–1341

Partridge, A. W., Therien, A. G. & Deber, C. M. (2002). Polar mutations in membrane proteins as a biophysical basis for disease. *Biopolymers* 66(5), 350–8

Popot, J. L. & Engelman, D. M. (1990). Membrane protein folding and oligomerization – The 2-stage model. *Biochemistry* 29(17), 4031–4037

Popot, J. L. & Engelman, D. M. (2000). Helical membrane protein folding, stability, and evolution. *Annu Rev Biochem* 69, 881–922

Sansom, M. S. & Weinstein, H. (2000). Hinges, swivels and switches: the role of prolines in signalling via transmembrane alpha-helices. *Trends Pharmacol Sci* 21(11), 445–51

Sansom, M. S. P. (1992). Proline residues in transmembrane helices of channel and transport proteins: a molecular modelling study. *Prot Engineer* 5, 53–60

Schneider, D. & Engelman, D. M. (2003). GALLEX, a Measurement of Heterologous Association of Transmembrane Helices in a Biological Membrane. *J. Biol. Chem.* 278(5), 3105–3111

Smith, S. O., Song, D., Shekar, S., Groesbeek, M., Ziliox, M. & Aimoto, S. (2001). Structure of the transmembrane dimer interface of glycophorin A in membrane bilayers. *Biochemistry* 40(22), 6553–6558

Tieleman, D., Shrivastava, I., Ulmschneider, M. & Sansom, M. (2001). Proline-induced hinges in transmembrane helices: possible roles in ion channel gating. *Proteins* 44(2), 63–72

Tomita, M. & Marchesi, V. T. (1975). Amino acid sequence and oligosaccharide attachment sites of human erythrocyte glycophorin. *Proc Natl Acad Sci U S A* 72(8), 2964–8

Ulmschneider, M. B. & Sansom, M. S. (2001). Amino acid distributions in integral membrane protein structures. *Biochim Biophys Acta* 2(1), 1–14

von Heijne, G. (1992). Membrane protein structure prediction – Hydrophobicity analysis and the positive-Inside rule. *J Mol Biol* 225(2), 487–494

Wallin, E. & von Heijne, G. (1998). Genome-wide analysis of integral membrane proteins from eubacterial, archaean, and eukaryotic organisms. *Protein Sci* 7, 1029–1038

White, S. H. & Wimley, W. C. (1999). Membrane protein folding and stability: physical principles. *Annu Rev Biophys Biomol Struct* 28, 319–65

White, S. H., Wimley, W. C., Ladokhin, A. S. & Hristova, K. (1998). Protein folding in membranes: determining energetics of peptide-bilayer interactions. *Methods Enzymol* 295, 62–87

Wigley, W. C., Corboy, M. J., Cutler, T. D., Thibodeau, P. H., Oldan, J., Lee, M. G., Rizo, J., Hunt, J. F. & Thomas, P. J. (2002). A protein sequence that can encode native structure by disfavoring alternate conformations. *Nat Struct Biol* 9(5), 381–8

Zhou, F. X., Cocco, M. J., Russ, W. P., Brunger, A. T. & Engelman, D. M. (2000). Interhelical hydrogen bonding drives strong interactions in membrane proteins. *Nat Struct Biol* 7(2), 154–60

2 Mobility Studies in Proteins by ^{15}N Nuclear Magnetic Resonance: Rusticyanin as an Example

Beatriz Jiménez, José María Moratal, Mario Piccioli and Antonio Donaire

Abstract

The knowledge of the molecular structure is the first step in comprehending how a protein works. The second level consists of understanding its mobility features. NMR is the only technique that allows the characterization of these two properties. Here,we will describe the relationships between the dynamics of a protein and some Nuclear Magnetic Resonance (NMR) parameters easily achievable, concretely, the ^{15}N relaxation times (T_1 and T_2), and the heteronuclear ^{1}H-^{15}N nuclear Overhauser effect. NMR also allows the detailing of the residues exposed to the solvent, i.e., to identify protein/water interactions. All this information together is essential for describing the functionality of a protein. As an example, we present here a ^{15}N heteronuclear NMR study on rusticyanin (Rc). Rc possesses a very high redox potential and is very stable at low pH values. Our study reveals that Rc is also very rigid and highly hydrophobic. The present study strongly indicates that both thermodynamic and mobility properties of Rc are correlated.

2.1 Introduction

Structure-function relationships are one of the main topics in biochemical related sciences. The characterization of the global and local tridimensional structure of a biomolecule is the first crucial step in order to understand its behavior. On one hand, protein functionality depends, in most cases, on the capability of interacting with a partner and, in turn, on how both molecules can be docked with each other and with the solvent. This docking would not be possible if proteins were rigid and fixed structures. Thus, the understanding of dynamic properties, in terms of both local fluctuations and domain–

Principles and Practice
Methods in Proteome and Protein Analysis
R.M. Kamp, J. J. Calvete, T. Choli-Papadopoulou (Eds.)

domain reorientations, is essential to get insights into protein functionality. On the other hand, protein function and stability depend on the way the protein is folded. This is a consequence not only of the intra-molecular protein interactions, but mainly, of interactions of the protein with the solvent (water molecules).

2.1.1 NMR Versus X-Ray for the Acquisition of Dynamic Information

X-ray diffraction was revealed as the most powerful technique to resolve tridimensional structures of proteins. Nevertheless, there are mainly three drawbacks with this technique. Firstly, the calculated structure belongs to crystallized molecules and not all molecules can be obtained in this form. Secondly, the conformation adopted by the protein under crystallization conditions (usually high salt concentrations) is not necessarily the same as the soluble (typically active) form of the protein. Thirdly, and most importantly with regard to this chapter, X-ray diffraction does not provide information on the dynamics properties of the system.

Although the so-called *temperature* or *B-factor* in a solved crystal structure is indicative of the degree of disorder of the corresponding region, this is not a direct measurement of the mobility itself. In any case, the time scale of the dynamic processes is not extracted from it. Moreover, X-ray diffraction details very accurately the position of the water molecules in the crystal cell, but, again, offers no information on the time scale of their exchange with the bulk solvent. The knowledge of these data can be relevant in understanding the role of the water molecules in the structure or the function of the molecule.

Nuclear magnetic resonance (NMR), on the contrary to this, is not only suitable for determining of the structure of proteins in the solution (ca. 20 % of the Protein Data Bank structures have been performed by NMR), but also can proportionate a detailed picture of the mobility of both the backbone and side chain of each individual residue (Peng and Wagner 1994, Dayie et al. 1996, Palmer 2001, Palmer et al. 2001, Bax 2003). NMR can also provide information on the time scale (from pico- to milliseconds) of the observed movements. Location of solvent molecules can also be detected by NMR (Brunne et al. 1993, Otting and Lieppinsh 1995, Bertini et al. 1997, Mesgarzadeh et al. 1998, Wider 1998, Bertini et al. 2000). More importantly, NMR can discriminate water molecules that remain in protein cavities longer than the correlation time of the molecule (probably structural or functional water molecules) from those that exchange fast with the bulk solvent (Dalvit and Hommel 1995, Dalvit 1996). Thus, NMR is a unique technique that permits the complete structural and dynamic characterization of a protein as well as of its hydration properties.

2.1.2 Dynamics of Proteins and NMR

2.1.2.1 Theoretical Considerations

When a molecule is placed into a magnetic field, the NMR active nuclei orient their magnetic moments and create a net magnetization in the direction of the applied field (by convention, *z*-axis), while the net magnetization perpendicular to that magnetic field (*xy*-plane) is zero. If we perturb (typically with a pulse) the system, it will tend to restore the equilibrium state via a process termed relaxation. The time that the net magnetization needs to recover the equilibrium will depend on the capability of each individual spin to exchange its excess of energy with the environment. Two relaxation times are currently defined: one related with the time that the magnetic moment takes to reach the equilibrium condition in the *z*-axis (longitudinal relaxation time, T_1); and the other related with the time that the magnetization takes to reach the equilibrium condition (net magnetization equal to zero) in the *xy*-plane (transversal relaxation time, T_2).

Relaxation is caused by fluctuation of magnetic dipoles around the concerned nucleus. In solution, the rate of these fluctuations depends on the overall molecular tumbling and also on the internal motions. Therefore, nuclear relaxation in solution is always intrinsically dependent on the dynamics of the nuclei and, hence, on the mobility of each residue.

Based on sensitivity considerations, ^{1}H would be the most appropriate nucleus to be investigated. However, proton relaxation is always the sum of many interactions of comparable magnitude and, hence, it is very complex to analyze in terms of internal dynamics. In contrast, the relaxation mechanisms of the ^{15}N nuclei in a protein uniformly enriched with ^{15}N are much simpler to predict and to account for. Indeed, the two mechanisms contributing to ^{15}N relaxation are the chemical shift anisotropy of ^{15}N nuclei (not discussed here) and the dipole–dipole coupling with its amide proton (Barbato et al. 1992; Peng and Wagner 1994, Tjandra et al. 1996). Among all backbone groups, both effects are essentially constant and only depend on the internal mobility of each individual backbone atom in the molecular frame. This makes ^{15}N an ideal nucleus to map internal dynamics in proteins.

The study of ^{13}C in a double (^{13}C and ^{15}N) labeled sample is a sort of intermediate state (Atkinson and Lefèvre 1999, Guenneugues et al. 1999, Walsh et al. 2001). Let us consider the case of a ^{13}Cα nucleus. Here, the main source of relaxation is still the dipolar coupling with Hα protons, but there are important couplings with the attached carbon nuclei (C′, Cβ) that cannot be ignored. The situation is simpler in the case of quaternary carbon atoms, like backbone ^{13}C′ carbonyl carbons. Relaxation is here essentially dominated by the strong $^1J_{C'-C\alpha}$ coupling and by C′ chemical shift anisotropy (Engelke and Rüterjans 1997). Therefore, the situation in principle is not different from the ^{15}N case and makes ^{13}C′ appropriate when studying protein dynamics. However, the ensemble of theoretical, experimental and economical reasons is such that the analysis of ^{15}N relaxation is by far the most common tool to address backbone dynamics in proteins.

From the experimental point of view, we need to collect both relaxation times T_1 and T_2 (usually expressed as the reciprocal quantities, R_1 and R_2, respectively) as well as the direct measurement of cross-relaxation occurring between ^1H and ^{15}N spins. This is also available on an experimental basis by means of the nuclear Overhauser effect (NOE) measurements.

Relaxation ^{15}N parameters (R_1, R_2, and ^1H-^{15}N NOE), without any further data analysis, provides direct indications on the relative mobility of each nucleus (Peng and Wagner 1994). For instance, R_1 or R_2 relaxation rates shorter than the average value in a protein (and, in general, in macromolecules with molecular weight larger than 3–5 kDa) denotes local mobility in the pico- to nanosecond time scale. Small (or even negative) ^1H-^{15}N NOE values are also indicative of this kind of dynamics in the studied region. In contrast, transversal relaxation rates, R_2, significantly larger than the average value indicates mobility in the micro- and millisecond time scale.

If we want to obtain not only qualitative information, we should analyze how backbone dynamics can be quantitatively extracted from ^{15}N relaxation parameters. A full exposition of the theory is reported elsewhere (Abragam 1961, Kowalewski 1987, Barbato et al. 1992; Mandel et al. 1995; Palmer 2001); here, we will comment on this analysis in a pictorial way. The extent of dipole-dipole ^1H-^{15}N relaxation as well as that of the ^1H-^{15}N NOE effect are described by the following equations (Mandel et al. 1995; Palmer 2001):

$$R_1\left({}^{15}N\right) = A r_{HN}^{-3} f_1\left(\omega_H, \omega_N\right) \tag{2.1}$$

$$R_2\left({}^{15}N\right) = B r_{HN}^{-3} f_2\left(\omega_0, \omega_H, \omega_N\right) + R_{ex} \tag{2.2}$$

$$NOE\left({}^1H-{}^{15}N\right) = 1 + C r_{HN}^{-3} f_3\left(\omega_H, \omega_N\right) \tag{2.3}$$

where A, B, and C are known constants, r_{HN} is the ^1H-^{15}N distance (0.91 Å), the term R_{ex} [Eq. (2.2)] accounts for chemical exchange processes (i.e. in the mico- and millisecond time scale) and the f_i functions (with i=1, 2, 3) are linear combinations of the so-called spectral density function, $J(\omega)$. Therefore, the measurement of these relaxation parameters (R_1, R_2, and ^1H-^{15}N NOE) can provide a detailed description of the spectral density function for each pair ^1H-^{15}N.

The magnetic field observed by a ^{15}N nucleus fluctuates due to its neighbor ^1H because of the whole protein tumbling and of the local movements. The spectral density functions describe how fluctuations at different frequencies affect the relaxation of a given nucleus. The contribution that a fluctuating field at frequency ω provides to the nuclear relaxation also depends on the global molecular motions from the following equation (Abragam 1961, Kowalewski 1987):

$$J(\omega)=\frac{\tau_c}{1+\omega^2\tau_c^2} \quad (2.4)$$

To factorize out internal motions from global reorientation motions due to molecular tumbling, the spectral density functions can be represented on the basis of the approach developed by Lipari and Szabo (Lipari and Szabo 1982, 1982), in the following form:

$$J(\omega)=\frac{2}{5}\left[\frac{S^2\tau_r}{1+(\omega\tau_r)^2}+\frac{(1-S^2)\tau_c}{1+(\omega\tau_c)^2}\right] \quad (2.5)$$

where τ_r is the correlation times for the overall tumbling of the molecule and $\tau_c^{-1}=\tau_r^{-1}+\tau_e^{-1}$, in which τ_e is the correlation time for internal motions. The correlation times account for the time scales in which dynamic processes occur. S^2 is defined as the generalized order parameter. It takes values from 0 to 1. A value of zero for the S^2 parameter indicates a completely free (isotropic) movement of the N-H pair (low degree of order). For an S^2 value of 1, the movement of that pair is completely restricted and anisotropic (maximum order). When an N-H pair rotates with the overall tumbling of the molecule (i.e. if this pair is fixed in the scaffold of the protein, as typically happens to protons present in a β-barrel or in an α-helix substructure; Fig. 2.1), $\tau_c=\tau_r$, and the second term of Eq. (2.5) vanishes. When the residue possesses its own local movements, then $\tau_c\neq\tau_r$. In this case, the S^2 provides the quantitative estimate of the degree of local motions while the time scale of this movement (τ_e) can be obtained by these measurements. This is usually the case of loops or less structured regions of proteins (see Fig. 2.1).

2.1.2.2 A Quantitative Analysis of the Model-Free Approach

According to the so-called model-free approach, developed by Lipari and Szabo (1982, 1982), fast internal motions can be described by two model independent quantities: a generalized order parameter, S^2, which provides a measure of the amplitude of the motion, and an effective correlation time, τ_e. The description of motions slower than the overall molecular tumbling are accounted for with the introduction of the parameter R_{ex} [Eq. (2.2)].

This three-parameter approach allows the possibility of using several different models to explain the experimental data. Palmer et al. (1991; Mandel et al. 1995; Palmer 2001) have developed a procedure based on the definition of a series of dynamical models and relying on the use of a model selection pro-

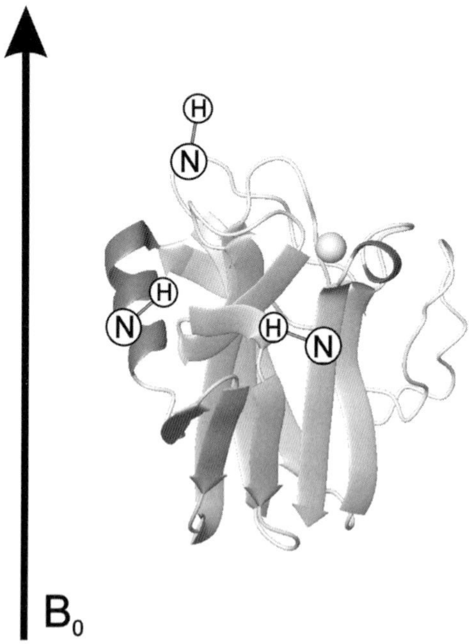

Fig. 2.1. Ribbon diagram of rusticyanin (1 cur.pdb; Botuyan et al. 1996); the copper ion is displayed as a ball at the top of the molecule. The β-barrel, and the α-helix are colored in *light* and *dark gray*, respectively. The *arrow* represents the orientation of the magnetic field. Three N-H pairs are specifically plotted: two of them belong to secondary structural elements of the protein (one in the α-helix, and the other one in the β-barrel), and hence, τ_c is equal to τ_r for them (see text); the third (at the *top*) is located in a loop. For this last pair, τ_c differs from τ_r

tocol derived from statistical methods. The steps of this protocol are briefly summarized here:

- *Analysis of the model for whole protein motions*: Local correlation times obtained for each individual N-H pair are statistically fitted versus an input structure of the protein to obtain an initial estimate of the diffusion tensor. Three different models were considered for the diffusion tensor: fully isotropic, axially isotropic or anisotropic movement. The best fit will provide the nature of the overall molecular motion experienced by the investigated system.
- *Analysis of the model for each backbone N-H pair*: Once the diffusion tensor is assumed, relaxation parameters of each N-H pair are individually examined and assigned to one of five different models. In model 1, the internal dynamics experienced by the N-H vector pair can be accounted for by using only a one-parameter fit, which is the generalized order parameter S^2. In model 2, two parameters are used, S^2 and τ_e, thus indicating that a fast correlation time (faster than the overall τ_c) is needed to account for relaxation parameters of individual residues. Models 3 and 4 are equivalent to models 1 and 2, respectively, but also assume the occurrence of chemical exchange rates in the micro- to millisecond time scale, i.e., the R_{ex} term [Eq. (2.2)] is fitted in both models. Finally, model 5 introduces a second order parameter S_f and uses an expression for the spectral density function more complex than Eq. (2.5). It accounts for those residues in which the

autocorrelation function cannot be approximated with a single exponential according to the Lipari and Szabo treatment, but it requires at least two exponential functions (Clore et al. 1990).

2.1.2.3 Practical Aspects

The sensitivity of an NMR experiment as a function of the acquired nuclei is proportional to (Ernst et al. 1987):

$$S/N \;\alpha\; \gamma_{exc}\gamma_{obs}^{3/2}\left\{1-\exp\left(-t/T_1^{(exc)}\right)\right\} \tag{2.6}$$

where *S/N* is the signal to noise ratio (i.e. the sensitivity of the experiment), γ_{exc} and γ_{obs} are the gyromagnetic ratios of the excited and observed nuclei, respectively, and *t* represents the time while the signal is being acquired (the acquisition time). Direct measurements of ^{15}N-NMR relaxation are performed by exciting and observing ^{15}N nuclei (γ_{15N}=–2.713×10^{7} T^{-1} s rad^{-1}). Inverse experiments are those heteronuclear experiments in which ^{1}H nuclei are excited and acquired (γ_{1H}=2.675×10^{8} T^{-1} s rad^{-1}). In these experiments, the magnetization is transferred from one nucleus to another by heteronuclear polarization transfer mechanisms and the low sensitivity nucleus is always frequency labeled in the indirect dimensions.

Equation (2.6) indicates that a ^{15}N-NMR experiment performed in the direct mode is ca. 350 times less sensitive than the corresponding inverse experiments. The basic inverse experiment is the heteronuclear ^{15}N single quantum coherence (HSQC) experiment (Bodenhausen and Ruben 1980), a 2D experiment that relates an amide ^{1}H with its attached ^{15}N. For a moderately large protein, this experiment allows the resolution of each individual N-H peak (fingerprint), to provide a very sensitive tool for a quick map of local structural rearrangements. In Fig. 2.2, the HSQC spectrum of rusticyanin is depicted. This protein possesses 155 amino acids, each of them (except the N-terminal and the proline residues) gives rise to a peak in the HSQC. As observed, most of the peaks are very well resolved. In the different experiments to measure the relaxation properties of the ^{15}N nuclei, a set of these experiments with a varying specific parameter (a delay time) of the corresponding pulse sequence is recorded. The intensity of the observed peak is modulated by this parameter according to the relaxation property of the ^{15}N nucleus. By fitting the intensity of the observed peak versus the value of that parameter, the relaxation property of the ^{15}N nucleus can be measured.

The typical experiment to measure ^{15}N R_1 relaxation (Peng and Wagner 1994) is shown in Fig. 2.3A. It is based on a polarization transfer from ^{1}H to ^{15}N magnetization whose frequency is labeled. Then, ^{15}N magnetization is rotated along the *z* axis and, during the period *T*, each individual ^{15}N nucleus is allowed to relax. The intensity of the acquired signal will depend on the ^{15}N T_1.

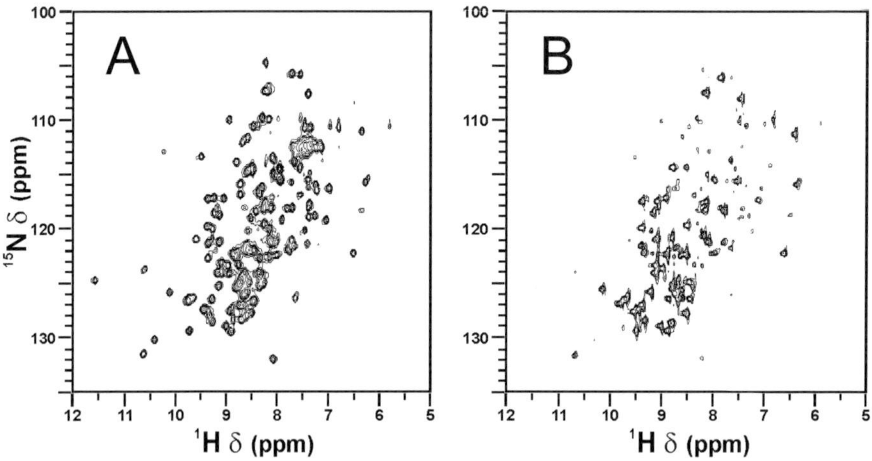

Fig. 2.2. ^{15}N-HSQC spectrum of copper(I) rusticyanin performed at 500 MHz (2.0 mM, pH 5.5, acetate buffer 0.1 M, 296 K): **A** in H_2O; **B** one week after having dissolved the sample in D_2O

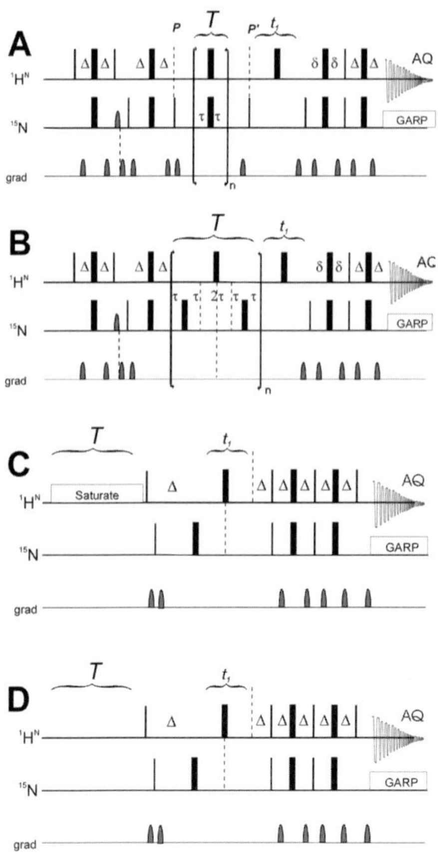

Fig. 2.3. Pulse sequences used to determine: **A** ^{15}N longitudinal relaxation rates, R_1 (Peng and Wagner 1994); **B** ^{15}N transversal relaxation rates, R_2 (Kay et al. 1992; Peng and Wagner 1994); **C** and **D** ^{1}H-^{15}N NOE values (Grzesiek and Bax 1993). In sequence **A**, the ^{15}N 90 pulse *P* takes ^{15}N magnetization to the *z*-axis. Then during the period *T* the ^{15}N nucleus relaxes (according to its T_1). The second 90 pulse (pulse *P′*) turns ^{15}N magnetization towards the *xy* plane. The rest of the sequence is a typical HSQC spectrum. In sequence **B**, the ^{15}N magnetization is always on the *xy* plane. During the period *T*, ^{15}N nuclei relax according to their T_2 values. For determining both T_1 and T_2 for each ^{15}N nuclei a series of this experiment, varying the delay *T* is acquired. Sequences **C** and **D** are almost the same: in the first, the amide protons are saturated (real NOE spectrum); while in sequence **D**, the amide protons are not saturated (NOE reference spectrum). ^{1}H-^{15}N NOE values are obtained from the intensity ratio observed in these two experiments (see text)

The experiment to measure ^{15}N R_2 relaxation (Kay et al. 1992; Peng and Wagner 1994) is shown in Fig. 2.3B. At variance with the previous one, each ^{15}N magnetization is not rotated onto the *z*-axis, and therefore, each signal is allowed to relax in the *x,y*-plane during the period *T*. Now, the intensity of the acquired signal will vary in accordance with its T_2. Pulses of 180° applied to both ^1H and ^{15}N during the variable relaxation delay *T* are required to prevent the evolution of chemical shifts and cross correlation effects (the latter will not be discussed here), which may affect the analysis of relaxation rates.

Figure 2.3C reports the experiment to collect ^1H-^{15}N NOE values (Grzesiek and Bax 1993). Basically, amide ^1H resonances are saturated before transferring the magnetization to the ^{15}N nucleus. Therefore, the intensity of each peak will depend on the cross relaxation operative between each ^1H-^{15}N pair. This intensity is then compared with reference experiment (reported in Fig. 2.3D) in which the ^1H amide spins are not excited.

2.1.3 The System: Rusticyanin

As an example of the information that NMR provides on protein dynamics and hydration, we report here the study of rusticyanin (Rc hereafter). Rc is the most abundant protein in the Gram negative bacterium *Thiobacillus ferrooxidans* (*Tf*), that lives in very acidic media by oxidizing the FeII ion to FeIII (Ingledew et al. 1977, Ronk et al. 1991). Thus, this protein is very stable at pH values lower than 2.5 (Cobley and Haddock 1975, Cox and Boxer 1978, Hall et al. 1996). Rc belongs to a class of proteins denominated blue copper proteins (BCPs) (Gray et al. 2000, Randall et al. 2000). The copper ion in these proteins presents a redox potential higher than that of the CuII/CuI pair in aqueous solution. Rc is the BCP with the highest redox potential (680 mV) (Ingledew and Cocco 1980, Blake and Shute 1987, Shoham 1992, Hall et al. 1999).

As all BCPs, Rc topology consists of a β-barrel structure with a copper ion in the so-called "north pole" of the molecule (Fig. 2.4A). These structural elements are, in the case of Rc, supplemented with an extension of 35 amino acids (N-35 extension, hereafter), not present in other BCPs (Djebli et al. 1992; Grossmann et al. 1995; Botuyan et al. 1996, Walter et al. 1996, Grossmann et al. 2002) (see Fig. 2.4A). This extension contains an α-helix and three β-strands, the first two being anti-parallel. For comparison, the tridimensional structure of plastocyanin from *Synechocystis* sp. PCC6803 (Bertini et al. 2001) is displayed in Fig. 2.4B. We will report here how ^{15}N relaxation data will account for the specific structural features of rusticyanin and how internal dynamics and protein hydration are important to address structure-function relationships in this protein.

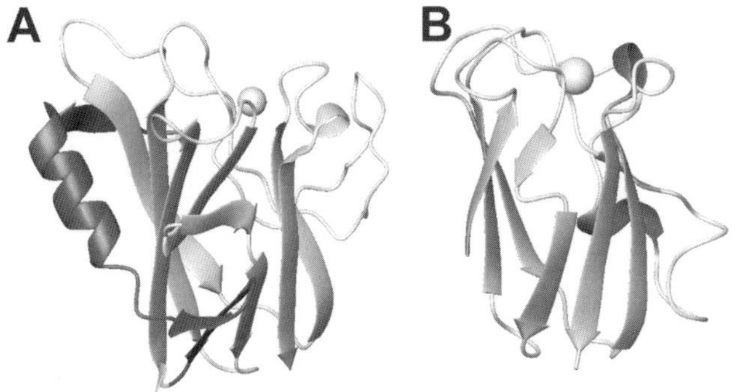

Fig. 2.4. Ribbon diagram of: **A** Rc; **B** Pc (Bertini et al. 2001). β-Barrels are displayed in *light gray*. Rusticyanin N-35 terminal extension is displayed in *dark gray*

2.2 Results and Discussion

2.2.1 Relaxation Properties of Rusticyanin

2.2.1.1 Relaxation Data

Figure 2.5A–C displays the relaxation data (R_1, R_2, and ^{1}H-^{15}N NOE) per residue for Cu(I)Rc (acetate buffer 50 mM, pH 5.5, 296 K). As outlined above, the direct analysis of relaxation data provides a quantitative indication of the occurrence of regions or domains characterized by peculiar features in terms of internal mobility. Large R_1, small R_2 and small ^{1}H-^{15}N NOE values are indicative of fast internal motions in the pico- or nanosecond time scale. The average value of the longitudinal relaxation rates (R_1) of the ^{15}N backbone nuclei was 1.00±0.06 s^{-1}. Six residues (Thr2, Leu3, Asp58, Ala70, Lys119, Trp127) exhibit R_1 values significantly larger than this average value. The average value of the transversal relaxation rates (R_2) was 17.4±1.2 s^{-1}. Five residues (Thr2, Leu3, Gly35, Lys36, and Gly93) have R_2 values smaller than the average. The average NOE value for Rc was 0.83±0.04. Eleven residues (Thr2, Leu3, Gly35, Lys36, Val38, Val56, Asp58, Ile66, Gly69, Tyr96, Ile102) have NOE values lower than this average NOE (remarkably, Leu3 NOE is negative). All these experimental data indicates that the most flexible region of the molecule is the N-terminal domain (residues Thr2, and Leu3, see Fig. 2.4). The regions encompassing amino acids 35–38, 66–70 and 93–96 are the other two loops of the protein with internal motions in the sub-nanosecond time scale.

The occurrence of conformational exchange, i.e., mobility in the micro- or millisecond time scale can be qualitatively detected from Fig. 2.5B. According to Eq. (2.2), chemical exchange (R_{ex}) may directly contribute to R_2. Therefore, residues with R_2 values, clearly above the average, experience an additional contribution to their average values expected in the absence of such motions. This is the case of the region 57–58, which is unambiguously above the general trend of the molecule (Fig. 2.5B). This is indicative of conformational

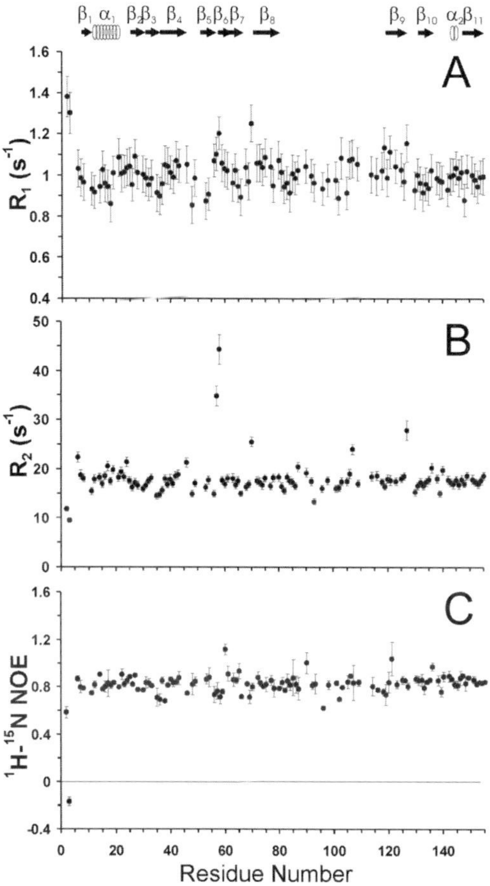

Fig. 2.5. Relaxation data for reduced rusticyanin (3.0 mM, pH 5.5, 296 K): **A** longitudinal relaxation rates, R_1; **B** transversal relaxation rates, R_2; **C** ^1H-^{15}N NOE values. The secondary structural elements of Rc are also displayed at the *top* of the figure

exchange phenomena occurring in time scales longer than the nanosecond time. Remarkably, His57 is the residue with the highest R_2 value. Since the pH value of our experiments (5.5) is coincident with the pK_a value of the imidazol side chain of a free histidine, it is reasonable to assume that protonation/deprotonation of this histidine produces two different conformations in our molecule. Its physiological relevance (if any) cannot be deduced from the present study.

To further understand the nature of the detected motions and for the exact knowledge of their time scales a model-free analysis was performed with the relaxation parameters.

2.2.1.2 An Analysis of the Generalized Order Parameter in Rc

Relaxation (R_1, and R_2) data of Rc were analyzed according to the well established procedure of Palmer et al. (Palmer et al. 1991; Mandel et al. 1995). They were satisfactorily fitted by using the isotropic model. The effective correlation time, τ_r, for Rc was 9.6 ns. The order parameter, S^2, was calculated for each residue by applying the model free analysis and according to the four dynamic models mentioned above. Figure 2.6A displays the S^2 values for the 111 residues whose relaxation properties have been determined. The average value is 0.93±0.03.

The overall trend of the order parameter reflects the secondary structure elements of the protein, with the highest values for the α-helix (0.93) and the β-strand (0.94) motifs and the largest flexibility in the short loop regions. The low values of residues 2–3 and 35–36 indicate that the mobility of the N-terminus and the 35–37 region is higher than that of the rest of the molecule. Other residues with low S^2 values are all located in loop regions of the protein (such as Gly48, Ile66, Gly93, Ile102, Thr130, Gln139).

Ninety-four out of 111 analyzed amino acids (84.7 %) fit within model 1 of the model-free analysis. Seven residues (Lys36, Val38, Ile66, Gly69, Tyr96, Ile102, Gln139) were fitted by taking into account fast internal motions in the sub-nanosecond time scale (model 2). Two of them, Lys36 and Val38, are located in a region that has on average a mobility higher than the rest of the protein. The same consideration also holds for Tyr96. Tyr96 is located in a

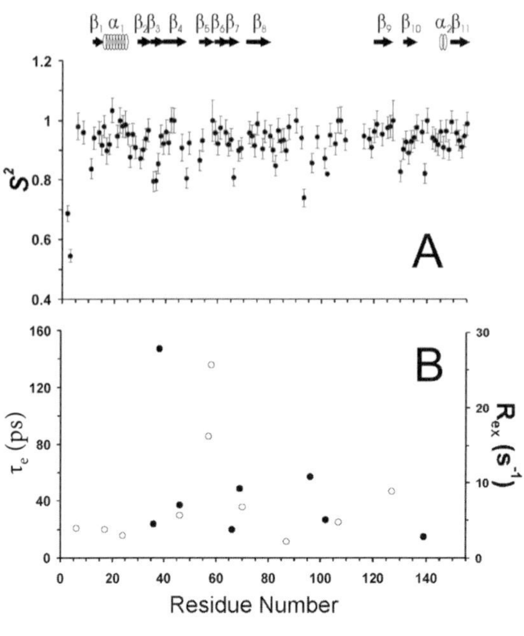

Fig. 2.6. A Generalized order parameter, S^2, versus the residue number for the reduced Rc (3.0 mM); **B** correlation times, τ_e, for residues with fast internal motions (these residues have been fitted according to models 2 or 4, *filled dots*, left y-axis); exchange rates, R_{ex}, for residues implicated in dynamics in the micro- or millisecond time scale (models 3 or 4, *open dots*, right y-axis). The secondary structural elements of Rc are also displayed at the *top* of the figure

Gly-Pro-Pro-Tyr stretch. The correlation times (τ_e) observed for these residues, shown in Fig. 2.6B, vary between 15 and 147 ps.

Nine residues (Fig. 2.6B) are only fitted when an exchange time constant in the millisecond time scale is introduced (model 3). The model-free analysis also reveals the existence of conformational exchange phenomena (R_{ex}) for these residues in time scales longer than the overall molecular tumbling. These exchange rates are in the 2.2–25.7 Hz range. While most of them are lower than 5 Hz, which is considered a threshold for the detectability of conformational exchange effects (Zinn-Justin et al. 1997), the region His57-Asp58 is unambiguously above that limit. Finally, Leu46 is the only residue which fits according to model 4, with τ_e and K_{ex} values of 37 ps, and 5.7 Hz, respectively.

2.2.2 D_2O/H_2O Exchange Experiments

In order to determine the degree of solvent accessibility, ^{15}N HSQC experiments performed with a fresh sample prepared in D_2O were collected between 3 h and 2 weeks after the H_2O/D_2O exchange. In Fig. 2.2B, one of these HSQC spectra is displayed. The 34.6 % of the assigned N-H groups does not appear in the experiment collected 3 hours after the D_2O exchange. Fifty residues (36.7 %) retain more than 50 % of their initial intensity after 2 weeks. The remaining amino acids have an intermediate behavior. These results are schematically shown in Fig. 2.7. The 47 rapid exchanging residues (in white in Fig. 2.7) are located either at the beginning of the sequence, in the first β-strand, in the 26–37 region, or distributed over the loops connecting the strands of the β-barrel. The longest hydrophilic region corresponds to residues encompassing amino acids 93–130. Forty out of the fifty slow exchanging residues (dark gray in Fig. 2.7) are mainly located in secondary structure elements characteristic of the classical BCP topology: 29 belong to the β-barrel structure; 3 are located in the small α-helix region in the northern part of the molecule close to the copper site; and 8 are located in loops.

2.2.3 Dynamics, Hydration, and Rusticyanin Stability

Mobility studies have been performed in three other BCPs: azurin (Az) from *Pseudomonas aeruginosa* (Kalverda et al. 1999), pseudoazurin (PsAz) from *Paracoccus pantotrophus* (Thompson et al. 2000), and plastocyanin (Pc) from *Synechocystis* sp. PCC6803 (Bertini et al. 2001). For all investigated cases, the average S^2 order parameters have relatively high values (0.87 in Pc, 0.86 in Az, 0.83 in PsAz). Thus, all BCPs have a rigid structure. This is due to the protein scaffold imposed by the Greek key topology of the β-barrel and to the high content in hydrogen-bonds (Gray et al. 2000). This feature allows electron transfer with a low re-organization energy (Gray et al. 2000, Randall et

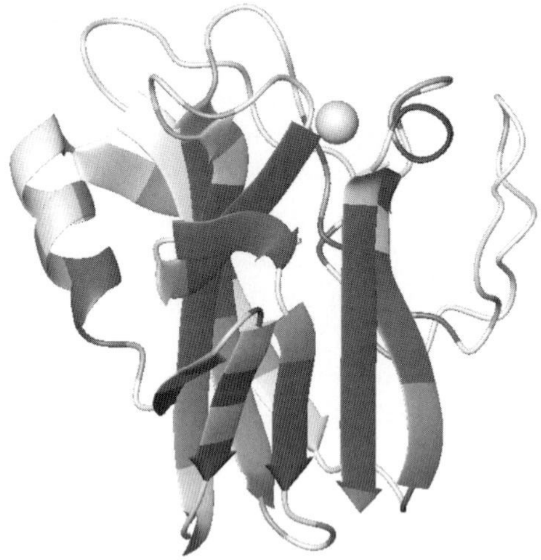

Fig. 2.7. Schematic view of Rc backbone displaying the amide protons with fast *(white)*, medium *(light gray)*, and slow *(dark gray)* D_2O/H_2O exchange pattern

al. 2000). In the case of Rc, the average S^2 (0.93±0.03) as well as the average NOE value (0.83 very close to the theoretical maximum, 0.835) indicate an unusually small degree of internal motions, even in comparison with other BCPs.

To get some insights into the specific features of Rc, we focused on those protein regions that are structurally different from other BCPs and those regions that have peculiar dynamic properties. The Rc elements characterized by some degree of internal motions are, in summary, regions 1–5, 35–38, 57–58, 66–70, 93–96 (see Figs. 2.5 and 2.6). Remarkably, residues 1–5 and 33–38 are the most mobile regions. By contrast, S^2 values indicate that the region 5–33 has a very restricted mobility. The long amphipatic helix (see Fig. 2.5A) seems to be (together with the C-terminal side, see below) the most rigid part of the molecule. Rusticyanin has 35 additional amino acids that extend the N-terminus, not present in the rest of BCPs (Grossmann et al. 2002; Fig. 2.4A). Thus, the mobile region (amino acids 33–38) connecting the classic β-barrel (the typical BCP topology) to such N-35 extension acts as a linker between two distinct domains. Crystallographic studies reveal that this extension acts as a belt with respect to the typical β-barrel fold (Walter et al. 1996, Hough et al. 2001, Kanbi et al. 2002). Given the peculiar features of Rc, this extension has been postulated to be the driving factor towards the increased acid stability of this protein. Hasnain and coworkers have recently shown that an N-35 depleted mutant is still soluble in an acidic environment (Grossmann et al. 2002). They also demonstrated that the N-terminal extension is responsible for the shielding of the hydrophobic core. The detailed analysis of hydration properties, that can be performed only by NMR, shows that the 36.7 % of Rc amide protons are hidden to the solvent.

Analogous studies performed on Pc (Fig. 2.4B) showed that the non-exchangeable amide protons are only the 17.2% of the residues (Bertini et al. 2001). This is a substantial difference that indicates how the N-terminus extension affects hydration properties of Rc.

Another region characterized by fast internal motions is the turn including amino acids 93–97 (Gly-Pro-Pro-Tyr-Ala). Of course the occurrence of two consecutive proline residues might be responsible for the high mobility. The equivalent residues in other BCPs (residues 46–49 in Pc (Bertini et al. 2001), 74–76 in Az (Kalverda et al. 1999), and 52–56 in PsAz (Thompson et al. 2000)), in which no prolines are present, also show a high degree of mobility in the same time scale. Indeed, this is the most mobile region for all these proteins (with the exception of the N-terminal residues). This turn is a part of a long loop connecting two strands in BCP. The first domain, encompassing residues 83–92, has essentially order parameter values (average S^2 in the region is 0.93) similar to those of the β-barrel regions even in the absence of defined secondary structure elements. Azurin, which has an α-helix in this region shows high S^2 values (Kalverda et al. 1999), although not as high as in Rc. In contrast, the degree of order in PsAz, with no defined secondary structure in this region, is lower than in Az and in Rc. Therefore, the high rigidity of Rc does not arise from a large content in secondary structure, but has to arise from tertiary interactions.

2.2.4 Mobility, Hydrophobicity, and High Redox Potential

Copper in Rc is bound to the side chain of four amino acids (Walter et al. 1996). Three of these copper ligands (Cys138, His143, and Met148) belongs to the two β-strands at the C-terminal end of Rc or to the loop interconnecting them. Our data shows that these two β-strands form one of the most rigid and highly shielded from solvent interaction regions of Rc. This could have two implications: (1) the larger protection of the backbone would provide an additional stabilizing effect, thus protecting against protein denaturation. (2) the dynamic, and thus, the structural properties of the copper ligands Cys138 and Met148 are driven by the interactions of these two β strands, as we have already proposed (Donaire et al. 2002). Therefore, the high rigidity and hydrophobicity of these two antiparallel β-strands provides the metal center with its unique features. Indeed, a highly hydrophobic environment of the copper ion would increase the redox potential of the protein, as previously appointed(Walter et al. 1996, Donaire et al. 2001). This would be in agreement with the entatic or rack state mechanism for the copper ion (Malmström 1994, Gray et al. 2000). It would also explain the facility of the interconversion copperI/copperII. The other two ligands, histidines 85 and 143, also have a very high S^2 value, indicating the high rigidity of the active center.

2.3 Conclusions

Our present study reveals that Rc is very rigid and highly hydrophobic. Not only secondary structural elements, but also tertiary interactions must be relevant in keeping the rigidity of Rc. The first N-35 domain, unique of this BCP, acts as an independent module of the whole protein and is partially responsible for the high degree of hidrophobicity of this protein. The hydrophobic environment is also present in copper ion surroundings. All these features are related (and, probably, are the crucial factors) in providing Rc its atypical thermodynamic properties (high stability at low pH values, and high redox potential).

This is another more example of how NMR can correlate structural-function relationships of a protein, via, in this case, its dynamic characterization.

Acknowledgements. Drs. S. Samar Hasnain (CCLRC Daresbury Laboratory) and John F. Hall (De Monfort University) are acknowledged for providing us with *E. coli* with the Rc plasmid. B.J. and M.P. would like to thank the Conselleria de Educación y Ciencia (Generalitat Valenciana) for a grant. This work was supported with financial aid from the DGICYT-Ministerio de Ciencia y Tecnología, Spain (BQU2002-02236). The support from the European Large Scale Facility PARABIO at the University of Florence, Italy (contract no. HPRI-CT-1999-00009) is acknowledged.

References

Abragam A (1961) The Principles of Nuclear Magnetism. Oxford University Press, Oxford

Atkinson RA, Lefèvre J-F (1999) Reduced spectral density mapping for proteins: validity for studies of 13C relaxation. J. Biomol. NMR 13:83–88

Barbato G, Ikura M, Kay LE, Pastor RW, Bax A (1992) Backbone dynamics of calmodulin studied by ^{15}N relaxation using inverse detected two-dimensional NMR spectroscopy; the central helix is flexible. Biochemistry 31:5269–5278

Bax A (2003) Weak alignment offers new NMR opportunities to study protein structure and dynamics. Protein. Sci. 12:1–16

Bertini I, Bryant DA, Ciurli S, Dikiy A, Fernandez CO, Luchinat C, Safarov N, Vila AJ, Zhao J (2001) Backbone dynamics of plastocyanin in both oxidation states. Solution structure of the reduced form and comparison with the oxidized state. J.Biol.Chem. 276:47217–47226

Bertini I, Dalvit C, Luchinat C, Huber JG, Piccioli M (1997) e-PHOGSY Experiments on a paramagnetic protein: location of the catalytic water molecule in the heme crevice of the oxidized form of Horse Heart Cytochrome c. FEBS Lett. 415:45–48

Bertini I, Huber JG, Luchinat C, Piccioli M (2000) Protein hydration and location of water molecules in oxidized horse heart cytochrome c by (1)H NMR. J.Magn.Reson. 147:1–8

Blake RC 2nd, Shute EA (1987) Respiratory enzymes of *Thiobacillus ferrooxidans*. A kinetic study of electron transfer between iron and rusticyanin in sulfate media. J.Biol.Chem. 262:14983–14983

Bodenhausen G, Ruben DJ (1980) Natural abundance nitrogen-15 NMR by enhanced heteronuclear spectroscopy. Chem. Phys. Lett. 69:185–189

Botuyan MA, Toy-Palmer A, Chung J, Blake RC 2nd, Beroza P, Case DA, Dyson HJ (1996) NMR Solution Structure of Cu(I) Rusticyanin from *Thiobacillus ferrooxidans:* Structural Basis for the Extreme Acid Stability and Redox Potential. J.Mol.Biol. 263: 752–767

Brunne RM, Lieppinsh E, Otting F, Wüthrich K, van Gunsteren WF (1993) Hydration of Proteins. A Comparison of Experimental residence times of Water Molecules Solvating the Bovine Pancreatic Trypsin Inhibitor with Theoretical Model Calculation. J.Mol.Biol. 231:1040–1048

Clore GM, Driscoll PC, Wingfield PT, Gronenborn AM (1990) Analysis of the backbone dynamics of interleukin-1 beta using two-dimensional inverse detected heteronuclear 15N-1H NMR spectroscopy. Biochemistry 29:7387–7401

Cobley JG, Haddock BA (1975) The respiratory chain of Thiobacillus ferrooxidans: the reduction of cytochromes by Fe2+ and the preliminary characterization of rusticyanin a novel "blue" copper protein. FEBS Lett. 60:29–33

Cox JC, Boxer DH (1978) The purification and some properties of rusticyanin, a blue copper protein involved in iron(II) oxidation from *Thiobacillus ferro-oxidans.* Biochem.J. 174:497–502

Dalvit C (1996) Homonuclear 1D and 2D NMR experiments for the observation of solvent-solute interactions. J. Magn. Reson. Ser. B 112:282–288

Dalvit C, Hommel U (1995) Sensitivity-improved detection of protein hydration and its extension to the assignment of fast-exchanging resonances. J.Magn.Reson.Ser.B 109: 334–338

Dayie KT, Wagner G, Lefevre JF (1996) Theory and practice of nuclear spin relaxation in proteins. Annu. Rev. Phys. Chem. 47:243–282

Djebli A, Proctor P, Blake RC 2nd, Shoham M (1992) Crystallization and preliminary X-ray crystallographic studies of rusticyanin from *Thiobacillus ferrooxidans.* J.Mol.Biol 227:581–582

Donaire A, Jiménez B, Fernández CO, Pierattelli R, Niizeki T, Moratal JM, Hall JF, Kohzuma T, Hasnain SS, Vila AJ (2002) Metal-Ligand Interplay in Blue Copper Proteins Studied by 1H NMR Spectroscopy: Cu(II)–Pseudoazurin and Cu(II)-Rusticyanin. J. Am.Chem. Soc. 124:13698–13708

Donaire A, Jiménez B, Moratal JM, Hall JF, Hasnain SS (2001) Electronic Characterization of the Oxidized State of the Blue Copper Protein Rusticyanin by 1H NMR: is the axial Methionine the Dominant Influence for the High Redox Potential? Biochemistry 40:837–846

Engelke J, Rüterjans H (1997) Backbone dynamics of proteins derived from carbonyl carbon relaxation times at 500, 600 and 800 MHz: Application to ribonuclease T1. J. Biomol. NMR 9:63–78

Ernst RR, Bodenhausen G, Wokaun A (1987) Principles of Nuclear Magnetic Resonance in one and two dimensions. Oxford University Press, London

Gray HB, Malmström BG, Williams RJP (2000) Copper coordination in blue proteins. JBIC 5:551–559

Grossmann JG, Hall JF, Kanbi LD, Hasnain SS (2002) The N-terminal extension of rusticyanin is not responsible for its acid stability. Biochemistry 41:3613–3619

Grossmann JG, Ingledew WJ, Harvey I, Strange RW, Hasnain SS (1995) X-ray absorption studies and homology modeling define the structural features that specify the nature of the copper site in rusticyanin. Biochemistry 34:8406–8414

Grzesiek S, Bax A (1993) The importance of not saturating water in protein NMR. Application to sensitivity enhancement and NOE measurements. J. Am. Chem. Soc. 115: 12593–12594

Guenneugues M, Gilquin B, Wolff N, Menez A, Zinn-Justin S (1999) Internal motion time scales of a small, highly stable and disulfide-rich protein: a 15N, 13C NMR and molecular dynamics study. J. Biomol NMR 14:47–66

Hall JF, Hasnain SS, Ingledew WJ (1996) The structural gene for rusticyanin from *Thiobacillus ferrooxidans:* cloning and sequencing of the rusticyanin gene. FEMS Microbiol.Lett. 137:85–89

Hall JF, Kanbi LD, Strange RW, Hasnain SS (1999) Role of the axial ligand in type 1 Cu centers studied by point mutations of Met148 in rusticyanin. Biochemistry 38:12675–12680

Hough MA, Hall JF, Kanbi LD, Hasnain SS (2001) Structure of the M148Q mutant of rusticyanin at 1.5 A: a model for the copper site of stellacyanin. Acta Cryst. D 57:355–360

Ingledew WJ, Cocco D (1980) A potentiometric and kinetic study on the respiratory chain of ferrous-iron-grown *Thiobacillus ferrooxidans.* Biochim.Biophys.Acta 590: 141–158

Ingledew WJ, Cox JC, Halling PJ (1977) A proposed mechanism for energy conservation during Fe^{2+} oxidation by *Thiobacillus ferroosidans* chemoosmotic coupling to net H^+ influx. FEMS Microbiol. Lett. 2:193–197

Kalverda AP, Ubbink M, Gilardi G, Wijmenga SS, Crawford A, Jeuken LJ, Canters GW (1999) Backbone dynamics of azurin in solution: slow conformational change associated with deprotonation of histidine 35. Biochemistry 38:12690–12697

Kanbi LD, Antonyuk S, Hough MA, Hall JF, Dodd FE, Hasnain SS (2002) Crystal Structures of the Met148Leu and Ser86Asp Mutants of Rusticyanin from *Thiobacillus ferrooxidans*: Insights into the Structural Relationship with the Cupredoxins and the Multi Copper Proteins. J. Mol. Biol. 320:263–275

Kay LE, Nicholson LK, Delaglio F, Bax A, Torchia DA (1992) Pulse sequences for removal of the effects of cross correlation between dipolar and chemical-shift anisotropyrelaxation mechanisms on the measurement of heteronuclear T1 and T2 values in proteins. J. Magn. Reson. 97:359–375

Kowalewski J (1987) Nuclear Spin Relaxation in Diamagnetic Fluids Part 2: Organic Systems and Solutions of Macromolecules and aggregates. Annu. Rep. NMR Spectrosc. 23:

Lipari G, Szabo A (1982) Model-Free Approach to the interpretation of Nuclear Magnetic Resonance Relaxation in Macromolecules. 1. Analysis of Experimental Results. J. Am. Chem. Soc. 104:4559–4570

Lipari G, Szabo A (1982) Model-Free Approach to the interpretation of Nuclear Magnetic Resonance Relaxation in Macromolecules. 1. Theory and Range of Validity. J.Am. Chem. Soc. 104:4546–4559

Malmström BG (1994) Rack-induced bonding in blue-copper proteins. Eur. J. Biochem. 233:711–718

Mandel AM, Akke M, Palmer AG 3rd(1995) Backbone dynamics of *Escherichia coli* ribonuclease HI: correlations with structure and function in an active enzyme. J.Mol.Biol. 246:144–163

Mesgarzadeh A, Pfeiffer S, Engelke J, Lassen D, Ruterjans H (1998) Bound water in apo and holo bovine heart fatty-acid-binding protein determined by heteronuclear NMR spectroscopy. Eur. J. Biochem. 251:781–786

Otting G, Lieppinsh E (1995) Protein Hydration viewed by high-resolution NMR Spectroscopy: implications for Magnetic Resonance Image Contrast. Acc. Chem. Res. 28:171–177

Palmer AG 3rd, Rance M, Wright JG (1991) Intramolecular motions of a zinc finger DNA-binding domain from Xfin characterized by proton-detected natural abundance carbon-13 heteronuclear NMR spectroscopy. J.Am.Chem.Soc. 113:4371–4380

Palmer AG 3rd (2001) NMR probes of molecular dynamics: overview and comparison with other techniques. Annu. Rev. Biophys. Biomol. Struct. 30:129–155

Palmer AG 3rd, Kroenke CD, Loria JP (2001) Nuclear magnetic resonance methods for quantifying microsecond-to-millisecond motions in biological macromolecules. Methods Enzymol. 339:204–238

Peng JW, Wagner G (1994) Investigation of protein motions via relaxation measurements. Methods Enzymol. 239:563–596

Randall DW, Gamelin DR, LaCroix LB, Solomon EI (2000) Electronic Structure contributions to electron transfer in blue Cu and CuA. JBIC 5:16–19

Ronk M, Shively JE, Shute EA, Blake RC 2nd (1991) Amino acid sequence of the blue copper protein rusticyanin from *Thiobacillus ferrooxidans*. Biochemistry 30:9435–9442

Shoham M (1992) Rusticyanin: Extremes in acid stability and redox potential explained by the crystal structure. J. Mol. Biol. 227:581–582

Thompson GS, Leung YC, Ferguson SJ, Radford SE, Redfield C (2000) The structure and dynamics in solution of Cu(I) pseudoazurin from *Paracoccus pantotrophus*. Protein Sci. 9:846–858

Tjandra N, Szabo A, Bax A (1996) Protein Backbone Dynamics and 15N Chemical Shift Anisotropy from Quantitative Measurement of Relaxation Interference Effects. J. Am. Chem. Soc. 118:6986–6991

Walsh ST, Lee AL, DeGrado WF, Wand AJ (2001) Dynamics of a de novo designed three-helix bundle protein studied by 15N, 13C, and 2H NMR relaxation methods. Biochemistry 40:9560–9569

Walter RL, Ealick SE, Friedman AM, Blake RC 2nd, Proctor P, Shoham M (1996) Multiple wavelength anomalous diffraction (MAD) crystal structure of rusticyanin: a highly oxidizing cupredoxin with extreme acid stability. J. Mol. Biol. 263:730–751

Wider G (1998) Technical aspects of NMR spectroscopy with biological macromolecules and studies of hydration in solution. Prog. NMR Spectrosc. 32:193–275

Zinn-Justin S, Berthault P, Guenneugues M, Desvaux H (1997) Off-resonance RF fields in heteronuclear NMR. Application to the study of slow motions. J. Biomol. NMR 10:363–372

3 Structure and Dynamics of Proteins in Crowded Media: A Time-Resolved Fluorescence Polarization Study

Silvia Zorrilla, German Rivas and Maria Pilar Lillo

3.1 Macromolecular Crowding in Physiological Media

The interior of cells in all living organisms, without exception, has a common feature: the high total concentration of macromolecules they contain, which occupy a considerable fraction (20–30 %) of the total volume. For example, macromolecular concentration of the *E. coli* intracellular media is around 400 mg/ml (Record et al. 1998, Zimmerman and Trach 1991), the concentration of hemoglobin in the cytoplasm of an erythrocyte is 330 mg/ml (Ralston 1990), and the total concentration of soluble proteins in blood plasma is around 80 mg/ml (Ellis 2001). In general, there is no single species whose concentration is so high, but the overall macromolecular content of the system gives rise to this high macromolecular concentration. For this reason physiological systems are referred to as crowded rather than concentrated (Minton 2001).

This characteristic of physiological environment is very often overlooked, and in typical biochemical/biophysical in vitro experiments the total concentration of macromolecules hardly ever surpasses 1 mg/ml (Ellis 2001, Minton 2001). There is consensus on the need to use experimental conditions that are as close as possible to the physiological ones in terms of pH, temperature and ionic strength, but in general, the crowded nature of biological fluids is not regarded. On the other hand, it has been recognized for a long time that the behavior of an individual protein in crowded media may be significantly different from the corresponding behavior in a diluted fluid. Moreover, crowding has been predicted (and in many cases experimentally tested) to have major influences on a broad range of biochemical reactions and processes physiologically relevant, including protein stability (chaperone-assisted) protein folding and folding pathways, amyloid formation, and the energetic and dynamics of protein interaction networks (protein–protein and protein-DNA complexes, supramolecular assemblies and fibers, reviewed in Ellis 2001; Minton 2001).

Principles and Practice
Methods in Proteome and Protein Analysis
R.M. Kamp, J. J. Calvete, T. Choli-Papadopoulou (Eds.)

3.1.1 Effect of Macromolecular Crowding on Chemical Equilibrium of Macromolecular Association Reactions

The high total concentration of macromolecules, present in almost all physiological media, can give rise to non-specific interactions among the species. These interactions can have a great effect upon the energetic and kinetics of macromolecular reactions taking place in such media. Non-specific interactions are those derived from global properties of macromolecules like the global charge and polarity, or the size and shape, more than from specific details of their primary or secondary structure. In dilute solutions, the intermolecular separation is really large, and the hydrodynamic interactions between macromolecules can be neglected. In highly concentrated solutions, the steric repulsion between macromolecules, which is always present at finite concentrations, is an important interaction to be examined (Minton 1997). As two macromolecules cannot occupy the same position in the space, the presence of a high concentration of macromolecules in a solution places restrictions in the positions that new macromolecules added can occupy. So, the position that a macromolecule can occupy in the solution is restricted to the volume from which it is not excluded (available volume). This has considerable energetic consequences (see Minton 2001). Excluded volume increases the chemical potential of all macrosolutes in the solution in a size and shape-dependent manner (Ellis 2001, Minton 2001). Moreover, crowding affects equilibria by preferentially stabilizing those states of the system excluding the least volume, relative to those excluding the most volume. Therefore, crowding provides a non-specific force for macromolecular compaction and association in crowded solutions.

3.1.2 Experimental Approaches to the Study of the Effect of Macromolecular Crowding Upon Biochemical Reactions

The quantitative characterization of excluded volume effects is essential for a better understanding of biochemical reactions and processes in physiological media. However, the study of the structure and dynamics of macromolecules in crowded media is a difficult experimental challenge. Most of the difficulties are mainly derived from the presence of a high number of species, with similar size to the *target labeled macromolecule*. Factors like high UV absorbance's values, refractive index changes, and the scattering and fluorescent backgrounds, present in these media, have to be really taken into account in the experimental design. On the other hand, the deviations from the ideal behavior observed from these samples make it difficult to gauge the correct interpretation of the experimental results. One important point in the application of hydrodynamic techniques in crowded media will be the characterization of the translational and rotational diffusion in these media, in terms of local viscosities, and sizes and shapes of the diffusing macromolecules.

Sedimentation equilibrium is among the very few methods which has been proven useful to study the behavior of proteins in crowded solutions. The theory of sedimentation equilibrium has been extended and successfully applied to characterize the state of association of dilute fibrinogen and tubulin (Rivas et al. 1999), and the bacterial cell division FtsZ protein (Rivas et al. 2001) in concentrated solutions of unrelated macromolecules. Recently, these methods have been also applied to characterize the state of association of ribonuclease A (RNase A) at concentrations up to 200 mg/ml (Zorrilla et al. 2003).

Fluorescence spectroscopy is a sensitive technique, widely used to quantify and characterize macromolecules in diluted solutions. It works like a molecular clock. It allows a study of all of those events which occur in times comparable to the fluorescence lifetime. Fluorescence anisotropy techniques provide information on the size, shape and flexibility of the macromolecules in diluted solutions, based on the depolarization induced by their rotational Brownian motions. The extension of the anisotropy methodologies to the characterization of both the rotation of the macromolecules as a whole and their segmental flexibility in highly concentrated solutions is a starting field (Zorrilla 2002). The use of extrinsic fluorophores overcomes most of the background problems and, in principle, allows the characterization of labeled proteins in the presence of other unrelated macromolecules. The only work to which fluorescence spectroscopy was previously applied was a study of interactions in a modeled crowded system by Wilf and Minton (1981). This is a steady-state anisotropy study that compares apomyoglobin (apoMb) and apohemoglobin (apoHb) hydrodynamic behavior in the presence of high concentrations of other unrelated proteins. ApoMb is a monomer and apoHb is a flexible dimer in diluted solutions (Sassaroli et al. 1986). The conclusion from Wilf and Minton study (1981) was that low concentrated ApoMb self-associates in crowded media, when the crowder species were proteins like RNase A, lysozime, etc. In a previous sedimentation equilibrium study, Minton and Lewis (1981) found that myoglobin also self-associates to a dimer and very probably to higher oligomers as well, in highly concentrated solutions of myoglobin (>150 mg/ml).

We have further extended the previous steady-state anisotropy study of apoMb, introducing time-resolved polarization methodologies to characterize the structure and dynamics of apoMb homodimer in these media.

3.2 Application of Time-Resolved Fluorescence Polarization Spectroscopy in Crowded Media

Fluorescence polarization spectroscopy has been widely used as a tool to study the structure, rotational dynamics and interactions of biomolecules in diluted solutions (Jameson and Sawyer 1995, Lakowicz 1999, Brown and Royer 1997). If the exciting light is made to be plane-polarized, it will result in a photoselection of those fluorophores attached to the protein, whose excitation transition moment has a component parallel to the plane of polarization. The emission anisotropy, measured as a function of the time, $r(t)$, will contain information about all the depolarization processes that happen during the fluorescence lifetime (electronic, energy transfer, motional effects, etc.). Dynamic processes in proteins occur on a wide range of time scales. It is important to note that only motions that change the orientation of the transition dipoles during the fluorescence time window will be detected in a fluorescence anisotropy decay. The use of extrinsic fluorophores permits a selection of this time window. In general, $r(t)$ will be analyzed in terms of a series of exponentials:

$$r(t) = \sum_j \beta_j \cdot \exp(-t/\phi_j) \tag{3.1}$$

where ϕ_j are the rotational correlation times.

A compact globular molecule in aqueous diluted solution, having a rigidly attached chromophore, is usually expected to show monoexponential anisotropy decay due to the overall global protein motion. The rotational correlation time, ϕ_G, is a function of the molecular hydrodynamic volume, V_H, the temperature, T, and the solvent viscosity, η, (Stokes-Einstein-Debye relationship):

$$\phi_G = (\eta V_H)/(k_B T) \tag{3.2}$$

In contrast to rigid particles, flexible particles will present a variety of conformations. In the simplest case, the anisotropy decays, $r(t)$, will be adequately described as a biexponential function:

$$r(t) = r(0)\left[\beta_S \cdot \exp(-t/\phi_S) + \beta_G \cdot \exp(-t/\phi_G)\right] \tag{3.3}$$

where β_S and β_G represent the amplitudes of the segmental motion (ϕ_S) and the overall global protein motion (ϕ_G).

The presence of highly concentrated unrelated proteins in the crowded media explored in this work, introduces complexity in the interpretation of the experimental anisotropy decays. The *solvent* is not homogeneous now and cannot be considered as a continuum. Therefore, the Stokes-Einstein-Debye hydrodynamic model cannot describe the diffusive behavior of proteins in these media. The rotational diffusion of the labeled macromolecules will be affected by collisions with the solvent molecules, and the unrelated molecules. In this case, the determined rotational correlation times may be described by a quasi-hydrodynamic relation: $\phi=(\eta_R V_H)/(k_B T)$, where η_R represents the "local rotational viscosity". Relative values of this apparent viscosity to the buffer viscosity can be obtained from the ratio of the rotational correlation times: $\phi/\phi_0=\eta_R/\eta_0$, where ϕ and ϕ_0 are the rotational correlation times of the protein in crowded media and buffer solution respectively, and η_0 is the buffer viscosity.

3.3 Volume Fraction and Intermolecular Separations in a Heterogeneous System

One important point in the study of biomolecular interactions in crowded media, using time-resolved anisotropy techniques, is to have an idea about the intermolecular separations between the different solute species. In this work, the crowded media was simulated by adding high concentrations of a non-related protein (250 mg/ml RNase A) to the diluted protein, the object of the self-association study (apoMb). Below, we propose an idealized simple microscopic model of the system that gives an estimation of the order of the collision times between macromolecules in highly concentrated solutions. This molecular view may help in the interpretation of the experimental rotational correlation times determined in different crowded media, in terms of hydrodynamic protein volumes, and the rotational viscosity.

3.3.1 Characterization of the Crowded Medium Itself

Highly concentrated solutions of RNase A alone were characterized by sedimentation equilibrium techniques (Zorrilla et al. 2003), and the conclusion was that RNase A presents weak self-association in concentrated solutions. The simplest models compatible with the experimental data assume RNase A monomer in equilibrium with a trimer, a tetramer, or a dimer and tetramer mixture. In the present study, for simplicity, we worked with the species distribution from the monomer-tetramer model, and we assumed spherical shape for all the proteins. In Table 3.1, we present the concentration, the number of molecules, the calculated radius, and the estimated occupied volume fraction for each protein, corresponding to a solution containing 1.7 mg/ml of apoMb dimer (10^{-4} M per monomeric subunit), and 250 mg/ml of RNase A.

3.3.2 Microscopic Model for Crowded Solutions

In order to be able to estimate intermolecular distances in highly concentrated RNase A solutions, we assumed that the microenvironment of individual apoMb, and RNase A proteins presented some *microscopic order* in the time window of our time-resolved fluorescence anisotropy experiments. In three-dimensions, one of the optimal dense sphere packing is the *cannon ball*

Table 3.1. Characterization of the different species for a solution containing apoMb dimer (1.7 mg/ml, 10^{-4} M per monomeric subunit), and 250 mg/ml of RNase A

	v_i (cm^3/g)	M_i (g/mol)	c_i (mg/cm^3)	N_i[a] (molecules/ $10^{-15}cm^3$)	R_{Hi}[b] (Å)	γ_i[c]
RNase momomer	0.703[d]	13,690	88	3870	17.6	0.092
RNase tetramer	(0.703)[c]	54,760	162	1780	27.9	0.168
ApoMb dimer	(0.743)[e]	34,000	1.7	30	24.1	0.002

[a] Number of molecules in a solution cubic element volume of $(100\ nm)^3$

[b] Calculated hydrodynamic radius of the three proteins assuming spherical shape. The corresponding hydrodynamic volume, $V_H = M(\bar{v}_i + h\bar{v}_0)/N_A$, was estimated assuming h=0.3 g H_2O/g protein, for hydration water (Cantor, 1980). $\bar{v}_i$ and $\bar{v}_0$ are the partial specific volume of protein i, and of hydration water, respectively. N_A is Avogadro's number

[c] Fractional volume occupied by protein i, $\gamma_i = c_i(\bar{v}_i + h\bar{v}_0)10^{-3}$, where c_i is the protein concentration

[d] Richards and Wyckhoff (1971); we assume the same value for RNase A tetramer

[e] Ehrenberg (1957) found the partial specific volume of apoMb to be identical to Mb. We assumed the same value for ApoMb dimer

packing where the centers of identical spheres are arranged on the face centered cubic (fcc) lattice (Conway and Sloane 1999). This is the packing usually found in fruit stands (note that in this packing the spheres will have a coordination number of 12 and they would occupy a fraction of the total volume, $\gamma_{max}=\pi\sqrt{2}/6=0.7405$, named as maximum packing density). We partitioned the 3-D space (total volume, V_T) in N_T spherical compartments, with volume V_c, where $V_C=(V_T \cdot \gamma_{max})/N_T$. The calculated radius of these compartments (see Table 1) for a solution containing 1.7 mg/ml of apoMb dimer (10^{-4} M per monomeric subunit), and 250 mg/ml of RNase A, is R_c=30.9 Å. If we assume that in one instant time t, each protein is enclosed in the center of each spherical compartment, the "instant" edge-to-edge separation between different adjacent molecules, d_{ij}, can be calculated as: $d_{ij}=(R_c-R_i)+(R_c-R_j)$, and d_{21} =20 Å, d_{24}=10 Å (where the subscripts *2*, *1*, and *4* refer to apoMb dimer, RNase A monomer, and RNase A tetramer, respectively). One of the biggest assumptions of this model is that the protein molecules are maximally and symmetrically distributed in solution, for protein concentrations near the maximal value. The macromolecular self association detected in highly concentrated RNase A solutions introduces size heterogeneity in the molecular distribution of our system. In principle, this heterogeneity in size would cause less compact packing, and the estimated intermolecular distances with the previous model would represent a lower limit for an average intermolecular distance in these solutions. The proposed microscopic model is an extension of the model by Endre and Kuchel (1986) for single species solutions. Our goal was to idealize the molecular microenvironment of apoMb in highly concentrated solutions of RNase A, proposing a certain short range order (at distances ~10–20 Å), to interpret the experimental diffusive behavior of apoMb dimer in complex solutions. A dimer of ApoMb in aqueous solution at 20 °C, with an approximate diffusion coefficient, $D_t \sim 8.6\times10^{-7}\,s^{-1}cm^2$ [calculated through the Stokes-Einstein equation, assuming spherical shape, $D_t=(k_B T)/(\eta\pi R_H)$] would be able to diffuse 10–20 Å (range of the estimated separations between ApoMb and the nearest RNase A molecule, monomer or tetramer), in about 2–8 ns. These are the estimated collision times for particles that have an average separation, d_{ij}, from the Einstein relationship: $t=(\langle d\rangle^2/6D_t)$, assuming $\langle d\rangle^2 \sim d_{ij}^2$.

3.4 Structure and Dynamics of apoMb Dimer in Crowded Protein Solutions

ApoMb is a protein which in diluted solution at pH 7 is a monomer, but in the presence of high concentrations of proteins like RNase A, β-lactoglobulin or lisozyme tends to self-associate to form dimers (Wilf and Minton 1981). The structure of apoMb monomer very closely resembles that of the hemoglobin subunits. Hemoglobin is a tetrameric protein, but the removal of the heme group changes its association state, so apoHb is a dimer (Sassaroli et al. 1984).

Wilf and Minton (1981) did a comparative study of apoMb and apoHb in crowded solutions. They showed that the steady state anisotropy of a solution of ApoMb labeled with 1,8-anilinonaphthalenesulfonic acid (ANS) increases with increased concentrations of unrelated proteins like RNase A. The steady state anisotropy of apoMb-ANS gradually approaches the anisotropy of a solution with the same concentration of apoHb-ANS. This happened only when the crowder was modeled with concentrated solutions of some proteins. At 250 mg/ml of RNase A, the steady-state anisotropy ApoMb-ANS coincides, within the error, with the determined for apoHb-ANS, which seems to indicate that at this concentration of RNase A all or almost all apoMb behaves hydrodynamically like apoHb dimers.

In this work, we have tried to study the structure and dynamics of the dimer of apoMb in the presence of high concentrations of a non-related protein, RNase A, by using time-resolved fluorescence anisotropy methodologies.

3.4.1 Preparation of Apomyoglobin and Labeling with ANS

ApoMb was prepared from commercial Mb by removing the heme group using a modification of the acid-acetone method (Rossi-Fanelli et al. 1958) as described in Wilf and Minton (1981). The efficiency of removal of the heme group was approximately 98 %, and the amount of the recovered apoMb was approximately 60 % of the initial myoglobin.

ApoMb was non-covalently labeled with ANS, which probably binds with apoMb in the hydrophobic heme pocket, previously occupied by the heme group (Stryer 1965). The ratio of labeling was 0.8 mol of ANS per mol of apoMb, calculated assuming a binding constant for ANS of 3.4×10^{-6} M (Stryer 1965). ApoMb samples (1.7 mg/ml) labeled with ANS, in phosphate buffer solutions (20 mM sodium phosphate, 150 mM NaCl, 0.1 mM EDTA, pH 7.4), were prepared in the presence, and in the absence of 250 mg/ml of RNase A.

3.4.2 Spectroscopic Properties of ANS Remain Essentially Unchanged Upon Dimer Formation

Before studying the rotational behavior of apoMb-ANS in crowded solutions, it is necessary to carefully determine the spectroscopic parameters of the fluorophore ANS when it is bound to apoMb. Corrected emission spectra of apoMb-ANS samples alone, and in the presence of 250 mg/ml of RNase A, were collected in an SLM 8000D spectrofluorometer at an excitation wavelength of 393 nm. ANS is a fluorophore whose fluorescence quantum yield increases over 200 fold upon binding to apoMb, and free ANS in aqueous solution does not contribute significantly to fluorescence emission. Besides, the fluorescence emission of a solution of 250 mg/ml of RNase A with or without ANS, were approximately the same, which indicates that ANS does not bind RNase A in the concentration range of this work. These two properties mentioned make ANS a suitable probe for the current study. The contribution of fluorescence background was always subtracted from the spectra, and it was less than 4 % of the total fluorescence intensity for a 250 mg/ml RNase A solution.

ApoMb-ANS in 250 mg/ml RNase A solutions showed a 20 % decrease in the total fluorescence intensity in relation to the determined in buffer solutions, maintaining essentially the same emission spectra shape. This intensity decrease is probably due to the inner filter effect of highly concentrated RNase A solutions with non-zero absorption at 393 nm.

Time-resolved fluorescence measurements were done by the time-correlated single photon technique with the instrument described in Organero et al. (2002). Fluorescence intensity decays were collected under magic angle conditions at excitation and emission wavelengths of 393 and 465 nm, respec-

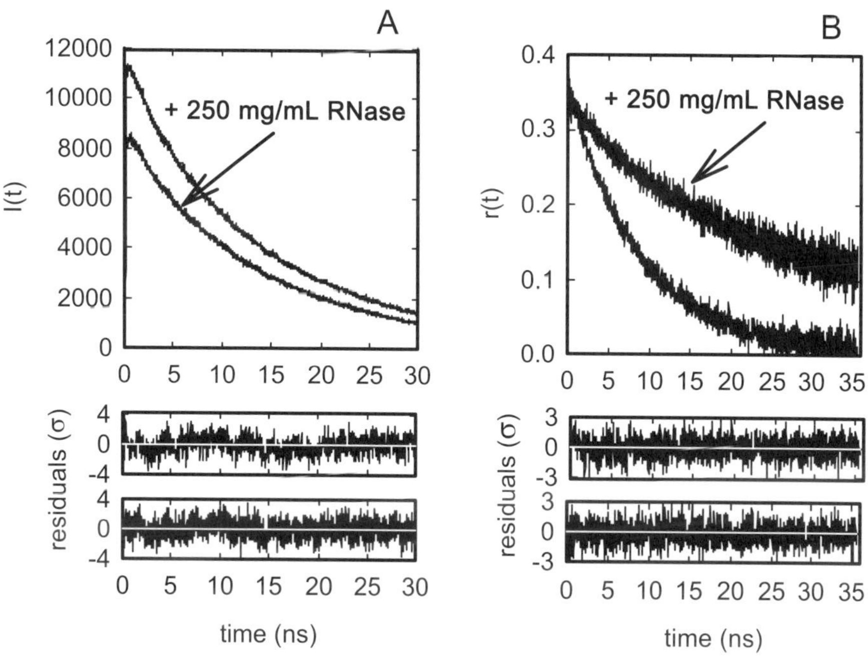

Fig. 3.1. ApoMb-ANS in 20 mM sodium phosphate buffer, 150 mM NaCl, 0.1 mM EDTA, pH 7.4, at 20 °C; protein concentration 10^{-4} M per monomeric subunit (0.8 mol of ANS/mol of apoMb monomer), in the absence and presence of 250 mg/ml RNase A. **A** Fluorescence decay. **B** Anisotropy decay. The line fits are superposed to the data, and the weighted residuals for the fits are presented below each decay. Fluorescence lifetimes and anisotropy parameters from the fits are reported in Tables 3.2 and 3.3

tively. Figure 3.1A shows the fluorescence intensity decays of apoMb-ANS in buffer, and in 250 mg/ml RNase A solutions. Both curves were biexponential, the corresponding decay parameters are presented in Table 3.2.

This biexponential behavior has been observed before for ANS labeled proteins (Robinson et al. 1978), and it could be related to the solvation of ANS in the heme pocket (Sassaroli et al. 1984).

As can be seen in Table 3.2 the lifetime of 15 ns of apoMb-ANS does not change in highly concentrated RNase A solutions. However, the shortest lifetime seems to increase with increasing concentrations of RNase A, while its contribution to the decay decreases. The contribution of this short lifetime to the total intensity is very low and the intensity-averaged, τ_m, and the amplitude-averaged, $<\tau>$, lifetimes were approximately the same within the error, for apoMb-ANS in buffer, and in 250 mg/ml RNase A solutions. Since there is not change in the photophysical properties of the fluorophore ANS bound to apoMb, the protein microenvironment of ANS may be maintained in apoMb

Table 3.2. Fluorescence intensity decay parameters of ApoMb-ANS buffer solutions and ApoMb-ANS (1.7 mg/ml, 10^{-4} M per monomeric subunit) in 250 mg/ml RNase A solutions (0.8 mol of ANS/mol of apoMb monomer). The measurements were performed at 20 °C in 20 mM phosphate, 150 mM NaCl, 0.1 mM EDTA, pH 7.4. λ_{exc}=393 nm; λ_{em}=465 nm

RNase A (mg/ml)	$a_1 \pm 0.02$[c]	τ_1(ns)± 0.3[c]	$a_2 \pm 0.02$[c]	τ_2(ns)± 0.1[c]	$<\tau>$[a] (ns)	τ_m[b] (ns)
0	0.16	3.4	0.84	15.2	13.3	14.7
250	0.15	5.5	0.85	15.2	13.7	14.6

[a] Amplitude average lifetime $<\tau>=\sum a_i \tau_i$
[b] Intensity average lifetime $\tau_m=(\sum a_i \tau_i^2)/(\sum a_i \tau_i)$
[c] Confidence intervals corresponding to 1 SD

dimer. On the other hand, the predominant fluorescence lifetime of apoMb-ANS of 15 ns establishes a good time window for conformational dynamic studies of molecules in the range of apoMb dimer size.

3.4.3 Conformational Dynamics of the Dimer of Apomyoglobin

Steady-state anisotropy measurements, $\bar{r}$, of apoMb-ANS samples in phosphate-buffered solution at 20 °C, were performed in an SLM 8000D spectrofluorometer at two excitation wavelengths, 375 and 393 nm. The anisotropy values (λ_{exc} 375 nm; λ_{em} 465 nm) agree within the error with values reported by Wilf and Minton (1981) for apoMb-ANS in RNase A solutions 0–250 mg/ml (Zorrilla 2002). The steady-state anisotropy increases from 0.14 to 0.22 (λ_{exc} 393 nm; λ_{em} 465 nm), when apoMb solutions contain 250 mg/ml of RNase A.

ApoMb-ANS in the absence of RNase A (λ_{exc}=393 nm, λ_{em} =465 nm) showed a monoexponential anisotropy decay (Fig. 3.1B), with a time cero anisotropy, $r(0)$=0.35, close to the maximum expected for ANS chromophore (Hudson and Weber 1973). The determined rotational correlation time, ϕ_G=9.0±0.2 ns at 20 °C (see Table 3.3), can be assigned to the global rotational motion of a monomer of apoMb. This value works in accordance with the determined from the tryptophan fluorescence anisotropy study of apoMb (Tcherkasskaya et al. 2000), and a NMR study of myoglobin (Wang et al. 1997). It is larger than the estimated value for a spherical monomer at 20 °C, with a molecular weight of 17,000, and a hydration of 0.3 g H_2O/g protein (7.4 ns). This difference indicates a slight deviation from the spherical shape of apoMb monomer (Tcherkasskaya et al. 2000). On the other hand, the determination of a single rotational correlation time shows that there are no local movements of ANS inside the heme pocket. This property is quite

Table 3.3. Fluorescence anisotropy decay parameters of ApoMb-ANS (1.7 mg/ml, 10^{-4} M per monomeric subunit; 0.8 mol ANS/mol apoMb monomer) in buffer and in 250 mg/ml RNase A solutions. The measurements were performed at 20 °C in 20 mM phosphate, 150 mM NaCl, 0.1 mM EDTA, pH 7.4. λ_{exc}=393 nm; λ_{em}=465 nm

RNase A (mg/ml)	$r(0)\pm0.01$[a]	β_S	ϕ_S (ns)	β_G	ϕ_G (ns)
0	0.35	-	-	1	9.0 (8.8–9.2)[a]
250	0.34	0.4 (0.35–0.47)[a]	12.4 (10.7–14.6)[a]	0.6 (0.53–0.65)[a]	54 (45–69)[a]

[a] Confidence intervals corresponding to 1 SD

suitable for hydrodynamic anisotropy studies, because the presence of local movements of the probe would decrease the resolution of the anisotropy decays at longer times, from which the global rotational motion parameters are determined.

Fluorescence anisotropy decays of apoMb-ANS in solutions where the concentration of RNase A was 250 mg/ml (Fig. 3.1B) were fitted to a biexponential function, with two well-separated correlation times (see Table 3.3). The fast one, ϕ_S=12.4 (10.7–14.6) ns, accounts for approximately 40 % of the depolarization. The remaining anisotropy decays through the slow rotational component, which is in the limit of our time window, ϕ_G=54 (45–69) ns. The interpretation of these two rotational correlation times, determined in highly concentrated RNase A solutions, in terms of rotational motions of apoMb-ANS is not straightforward. As we showed before, in diluted solutions the rotational correlation times are a function of the viscosity and the size of the rotating unit. However, the microscopic viscosity that acts over the rotational diffusion of a macromolecule like apoMb-ANS in this crowded solution is not the actual macroviscosity (bulk viscosity) of the solution, and in fact it can be very different (Gavish 1980, Lavalette et al. 1999, Endre and Kuchel 1986). For this reason, a detailed time-resolved anisotropy study of apoMb-ANS was performed, in the presence of various RNase A concentrations, up to 250 mg/ml (Zorrilla 2002). Two rotational correlation times were determined for all the apoMb-ANS/RNase A solutions, and their values were represented as a function of RNase A concentration. The extrapolation of the fast rotational correlation time to cero concentration of RNase A, returns a value that agrees, within the error, with the global movement of a monomer of apoMb. A similar extrapolation for the slow rotational correlation time, returns a value of 22±2 ns that is in good agreement with the rotational correlation time reported for the global motion of apoHb, 22.3±0.8 ns (Sassaroli et al. 1986).

The simplest model that supports our time-resolved anisotropy results would associate the slow rotational correlation time, ϕ_G, to the global rotation of a dimer of apoMb-ANS and the fast one, ϕ_S, to the global rotation of a monomer of apoMb, both species coexisting in the solution. From the anisotropy fractional amplitudes, β_i, the proportion of apoMb monomer in 250 mg/ml RNase A solutions would be 40%. This simple model assumes that ApoMb monomer and dimer behave as rigid particles in a solution.

The macroscopic viscosity of a 250 mg/ml RNase A solution, at 20 °C, measured with a capillary Ostwald microviscometer, was nearly 4.5 cP. The calculated global rotational correlation times for a monomer and a dimer of apoMb, assuming spherical shape, in a glycerol solution with a viscosity of 4.5 cP, would be 33 ns and 66 ns respectively (note that in glycerol, the bulk viscosity and the rotational viscosity would be the same). These are approximated values, since the separation from the spherical shape, especially for the dimer molecules, would give rise to longer rotational times.

The assumption of monomer and dimer species of apoMb coexisting in 250 mg/ml RNase A solutions is not compatible with the previous results from Wilf and Minton (1981). At this RNase A concentration, the hydrodynamic behavior of apoMb-ANS is the same as the corresponding to apoHb-ANS, so all or almost all apoMb-ANS should be forming dimers in the presence of 250 mg/ml of RNase A. The only reasonable explanation for this discrepancy is that the dimer of apoMb-ANS is not a rigid specie, but presents internal flexibility characterized by local motions of the two monomeric subunits of apoMb-ANS. This hypothesis is supported by the fact that internal flexibility that involves motions of the monomeric subunits has previously been found for apoHb (Sassaroli et al. 1986). According to this model, the motions of the monomeric subunits of apoMb dimer will be less affected by the presence of RNase A molecules at distances of about 10–20 Å, than the global rotational motion, and the estimated rotational viscosity ratio, ϕ_i/ϕ_i^0, will be 1.4 and 2.5 for the internal flexibility and the global motion respectively.

Time-resolved fluorescence anisotropy data are obtained from a large number of excited apoMb-ANS molecules, therefore the fast rotational time, ϕ_S, and the global rotation time, ϕ_G, determined from this methodology, would represent an average of the local and overall dynamic processes of apoMb that happens during the fluorescence lifetime of ANS. Going back to the microscopic model of this system proposed beforehand, the fluorescence lifetime of ANS is in the same order as the estimated collision times between the different species, so that would explain the observed behavior of the rotational viscosity associated with the different motions.

3.5 Conclusions and Outlook

The studies presented in this article illustrate how time-resolved fluorescence polarization techniques can be used to study the hydrodynamic behavior of dilute labeled proteins in highly concentrated solutions of unrelated proteins resembling the physiological milieus. The extension of these studies to different crowder systems, varying the relative sizes and concentrations of the tracer and the crowder molecules, is one of the interests of our laboratory. These studies will allow us to obtain general laws which relate diffusion parameters such as the rotational correlation times, with measurable magnitudes like the macroscopic viscosity. These studies will help the identification and characterization of interactions in crowded media. It will be important also to optimize the design of new high throughput interaction assays in crowded media, based on fluorescence anisotropy methodologies.

Acknowledgements. We wish to thank Dr. A. Douhal for the use of the time-resolved instrument. We would also like to thank Dr. A.U. Acuña, Dr. J. Garcia de la Torre, and Dr. A.P. Minton for stimulating discussions. This work was financed by grants BIO99-0859-C03-03 and BQU/2000-1500 from the Spanish Direccion General de Enseñanza Superior e Investigación (DGESI). S.Z. was supported by a fellowship from the Comunidad de Madrid (CAM)

References

Brown MP, Royer C (1997) Fluorescence spectroscopy as a tool to investigate protein interactions. Current Opinion in Biotechnology 8:45–49

Cantor CR, Schimmel PR (1980) Size and Shape of macromolecules. In: Biophysical Chemistry. Part II: Techniques for the study of biological structure and function. WH Freeman and Company, San Francisco, pp 539–590

Conway JH, Sloane NJA (1999) Sphere packing and kissing numbers. In: Sphere packings, lattices and groups. Springer-Verlag, New York, pp 1–30

Ehrenberg A (1957) Determination of molecular weight and diffussion coefficients in the ultracentrifugue. Acta Chem Scand 11:1257–1270

Ellis RJ (2001) Macromolecular crowding: an important but neglected aspect of the intracellular environment. Curr. Opin. Struct. Biol. 11:114–9

Endre ZH, Kuchel PW (1986) Viscosity of concentrated solutions of human erythrocyte cytoplasm determined from NMR measurement of molecular correlation times. The dependence of viscosity on cell volume. Biophys Chem 24:337–356

Gavish B (1980) Possition /dependent viscosity effects on rate coefficients. Phys. Rev. Lett. 1160–1163

Hudson EN, Weber G (1973) Synthesis and characterization of two fluorescent sulfhydryl reagents. Biochemistry. 12:4154–4161

Jameson DJ, Sawyer WH (1995) Fluorescence anisotropy applied to biomolecular interactions. Meth Enzymol 246:283–300

Lakowicz JR (1999) In: Lakowicz JR (ed) Principles of fluorescence spectroscopy. 2nd Edition. Kluwer Academic-Plenum Publishers, New York

Lavalette D, Tetreau C, Tourbez M and Blouquit Y (1999) Microscopic viscosity and rotational diffusion of proteins in a macromolecular environment. *Biophys. J.* 76:2744–2751

Minton AP (1997) Influence of excluded volume upon macromolecular structure and associations in crowded media. Curr. Opin. Biotechnol. 8:65–69

Minton AP (2001) The influence of macromolecular crowding and macromolecular confinement on biochemical reactions in physiological media. J. Bio l. Chem. 276:10577–10580

Minton AP, Lewis MS (1981) Self association in highly concentrated solutions of myoglobin: a novel analysis of sedimentation equilibrium of highly non ideal solutions. Biophys. Chem. 14:317–324

Organero JA, Tormo L and Douhal A (2002) Caging ultrafast proton transfer and twisting motion of 1-hidroxi-2-acetonaphthona. Chem. Physics letters. 363:409–414

Ralston GB (1990) Effects of crowding in protein solutions. Journal of chemical education. 67:857–860

Record MT Jr, Courtenay ES, Cayley S and Guttman HJ (1998). Biophysical compensation mechanisms buffering *E. coli* protein-nucleic acid interactions against changing environments. Trends. Biochem. Sci. 23:190–194

Richards FM, Wyckhoff HW (1971) Bovine pancreatic ribonuclease. In: Boyer PD (ed) The Enzymes, Vol 4, Academic Press, New York, pp 647–806

Rivas G, Fernandez JA, Minton AP (1999) Direct observation of the self-association of dilute proteins in the presence of inert macromolecules at high concentration via tracer sedimentation equilibrium: theory, experiment, and biological significance. Biochemistry. 38:9379–9388

Rivas G, Fernandez JA, Minton AP (2001) Direct observation of the enhancement of non cooperative protein self-assembly by macromolecular crowding: independent linear self-association of bacterial cell division protein FtsZ. Proc. Natl. Acad. Sci. USA 98:3150–3155

Robinson, GW, Caughney, TA, Auerbach RA (1978). In: Zewail AH (ed) Advances in Laser Chemistry. Springer, New York, pp 108–125

Rossi-Fanelli A, Antonini E, Caputo A (1958) Structure of hemoglobin. I. Physicochemical properties of human globin. Biochim. et Biophys. Acta 30: 608–615

Sassaroli M, Bucci E, Liesegang J, Fronticelli C, Steiner RF (1984) Specialized functional domains in hemoglobin: dimensions in solution of the apohemoglobin dimer labeled with fluorescein iodoacetamide. Biochemistry. 23:2487–2491

Sassaroli M, Kowalczyk J, Bucci E (1986) Probe dependence of correlation times in heme-free extrinsically labeled human hemoglobin. Arch. Biochem. Biophys. 251:624–628

Stryer L (1965) The interaction of a naphthalene dye with apomyoglobin and apohemoglobin. A fluorescent probe of non-polar binding sites. J. Mol. Biol. 13:482–495

Tcherkasskaya O, Ptitsyn OB, Knutson JR (2000) Nanosecond dynamics of tryptophans in different conformational states of apomyoglobin proteins. Biochemistry. 39:1879–1889

Wang D, Kreutzer U, Chung Y, Jue T (1997) Myoglobin and hemoglobin rotational diffusion in the cell. Biophys. J. 73:2764–2770

Wilf J, Minton AP (1981) Evidence for protein self-association induced by excluded volume. Myoglobin in the presence of globular proteins. Biochim. Biophys. Acta 670: 316–322

Zimmerman SB and Trach SO (1991) Estimation of macromolecule concentrations and excluded volume effects for the cytoplasm of *Escherichia coli*. J. Mol. Biol. 222:599–620

Zorrilla S (2002) Structure and dynamic of proteins in crowded media. Doctoral Thesis. Univ. Complutense Madrid

Zorrilla S, Jiménez M, Lillo MP, Rivas G, Minton AP (2003) Sedimentation equilibrium in a solution containing an arbitrary number of solute species at arbitrary concentrations: Theory and application to concentrated solutions of ribonuclease. Biophys Chem (in press)

4 Analyses of Wheat Seed Proteome: Exploring Protein-Protein Interactions by Manipulating Genome Composition

NAZRUL ISLAM and HISASHI HIRANO

4.1 Summary

Current knowledge of protein–protein interactions in proteome is very limited due to the lack of high throughput and efficient experimental techniques at the protein levels. In this chapter, we briefly discuss the techniques currently available to determine protein-protein interactions and then focus on a novel technique developed by manipulating genome composition in common wheat. Genome manipulation has been carried out either by deleting a chromosomal arm, ditelocentric lines (DT), or a part of a chromosome, fine deletion line. The network of protein-protein interactions in the manipulated genome was then investigated by quantitative analysis of the expressed proteins. Out of the 1755 major spots detected in 39 DT lines: 1372 (78 %) spots were found variable in different spot parameters, 147 (11 %) disappeared, 978 (71 %) up-regulated and 247 (18 %) down-regulated. High correlations in changes of protein intensities among the proteins were observed. Analysis of proteome in a fine deletion line by isotope-coded affinity tag labeling (ICAT) of peptides in tryptic digest followed by electrospray ionization quadrupole time of flight mass spectrometry method also showed an imbalanced in protein expression. Taking an example of manipulated wheat genome, it has been demonstrated that the expression of proteins in proteomes is not totally independent; rather it is the product of interactions among all other proteins in the proteome.

4.2 Introduction

The complete genomic sequencing of prokaryotes, such as bacteria, viruses, and eukaryotes, such as human, rice, *Arabidopsis* and yeast, has provided an unprecedented amount of genetic information, nucleotide sequences, genetic maps, and DNA markers. The genetic information produced by genome

Principles and Practice
Methods in Proteome and Protein Analysis
R.M. Kamp, J. J. Calvete, T. Choli-Papadopoulou (Eds.)

analysis does not necessarily match quantitatively or qualitatively at the protein levels. Biological influences including the stability, half-life, post-transcriptional, co-translational and degradative modifications of proteins along with the environmental stimulatory factors affect the gene-products (Cordwell et al. 2001). This has led to a conclusion that there is no strict linear relationship between genes and protein expression of a cell (Pandy and Mann 2000). Therefore, to make the best use of genome information and to gain a comprehensive understanding of complex biological processes, a global analysis of the gene products, a study of the proteome, is essential.

The proteome analysis, in general, is performed by (1) separation of proteins in two-dimensional electrophoresis (2-DE), (2) determination of peptide mass fingerprints/amino acid sequences by mass spectrometry (MS), (3) identification of protein/protein homologues using databases and (4) characterization of proteins without known function by amount, localization, structure, post-translational modification, enzyme activity, etc. (Gygi et al. 2000; Woo et al. 2002; Fukuda et al. 2003). Recent developments in MS techniques such as matrix-assisted laser desorption/ionization time of flight (MALDI TOF) and electrospray mass spectrometry (ESI-MS) have provided a unique opportunity to look at the protein complement of whole genome within a very short time. MALDI TOF analysis of proteins, although considered fast and specific, does however present a high mass resolution, which is still poor compared with the ESI-MS. The direct analysis of the total digest of a complex protein mixture by nanoscale capillary high performance liquid chromatography (LC)-ESI-MS has recently become a cutting edge tool for rapid and accurate protein identification. The accuracy of protein determination particularly for complex proteins by LC-ESI-MS has further increased with the recent development and use of isotope based techniques (Smolka et al. 2002; Gygi et al. 1999).

Although the tremendous development of proteomic techniques in the last few years has provided a unique platform of large scale protein identification, a key question about proteins is yet to be resolved, how proteins interact with other proteins to perform particular cellular tasks. In this chapter, we provide an overview of recent approaches used to determine the protein–protein interaction, and demonstrate a novel technique used to explore the network of protein–protein interactions by using manipulated genomes of common wheat.

4.3 Techniques of Protein–Protein Interactions

Most cellular processes are regulated by multiprotein complexes, where the interactions among proteins play a vital role in determining all the biological events in organisms. Several approaches have so far been made to understand these processes. One of the generic ways of identifying the interaction partners of a new protein is to tag it with an epitope. In this technique, the clues to the function of the unknown protein are determined by investigating its reaction with proteins of known functions. The entire protein complex is then purified by affinity based methods (Eisenberg et al. 2000).

The yeast two-hybrid system is one of the most popular techniques which has been adapted to identify and characterize protein–protein interactions. In this system, the eukaryotic transcriptional factors are divided into (1) an activation domain (AD) and (2) a DNA-binding domain (DBD). The activity of these factors is restored by reconstitution. One of the proteins of interest is expressed as a hybrid with AD and the other is expressed as a hybrid with DBD. An interaction between two proteins is measured by a reporter plasmid fused with DBD. Normally, this two-hybrid experiment is performed in yeast, although there are some recent reports of using mammalian cells instead of yeast.

In addition, some other techniques, such as tandem affinity purification (TAP), surface plasmon resonance (eg BIA/MAS), immunoprecipitation and the G protein based screening system, are also available to study protein-protein interactions (Auerbach et al. 2002; Gavin et al. 2002). Recently, Gavin et al. (2002) have studied the large-scale functional organization of yeast protein complex by using TAP. In this technique, the target protein is attached with calmodulin-binding peptides containing TEV mediated proteins. The protein is then allowed to interact with associated proteins. Successive purification is then carried out using two affinity-based columns (IgG beads and calmodulin beads), and the eluted proteins are separated in SDS PAGE and characterized by MS.

4.4 Chromosome Manipulation: An Alternative Approach

4.4.1 Principle

Bread wheat is a hexaploid species with three diploid genomes denoted by A, B and D and is composed of seven pairs of chromosomes in each genome. Because of genetic triplication in wheat, whole or partial deletions of chromosomes tend not to be as lethal as in diploids. Taking this as one of the advantages of common wheat, geneticists have manipulated the genome composition, either by deleting the whole or a part of the chromosome or by introducing a whole chromosome from an alien species. Endo and Gill (1996) produced 436 deletion stock in wheat; one of the important stocks is the ditelocentric lines (DT) which carry all the normal chromosome complements of wheat chromosomes except for one chromosome for which one arm is missing. Recently, using gametocidal genes (*Gc*) of alien wheat species, Tsujimoto et al. (2001) have reported the construction of a fine deletion map of wheat chromosome 1B. The *Gc*, also known as selfish genes, are introduced into the common wheat through backcrossing of hybrids with related alien species, *Aegilops* (Endo 1990). The *Gc* genes, once introduced into the host cell, can cause chromosomal breakage, which can be visualized by C-banding (Tsujimoto et al. 1989; Ogihara et al. 1994; Endo and Gill 1996). This uniqueness of common wheat has enabled cereal scientists to study the mechanisms involving in controlling protein expression or suppression in the manipulated genomes. The deletion of chromosomes or a part of a chromosome gene(s) removed from the genome might cause a serious imbalance in the network of protein expressions in the proteome. We believe an attempt to explore the imbalance of protein expression in the manipulated genome would provide a better understanding of the network of interacting proteins. Using proteomic techniques, we investigate the quantitative changes in protein spots of wheat-endosperm proteome in euploid (hexaploid) and 39 DT lines by analyzing 42 gels in computer assisted image analyzer. A correlation analysis of changes in protein intensities of 24 protein spots across the DT lines was performed. The network of protein interaction has also been studied in fine deletion line by using isotope coded affinity tagging followed ESI-MS/MS technique.

4.4.2 Experimentation

4.4.2.1 Plant Materials

Common wheat (*Triticum aestivum* L.) cv. Chinese Spring and all possible ditelocentric (DT) lines except DT-5DS, DT-5DS and DT-7DL were used in this study. These DT lines were originally produced by Sears (1954), and have been maintained in the gene bank of the Kihara Institute for Biological Research. The codes designated in DT lines are the chromosomal arm present in the lines, and thus, DT-1BS, for example, indicates the plant is lacking the whole long arm of chromosome 1B. Of these 39 lines used, 4 were maintained as ditelocentric for the arm designated and in addition monotelocentric for the opposite arm; these are DT-2AL, DT4BL, 5AS, and 5BS, which were self-pollinated progeny of the ditelo-monotelosomic lines. These seeds were harvested in four different years (1992, 1997, 1996, and 1999) and kept in a desiccator at −20 °C.

The fine deletion line of chromosome1B used in this experiment was produced in the experimental field of the Kihara Institute for Biological Research. A momosomic alien chromosome additional line of common wheat (*T. aestivum* L.) cv. Chinese Spring carrying chromosome 2C of *Aegilops cylindrica* was used to produce fine deletion lines (Tsujimoto et al. 2001). The alien gene causing chromosomal breakage during gametogenesis is known as gametocidal gene (*Gc*). As the deletion of chromosomes is physical, the genome composition between the euploid (full set of chromosomes) and deletion lines is totally identical except in the site of cytologically observed

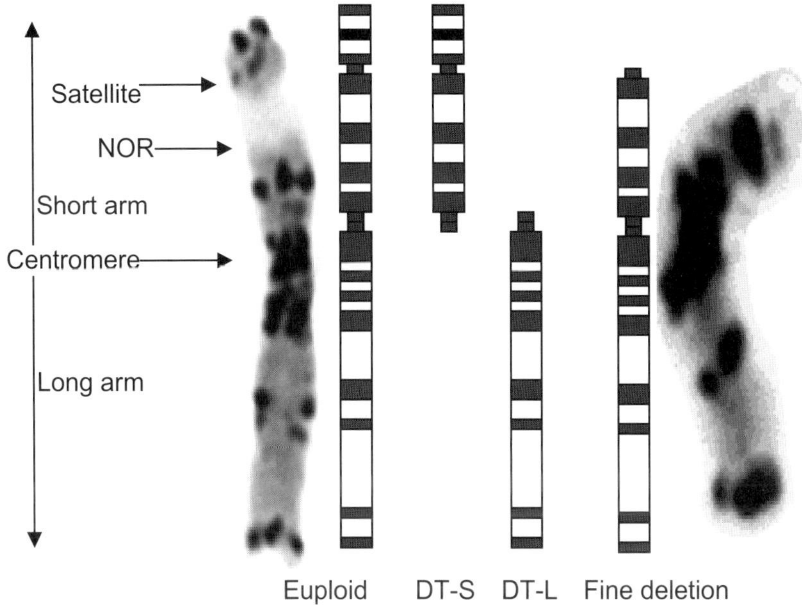

Fig. 4.1. Parameters used to manipulate genome composition in common wheat. Ditelocentric *(DT)* lines carry all the normal chromosome complements of hexaploid wheat (euploid) except for one chromosome for which one arm. In fine deletion a part of the chromosome is missing. *S* denotes short and *L* long arms

chromosome aberration. Parameters used to manipulate genome composition is shown in Fig. 4.1.

4.4.2.2 Two-Dimensional Electrophoresis

Endosperm was dissected from mature grains by a sharp blade; ground and sieved to obtain flour. A portion of the flour (10 mg), was digested for 1 hr with lysis buffer (30 µl per mg of flour) containing 8 M urea, 2 % (v/v) of Nonidet P-40 (NP-40), 0.8 % (v/v) ampholine of pH 3.5 to10, 5 % (v/v) 2-mercaptoethanol and 5 % (w/v) polyvinylpyrrolidone-40 (O Farrell 1975). After digestion, the samples were centrifuged at 15,000×*g* for 10 min. and the supernatant was applied to an isoelectric focusing (IEF) (pH 4–10) rod gel. Sample-solutions (40 µl) were applied to the acidic side of the IEF gels for the first dimension, and anodic and cathodic electrode solutions were filled in the upper and lower electrode chambers, respectively (Ito et al. 2000). Sodium dodecyl sulfate-polyacrylamide gel electrophoresis (SDS-PAGE) in the second dimension was performed with 17 % separation and 5 % stacking gels. Protein spots in 2-DE gels were visualized by Coomassie Brilliant Blue (CBB) R250 staining (Hirano and Watanabe 1990).

4.4.2.3 Quantitative Analysis of Electrophoresis Patterns

Following staining, the 2-DE gel patterns were scanned using a flatbed scanner, and analyzed using PDQuest software version 6.2 (Bio-Rad discovery series, Bio-Rad Laboratories, USA). After scanning, spots in 39 gels were detected using the same parameters and quantified by two-dimensional Gaussian modelling. Taking a full complement of genes in Chinese Spring as standard, we made a match set to compare protein-spots of 39 gels of DT lines. Thirty landmarks were used to align and position all the members of the match set. Gels were then normalised and data were exported to Excel (Microsoft). A specific spot present in all the gels was selected as the internal standard and the intensity of all other spots was expressed as the ratio of the internal standard. Each deletion line was run at least twice and conclusions for the gene location of protein spots were drawn from three gels of the deletion lines.

4.4.2.4 Statistical Analysis

To determine the patterns of protein expression, especially up and down regulation, a correlation analysis was performed on 24 major spots of 39 DT lines. An interactome map was then made using the correlation coefficient.

4.4.2.5 Sample Preparation for ICAT-ESI

Proteins from wheat endosperm were extracted by Tris-HCl buffer (10 mM, pH 7.5) containing 0.1 % SDS. Flour (7 mg) from each of the euploid and fine deletion line was centrifuged at 15,000×*g* for 10 min. To remove the non-protein contamination from the crude extract, proteins were subjected to acetone precipitation with 80 % acetone cooled down to –30 °C for more than 2 h. The pellet collected after centrifugation at 15,000×*g* for 10 min was digested with 150 µl of ICAT denaturing buffer, Tris-HCl (50 mM, pH 8.5) containing 0.1 % SDS. After centrifugation at 15,000×*g* for 10 min, 100 µl of the supernatant was treated with 2 µl of reducing agent (50 mM TCEP). The samples were then boiled for 10 min, followed by a cooling period at room temperature. After centrifugation at 15,000×*g* for 10 min, 90 µl of each of the euploid and chromosome 1B deletion lines (K64) were transferred to the vial of ICAT-labeling reagents, light (D0) and heavy (D8), respectively, containing sulfhydryl-modifying biotinylation reagent and then allowed to incubate for 1 h at room temperature. After centrifugation at 15,000×*g* for 10 min, 80 µl from each of the euploid and chromosome1B deletion line were combined, mixed thoroughly and treated with trypsin (ICAT Kit Applied Biosystem, USA) and incubated at 37 °C for 16 h. The resultant peptides were then cleaned by cation exchange column and purified by avidin affinity column (for details, see ICAT kit instructions). This analytical procedure was repeated twice, and the variation of peak height between the analyses was less than 15 %.

4.4.2.6 *Protein Analysis by ESI-MS/MS*

The purified peptide fragments were injected into a capillary liquid chromatography (Waters) equipped with a pre-column, LC packing Pep map (320 μm i.d.×1 mm), and an analytical column, LC Packing Pepmap (75 μm i.d.×150 mm). After desalting, the peptides were eluted by gradient flow of acetonitrile and water containing 0.1 % formic acid and then injected into Q-TOF MS (Q-Tof Ultima, Micromass, UK) through nano LC probe (ESI) under the following analytical conditions: cycle time (s), 2.10; scan duration (s), 2.00; retention window, 0.00 to 70.00; ionization mode, ES+; mass range, 400 to 1800. Five parent masses were selected for MS/MS analyses with collision energy of 20–30 eV. Data acquisition was performed through MassLynx, and peptide sequence analyses were done by Biolynx (Micromass, UK). The sequence information was submitted to Swiss-Prot and EMBL protein and peptide databases for identification.

4.4.3 Results and Discussion

4.4.3.1 *Localization of Structural Genes*

The mature wheat grains contain several types of proteins including the storage proteins, gluten, and non-storage proteins which include enzymes like alpha amylase inhibitors and structural proteins such as those in membranes. The gluten can be classified into gliadin, a monomeric protein with intramolecular disulfide bonds, and glutenin, a polymeric protein with intra and intermolecular disulfide bonds. Based on lactate PAGE (Acid PAGE), gliadin proteins are classified into alpha, beta, gamma and omega gliadins The glutenin proteins are again classified into two groups based on their mobility in electric filed, HMW and low molecular weight (LMW) glutenins. Upon reduction, the glutenin dissociates into several subunits, denoted as subunit 1, subunit 2 and so on. Payne et al. (1985) first attempted to divide the wheat endosperm proteome, the total proteins expressed by genome, into several areas based on different types of protein expressed. Recently, Skylas et al.(2001) have divided the wheat endosperm proteome into three regions; glutenin protein regions 1, glutenin protein region 2 and non-glutenin protein region 3. To ease the interpretation of protein–protein interactions, we divided the 2-DE map into four regions; HMW glutenin region, omega gliadin regions, LMW glutenin and alpha, beta and gamma region and non-storage protein region (Fig. 4.2)

To estimate variations in spot parameters quantitatively and to identify their gene locations, image analysis of protein-spots was performed by PDQuest

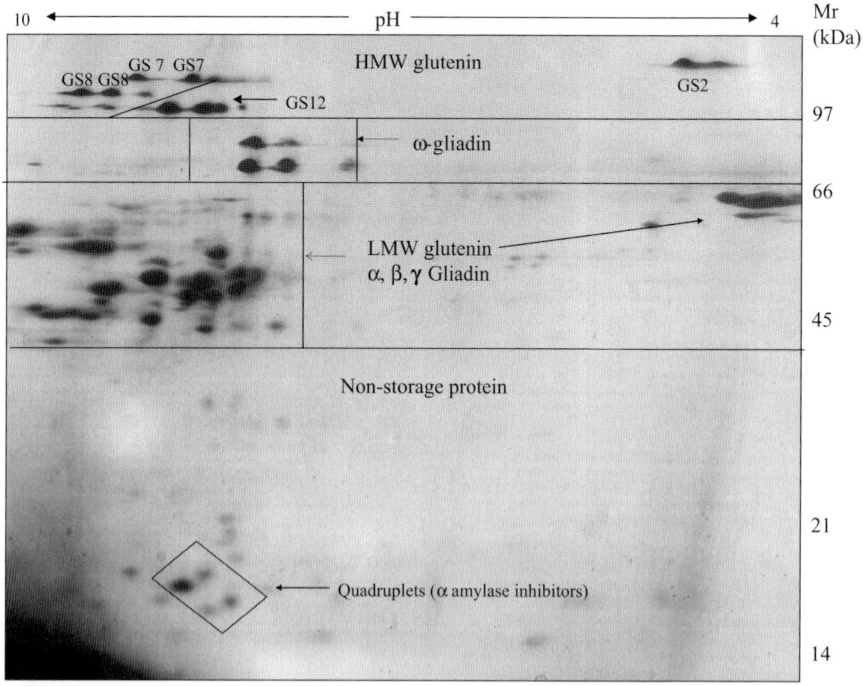

Fig. 4.2. Endosperm proteome of wheat var. Chinese Spring *(CS)*. *Boxes* indicate region of different classes of proteins with their subunits. Endosperm proteins were extracted by urea-based lysis buffer and applied in IEF rod gel (pH 3–11) and run in large size SDS PAGE (17 %). Proteins were detected by CBB staining and gels were scanned by a flatbed scanner

(The discovery series, Version 6.2, Bio-Rad) (Fig. 4.3). A total of 105 spots were detected in a CBB staining gel, of which 26 were found to be directly related to the ten missing chromosomal arms (Fig. 4.4). When a spot disappears due to the lack of only one chromosomal arm, it is assumed that the structural gene controlling the protein is located in that arm (Thiellement et al. 1999; Payne et al 1980, 1982, 1985). The same principle is also applied to localize enzyme loci [20]. Although the studies of Payne et al. (1985) have revealed the locations of most of the storage proteins resolved in 2-DE, in our studies, the assignment for a few major spots was not possible, probably because of interactions among regulators controlling the same protein. Studying the chromosomal location of structural genes and regulators in wheat, Colas des Francs and Thiellement (1985) were able to assign gene locations for only 26 out of 477 spots resolved in 2-DE. The allohexaploid nature of wheat where most polypeptides exist in two or three doses and the various locations of the regulator controlling the same polypeptide were claimed to be a major problem for localization of structural genes controlling the proteins.

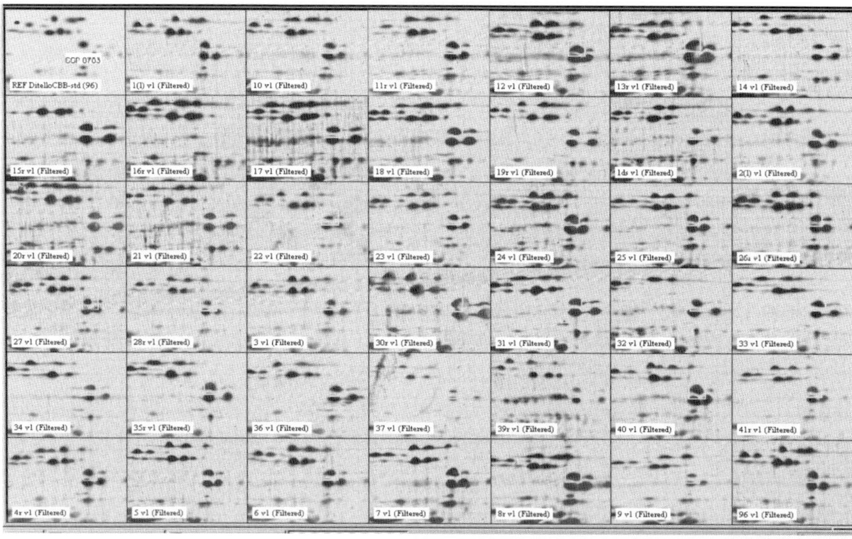

Fig. 4.3. A match set used to compare protein spots between the euploid and 39 DT lines. Following staining and scanning, the 2-DE gel patterns were analyzed using PDQuest software version 6.2 (Bio-Rad discovery series, Bio-Rad Laboratories, USA). Spots in 39 gels were detected using the same parameters and quantified by two-dimensional Gaussian modeling

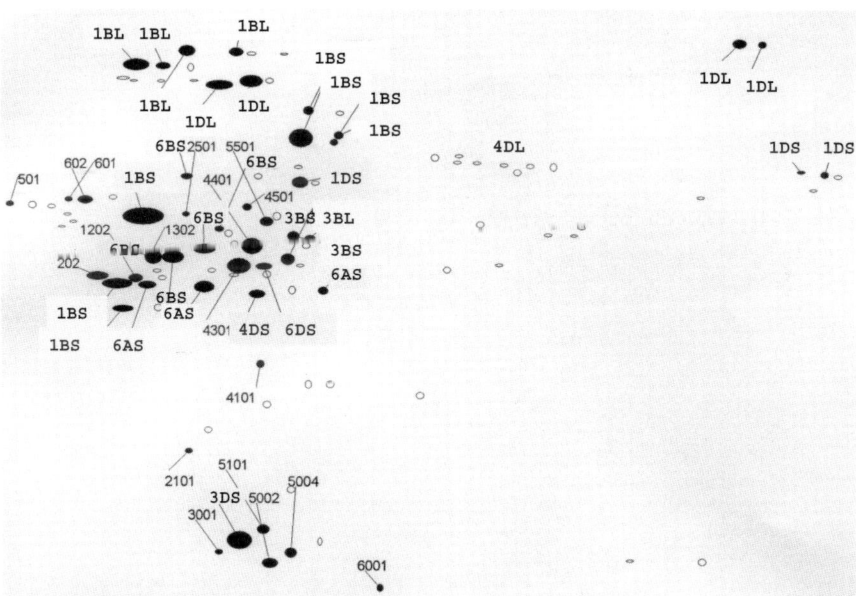

Fig. 4.4. Chromosomal locations of structural genes encoding protein in common wheat shown on a synthetic image created in PDQuest. *L* Long arm and *S* short arm of A, B and D genomes and their chromosomes

4.4.3.2 Exploring Protein-Protein Interactions

In ditelocentric lines using computer-assisted image analyser:
The quantitative image analysis of protein-spots revealed that, out of the 1755 major spots detected in 39 DT lines, 1372(78 %) spots were found variable in different spot parameters, 147 (11 %) disappeared, 978 (71 %) up-regulated and 247 (18 %) down-regulated. To understand the network of protein expression in wheat proteome, especially up and down regulation, correlation analyses were performed among 24 major spots of 39 DT lines. Results revealed that significant correlations in the protein intensity of 24 major spots across the 39 DT lines was observed (Table 4.1). For example, the regulation of spot 1803 was highly correlated with 3801 (0.94) followed by 2902 (0.87), 6707 (0.85), 4201 (0.85). Similarly a significant correlation of spot 1501 was observed with 1102 (0.88) followed by 6707 (0.81), 9608 (0.78) and 3201 (0.71). Based on the results of correlation, an interactome map was drawn (Fig. 4.5). It can be seen in the figure that there is a strong network of interactions of glutenin subunits 7 and 8 with all other proteins indicating that these two subunits might play a vital role in the formation of gluten backbone. The gluten polymers, sized into the millions of Daltons, are among the largest molecules in nature (Wrigley 1996). Understanding how the mechanism of glutenin for-

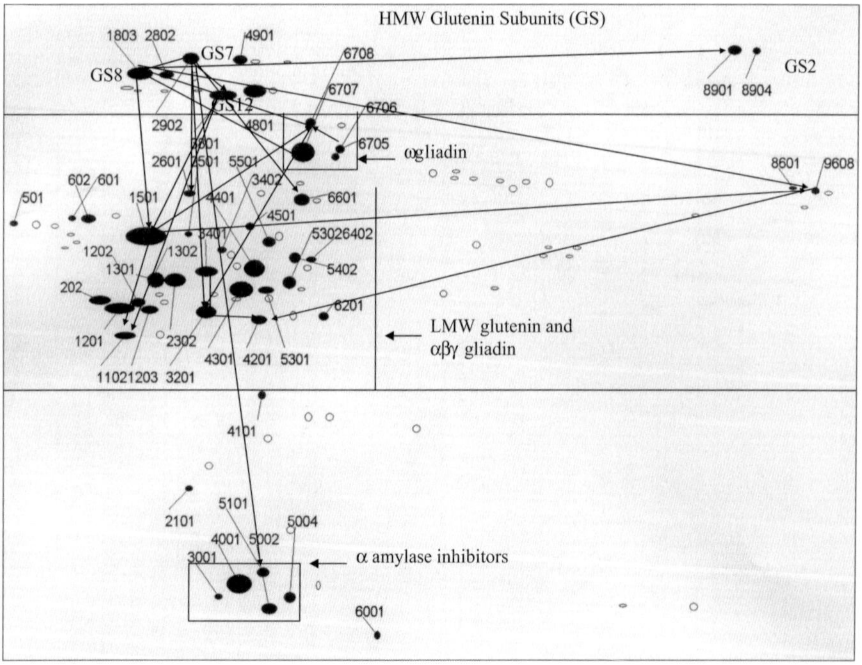

Fig. 4.5. The complex network of gluten protein. Links were made based on a correlation coefficient of more than 0.74

Table 4.1. Correlation coefficients of 24 major spots. High correlation coefficients (above 0.74) are indicated in *bold*

Spot no.	1803	2802	2902	4901	3801	4801	8901	6705	8904	6706	1501	6707	9608	3201	3401	6601	1102	5101	4201	5402	5301	2601	3402	6402
2802	0.68	1																						
2902	**0.87**	0.61	1																					
4901	0.48	0.52	0.43	1																				
3801	**0.94**	0.65	**0.86**	0.43	1																			
4801	0.60	**0.79**	0.50	0.66	0.59	1																		
8901	**0.75**	0.51	0.67	0.64	0.80	0.58	1																	
6705	0.55	0.21	0.40	0.25	0.57	0.17	0.49	1																
8904	0.20	0.41	0.07	0.54	0.13	0.53	0.35	0.03	1															
6706	0.39	0.52	0.31	0.27	0.24	0.33	0.19	-0	0.09	1														
1501	**0.84**	0.49	0.74	0.17	**0.82**	0.39	0.52	0.53	-0.10	0.32	1													
6707	**0.85**	0.46	0.67	0.33	**0.81**	0.41	0.63	**0.74**	0.09	0.35	**0.81**	1												
9608	**0.77**	0.44	0.67	0.19	0.74	0.47	0.50	0.4	0.11	0.21	**0.78**	0.67	1											
3201	**0.84**	0.56	**0.88**	0.36	**0.87**	0.45	0.67	0.45	0.14	0.34	0.71	**0.74**	0.68	1										
3401	0.65	0.20	0.58	0.29	0.63	0.27	0.53	0.46	0.03	0.20	0.65	0.64	0.51	0.62	1									
6601	0.66	0.59	**0.77**	0.17	0.70	0.40	0.42	0.41	0.05	0.25	0.65	0.56	0.47	**0.74**	0.36	1								
1102	**0.77**	0.53	0.79	0.13	**0.78**	0.37	0.47	0.41	-0.10	0.28	**0.88**	0.64	0.69	0.72	0.45	0.80	1							
5101	0.72	0.39	**0.74**	0.03	0.68	0.30	0.45	0.19	-0.10	0.30	0.74	0.52	0.69	0.73	0.48	0.63	0.77	1						
4201	**0.85**	0.63	0.69	0.34	0.80	0.62	0.61	0.40	0.28	0.37	0.71	0.73	**0.75**	**0.77**	0.52	0.59	0.67	0.75	1					
5402	**0.74**	0.63	0.68	0.51	0.66	0.64	0.55	0.25	0.19	0.42	0.61	0.59	0.67	0.67	0.40	0.45	0.55	0.61	0.76	1				
5301	0.72	0.49	0.69	0.60	0.71	0.61	0.72	0.33	0.42	0.14	0.55	0.60	0.63	0.67	0.50	0.46	0.45	0.54	0.67	0.69	1			
2601	0.52	0.40	**0.75**	0.21	0.58	0.25	0.40	0.22	-0	0.16	0.48	0.35	0.41	0.72	0.46	0.68	0.61	0.55	0.43	0.47	0.48	1		
3402	0.54	0.20	0.56	0.21	0.52	0.19	0.39	0.22	-0	0.26	0.52	0.46	0.48	0.59	0.43	0.27	0.44	0.57	0.50	0.52	0.45	0.49	1	
6402	0.31	0.15	0.18	0.25	0.22	0.27	0.31	0.14	0.57	0.08	0.17	0.23	0.29	0.24	0.41	0.15	0.11	0.30	0.40	0.18	0.48	0.15	0.15	1
8601	0.28	0.40	0.26	0.44	0.20	0.52	0.31	0.25	0.33	0.28	0.18	0.25	0.31	0.18	0.17	0.25	0.12	0.04	0.31	0.26	0.39	0.18	0.08	0.23

mation works and what components and interactions are involved has been the object of increasing research. A series of models were proposed (Wrigley 1996; Islam et al. 1999). Research from our experiment clearly demonstrated that the subunits 7 and 8 provide an important role in the framework of the gluten polymer.

In fine deletion lines using ICAT ESI MS/MS:
ICAT facilitates quantitative identification of proteins, particularly those having very similar Mr and *pI* which prevent them from separating by the conventional 2-DE. ICAT coupled with ESI-MS/MS provides a unique opportunity to quantify proteins even from nano to femtomole level. In this technique, ICAT reagents are added to two different samples, one with light ICAT reagent (D0) and another with heavy ICAT reagent (D8). The samples are then derivatised, combined, and then digested with trypsin. Excess reagent and trypsin are removed using cation exchange chromatography. Biotinylated, cysteine-containing tryptic fragments are captured on an avidin affinity column, and labeled peptides are separated from non-labeled material. After elution from the elution column, labeled peptides are further separated using capillary reversed phase HPLC, then analyzed by mass spectrometry. Relative quantitation between the control and test samples is determined from the mass ratio of D8-labeled peptide to D0-labeled peptides (ICAT Kit, Applied Biosystems, USA).

In our study, we treated the euploid sample with D0 and 1B deletion sample with D8, and achieved a clear separation of peptides between these two samples by using ESI-MS (Fig. 4.6), suggesting that peptides with a very similar *pI* and Mr can easily be separated by using ICAT-ESI MS technique. In wheat, separation of proteins is reported to be difficult due to the interactions among three genomes, A, B and D (Islam et al. 2002; Holt et al. 1981). It has also been reported that about 70 to 80 % of protein-spots in the wheat proteome exist in two or three doses. Poor separation of protein spots and interpretation difficulties for gene-locations of the protein spots due to the allohexaploid nature of the wheat genome were also reported by Zivy et al. (1984). Similar to Zivy et al.(1984), we could not identify the gene location of 70 % protein spots separated in 2-DE (Islam et al. 2002). Studying the chromosomal location of structural genes and regulators in wheat, Colas des Francs and Thiellement (1985) also found that the expression of most proteins in wheat are controlled by more than one regulator. However, results from our present study have revealed that these complexities of protein identification in wheat can be avoided, and more sensitive and accurate separation can be achieved by using ICAT-ESI-MS/MS.

The relative abundance of proteins was quantified by taking peak height ratio of monoisotopic peak of the light ICAT-labeled peptide to that of the heavy-ICAT-labeled peptide. We repeated the same procedure twice and found less than a 10–15 % variation of peak height between the replicates. As

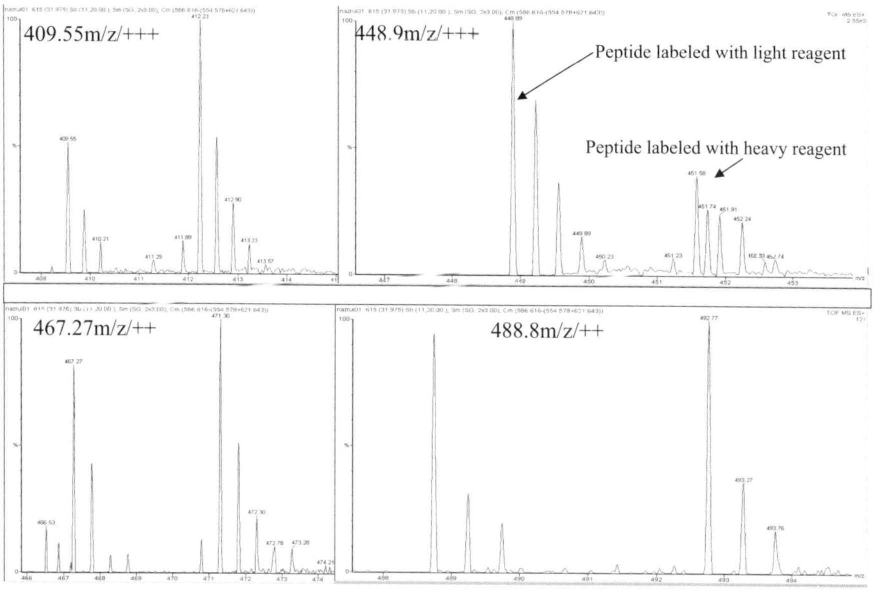

Fig. 4.6. Peak intensity of peptide labeled with light and heavy ICAT reagent. After separately labeling with light and heavy ICAT reagent, the two samples were combined, cleaned and purified by cation exchange and avidin column. The purified peptides were injected into a capillary liquid chromatography(Waters) equipped with a pre-column, LC packing Pep map (320 μm i.d × 1 mm), and an analytical column, LC Packing Pepmap (75 μm i.d. × 150 mm). After desalting, the peptides were eluted by a gradient flow of acetonitrile and water containing 0.1 % formic acid and then injected into Q-TOF MS (Q-Tof Ultima, Micromass, UK) through nano-LC probe (ESI)

shown in Fig. 4.6 and Table 4.2, the deletion of chromosomal segment resulted appreciable variations in relative intensity of peptide fragments. For example, compared to the euploid, a 35 % decrease in peptide intensity of mass 728.30 was observed in fine deletion lines. Similarly, a 279 % increase observed for mass fragment of 774.40 compared to the euploid, which confirmed our previous finding, that there is a network of protein that ultimately determines the intensity of protein expression in wheat proteome (Islam et al 2002). Of the 14 peptides identified by mass fragments analysis of the euploid, three down-regulated, ten up-regulated and one did not show appreciable changes due to the deletion of chromosome segment (Table 4.2). The substantial variations of peptides intensities due to the deletion of chromosomal segment or arm, hence genes, indicate that the magnitude of protein expression in a proteome depends on the context of its interactions with other proteins under a set of conditions. Although this phenomenon contradicts the traditional concept of protein function where a protein function is defined as its action in the presence of a substrate and a catalyst, it is quite supportive to the post-genomic

Table 2. Mass fragment analyzed by ESI MS/MS for sequence information, and proteins were identified by SWISS PROT database

Mass fragment D0	D8	Charge (z)	Peak height height D0	D8	Percent over D0	Sequence	Protein identified
503.76	511.81	+1	16.1	8.3	52		
510.25	514.27	+2	4.2	16.2	386		
568.24	576.29	+1	15.1	16.2	107	GNRDG	
576.26	584.31	+1	9.5	16.3	172		
600.31	604.33	+2	3.5	16.1	460		
632.32	636.34	+2	16.1	6.0	37		
664.04	669.40	+3	8.1	16.2	200		
705.86	713.90	+1	8.3	16.2	195		
728.30	740.38	+2	15.9	5.5	35	YTDKPA	Proteasome α7 subunit (rice)
731.36	739.41	+1	6.4	16.0	250		
774.40	778.45	+2	5.8	16.2	279	AASITAV	α-Amylase inhibitor (wheat)
789.38	801.46	+2	11.2	15.9	142	GGLTMA	α-Amylase inhibitor (wheat)
800.09	805.45	+3	6.5	14.8	228		
978.50	982.53	+2	5.9	16.2	275	AKVSGI	1,4-α-D-Glucan mathohydrolase (barley)

view of protein expression, where a protein function is defined by its interaction with all other functionally interacting proteins (Pandy and Mann 2000). The imbalance of the whole network of protein expression as revealed from our studies (Islam et al. 2002, 2003a, b) by up- and down-regulation due to the chromosomal deletion could also be a result of some of the functional linkages in the metabolic or signaling pathways responsible for the protein expression. For example, if there are three proteins, namely A, B and C where the expression of B depends on A and that C on B. Then if B was removed from the system, A would be up-regulated and C would be down-regulated.

In the present study, we used deletion lines produced by the *Gc* gene, and the chromosomal deletion included many genes, hence the gene products (protein). Therefore the differential expression of proteins in our study cannot absolutely be claimed as protein-protein interaction as many genes and hence gene products were deleted, but it is the first step in understanding interacting proteins in wheat proteome. We are now producing lines with a micro-level deletion of chromosomes, which could result in mutants with a single gene deletion (Tsujimoto et al 2001). These lines may be useful in future to determine the interaction of one protein with all other proteins in the proteome. Unlike the other techniques of interactive proteomics, such as yeast two- hybrid and other affinity-based techniques where two to three proteins are taken at a time to study their interaction, the genome manipulation technique accounts for the influence of all proteins in the proteome. Another major concern of the yeast two-hybrid (Y2H) system is the false-positive and false-negative signals, which may lead to ambiguous results of protein-protein interaction. Recently Ito et al. (2002) performed a comprehensive study on the Y2H system and reported that 90 % of the known interactions came from a false negative. The reasons for false signaling are not known yet. As Y2H produces binary interaction, it is difficult to deduce meaningful comparative information for the Y2H based technique. Moreover, detection of interaction in theY2H system does not consider post-translational modification or the folding/structure of the protein complex. These problems are reported to be associated with affinity-based interactome and other techniques. On the other hand, in the genome manipulation technique, there is less chance of false signaling as the study is performed in vitro

Eight well-separated peptide fragments were selected for MS/MS analysis. Clear peptide mass fingerprints of the amino acid sequence were found when peptide fragments were analyzed by MS. Sequence information of five peptide fragments was submitted to Swiss-Prot database for protein identification (Table 4.2). Of the five proteins submitted, four were identified as alpha-amylase inhibitor, alpha-amylase/subtilisin inhibitor precursor, proteasome subunit alpha type 7 and 1,4 alpha-glucan-D-mathohydrolase. Most of the proteins identified by ICAT ESI MS/MS technique were non-storage proteins with molecular weight less than 30 kDa, except for 1,4 alpha-glucan-D-mathohydrolase, suggesting that the proteins extracted by 0.1 % SDS are of a low molecular weight. Results from our experiment revealed that the ICAT based technique is highly sensitive, as in this approach minor changes between two peptides such as one or two amino acids which can not cause big variation in p*I* and thus remain undetectable by 2-DE, are amplified by heavy ICAT reagent (deuterium) and can easily be detected by mass spectrometry.

A quick analysis of all seed proteins is likely to be revealed in the future, especially by combining ESI-MS/MS and software with automatic protein identification through direct access to different databases information, which will lead to a dramatic increase in the knowledge of complex protein network in wheat. Although as an initial step to investigate the feasibility of quick protein identification in the crude extract of wheat endosperm using ICAT-ESI MS/MS has been achieved in our study, several challenges are also apparent. For example, it was often difficult to identify C terminal fragments due to signals resulted from the fragmentation of ICAT reagents. In addition, in this procedure, the cleaning and purification of samples from salts and detergent by columns limits the use of urea-based solvent considered to be ideal for wheat protein extraction. In our research, we have attempted to improve the protein extraction by lysis buffer we used for 2-DE followed by acetone precipitation and digestion in ICAT denaturing buffer. Unfortunately, a major portion of the proteins remained insoluble in the ICAT-denaturing buffer which has limited us in making a useful correlation between the quantitative ICAT and 2-DE data. Therefore, sample extraction procedure of ICAT has to be improved to make a valid comparison between these two methods.

4.5 Concluding Remarks

The genome manipulation approach was found efficient in determining protein-protein interactions in proteome. Taking an example of wheat genome, we have demonstrated clearly that the interactome information obtained from the genome manipulation technique is more reliable compared to the Y2H system and other affinity based techniques. We believe that a comprehensive map of wheat proteins interactome is possible by the genome manipulation technique when sufficient information on wheat proteins is available in the wheat protein databases.

References

Auerbach D, Thaminy S, Hottiger MO, and Stagljar I (2002) The post-genomic era of interactive proteomics: Facts and perspective. *Proteomics* 2:611–623

Colas des Francs C and Thiellemen H (1985) Chromosomal localization of structural genes and regulators in wheat by 2D electrophoresis of ditelosomic lines. *Theor. Appl. Genet.* 71:31–38

Cordwell SJ, Nouwens AS and Walsh BJ (2001) Comparative proteomics of bacterial pathogens. *Proteomics* 1:461–472

Endo TR (1990) Gametocidal chromosomes and their induction of chromosome mutations in wheat *Jpn. J. Genet.* 65:135–152

Endo TR and Gill BS (1996) The deletion stocks of common wheat. *J. Hered.* 87:295–307

Eisenberg D, Marcotte M, Xenarios I, and Yeates TO (2000) Protein function in the post-genomic era. *Nature* 12:823–826

Fukuda M, Islam N, Woo SH, Yamagishi A, Takaoka M, and Hirano H (2003) Assessing matrix assisted laser desorption/ionization-time of flight-mass spectrometry as a means of rapid embryo protein identification in rice. *Electrophoresis* 24:1319–1329

Gygi SP, Rist B, Gerber SA, Turecek F, Gelb MH, and Aebersold R (1999) Quantitative analysis of complex protein mixtures using isotope-coded affinity tags. *Nat. Biotechnol.* 17:994–999

Gygi SP, Corthals GL, Zhang Y, Rochon Y, and Aebersold R (2000) Evaluation of two-dimensional gel electrophoresis based proteome analysis technology. Proc. Natl. Acad. Sci. U.S.A. 97:9390–9395

Gavin A, Bosche M, Krause R, Grandi P, Marzioch M, Bauer A, Schultz J, Rick JM, Michon A, Cruciat C, Remor M, Hofert C, Schelder M, Brajenovic M, Ruffner H, Merino A, Klein K, Hudak M, Dickson D, Rudi T, Gnau V, Bauch A, Bastuck S, Huhse B, Leutwein C, Heurtier M, Copley RR, Edelmann A, Querfurth E, Rybin V, Drewes G, Raida M, Bouwmeester T, Bork P, Seraphin B, Kuster B, Neubauer G, and Superti-Furga G (2002) Functional organization of the yeast proteome by systemic analysis of protein complexes. *Nature* 415:141–146

Holt LM, Astin R and Payne PI (1981) Structural and genetical studies on the high-molecular-weight subunits of wheat glutenin. *Theor. Appl. Genet.* 60:237–243

Hirano H and Watanabe T (1990) Microsequencing of proteins electrotransferred onto immobilizing matrices from polyacrylamide gel electrophoresis: Application to an insoluble protein. *Electrophoresis* 11:573–580

Hart GE (1983) Hexaploid wheat In Tanksley SD, Orton,TJ (eds) Isozymes in plant genetics and breeding, part B. Elsevier, Amsterdam, pp 35–56

Islam N, Larroque OR, Clarke BC, and Bekes F (1999) Accumulation of wheat storage proteins during the early stages of endosperm development In: Cereals, Proc. 49th Australian Cereal Chem. Conf., Royal Aust. Chem. Isnt, Melbourne 1999, pp. 26–31

Islam N, Woo SH, Tsujimoto H, Kawasaki H and Hirano H (2002) Proteome approaches to characterize seed storage proteins related to ditelocentric chromosomes in common wheat (*Triticum aestivum* L.). *Proteomics* 2, 2:1146–1155

Islam N, Tsujimoto H, and Hirano H (2003a). Wheat proteomics: Relation between fine chromosome deletion and protein expression in common wheat. *Proteomics* 3:307–316

Islam N, Tsujimoto H, and Hirano H (2003b) Proteome analyses of diplid, tetraploid and hexaploid wheat: Towards understanding genome interaction in protein expression. *Proteomics* 3:549–559

Ito K, Kimura Y, Sassa H and Hirano H (2000) Identification of the rice 20S proteasome subunits and analysis of their N-terminal modification. *Jpn. J. Electrophoresis* 44:205–210

O`Farrell P H (1975) High resolution two-dimensional electrophoresis of proteins. *J. Biol. Chem.* 250:4007–4021

Ogihara Y, Hasegawa K, and Tsujimoto H (1994) High-resolution cytological mapping of the long arm of chromosome 5A in common wheat using a series of deletion lines induced by gametocidal (*Gc*) genes of *Aegilops speltoides. Mol. Gen. Genet.* 244:253–259

Payne P I, Holt L M, Worland, AJ, and Law C N (1982) Structural and genetical studies on the high-molecular-weight subunits of wheat glutein. *Theor. Appl. Genet.* 63:129–138

Payne P I, Law C N, and Mudd E E (1980) Control by homoeologous group 1 chromosomes of the high-molecular-weight subunits of glutenin, a major protein of wheat endosperm. *Theor. Appl. Genet.* 58:113–120

Pandy A, and Mann M (2000) Proteomics to study genes and genomes. *Nature* 405: 837–846

Payne PI, Holt LM, Jarvis MG, and Jackson EA (1985) Two-dimensional fractionation of the endosperm proteins of bread wheat (*Triticum aestivum*): Biochemical and genetic studies. *Cereal Chem.* 62:319–326

Sears ER (1954) The aneuploids of common wheat. Research Bull. Univ. Missouri Agric.Exper. Station 572:1–59

Skylas DJ, Copeland L, Rathmell WG and Wrigley CW (2001) The wheat-grain proteome as a basis for more efficient cultivar identification. *Proteomics* 1:1542–1546

Smolka M, Zhou H and Aebersold R (2002) Quantitative protein profiling using two-dimensional gel electrophoresis, isotope-coded affinity tag labeling, and mass spectrometry. *Mol. Cell. Proteomics* 1:19–29

Thiellement H, Bahrman N, Damerval C, Plomion C, Rossignol M, Santoni V, Vienne D de and Zivy M (1999) Proteomics for genetic and physiological studies in plants. *Electrophoresis* 20:2013–2026

Thiellement H, Bahrman N, and Colas des Francs C (1986) Regulatory effects of homoeologous chromosome arms on wheat proteins at two developmental stages. *Theor. Appl. Genet.* 73:246–251

Tsujimoto H, Yamada T, Hasegawa K, Usami N, Kojima T, Endo TR, Ogihara Y and Sasakuma T (2001) Large-scale selection of lines with deletions in chromosome 1B in wheat and applications for fine deletion mapping. *Genome* 44:501–508

Tsujimoto H and Noda K (1989) Structure of chromosomes 5A of wheat speltoid mutants induced by the gametocidal genes of *Aegilops speltoides. Genome* 32:1085–1090

Tsujimoto H, Yamada T, and Abe T (2001) Chromosome breakage in wheat induced by heavy ion-beam irradiation. Riken Reports 34:172

Wrigley CW (1996) Giant proteins with flour power *Nature* 381:738–739

Woo SH, Fukuda M, Islam N, Takaoka M, Kawasaki H, Hirano H (2002) Efficient peptide mapping and its application to identification of embryo proteins in the rice proteome analysis Electrophoresis 23: 647–654

Zivy M, Thiellement H, de Vienne D and Hofmann JP (1984) Study on nuclear and cytoplasmic genome expression in wheat by two-dimensional gel electrophoresis. 2. genetic differences between two lines and two groups of cytoplasm at five developmental stages or organs. *Theor. Appl. Genet.* 68:335–345

5 Modification-Specific Proteomic Strategy for Identification of Glycosyl-Phosphatidylinositol Anchored Membrane Proteins

FELIX ELORTZA, LEONARD J. FOSTER, ALLAN STENSBALLE
and OLE N. JENSEN

5.1 Summary

Post-translational modification of proteins is crucial for dynamic regulation and control of cellular processes. In eukaryotic cells, a subset of post-translationally modified membrane proteins is attached to the external leaflet of the plasma membrane by a glycosyl-phosphatidylinositol (GPI) anchor. There is substantial evidence suggesting that GPI-anchored proteins are clustered in sphingolipid-sterol microdomains or rafts. Lipid rafts are involved in numerous cellular functions including membrane trafficking and cell signalling. Due to the important role of membrane proteins in health and disease there is a need for a better understanding of membrane microdomains and the proteins they contain, including GPI-anchored proteins. GPI-anchored proteins contain special amino acid sequence features and they can be classified by genome wide sequence analysis by bioinformatics methods. However, direct experimental evidence is required in order to validate such sequence-based predictions and for improving the bioinformatic data-mining procedures. In the present study we have combined a detergent (Triton X-114) based phase separation method with enzyme treatment by phospatidylinositol-phospholipase C (PI-PLC) to selectively isolate GPI anchored proteins from purified lipid rafts from HeLa cells. This approach resulted in the recovery and identification by mass spectrometry of six known human GPI-anchored proteins. This modification-specific proteomic strategy is general and applicable to the systematic study of GPI-anchored membrane proteins in a variety of eukaryotic cell type.

Principles and Practice
Methods in Proteome and Protein Analysis
R.M. Kamp, J. J. Calvete, T. Choli-Papadopoulou (Eds.)

5.2 Introduction

Membrane proteins constitute an important class of biomolecules in living cells as they are at the interface to the surrounding environment. They are highly interesting for biomedical and pharmaceutical purposes, as they act as targets for most of the pharmaceuticals in use nowadays (Hopkins and Groom, 2002). The systematic study of membrane proteins may reveal new potential drug targets and novel molecular mechanisms for signal transduction. It has been estimated that 20–30 % of the human genome encodes membrane proteins. However, due to the inherent hydrophobic nature of most membrane proteins, it is difficult to analyse them using the analytical techniques that are typically applied to study soluble proteins. In scientific literature, studies of membrane proteins are significantly underrepresented in comparison to reports on soluble cytoplasmic or nuclear proteins. However, a number of novel proteomic approaches are directed towards the systematic study of membrane proteins (Wu and Yates, 2003).

5.2.1 Glycosyl-Phosphatidylinositol Anchored Proteins

A majority of membrane proteins are post-translationally modified, and a subset of them are modified by a covalent attachment of a glycosyl-phosphatidylinositol (GPI) moiety at the C terminus of the protein (Ferguson et al., 1988; Fig. 5.1). This hydrophobic group anchors the protein in the lipid bilayer, conferring on the GPI-AP many of the physichochemical properties of intrinsic plasma membrane proteins thereby allowing them to act as surface coat proteins, receptors, adhesion molecules, ectoenzymes and differentiation antigens (Angst et al., 2001; Ferguson, 1999; Hooper, 1997). Additional functions have been proposed for GPI anchored proteins, e.g. their involvement in intracellular sorting and in transmembrane signalling (Horejsi et al., 1999; Muniz and Riezman, 2000).

In contrast to integral membrane proteins, membrane-associated GPI-anchored proteins can become water-soluble by cleavage of the GPI anchor by various phosphatidyl-inositol specific phospholipases. As a result of this cleavage, the protein moiety is released to the extracellular environment. Thus, the GPI-anchor confers on the cell the ability to modulate the activity of the GPI-anchored protein in the surrounding intercellular space. The remaining lipidic part of the GPI-anchor possibly acts as a secondary messenger (Nishizuka, 1995).

All known GPI-anchored proteins share a number of common features (Hooper, 2001). Firstly, they have a cleavable N-terminal secretion signal for translocation into the endoplasmic reticulum. Secondly, they lack any transmembrane polypeptide domain in the mature protein. Thirdly, they contain a predominantly hydrophobic region in the C-terminus, which most likely

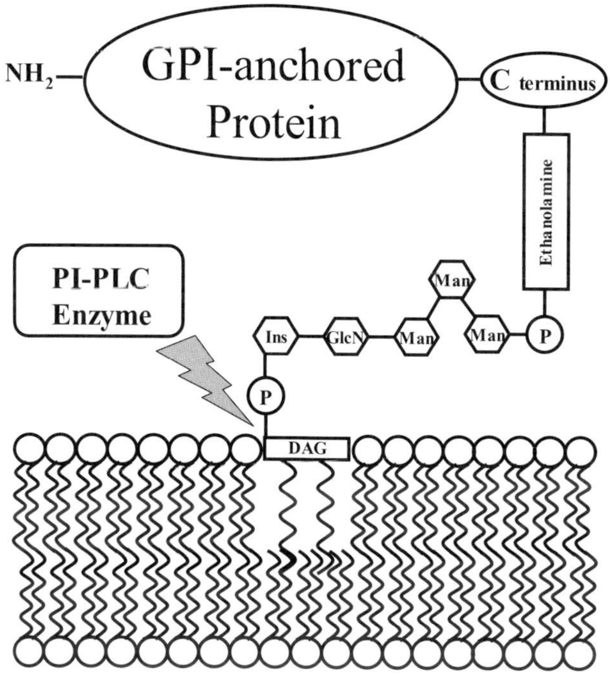

Fig. 5.1. Schematic representation of a glycosyl-phosphatidylinositol anchored protein and the phosphatidylinositol-phospholipase C (*PI-PLC*) enzyme cleavage site

forms a transient transmembrane domain and is thought to function as a recognition signal for a transamidase. The latter enzyme cleaves the C-terminal hydrophobic tail of the nascent protein at the so-called *ω-site*, and transfers the truncated protein to a pre-synthesised GPI anchor. Analysis of native GPI-anchored proteins and site-directed mutagenesis studies have shown that there are certain sequence constraints for the ω-site (Eisenhaber et al., 1998; Udenfriend and Kodukula, 1995).

A number of bioinformatic tools have been developed for the detection of GPI-anchored proteins based on the above-mentioned characteristic sequence features (Borner et al., 2002; Eisenhaber et al., 2001; Kronegg and Buloz, 1999).

In mammalian cells GPI-anchored proteins are thought to be located mainly in sphingolipid- and cholesterol-enriched domains (Harder et al., 1998) known as lipid rafts, detergent-resistant membranes (DRMs) or detergent insoluble material (DIGs). GPI-anchored proteins are thought to concentrate in rafts due to interactions between sphingolipids, cholesterol and the long saturated acyl chains from the PI moiety of the GPI anchor (Schroeder et al., 1994). As lipid rafts seem to be a critical component of many processes there is abundant interest in understanding such membrane microdomains and the proteins they contain.

Few GPI-anchored proteins have been experimentally determined or characterized in lipid raft preparations. In a study of detergent-resistant membrane fractions from baby hamster kidney cells, six GPI anchored proteins were detected by comparing PI-PLC treated and untreated DRM fractions using 2D gels but only the Thy-1 protein was identified (Fivaz et al., 2000). In another study, Jurkat T cell detergent-resistant membranes were analyzed and three GPI-anchored proteins (GDNFR-a, prion-protein and CD48) were detected among more than 70 proteins. Only CD48 was expected to be present based on the current knowledge of T cell biology (von Haller et al., 2001).

Here, we can demonstrate a general proteomic approach directed towards the selective isolation and identification of GPI-anchored proteins. Using the

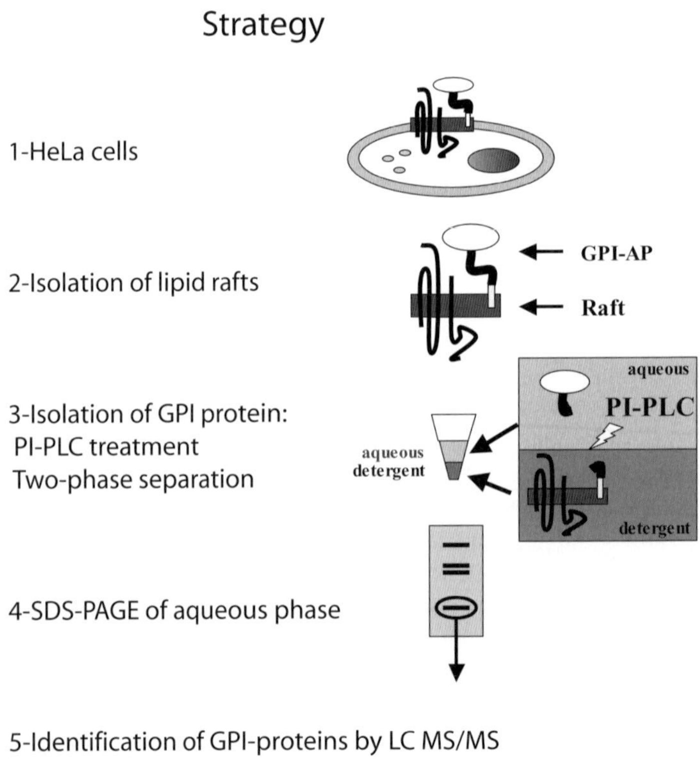

Fig. 5.2. Modification-specific proteomic strategy for isolation and identification of GPI-anchored proteins by mass spectrometry. *Stage 1* HeLa cells; *2* rafts isolated based on insolubility at a high pH and after sucrose gradient centrifugation; *3* isolation of GPI-anchored proteins by combination of phospatidylinositol-specific phospholipase C (*PI-PLC*) and Triton X-114 based two-phase separation; *4* SDS-PAGE of aqueous GPI-anchored protein enriched fraction after acetone precipitation; *5* identification of GPI-anchored proteins by tryptic in-gel digestion of excised bands and peptide analysis by nanoscale high performance liquid chromatography (HPLC) interfaced to electrospray quadrupole time-of-flight (QTOF) tandem mass spectrometry (nLC-MS/MS)

concept of 'modification-specific proteomics' (Jensen, 2000), we combined membrane protein fractionation methods, a specific biochemical technique for isolation of GPI-anchored proteins and advanced mass spectrometry for protein identification (Fig. 5.2).

5.3 Results

5.3.1 Selective Isolation of GPI-Anchored Proteins

Cells exhibit a broad range of protein expression. Thus, in any proteomic study it is useful to enrich for the subset or class of proteins to be investigated. The importance of reducing the sample complexity is especially important in order to be able to detect and study low abundance proteins, such as membrane proteins.

A raft-enriched fraction was isolated based on insolubility at a high pH. The sphingolipid and cholesterol content confers lipid rafts with distinct buoyancy compared to other membrane fractions and they can be isolated by sucrose gradient centrifugation.

Triton X-114 detergent-based two-phase separation was used to partition membrane proteins from soluble proteins present in the raft preparation (Bordier, 1981; Fig. 5.2). After removal of membrane-associated proteins by several cycles of two-phase separation the isolated membranes were treated with phosphatidylinositol-phospholipase C (*Bacillus cereus*, PI-PLC) enzyme in the presence of Triton X-114 by adapting the method of Hooper *et al.* (Hooper et al., 1987). PI-PLC hydrolyzes the phosphatidylinositol, thereby releasing the soluble protein moiety from the membrane/detergent phase and enabling its recovery in the aqueous phase (Figs. 5.1 and 5.2).

Next, the isolated proteins were separated by gel electrophoresis. In classical proteomic studies, 2D polyacrylamide gel electrophoresis is a widely used method for resolving complex mixtures of proteins. However, proteins which are very hydrophobic are often underrepresented due to solubility problems during the isoelectrofocusing process, a problem also observed with GPI-anchored proteins (Fivaz et al., 2000; Santoni et al., 2000). GPI proteins are often highly *N*-glycosylated, leading to numerous isoforms and heterogeneity that generate poorly resolved trails of protein spots or smears in 2D PAGE (Kuster et al., 2001). This problem can be addressed by treating the sample with exoglycosidases to reduce the heterogeneity of protein. However, to avoid these complicating sample handling procedures and to gain sensitivity, we used regular one-dimensional SDS-PAGE. This combination of methods, i.e. detergent-based two-phase separation combined with PI-PLC treatment and SDS-PAGE, makes it possible to generate and separate a highly enriched GPI anchored protein fraction (Fig. 5.3). As shown in Fig. 5.3, sample complexity is significantly reduced when going from a total raft protein preparation (lane 1)

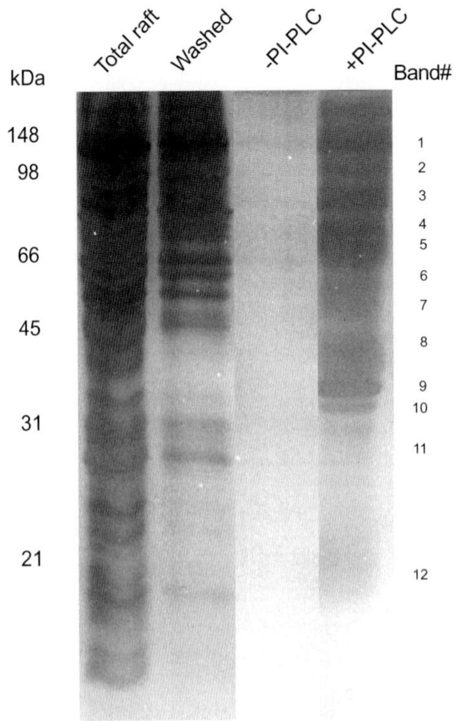

Fig. 5.3. Isolation of GPI-APs for mass spectrometry analysis. SDS-PAGE separation of proteins. *Total raft* Crude protein extract of HeLa rafts; *Washed* proteins in aqueous phase after Triton X 114 solubilization of rafts and before adding PI-PLC enzyme; *–PI-PLC* analysis of the aqueous fractions after the two-phase separation without PI-PLC treatment; *+PI-PLC* analysis of the aqueous fractions after the two-phase separation with PI-PLC treatment; *Band #* excised bands

to the fraction enriched for GPI-anchored proteins (lane 4). The selectivity of the enzymatic PI-PLC treatment is obvious when comparing lane 3 (untreated) and lane 4 (treated) protein sample in Fig. 5.3. Subsequently, we rely on peptide analysis by LC-MS/MS to resolve and identify GPI-anchored proteins isolated by SDS-PAGE.

5.3.2 Identification of GPI-Anchored Proteins by Mass Spectrometry

When starting from a complex sample there is often more than one protein in each of the excised protein bands from SDS-PAGE. In-gel digestion (Shevchenko et al., 1996) of proteins leads to complex peptide mixtures, which are subsequently analyzed and sequenced by mass spectrometry, leading to unambiguous identification of the different proteins present in the same gel band.

Protein samples in twelve individual gel slices were in-gel digested with trypsin and the recovered peptides were separated and sequenced by nanoscale high performance liquid chromatography (HPLC) interfaced to electrospray quadrupole time-of-flight (QTOF) tandem mass spectrometry

(LC-MS/MS). For each LC-MS/MS run, the complete set of peptide tandem mass spectra was submitted for sequence database searching using MASCOT software (Perkins et al., 1999). In this way a list of 17 proteins was produced by LC-MS/MS analysis of the twelve protein bands cut from the SDS-PAGE gel (Fig. 5.3 and Table 5.1).

Among the 17 identified proteins, six were characterised as known GPI-anchored proteins. A majority of the non GPI-anchored proteins were transmembrane proteins whose presence may be due to their partial solubilisation during the two-phase separation process (Table 5.1).

5.3.3 Protein Sequence Analysis

The amino acid sequences of the identified proteins were inspected for features that are specific for GPI-anchored proteins. The predicted modification site or "ω-site" for each of these six GPI-anchored proteins are shown in Table 5.1. Five out of the six known GPI-APs were categorized as GPI-AP by two sequence analysis tools: Big-PI (http://mendel.imp.univie.ac.at/gpi/gpi_server.html) and DGPI (http://129.194.185.165/dgpi/index_en.html). Carboxypeptidase M was only recognized as GPI-AP by the latter method (DGPI). The remaining 11 proteins were categorized as non-GPI-APs by both of these sequence analysis tools. Thus, based on this limited experimental data set the GPI-AP prediction servers seem more likely to produce false negatives than false positives. Bioinformatic methods can provide a very useful starting point for identifying GPI-APs in a variety of model organisms (Borner et al., 2002), but they have to be used with caution and experimental validation is required (Elortza et al., 2003).

5.4 Discussion

A modification specific proteomic strategy for detection of GPI-anchored proteins was used to isolate and identify 17 proteins in a lipid raft preparation from HeLa cells. Six of these 17 proteins were known GPI-anchored proteins, demonstrating the high selectivity and specificity of the technique. GPI anchored proteins include a wide variety of functional classes. This is also evident from this study where we detected six GPI-anchored proteins acting as receptors, enzymes and surface antigens (Table 5.1). Non-tissue-specific alkaline phosphatase is a hydrolase that may play a role in skeletal mineralisation (Weiss et al., 1988). Carboxypeptidase M enzyme is an exoprotease that acts by cleaving at arginine or lysine residues at the carboxy terminus of polypeptides (Tan et al., 1995). Decay acceleration factor (CD55) plays a role in host-pathogen interactions (Ward et al., 1994). CD59 is involved in signal transduction for T-cell activation via a complex with a protein tyrosine kinase (Stefanova et al., 1991). Urokinase plasminogen activator receptor (UPAR) is

Table 5.1. GPI-APs identified by modification-specific proteomic analysis

Swiss Prot[a]	Name of GPI-anchored proteins	No.[b]
P05186	Alkaline phosphatase	11
P08174	Decay acceleration factor, CD55	9
P15328	CD59 glycoprotein	8
P13987	Folate receptor 1	3
P14384	Carboxypeptidase M	2
Q03405	Urokinase plasminogen activator receptor (UPAR)	1
	Name of non-GPI-anchored proteins	
P43121	Melanoma adhesion molecule, CD146	2
P11912	B-cell antigen receptor complex ass. Prot. α-chain, CD79	3
P15999	ATP synthase alpha chain	3
P08195	4F2 heavy chain antigen, CD98	4
P21796	Voltage-dependent anion channel	4
P05556	Fibronectin receptor, CD29	13
P16070	Epican, CD44	2
Q9UK57	Mesotheline/megakaryocyte potentiation factor	3
Q9NZS6	Glucocorticoid receptor AF-1 specific elongation factor	3
P02571/P02572	Actin, beta and/or gamma actin	4
P11021	78-kDa glucose-regulated protein	15

[a] Swiss-Prot accession number

[b] The number of peptide tandem mass spectra that were matched to the protein sequence

[c] The score obtained in Mascot database search (significance threshold 54). The mascot score is $-10\times\text{Log}(P)$, where P is the probability that the observed match is a random event. Individual ions scores >50 indicate identity or extensive homology ($p<0.05$)

[d] Omega site: ω-site predicted for each protein

[e] Big Pi (http://mendel.imp.univie.ac.at/gpi/gpi_server.html)

[f] DGPI (http://129.194.185.165/dgpi/index_en.html); * means recognized as GPI-AP and ** means not recognized as GPI-AP

the receptor for urokinase plasminogen activator (uPA) and plays a role in targeting and promoting plasmin formation (Preissner et al., 2000). Folate receptor 1 binds folate and reduced folic acid derivatives, mediating the delivery of 5-methyltetrahydrofolate to the interior of the cells (Wang et al., 1996). The only common feature of these proteins is the glycosyl-phosphatidylinositol anchor that locates the proteins in lipid rafts at the extra cellular leaflet of the plasma membrane, where they are able to perform their biological function.

Mascot score[c]	Omega site[d]	Big Pi[e]	DGPI[f]
322	SSAGSLAAGPLLLALALYPLSVLF	*	*
376	SGTTRLLSGHTCFTLTGLLGTLVTMGLLT	*	*
279	SGAGPWAAWPFLLSLALMLLWLLS	*	*
131	NGGTSLSEKTVLLLVTPFLAAAWSLHP	*	*
79	SAATKPSLFLFLVSLLHIFFK	**	*
62	SGAAPQPGPAHLSLTITLLMTARLWGGTLLWT	*	*
	Transmembrane		
80	Yes	**	**
135	Yes	**	**
159	Yes	**	**
181	Yes	**	**
191	Yes	**	**
585	Yes	**	**
75	No	**	**
91	No	**	**
104	No	**	**
147	No	**	**
768	No	**	**

5.5 Conclusion

A novel proteomic strategy targeted towards the identification of glycosylphosphatidylinositol anchored membrane proteins successfully recovered and identified six known GPI-anchored proteins. The identification of these six known GPI-APs in human lipid rafts validates our strategy and our preliminary result supports the hypothesis that GPI-APs are enriched in lipid raft preparations. The number of GPI-APs recovered from the lipid raft preparation and the sequence coverage obtained for these proteins by mass spectrometry demonstrate the selectivity, specificity and the sensitivity of the method, and suggests that this modification specific strategy is generally applicable for proteomic analysis of GPI anchored proteins. It generates experimental data which validates *in silico* based predictions of GPI-anchored proteins. The method has the potential to systematically determine the composition of GPI-anchored proteins in cell membranes for classification of cell types and differentiation stages and for detection of pathogenic or perturbed conditions.

5.6 Material and Methods

5.6.1 Lipid Raft Preparation

Rafts were isolated based on their ability to resist solubilization at a high pH. The cells were scraped into 100 mM Na_2CO_3, pH 11, passed 10x through a Dounce homogenizer and sonicated for three 20-s pulses with a 3 mm probe sonicator. Following a low speed centrifugation to clarify the lysates, the solution was put into 45% sucrose, placed in a Sorvall SureSpin 630 ultracentrifuge tube, and a discontinuous gradient (5/35% sucrose) was layered on top. The gradient was centrifuged for 18 h at ~170,000×*g* and the layer containing the rafts (near 25% sucrose) was subsequently extracted, diluted in 100 mM Na_2CO_3, pH 11 and centrifuged for an additional 2 h (~170,000×*g*) to pellet the rafts.

5.6.2 Two-Phase Separation and Phosphoinositol-Phospholipase C Treatment

Membranes were equilibrated by resuspending the pellet in buffer A (20 mM Hepes pH 7.5, 0.2 mM Phenylmethylsulfonylfluoride (PMSF) and 0.5 tablet of protease inhibitor per mL) and pelleted again at 20,000×*g* for 20 min. The membrane fraction was resuspended in 100 µl buffer A, then the same volume of Triton X 114 (Glyko) was added and mixed to homogeneity. The mixture was chilled on ice for 5 minutes, and then transferred to 37 °C for 20 min for phase separation. The aqueous supernatant was discarded and the extraction procedure was repeated. The detergent phase was recovered and 100 µL of buffer A with two units of PI-PLC (Molecular Probes) were added and the entire mixture was incubated at 37 °C with shaking. After 1 h, phase separation was performed again and the aqueous supernatant was recovered. Buffer and enzyme were added again to the detergent phase and the procedure was repeated. The two resulting aqueous supernatant fractions were pooled and the proteins were recovered by acetone precipitation, separated by SDS-PAGE and visualized by silver staining. Protein bands were cut out and in-gel digested with trypsin (Shevchenko et al., 1996).

5.6.3 Mass Spectrometry

Automated nanoflow liquid chromatography/tandem mass spectrometry analysis was performed using a electrospray ionization QTOF Ultima mass spectrometer (Waters/Micromass UK Ltd., Manchester, UK) employing automated data dependent acquisition. A nanoflow-HPLC system (Ultimate; Switchos2; Famos; LC Packings, Amstersdam, The Netherlands) was used to deliver a flow rate of 175 nl min^{-1} to the mass spectrometer. Chromatographic separation was accomplished by using a 2-cm fused silica precolumn (75 mm i.d.; 360 mm o.d.; Zorbax® SB-C18,5 μm (Agilent, Wilmington, DE)) connected to a 8 cm analytical column (75 mm i.d.; 360 mm o.d.; Agilent Zorbax SB-C18 3.5 μm). Peptides were eluted by a gradient of 5–32% ACN in 30 min.

The mass spectrometer was operated in positive ion mode with a source temperature of 80 C and a countercurrent gas flow rate of 150 l h^{-1}. Data dependent analysis was employed (three most abundant ions in each cycle): at 1 s MS m/z 350–1500 and a maximum of 4 s MSMS m/z 50–2000 (continuum mode), 30 s dynamic exclusion.

Raw data were processed using MassLynx 3.5 ProteinLynx and the resulting MS/MS data set exported in the Micromass pkl format. Automated peptide identification from raw data was performed using an in-house MASCOT server (v. 1.8) (Matrix Sciences, London, UK). External mass calibration using NaI resulted in mass errors of less than 50 ppm, typically 5–15 ppm in the m/z range 50–2000.

5.6.4 Bioinformatics

Two sequence-based GPI-AP prediction tools, Big-PI (http://mendel.imp.univie.ac.at/gpi/gpi_server.html) and DGPI (http://129.194.185.165/dgpi/index_en.html), were applied to analyze the identified protein sequences.

Acknowledgements. F.E. was supported by a post-doctoral fellowship from the Basque Government. L.J.F. was supported by an EMBO long-term fellowship. This project was supported by a grant from the Danish Natural Sciences Research Council (O.N.J.).

References

Angst, B.D., C. Marcozzi, and A.I. Magee. 2001. The cadherin superfamily: diversity in form and function. *J Cell Sci.* 114:629–41

Bordier, C. 1981. Phase separation of integral membrane proteins in Triton X-114 solution. *J Biol Chem.* 256:1604–7

Borner, G.H., D.J. Sherrier, T.J. Stevens, I.T. Arkin, and P. Dupree. 2002. Prediction of glycosylphosphatidylinositol-anchored proteins in *Arabidopsis*. A genomic analysis. *Plant Physiol.* 129:486–99

Eisenhaber, B., P. Bork, and F. Eisenhaber. 1998. Sequence properties of GPI-anchored proteins near the omega-site: constraints for the polypeptide binding site of the putative transamidase. *Protein Eng.* 11:1155–61

Eisenhaber, B., P. Bork, and F. Eisenhaber. 2001. Post-translational GPI lipid anchor modification of proteins in kingdoms of life: analysis of protein sequence data from complete genomes. *Protein Eng.* 14:17–25

Elortza, F., T.S. Nühse, A. Stensballe, S.C. Peck, and O.N. Jensen. 2003. Proteomic analysis of glycosylphosphatidylinositol-anchored proteins in *Arabidopsis thaliana* by mass spectrometry. *Moll Cel. Proteomics* 2 (II) (in press)

Ferguson, M.A. 1999. The structure, biosynthesis and functions of glycosylphosphatidylinositol anchors, and the contributions of trypanosome research. *J Cell Sci.* 112 (Pt 17):2799–809

Ferguson, M.A., S.W. Homans, R.A. Dwek, and T.W. Rademacher. 1988. Glycosyl-phosphatidylinositol moiety that anchors *Trypanosoma brucei* variant surface glycoprotein to the membrane. *Science.* 239:753–9

Fivaz, M., F. Vilbois, C. Pasquali, and F.G. van der Goot. 2000. Analysis of glycosyl phosphatidylinositol-anchored proteins by two-dimensional gel electrophoresis. *Electrophoresis.* 21:3351–6

Harder, T., P. Scheiffele, P. Verkade, and K. Simons. 1998. Lipid domain structure of the plasma membrane revealed by patching of membrane components. *J Cell Biol.* 141:929–42

Hooper, N.M. 1997. Glycosyl-phosphatidylinositol anchored membrane enzymes. *Clin Chim Acta.* 266:3–12

Hooper, N. M. 2001. Determination of glycosyl-phosphatidylinositol membrane protein anchorage. *Proteomics.* 1:748–55

Hooper, N.M., M.G. Low, and A.J. Turner. 1987. Renal dipeptidase is one of the membrane proteins released by phosphatidylinositol-specific phospholipase C. *Biochem J.* 244:465–9

Hopkins, A.L., and C.R. Groom. 2002. The druggable genome. *Nat Rev Drug Discov.* 1: 727–30

Horejsi, V., K. Drbal, M. Cebecauer, J. Cerny, T. Brdicka, P. Angelisova, and H. Stockinger. 1999. GPI-microdomains: a role in signalling via immunoreceptors. *Immunol Today.* 20:356–61

Jensen, O.N. 2000. Modification-specific proteomics: Strategies for systematic studies of post-translationally modified proteins. *Proteomics: a Trends Guide*:36–42

Kronegg, D., and D. Buloz. 1999. DGPI: detecting GPI-anchor proteins.

Kuster, B., T.N. Krogh, E. Mortz, and D.J. Harvey. 2001. Glycosylation analysis of gel-separated proteins. *Proteomics.* 1:350–61

Muniz, M., and H. Riezman. 2000. Intracellular transport of GPI-anchored proteins. *Embo J.* 19:10–5

Nishizuka, Y. 1995. Protein kinase C and lipid signaling for sustained cellular responses. *Faseb J.* 9:484–96

Perkins, D.N., D.J. Pappin, D.M. Creasy, and J.S. Cottrell. 1999. Probability-based protein identification by searching sequence databases using mass spectrometry data. *Electrophoresis.* 20:3551–67

Preissner, K.T., S.M. Kanse, and A.E. May. 2000. Urokinase receptor: a molecular organizer in cellular communication. *Curr Opin Cell Biol.* 12:621–8

Santoni, V., M. Molloy, and T. Rabilloud. 2000. Membrane proteins and proteomics: un amour impossible? *Electrophoresis.* 21:1054–70

Schroeder, R., E. London, and D. Brown. 1994. Interactions between saturated acyl chains confer detergent resistance on lipids and glycosylphosphatidylinositol (GPI)-anchored proteins: GPI-anchored proteins in liposomes and cells show similar behavior. *Proc Natl Acad Sci U S A.* 91:12130–4

Shevchenko, A., M. Wilm, O. Vorm, and M. Mann. 1996. Mass spectrometric sequencing of proteins silver-stained polyacrylamide gels. *Anal Chem.* 68:850–8

Stefanova, I., V. Horejsi, I.J. Ansotegui, W. Knapp, and H. Stockinger. 1991. GPI-anchored cell-surface molecules complexed to protein tyrosine kinases. *Science.* 254:1016–9

Tan, F., P.A. Deddish, and R.A. Skidgel. 1995. Human carboxypeptidase M. *Methods Enzymol.* 248:663–75

Udenfriend, S., and K. Kodukula. 1995. How glycosylphosphatidylinositol-anchored membrane proteins are made. *Annu Rev Biochem.* 64:563–91

von Haller, P.D., S. Donohoe, D.R. Goodlett, R. Aebersold, and J.D. Watts. 2001. Mass spectrometric characterization of proteins extracted from Jurkat T cell detergent-resistant membrane domains. *Proteomics.* 1:1010–21

Wang, X., G. Jansen, J. Fan, W.J. Kohler, J.F. Ross, J. Schornagel, and M. Ratnam. 1996. Variant GPI structure in relation to membrane-associated functions of a murine folate receptor. *Biochemistry.* 35:16305–12

Ward, T., P.A. Pipkin, N.A. Clarkson, D.M. Stone, P.D. Minor, and J.W. Almond. 1994. Decay-accelerating factor CD55 is identified as the receptor for echovirus 7 using CELICS, a rapid immuno-focal cloning method. *Embo J.* 13:5070–4

Weiss, M.J., D.E. Cole, K. Ray, M.P. Whyte, M.A. Lafferty, R.A. Mulivor, and H. Harris. 1988. A missense mutation in the human liver/bone/kidney alkaline phosphatase gene causing a lethal form of hypophosphatasia. *Proc Natl Acad Sci U S A.* 85:7666–9

Wu, C.C., and J.R. Yates. 2003. The application of mass spectrometry to membrane proteomics. *Nat Biotechnol.* 21:262–7

6 Diocleinae Lectins: Clues to Delineate Structure/Function Correlations

Francisca Gallego Del Sol, Vania M. Ceccatto, Celso S. Nagano, Frederico B.M.B. Moreno, Alexandre H. Sampaio, Thalles B. Grangeiro, Benildo S. Cavada and Juan J. Calvete

6.1 Introduction

Lectins are a structurally heterogeneous group of carbohydrate-binding proteins of non-immune origin comprising distinct families of evolutionarily related proteins (van Damme et al. 1998) (consult also the 3D Lectin Database at http://www.cermav.cnrs.fr/databank/lectine). Sugar recognition mechanisms have evolved independently in diverse protein frameworks, and lectins are ubiquitous in animals, plants and microorganisms. Lectins play biological roles in many cellular processes, such as: cell communication, host defense, fertilisation, development, parasitic infection and tumour metastasis, by deciphering the glycocodes encoded in the structure of glycans attached to soluble and integral cell membrane glycoconjugates (Gabius and Gabius 1997).

Though in many instances their exact biological roles remain elusive, lectins from terrestrial plants have been extensively exploited as biochemical tools in biotechnology and biomedical research due to their exquisite carbohydrate recognition properties. Moreover, legume lectins have traditionally been represented as a paradigm for studying protein-carbohydrate interactions.

Plant lectins from the Leguminosae family remain the most thoroughly investigated group of sugar-binding proteins. Despite large conservation in their primary structures, Leguminosae lectins exhibit considerable diversity in their carbohydrate binding properties. This diversity is observed not only in lectins that recognise different monosaccharides but also in lectins that share the same nominal monosaccharide specificity. The structural basis and thermodynamics of selective sugar recognition have been assessed by X-ray crystallography (Weis and Drickamer 1996; Elgavish and Shaanan 1997; Loris et al. 1998; Bouckaert et al. 1999) and isothermal titration calorimetry (Chervenak and Toone 1995; Dam et al. 1998; 2000a, b). Low affinity primary bind-

Principles and Practice
Methods in Proteome and Protein Analysis
R.M. Kamp, J. J. Calvete, T. Choli-Papadopoulou (Eds.)

ing sites for monosaccharides are pre-formed in shallow grooves on the protein surface. Carbohydrate recognition and binding selectivity is achieved through a combination of hydrogen bonds between sugar hydroxyls and protein main- and side-chain groups, water-mediated contacts, van der Waals packing of the hydrophobic sugar ring face against an aromatic amino acid residue, and hydrophobic interactions. Higher selectivity and affinity for complex carbohydrate ligands are achieved both by the existence of subsites in the lectin monomer that extend the monosaccharide binding pocket and by the clustering of these carbohydrate-binding sites in oligomeric lectins, i.e. subunit multivalency.

Legume lectin subunits display a high degree of tertiary fold conservation. The architecture of the *legume lectin fold* is structurally related to a jelly roll motif built by a curved seven-stranded 'front' β-sheet, almost a flat six-stranded *back* β-sheet, and a smaller five-stranded *top* β-sheet (reviewed by Loris et al. 1998). The front and back sheets form a sandwich, which is bridged on one side by the top sheet. The sugar-binding site lies in a shallow cleft against the front sheet and involves residues originating from three loops connecting the strands of the β-sheet.

6.2 Quaternary Structure Variability

Leguminosae lectins exhibit considerable diversity in the modes in which their monomers oligomerise (Prabu et al. 1999). Quaternary structure relates to biological activity because the spatial distribution of carbohydrate-binding sites in oligomeric lectins determine the ability of these lectins to distinguish and cross-link multivalent saccharide ligands or specific cell surface arrays of glycoconjugates. Thus, there is a longstanding correlation between the ability of lectins, such as ConA, the *Canavalia ensiformis* seed lectin, to aggregate cell surface glycoproteins and their mitogenic activity (Lis and Sharon 1986).

Lectins isolated from seeds of the Diocleinae subtribe of the Phaseoleae tribe of leguminous vines have amino acid sequences which are closely related to ConA but exhibit distinct pH-dependent dimer-tetramer equilibria (Calvete et al. 1999) and show different biological activities, such as induction of lymphocyte proliferation and interferon γ production (Barral-Netto et al. 1992), pro-inflammatory activities (Rodriguez et al. 1992), and histamine release from rat peritoneal mast cells (Ferreira et al. 1996; Barbosa et al. 2001; Cavada et al. 2001). Accumulating evidence indicates that several factors may contribute to the distinct biological effects exerted by these closely related lectins. For example, the histamine releasing activity appears to correlate with the relative affinities of the Diocleinae lectins for a biantennary complex oligosaccharide (Dam et al. 1998). On the other hand, a single amino acid replacement at position 58 (Asp in ConA; Gly in *Canavalia brasiliensis* seed lectin, ConBr) disrupts a hydrogen bond at the dimer-dimer interface leading to significant differences in the relative orientation of the carbohydrate binding sites in the quaternary structures of ConA and ConBr (Sanz-Aparicio et al. 1997). Thus, ConA and ConBr may bind to similar or identical carbohydrate structures that are differently exposed on cell surfaces, thereby triggering the response of different cell populations or different quantitative effects on the same cells. In addition, since only the tetravalent form is able to cause cross-linking of the receptor on the cell surface, the different ratios between divalent and tetravalent lectin species at a given pH may also contribute to the variability of the biological function of Diocleinae lectins.

6.3 Structural Basis of pH-Dependent Oligomerisation: The Crystal Structures of the Lectins from *Dioclea grandiflora* and *Dioclea guianensis*

The seed lectin from *Dioclea guianensis* (Dguia) exhibits pH-dependent dimer-tetramer equilibrium above pH 5.5, while that of *Dioclea grandiflora* (Dgran), whose crystal structure has been recently reported (PDB accession code 1DGL) (Rozwarski et al. 1998), exists as a tetramer above pH 4.5 (Calvete et al. 1999). We have recently solved the $P4_32_12$ crystal structure of *Dioclea guianensis* seed lectin at 2.0-Å resolution from diffraction data collected at cryogenic temperature at the Elettra Synchrotron Light Source (Italy) on beamline 5.2R and at the EMBL-DESY Outstation (Hamburg) on beamline BW7B (PDB code 1H9P) (Wah et al. 2001). In the crystals of native *Dguia* lectin, a canonical legume lectin dimer occupies the asymmetric unit and a tetramer is formed from two symmetry-related dimers in a similar manner as has been observed in every ConA crystal space group reported to date. Bouckaert and colleagues (1996) have also noticed that in crystals grown in conditions (pH 5.0), where the lectin is known to be dimeric in solution, ConA dimers change into a tetramer-like oligomer resembling the native ConA tetramer, but with a drastically smaller number of interdimer interactions. Thus, the tetramer appears to represent the most favourable oligomeric association in the crystal nucleation process, although the lectin might be regarded as dimeric in the crystal just as in the solution at the same pH.

Two dimers of crystallised *Dguia* lectin range cross-wise back to back to form the tetramer. At the interface, the two concave, 12-stranded β-sheets of each dimer are offset from one another by roughly 75° so that the dimer-dimer interactions are made on the periphery of the sheets. The center of the tetramer is a large (25×8 Å^2) water-filled cavity. The loops comprising residues 117–123 of each monomer extend deep into this cavity towards the opposite β-sheet. In the crystal structure of *Dioclea guianensis* lectin, these loops are fairly disordered (Fig. 6.1d) and poor density for this region has been also reported in different crystal forms of Con A and ConBr (Loris et al. 1998; Sanz-Aparicio et al. 1997). On the other hand, residues 114–125 are

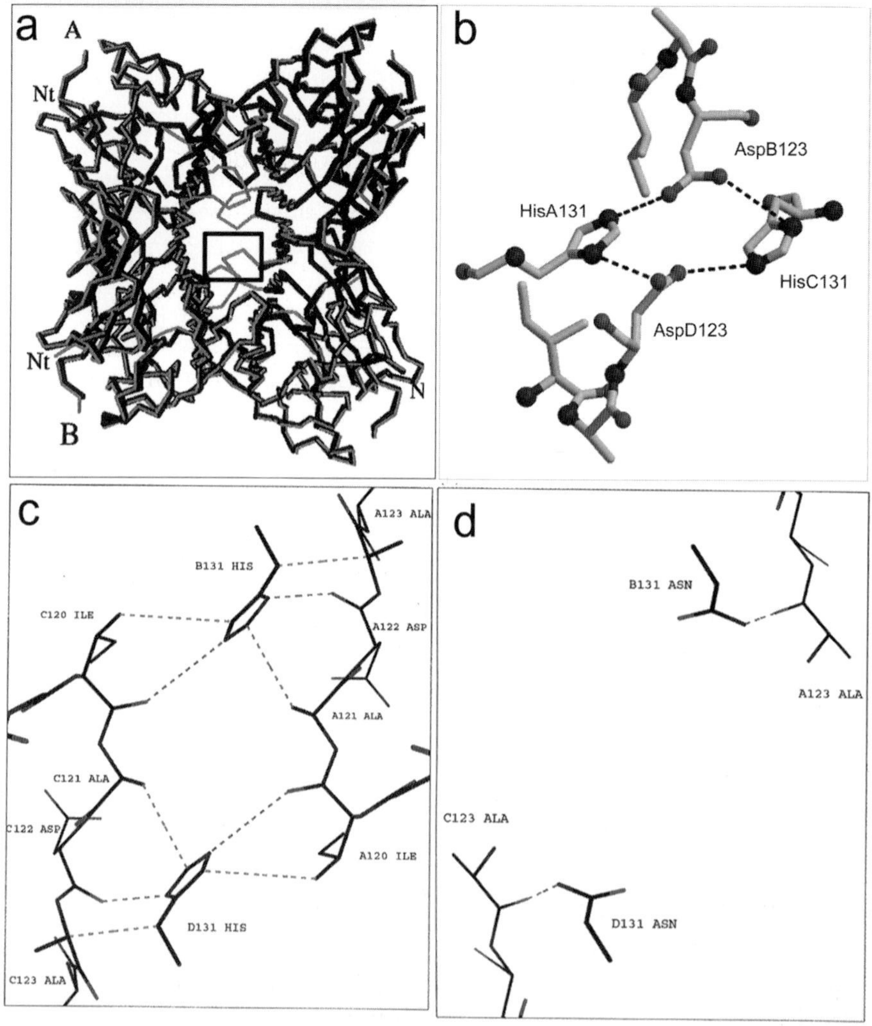

Fig. 6.1. **a** Ca superposition of the structures of *Dioclea guianensis* lectin and *Dioclea grandiflora* lectin. Two monomers (*labelled A* and *B*) change into a canonical dimer along their β-sheets to form an extended 12-stranded β-sheet. Tetramers are formed by face-to-face association of symmetry-related dimers. Note that the loop comprising of residues 114–125, extending deep into the central cavity of the tetramer, is ordered in the crystal structures of the seed lectins from *Dioclea grandiflora* and *Dioclea violacea* (*pale gray*) but is disordered in the *Dioclea guianensis* lectin structure. The N-termini of the monomers are labelled *N-t.* **b, c** show details of the interactions at the dimer-dimer interface between the 117–123 loop and residue 131 in the *Dioclea violacea* and *Dioclea grandiflora* lectin (1DGL) tetramers, respectively, which are absent in the *Dioclea guianensis* tetramer (1H9P) (**d**). Hydrogen bonds and van der Waals interactions are represented by *dashed lines*

ordered at the dimer subunit interface in the crystal structure of the *Dioclea grandiflora* seed lectin (Rozwarski et al. 1998; Fig. 6.1a). In this latter structure, a network of hydrogen bonds involving residues from all four tetramer subunits contribute to order the conformation of loop 114–125 (Fig. 6.1 c). For instance, the backbone carbonyl of Ile 120 hydrogen bonds to the side chain of His 131 in a different subunit, and this interaction is duplicated within the dimer–dimer interface by a two-fold symmetry axis. Moreover, the ND1 of His B131 hydrogen bonds with the main-chain oxygen of Asp A122. Van der Waals contacts are made between Cβ and the Cα of Ala A123 and between CE1 and the Cβ of Ala A121. His B131 makes additional contacts with the loop of monomer C from the other dimer. NE2 donates a hydrogen bond to the main-chain oxygen of Ile C120, and a van der Waals interaction occurs between CD2 of His B131 and CG2 from the side chain of Ile C120.

Residue 131 is an asparagine in *Dguia* lectin, and in striking contrast to *Dgran* lectin, many fewer interactions are observed at the dimer–dimer interface. Asn B131 has only one contact with the loop of its dimer partner monomer A, that is a hydrogen bond between ND2 and the main-chain oxygen of residue A123 (Fig. 6.1d).

Similarly to *Dguia* lectin, ConA behaves as a dimer in buffers below pH 5.0 and forms tetramers above pH 7.0. The effects of temperature and pH on the reversible dimer-tetramer association of ConA has been studied by sedimentation equilibrium, the authors proposed that the association is governed by the ionisation of a histidine imidazole group on each subunit, either His51 or His121. In this model, protonation of His121 of each protomer would cause charge repulsion and dissociation of the dimers (Senear and Teller 1981). However, our results suggest an alternative explanation.

A comparison of the three-dimensional structures of *Dgran* and *Dguia* lectins clearly shows that except for the residue at position 131, none of the amino acid replacements significantly affect tertiary or quaternary interactions. On the other hand, replacement of His for Asn at position 131 in the *Dguia* lectin accounts for the drastically reduced number of interdimer interactions compared to the *Dgran* structure. This is a possible explanation for the existence of pH-dependent dimer-tetramer equilibrium in the *Dguia* lectin, but not in the *Dgran* lectin,. We hypothesise that the ordering of the 114–125 loop structure by His131 stabilises the tetrameric association.

6.3.1 The Key Role of His-131: The Crystal Structure of *Dioclea violacea* (Dviol) Seed Lectin

To validate our hypothesis that small differences in key positions of the primary structure of close phylogenetically related lectins, in particular at position 131, may have an impact in oligomer formation and this, in turn, may have profound functional consequences, we have analysed by analytical centrifugation equilibrium sedimentation a set of *Diocleinae* lectins (Fig. 6.2), and determined their primary structures by a combination of automated Edman degradation of HPLC-isolated proteolytic peptides and mass spectrometry. In keeping with our hypothesis, all Diocleinae lectins exhibiting pH-dependent tetrameric association possess Asn at position 131. These include the seed lectins from *Dioclea guianensis* (^{119}SIADANSLHFSF**N**QFSQ^{135}), *Canavalia brasiliensis* (^{119}STHETNALHFMF**N**QFSK^{135}), *Canavalia bonaerensis* (^{119}STADANSLHFTF**N**QFSQ^{135}), *Cratylia floribunda* (^{119}STADAQSLHFTF**N**QFSQ^{135}), *Dioclea rostrata* (^{119}SIADANSLHFTF**N**QFSQ^{135}), and *Dioclea virgata* (^{119}SIADANSLHFSF**N**QFSQ^{135}). On the other hand, the lectin from *Dioclea violacea*, which like that of *Dioclea grandiflora* is tetrameric in the pH range 4.5–8.5, possesses His131 [15] (^{119}SIADENSLHFSF**HK**FSQ^{135}). The primary structure of the *Dioclea violacea* lectin was established by mass peptide fingerprint (Fig. 6.3A), PSD analysis of selected ions (Fig. 6.3B), and N-terminal sequence analysis (Fig. 6.4B). It differs in only nine residues from that of *Dioclea guianensis*. These residues (Dguia/Dviol) are: Ser21/Asn, Ala50/Val, Thr68/Ser, Ala123/Glu, Asn131/His, Gln132/Lys, Thr147/Phe, Ser168/Asn, and Asp205/Glu.

The *Dioclea violacea* seed lectin was crystallised at 22 °C using the vapour-diffusion method in hanging drops equilibrated against a mixture of 10 % PEG 8000 and 10 % PEG 1000 as precipitant. X-ray diffraction data to a maxi-

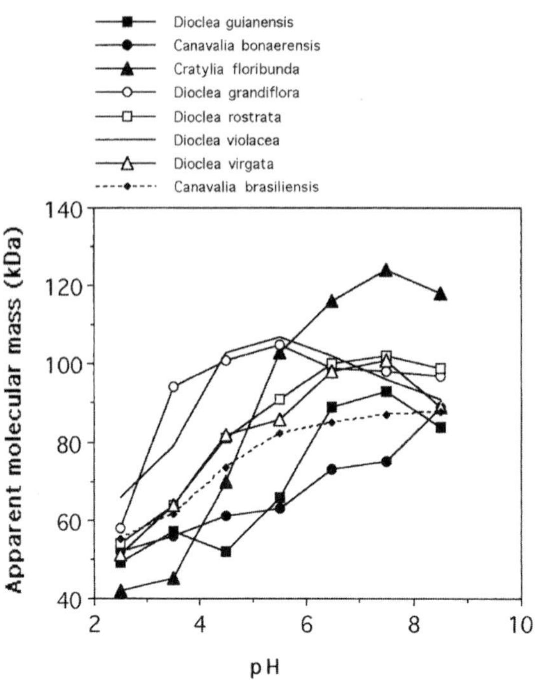

Fig. 6.2. pH-dependent oligomerisation of Diocleinae seed lectins determined by analytical ultracentrifugation equilibrium sedimentation

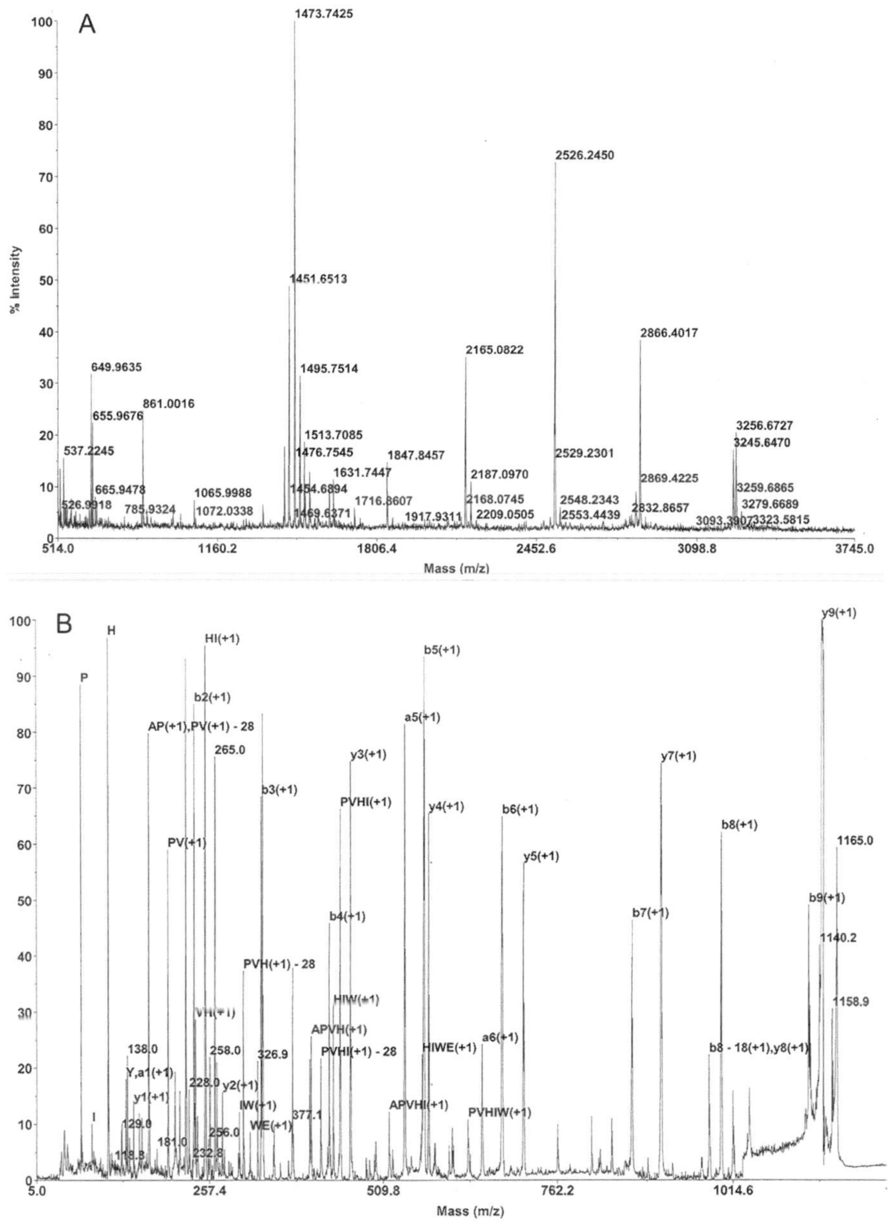

Fig. 6.3. A. Mass tryptic peptide fingerprinting of the seed lectin from *Dioclea violacea* recorded by MALDI-TOF mass spectrometry. **B** displays a typical PSD mass spectrum (from peptide ^{176}APVHIWEK^{184}, parent ion of m/z 1142.72) used for elucidation of the primary structure of the *Dioclea violacea* seed lectin shown in Fig. 6.4B

A

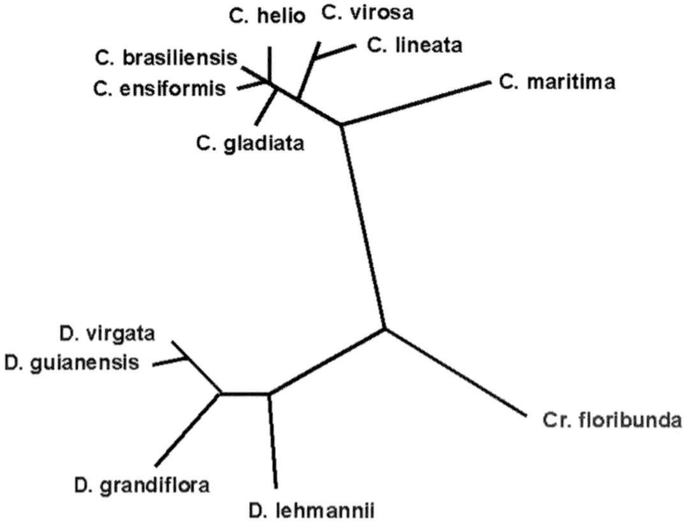

B

```
1        10        20        30        40        50
ADTIVAVELDSYPNTDIGDPNYPHIGIDIKSIRSKSTARWNMQTGKVGTV
|--------------- N-terminal sequencing -------------
|------------------------- β (12872 ± 2 Da) --------
|--------- 3256.6 ----------|  |-649.9-|--- 865 --|-----

51       60        70        80        90        100
HISYNSVAKRLSAVVSYSGSSSTTVSYDVDLNNVLPEWVRVGLSATTGLY
------------------------- β -------------------------
- 1374.7 ----| |---------- 3244.6 ----------|---- 1109.6 --

101      110       120       130       140       150
KETNTILSWSFTSKLKTNSIADENSLHFSFHKFSQNPKDLILQGDAFTDS
-||------- 1754.9 -------| |------- N-terminal sequencing --------
---------- β -------|------------- γ (12750 ± 2 Da) -----
                  |------ 1631.8 --------|    |---- 2166.3 ------
               |-------- 1841.8 ---------|

151      160       170       180       190       200
DGNLELTKVSSSGDPQGNSVGRALFYAPVHIWEKSAVVASFDATFTFLIK
-----------
-----------||------- 1346.6 -----||----- 1473.8 ----|-------- 1716.9 ------|
------------------------- γ -------------------------

201      210       220       230
SPDREPADGITFFIANTDTSIPSGSGGRLLGLFPDAN
------------- γ ---------------------|
|--------- 2865.3 -----------|--- 959.5---|
```

Fig. 6.4. A. The package of programs PHYLIP (the PHYLogeny Inference Package; obtained at: http://evolution.genetics.washington.edu/phylip.html) and MEGA (Molecular Evolutionary Genetic Analysis) (http://www.megasoftware.net) were employed for inferring a phylogenetic tree from a multiple alignment of Diocleinae lectins from the genera *Canavalia* (*C. ensiformis* [P02866], *C. brasiliensis* [P55915], *C. gladiata* [P14894], *C. virosa* [P81461], *C. lineata* [P81460], *C. maritima* [P81364], and *C. helio* [u.r., unpublished results]), *Dioclea* (*D. virgata* [u.r.], *D. guianensis* [P81637], *D. grandiflora* [P08902], and *D. lehmannii* [A45587]) and *Cratylia* (*Cr. floribunda* [P81517]). The tree represents the minimum evolutionary distance estimated through neighbour joining using maximum likelihood distances. **B** Primary structure of the *Dioclea violacea* seed lectin obtained by combination of mass spectrometry (as in Fig. 6.3) and N-terminal sequencing. *Dioclea*-specific amino acid residues are highlighted

mum resolution of 3.2 Å were collected at the beamline BM14 (ESRF, Grenoble). The asymmetric unit (a=55.25, b=113.33, c=123.27) contained a tetramer and belonged to space group $P2_12_12_1$. The *Dioclea violacea* seed lectin was solved by molecular replacement using the coordinates of *Dioclea grandiflora* seed lectin (PDB code 1DGL) as the search model. The structure is being currently refined and will be reported in detail elsewhere. For the purpose of discussion here it is suffice to notice that the loop 114–25 is ordered owing to interactions with His131 of one monomer with aspartate residues 123 from neighbouring monomers (Fig. 6.1b). It is worth noting that this network of quaternary contacts is different from the interactions in the dimer-dimer interface of the *Dioclea grandiflora* lectin. Nevertheless, the data support the view that His131 plays a critical role in stabilising the pH-independent tetrameric association through the ordering of the 114–125 loop structure. Molecular cloning of native and mutant Diocleinae lectins is underway in our laboratories to further elucidate structural requirements that modulate the dimer-tetramer equilibrium.

6.4 Diocleinae Lectin Sequence Characteristics as Phylogenetic Markers

A dendogram generated from a multiple sequence alignment of Diocleinae lectins clearly segregates the genera *Diocleae*, *Canavalia*, and *Cratylia* (Fig. 6.4A). A close examination of the aligned sequences revealed that certain positions are occupied by *genera-specific* residues. Those amino acids that are invariant in all *Dioclea* lectins are highlighted in Fig. 6.4B. In *Canavalia* lectins these positions are also conserved (T^{11}, N^{21}, V^{32}, K^{36}, K^{39}, N^{44}, I^{53}, $^{69}GDS^{71}$, D^{82}, S^{96}, S^{117}, $^{120}THET^{123}$, A^{125}, M^{129}, $^{135}KDQ^{137}$, G^{149}, R^{158}, N^{162}, S^{163}, S^{184}, S^{204}, A^{211}, S^{215}, I^{217}, and T^{226}). The seed lectin of *Cratylia floribunda*, the only lectin of this genus whose sequence has been reported (Cavada et al. 1999), represents an intermediate situation: some positions (21, 39, 53, 69–71, 121–123, 135–137, 184, 211, 215, 217, 226) display *Dioclea*-specific residues, while a few other positions (32, 96, 204) exhibit amino acids characteristic of *Canavalia* lectins. This evidence suggests that defined amino acids have been distinctly conserved in the evolution of *Diocleinae* lectins. Though the physiological significance of this genus-specific conservation of defined amino acids remains to be disclosed, these primary structural features might be exploited as phylogenetic markers to aid in cases of conflicting taxonomical classification.

Acknowledgements. This work has been financed by Conselho Nacional de Desenvolvimento Científico e Tecnológico (CNPq), CAPES, PADCT, FUNCAP, and Banco do Nordeste do Brasil (BNB), and the Dirección General de Investigación Científica y Técnica (BMC2001-3337), Madrid, Spain. A.H.S., T.B.G. and B.S.C are senior investigators of CNPq/Brazil.

References

Barbosa T, Arruda S, Cavada BS, Grangeiro TB, Freitas LAR, Barral-Netto M (2001) In vivo lymphocyte activation and apoptosis by lectins of the Diocleinae subtribe. Mem Inst Oswaldo Cruz 96:673–678

Barral-Netto M, Santos SB, Barral A, Moreira LIM, Santos CF, Moreira RA, Oliveira JTA, Cavada BS (1992) Human lymphocyte stimulation by legume lectins from the Diocleinae tribe. Immunol Invest 21:297–303

Bouckaert J, Loris R, Wyns L (1996) Change of the crystal space group and of the oligomeric structure of concanavalin A as a function of metal ion binding and pH. In: Van Driessche E, Rougé P, Beeckmans S, Bøg-Hansen TC (eds.) Lectins. Biology, Biochemistry, Clinical Biochemistry vol.11. Textop, Hellerup, Denmark, pp. 50–60

Bouckaert J, Hamelryck T, Wyns L, Loris R (1999) Novel structures of plant lectins and their complexes with carbohydrates. Curr Op Struct Biol 9:572–577

Calvete JJ, Thole HH, Raida M, Urbanke C, Romero A, Grangeiro TB, Ramos MV, Almeida da Rocha IM, Guimarães FN, Cavada BS (1999) Molecular characterization and crystallization of Diocleinae lectins. Biochim Biophys Acta 1430:367–375

Cavada BS, Nogueira NAP, Farias CMAS, Grangeiro TB, Ramos MV, Thole HH, Raida M, Rouge P, Calvete JJ (1999) Primary structure and kinetic interaction with glycoproteins of the lectin from seeds of *Cratylia floribunda*. Protein Pept Lett 6:27–34

Cavada BS, Barbosa T, Arruda S, Grangeiro TB, Barral-Netto M (2001) Revisiting Proteus: do minor changes in lectin structure matter in biological activities? Lessons from, and potential biotechnological uses of, *Diocleinae* subtribe lectins. Curr Prot Pept Sci 2:123–135

Chervenak MC, Toone EJ (1995) Calorimetric analysis of the binding of lectins with overlapping carbohydrate-binding ligand specificities. Biochemistry 34: 5685–5695

Dam TK, Cavada BS, Grangeiro TB, Santos CF, de Sousa FAM, Oscarson S, Brewer CF (1998) Diocleinae lectins are a group of proteins with conserved binding sites for the core trimannoside of asparagine-linked oligosaccharides and differential specificities for complex carbohydrates. J Biol Chem 273:12082–12088

Dam TK, Cavada BS, Grangeiro TB, Santos CF, Ceccatto VM, de Sousa FAM, Oscarson S, Brewer CF (2000a) Thermodynamic binding studies of lectins from the *Diocleinae* subtribe to deoxy analogs of the core trimannoside of asparagine-linked oligosaccharides. J Biol Chem 275:16119–16126

Dam TK, Roy R, Das SK, Oscarson S, Brewer CF (2000) Binding of multivalent carbohydrates to concanavalin A and *Dioclea grandiflora* lectin. Thermodynamic analysis of the "multivalency effect". J Biol Chem 275:14223–14230

Elgavish S, Shaanan B (1997) Lectin-carbohydrate interactions: different folds, common recognition principles. Trends Biochem Sci 22:462–467

Ferreira RR, Cavada BS, Moreira RA, Oliveira JTA, Gomes JC (1996) Characteristics of the histamine release from hamster cheek pouch mast cells stimulated by lectins from Brazilian beans and concanavalin A. Inflamm Res 45:442–447

Gabius H-J, Gabius, S (1997) Glycoscience. Status and perspectives. Chapman and Hall, Weinheim

Lis H, Sharon N (1986) Lectins as molecules and as tools. Annu Rev Biochem 55:35–67

Loris R, Hamelryck T, Bouckaert J, Wyns L (1998) Legume lectin structure. Biochim Biophys Acta 1383:9–36

Prabu MM, Suguna K, Vijayan M (1999) Variability in quaternary association of proteins with the same tertiary fold: a case study and rationalization involving legume lectins. Proteins: Struct Funct Genet 35:58–69

Rodriguez D, Cavada BS, OLiveira JTA, Moreira RA, Russo M (1992) Differences in macrophage stimulation and leukocyte accumulation in response to intraperitoneal administration of glucose/mannose-binding plant lectins. Braz J Med Biol Res 25: 823–826

Rozwarski, DA, Swami BM, Brewer CF, Sacchettini JC (1998) Crystal structure of the lectin from Dioclea grandiflora complexed with core trimannoside of asparagine-linked carbohydrates. J Biol Chem 273:32818–32825

Sanz-Aparicio J, Hermoso J, Grangeiro TB, Calvete JJ, Cavada BS (1997) The crystal structure of *Canavalia brasiliensis* lectin suggests a correlation between its quaternary conformation and its distinct biological properties from concanavalin A. FEBS Lett 405:114–118

Senear DF, Teller DC (1981) Thermodynamics of concanavalin A dimer-tetramer self-association: sedimentation equilibrium studies. Biochemistry 20:3076–3083

Van Damme EJM, Peumans WJ, Barre A, Rougé, P (1998). Plant lectins: a composite of several distinct families of structurally and evolutionary related proteins with diverse biological roles. Crit Rev Plant Sci 17:575–692

Wah DA, Romer A, Gallego del Sol F, Cavada BS, Ramos MV, Grangeiro TB, Sampaio AH, Calvete JJ (2001) Crystal structure of native and Cd/Cd-substituted *Dioclea guianensis* seed lectin. A novel manganese-binding site and structural basis of dimer-tetramer association. J Mol Biol 310:885–894

Weis WI, Drickamer K (1996) Structural basis of lectin-carbohydrate recognition. Annu Rev Biochem 65:441–473

7 The Contribution of Optical Biosensors to the Analysis of Structure-Function Relationships in Proteins

Marc H.V. Van Regenmortel

7.1 Introduction

Elucidating the nature of the relationships between the structure and function of biomolecules remains the key problem that molecular biologists need to solve in order to complete their reductionist agenda. According to Francis Crick (1966): "The aim of the modern movement in biology is to explain all biology in terms of physics and chemistry ". This reductionist credo disregards the fact that biological systems possess so-called emergent properties that arise through the innumerable interconnections that exist between the individual constituents of each system. These emergent or relational properties do not exist in the constituent parts and cannot be deduced or predicted from the properties of the individual's isolated components (Holland 1994, Van Regenmortel and Hull, 2002). The beauty of a melody is not present in the individual notes but in the melody as a whole.

In recent years, an increasing number of biologists and philosophers have pointed out the fallacy of attempting to reduce living organisms to a simple juxtaposition of non-living constituents. It has become clear that biological systems can only be understood in terms of their evolutionary history on Earth and that biology is an autonomous discipline requiring its own explanatory concepts not found in chemistry and physics (Dupré 1993; Rosenberg 1994; Bhalla and Iyengar 1999; Weng et al 1999; Van Regenmortel and Hull 2002). Concepts like replication error, adaptation, natural selection, fitness or protection against infection are only meaningful in a biological context. These concepts require the use of a functional language that refers to the roles that molecules, organelles, cells and organs play in meeting the needs of an organism and keeping it alive (Hull and Ruse 1998; Van Regenmortel 2002a).

For the last 50 years molecular biologists have been guided by the reductionist paradigm: "structure determines function". This paradigm assumes that functions, which are always necessarily part of the biological realm, can

Principles and Practice
Methods in Proteome and Protein Analysis
R.M. Kamp, J. J. Calvete, T. Choli-Papadopoulou (Eds.)

be analysed and explained solely in terms of the chemical structure of biomolecules. In effect, it assumes that biological questions and problems can be translated in the language of physics and chemistry. If this were feasible, it would transform biologists into physical chemists and would, at long last, succeed in making biology truly molecular. I believe that such an alleged transformation of biology into genuine *molecular* biology is an example of figurative or metaphorical speech. Crystallographers, when they study the structure of antibodies or enzymes do not, as a result, automatically become immunologists or enzymologists. Today, a considerable part of what is described, as research in molecular biology belongs to fields like physical chemistry or protein chemistry and is aimed at solving chemical rather than biological problems. As pointed out by Sterelny and Griffiths (1999, p. 147): "Molecular biology is the study of how biochemical and other physical laws operate in the complex and varied cellular contexts that evolution has produced". If the cellular and organismic context is ignored, the biological enquiry becomes a purely chemical investigation and its relevance to biology may be questionable.

7.2 Structures Do Not Cause Function

Although a biological function cannot take place in the absence of an underlying material structure, the structure only sets constraints for the type of biological activity that is possible within the context of a given system. A structure does not possess a necessary and sufficient causal efficacy in bringing about a particular function. Causal relations are actually relations between successive events and not between two material objects or between an object and an event. Biological functions result from the integrated interactions of many individual biomolecules and macromolecular assemblies. A single protein is able, on average, to interact with as many as five different partners through the various binding sites it harbours and, furthermore, the same biological activity can be generated by a variety of structures (Martin et al 1998). It is thus illusory to look for a single cause as the explanation of any biological phenomenon. Biological systems are complex and any observed effect is always the result of a network of interactions. Instead of invoking causes, it is preferable to refer to the many factors that simultaneously influence the state of a biological system (Van Regenmortel 2002a).

Another difficulty with the paradigm *structure determines function* is that it is rarely made clear what exact meaning is given to the terms *structure* and *function*. Obviously, a close link between a protein structure and binding activity can be expected only if attention is restricted to the binding site recognized by a specific ligand rather than to the whole protein.

The term function is even more problematic since it is used in various ways depending on the level of biological organization under consideration (Hull and Ruse 1998). Biochemists tend to focus only on the molecular level and restrict their attention to functions like binding, catalysis or signaling. In many cases, the only function that biochemists discuss is binding activity and as a result, the term function is often used as if it were synonymous with binding. It should be emphasized that it is not possible to deduce binding activity from the structure of a protein unless a particular relationship with a specific partner has first been identified. The reason for this is that a binding site is a relational entity defined by the ligand and not by intrinsic structural features of the protein identifiable independently of the ligand. The structure of a binding site as opposed to the structure of a molecule cannot be described without considering the binding partner (Van Regenmortel 2002a).

Since structure is not the *cause* of a function like binding, any analysis of structure-function relationships must be limited to a search for correlations rather than for causal relations. Instead of looking for the cause of binding, the aim should be to investigate the various factors and environmental parameters that can influence binding activity.

Most biological functions have a meaning only at the cellular and organismic level (Bork et al 1998). Protein functions, for instance, have been differentiated by the roles they play at the level of the whole organism and this has led to classification in three categories according to energy, information and communication-associated proteins (Tamames et al 1996; Patthy 1999). The link between such functions and protein structure is less direct than between binding and structure since these functions result from the integrated interactions of many individual proteins. It is to be expected that any correlation between structure and function will then be even more difficult to unravel.

7.3 Can Protein Functions Be Predicted from Structure or Should They Be Determined Experimentally?

Since there is no unique, causal relationship between the structure of a protein and its function, it is not astonishing that the prediction of biological functions of proteins de novo and *in silico* has until now met with little success. Most attempts at predicting protein functions are based on sequence comparisons of proteins of known and unknown function, and generally homology searches succeed only in placing a protein in a broad functional class (Fields 1997). When the degree of sequence difference between two proteins is extensive, any functional prediction becomes even less reliable.

Our ability to predict the 3D structure of a protein from its sequence is still too limited to allow one to locate the position of a binding site in more than about 50% of the cases (Russell et al 1998). Furthermore, when a potential binding site has been located, this does not necessarily indicate the nature of the ligand since similar binding sites may bind ligands that differ markedly in chemical properties. As pointed out by Moodie et al (1996) binding motifs are fuzzy in the sense that the shape and chemical complementarity between two partners can be achieved by a large number of alternative amino acid arrangements. In view of the low success rate of predicting the function of proteins from their structure, it has been argued that reliable information on function is more likely to be obtained by direct experimental determination rather than by the use of prediction algorithms. In the field of functional proteomics, this means that research efforts may be more successful if they were guided by the paradigm: *binding determines function* instead of by the paradigm: *structure determines function* (Van Regenmortel 2002b).

7.4 Analyzing Structure-Activity Correlations with Biosensors

The biological activity of a protein always involves a process of molecular recognition which depends on an initial, specific binding step with a ligand. Therefore, the primary criterion for assessing biological activity is the ability of the protein to bind specifically in a binding assay. Although the presence of biological activity in a protein can be demonstrated by various in vitro and in vivo biological assays, such assays are less accurate and reproducible than binding assays. Therefore, assays that measure the initial binding step of the protein to a ligand are a useful surrogate method for quantifying biologically active proteins compared with cellular and other bioassays.

In recent years, the development of biosensor instruments based on surface plasmon resonance (SPR) has become the method of choice for measuring the binding characteristics of biomolecules. These instruments measure the binding between a ligand immobilized on a sensor chip and an analyte introduced in a flow passing over the surface (Wilson 2002). The binding process is followed on a computer screen as a function of time, in terms of change in mass concentration occurring at the surface of the chip. None of the reactants are labeled, which avoids the artefactual changes in binding properties that often occur when molecules are labeled. Since the interaction is measured in real time, it is possible to determine kinetic rate constants and equilibrium affinity constants (Karlsson and Roos 1997, Morton and Myszka 1998).

The most commonly used biosensor instrument is the BIACORE, which exists in several versions and has been used in about 90 % of the studies published so far (Myszka 1999 a, Rich and Myszka 2000, 2001, 2002). This instrument consists of an optical detector system, exchangeable sensor chips, a processing unit and a personal computer for control and evaluation. The sensor chip is a glass slide coated on one side with a gold film and covered with a dextran layer extending about 100 nm from the surface. The dextran is usually carboxymethylated which allows the immobilization of molecules containing primary amines although other immobilization strategies like aldehyde coupling, hydrazide group coupling, sulfhydryl group coupling and chelate linkage of oligohistidine tags can also be used (Jöhnsson et al 1995). The hydrophilic dextran matrix provides a flexible anchor for ligand immobilization and allows interactions to occur very much like in solution. Four independent flow cells are present on each sensor chip. Normally one of the cells is used as a reference surface, which must be submitted to the same activation and deactivation steps as the reaction surface. This reference surface is used to check for bulk refractive index changes, nonspecific binding, detector drift and injection noise and is essential for collecting high quality kinetic data (Myszka 1999 b). Changes in SPR signal corresponding to the mass of bound species are monitored continuously and are visualized on the computer screen as a plot of resonance units (RU) versus time, known as a sensorgram. For proteins, a signal of 1 RU corresponds to a surface concentration change of 1 pg/mm^2. After each analysis, the sensor surface can be regenerated by introducing a small volume of a suitable dissociating agent which removes the analyte from the immobilized ligand. At least 100 analytical cycles can be performed on the same ligand surface.

In order to obtain reliable kinetic data, it is important to avoid various system-dependent artefacts and to use low concentrations of immobilized ligand (Day et al 2002). Possible artefacts arise from mass transport effects, aggregation, heterogeneity induced by surface immobilization, avidity and matrix effects and nonspecific binding (Myszka 1999 b). The validity of BIACORE data is sometimes questioned because it is believed that the normal kinetics of interaction are altered by the use of an immobilized ligand. On the basis of computer simulations, it was suggested that the rate constants would be affected by the slow diffusion of analyte into the dextran matrix (Schuck 1997). There is evidence, however, that under the conditions recommended for kinetic measurements, the hydrodynamic flow into the dextran is high enough not to significantly affect the observed kinetic rates (Witz 1999). It has been shown experimentally that identical kinetics are observed when ligands are immobilized to surfaces in the presence or absence of a dextran layer (Karlsson and Fält 1997). Furthermore, several recent studies have shown that the use of an immobilized ligand in the BIACORE leads to equilibrium constants that do not differ significantly from those measured in solution by titration calorimetry or analytical ultracentrifugation (Kortt et al 1999 Rich and Myszka 2000, Day et al 2002). The binding data are analyzed using global curve fitting and numerical integration and a theoretical best-fit curve to the primary data should be presented to demonstrate that the interaction model used to interpret the data is applicable. The range of kinetic parameters that can be measured reliably is 10^3–10^7 M^{-1} s^{-1} for the on-rate and 10^{-5} to 10^{-1} s^{-1} for the off-rate (Karlsson 1999).

One of the advantages of using a biosensor for detecting and quantifying a protein is that the protein that is measured is most probably in its native conformation since it has to be recognized by a specific ligand. When physical and chemical techniques like chromatography, electrophoresis or spectroscopy are used to monitor the presence of a protein, only the chemical composition and the primary structure are assessed and there is no guarantee that the protein has the correct tertiary structure and is biologically active. On the other hand, when the concentration of a protein is measured by its ability to interact specifically with a ligand in a biosensor, it is the biologically active protein that is measured and not the amount of chemical substance that may consist of correctly as well as incorrectly folded molecules. In most biological investigations, it is of course the concentration of biologically active molecules that is relevant. For this reason the best method available at present for measuring active concentrations is by means of biosensors. One particular biosensor procedure which utilizes measurements under various flows is especially attractive because it obviates the usual need for a standard calibration curve (Richalet-Secordel et al 1997). Such a curve requires that a preparation of known active concentration be available, which by itself begs the question. The active concentration of a protein measured with a biosensor is nearly always lower than the nominal concentration determined by conventional physico-chemical methods. In the case of recombinant proteins it is not rare to observe that the active concentration is less than 10% of the nominal concentration measured by spectrophotometry (Zeder-Lutz et al 1999).

When searching for correlations between the structure and activity of proteins, it is customary to modify the proteins by mutagenesis and to assess the effect of mutations on the binding affinity (Van Regenmortel 2001a). In view of their reliability and ease of utilization, biosensors have replaced most other techniques for measuring binding affinity in this type of study.

When residues present at the binding interface are mutated, it is usually found that only a few of the changes lead to an affinity that is reduced by more than 100-fold. This has led to the conclusion that only a small fraction of the residues that are assigned to binding sites because they make contact with the ligand are actually contributing to the binding free energy (Jin and Wells 1996). On the other hand, substitutions of residues that are not in contact at the interface between two proteins are also frequently found to affect binding affinity, presumably because these substitutions produce structural changes that propagate far beyond the mutated region (Ben Khalifa et al 2000). Furthermore, multiple substitutions often have non-additive or cooperative effects on the binding kinetics (Rauffer-Bruyère et al 1997). In view of the difficulty of predicting the effect of mutations on binding affinity, it is not astonishing that the so-called rational design of improved proteins by site-directed mutagenesis has so far met with little success (Tobin et al 2000; Van Regenmortel 2000). Although some success has been reported in predicting the kinetics of an interaction measured with BIACORE by a multivariate QSAR approach involving modifications in sequence and buffer composition, the predictions are valid only for the particular protein under study and cannot be generalized for other systems (Andersson et al 2001; Choulier et al 2002).

The neutralization of virus infectivity by antibodies achieved by vaccination is an example of a functional activity that is only meaningful at the level of the organism since only whole organisms can be protected against infection. In view of the considerable practical importance of infectivity neutralization by antibodies, many studies have tried to define which structural features differentiate antibodies that neutralize the infectivity of infectious agents from those that do not. In the case of viruses, it has been shown that the capacity of antibodies to neutralize virus infectivity is correlated with the kinetic dissociation rates of antibodies rather than with their equilibrium affinity constants (Van Cott et al 1994). However, this knowledge is of little use for designing improved vaccines, since immunologists do not know how to stimulate the immune system to preferentially produce antibodies with slow off-rates. For the same reason, X-ray crystallographic studies of complexes between viral antigens and neutralizing monoclonal antibodies do not provide information that facilitates the design of new vaccines (Van Regenmortel 2002a). Knowledge of the structure-activity correlates in a neutralizing antibody combining site only gives insight into the structural complementarity between one antigen binding site and a particular antibody. Such information may be useful for improving the binding properties of a given recombinant antibody intended for use in passive immunotherapy (Gauduin et al 1997). However, the development of a vaccine requires a control over the immunogenicity of the antigen used for immunization, and immunogenicity depends on the potentialities of the host being immunized. Parameters like the immunoglobulin gene repertoire of the host and various cellular and regulatory mechanism which determine the type of antibodies that will be produced after vaccination cannot be controlled because one understands the structural correlates of a single antigen-antibody interaction (Van Regenmortel 2001 b).

The immunogenic capacity of an antigen to induce neutralizing antibodies that protect the organism against infection is a property that belongs to the biological realm and which cannot be reduced to the chemical level of a specific recognition between the binding site of an antigen and a particular antibody. The structural analysis of a virus-antibody complex clarifies the nature of the binding reaction with one neutralizing monoclonal antibody but it cannot identify which vaccine immunogen will be able, in a biological context, to induce antibodies that protect against disease. The development of new vaccines, which belongs to the biological discipline of immunology, will therefore continue to be based on the empirical testing rather than on the so-called rational design of vaccine candidates

References

Andersson K, Choulier L, Hämäläinen MD, Van Regenmortel MHV, Altschuh D, Malmqvist M. 2001. Predicting the kinetics of peptide-antibody interactions using a multivariate experimental design of sequence and chemical space. J.Mol.Recognit. 14: 62–71

Ben Khalifa M, Weidenhaupt M, Choulier L, Chatellier J, Rauffer-Bruyère N, Altschuh D, Vernet T. 2000. Effects of interaction kinetics of mutations at the VH-VL interface of Fabs depend on the structural context. J.Mol. Recognit. 13: 127–139

Bhalla US, Iyengar R. 1999. Emergent properties of networks of biological signaling pathways. Science 283: 381–387

Bork P, Dandekar T, Diaz-Lazcoz Y, Eisenhaber F, Huynen M, Yuan Y. 1998. Predicting function: from genes to genomes and back. J. Mol. Biol. 283: 707–725

Choulier L, Andersson K, Hämäläinen MD, Van Regenmortel MHV, Malmqvist M, Altschuh D. 2002. QSAR studies applied to the prediction of antigen-antibody interaction kinetics as measured by BIACORE. Prot. Eng. 15:101–110

Crick FHC. 1966. Of Molecules and Men. University of Washington Press, Seattle

Day YS, Baird CL, Rich RL, Myszka DG. 2002. Direct comparison of binding equilibrium, thermodynamic and rate constants determined by surface and solution based biophysical methods, Protein Sci. 11:1017–1025

Dupré J. 1993. The Disorder of Things. Metaphysical Foundations of the Disunity of Science. Harvard University Press, Cambridge MA

Fields S, 1997. The future is function. Nat. Genet 15: 325–327

Gauduin MC, Parren PW, Weir R, Barbas CF, Burton DR, Koup RA. 1997. Passive immunization with a human monoclonal antibody protects hu-PBL-SCID mice against challenge by primary isolates of HIV-1. Nat. med. 3: 1389–1393

Holland JH. 1994. Emergence. Perseus Books: Reading, MA

Hull DL, Ruse M (eds) 1998. The Philosophy of Biology. Oxford University Press, New York

Jöhnsson B, Löfas S, Linquist G, Edström A, Muller-Hillgren RM., Hansson A 1995: Comparison of methods for immobilisation to carboxymethyl dextran sensor surface by analysis of the specific activity of monoclonal antibodies. J.Mol.Recogn. 8: 125–131

Karlsson R. 1999 Affinity analysis of non-steady-state data obtained under mass transport limited conditions using BIAcore technology. J.Mol.Recognit. 12: 285–292

Karlsson R, Fält A 1997: Experimental design for kinetic analysis of protein-protein interactions with surface plasmon resonance biosensors. J Immun Meth. 200:121–133

Karlsson R, Roos H 1997: Reaction kinetics; in Price CP, Newman D.J. (eds): Principles and Practice of Immunoassay, 2nd edition, London, MacMillan, pp 101–122

Kortt AA, Nice E, Guen LC 1999: Analysis of the binding of the Fab fragment of monoclonal antibody NC10 to influenza virus N9 neuraminidase from tern and whale using the BIACORE biosensor: Effect of immobilisation level and flow rate on kinetic analysis. Anal Biochem. 273: 133–141

Martin ACR, Orengo CA, Hutchinson EG, Jones S, Karmirantzou M, Laskowski R.A, Mitchell JBO, Taroni C, Thornton J. 1998 Protein folds and functions. Structure 6: 874–8

Moodie SL, Mitchell JBO, Thornton JM 1996 Protein recognition of adenylate: an example of a fuzzy recognition template J.Mol. Biol. 263: 486–500

Morton TA, Myszka DG 1998: Kinetic analysis of macromolecular interactions using surface plasmon resonance biosensors. Methods Enzymol. 295: 268–294

Myszka DG1999 a: Survey of the 1998 optical biosensor literature. J.Mol. Recogn. 12: 390–408

Myszka DG1999 b: Improving biosensor analysis. J.Mol. Recogn. 12: 279–284

Rich RL, Myszka DG. 2000. Survey of the 1999 surface plasmon resonance biosensor literature J.Mol. Recognit. 13: 388–407
Rich RL, Myszka DG. 2001. Survey of the year 2000 commercial optical biosensor literature. J.Mol. Recognit. 14: 273–294
Rich RL, Myszka DG 2002. Survey of the year 2001 commercial optical biosensor literature. J.Mol.Recognit. 15: 352–376
Patthy L 1999 Protein Evolution. Blackwell Science, Oxford
Rauffer N, Zeder-Lutz G, Wenger R, Van Regenmortel MHV and Altschuh D. 1994 Structure-activity relationship for the interaction between cyclosporin A derivatives and the Fab fragment of a monoclonal antibody. Mol. Immunol. 31: 913–922
Richalet-Secordel PM, Rauffer-Bruyère N, Christensen LL, Ofenloch-Haehnle B, Seidel C, Van Regenmortel MHV1997: Concentration measurement of unpurified proteins using biosensor technology under conditions of partial mass transport limitation. Anal Biochem. 249: 165–173
Rosenberg A 1994. Instrumental Biology or the Disunity of Science. University of Chicago Press: Chicago, IL
Russell RB, Sasieni PD, Sternberg MJE 1998. Supersites within superfolds. Binding similarity in the absence of homology. J.Mol.Biol 282: 903–918
Schuck P 1997: Use of surface plasmon resonance to probe the equilibrium and dynamic aspects of interactions between biological macromolecules. Ann Rev Biophys Biomol Struct. 26: 541–566
Sterelny K, Griffiths PE 1999. Sex and Death. An Introduction to Philosophy of Biology. University of Chicago Press, Chicago
Tamames J, Ouzounis C, Sander C, Valencia A 1996. Genomes with distinct function composition. FEBS Lett 389: 96–101
Tobin MB, Gustafson C, Huisman GW. 2000. Directed evolution: the'rational' basis for'irrational' design. Curr. Opin. Struct. Biol.10: 421–427
Van Cott TC, Bethke FR, Polonis AR, Gorny MK, Zolla-Pazner S, Redfield RR, Birx DL, 1994 Dissociation rate of antibody-gp120 interactions is predictive of V3-mediated neutralization of HIV-1. J. Immunol.153: 449–458
Van Regenmortel MHV (2000) Are there two distinct research strategies for developing biologically active molecules: rational design and empirical selection? J.Mol.Recognit. 13: 1–4
Van Regenmortel MHV. 2001a. Analysing structure-function-relationships with biosensors. Cell.Mol.Life Sci. 58: 794–800
Van Regenmortel MHV, 2001b. Pitfalls of reductionism in the design of peptide-based vaccines. Vaccine 19: 2369–2374
Van Regenmortel MHV 2002a. Reductionism and the search for structure-function relationships in antibody molecules J. Mol. Recognit. 15: 241–248
Van Regenmortel MHV 2002b A paradigm shift is needed in proteomics: "structure determines function" should be replaced by "binding determines function" J.Mol. Recognit. 15: 349–351
Van Regenmortel MHV, Hull DL (eds) 2002. Promises and Limits of Reductionism in the Biomedical Sciences. John Wiley, Chichester
Weng G, Bhalla US, Iyengar R. 1999. Complexity in biologically signaling systems. Science 284: 92–98
Wilson WD. 2002. Analysing biomolecular interactions. Science 295: 2103–2104
Witz J 1999: Kinetic analysis of analyte binding by optical biosensors: hydrodynamic penetration of the analyte flow into the polymer matrix reduces the influence of mass transport, Anal Biochem. 270: 201–206
Zeder-Lutz G, Benito A, Van Regenmortel MHV. 1999. Active concentration measurements of recombinant biomolecules using biosensor technology. J.Mol Recognit. 12:300–309

8 The Use of Protein–protein Interaction Networks for Genome Wide Protein Function Comparisons and Predictions

CHRISTINE BRUN, ANAÏS BAUDOT, ALAIN GUÉNOCHE
and BERNARD JACQ

Abstract

The concept of protein function is widely used by biologists. However, the means of the concept and its understanding can vary largely depending on the functional level under consideration (molecular, cellular, physiological, etc.) Function is therefore a complex notion and the development of efficient ways of representing function which can be computer-tractable is presently the goal of many research efforts. Moreover, genomic studies and new high-throughput methods of the post-genomic era provide the opportunity to shed a new light on the concept of protein function. Among them, the analysis of large protein–protein networks will permit the emergence of a more integrated view of protein function.

In this context, we have proposed a new method for protein function comparison and classification which, unlike usual methods based on sequence homology, permits the definition of functional classes of protein based solely on the identity of their interacting partners, thus giving access for the first time to function at the cellular level. This method, named PRODISTIN for Protein Distance based on Interactions, has been first applied to the *Saccharomyces cerevisiae* interactome (proteome-wide protein–protein interactions). An example of a classification/comparison is shown and discussed for a subset of *S. cerevisiae* proteins, accounting for 10 % of its proteome (600 proteins). Functional classification trees have also been made for the *Helicobacter pylori* proteome, confirming the generic aspect of the method. We demonstrated that the method is robust (biologically and statistically) and can be used to predict function for unknown proteins and groups of proteins.

Finally, the potential use of protein–protein interaction data and of the PRODISTIN method in structural biology projects is presented and dis-

Principles and Practice
Methods in Proteome and Protein Analysis
R.M. Kamp, J. J. Calvete, T. Choli-Papadopoulou (Eds.)

cussed. In the future, this method could also be potentially applied to other types of networks such as transcriptional and genetic networks.

8.1 Introduction

It is now widely accepted that the elucidation of gene/protein function is the next big challenge in biology after the determination of many complete prokaryotic and eukaryotic genome sequences. Although it was anticipated that comparisons of newly determined protein sequences with sequences of proteins of known functions could rapidly help to decipher functions, 30–35 % of encoded proteins per completely sequenced genome still have an unknown predictable function [15]: each newly sequenced genome exhibits a significant percentage of translated protein sequences which do not show any or only a marginal sequence similarity with sequences already stored in databases. In addition, it has to be emphasized that even when sequence similarity is detected, the path to the function is not straightforward: many detected similarities are found between putative proteins of unknown function. When a search identifies a similarity with a protein domain of a known function, this is only indicative of a general biochemical function, and is insufficient in specifying a cellular function. Thus, in order to help elucidate protein function, two key factors fostered the need for new high-throughput technologies and bioinformatic tools adapted to proteins: (i) the partial inability of sequence comparison programs to help in protein function prediction as said above and (ii) the fact that knowledge of the genome and of the transcriptome are not sufficient to give comprehensive access to the functional proteome. As a result, a shift of many research efforts from genomics to proteomics and proteome bioinformatics has clearly been visible in the last two years.

Proteomics is the necessary complement of genomics and can be defined as the study of proteins using high-throughput methods. It can be further divided into structural proteomics (also misleadingly called structural genomics), which is defined as the high-throughput study of protein structures, and functional proteomics, which is the study of protein function on a large scale. Functional proteomics studies can be performed at different levels of increasing integration, from the molecular level to the cellular and physiological levels. Since high-throughput methods have allowed us to unravel the composition of large protein complexes and to identify hundreds of pairs of interacting proteins which have been developed, the study of protein–protein interactions is becoming increasingly used to understand and describe protein function at the molecular level.

In this chapter, we will focus on the notion of protein function and on the use of protein–protein interaction data to obtain a new, integrated view of function. We will first discuss the notion of the function(s) of a protein and the different possible ways which can be used to obtain a representation of a function in the computer. Then, molecular interactions will be described, with a particular interest in protein–protein interactions. We will present our own work, which grounded a method for functional classifications of proteins on such interactions. Examples of such classifications will be described for different organisms and the potential uses of the method will be discussed. Finally, the possible use of our method and of interaction data in the context of structural biology will be introduced.

8.2 How Is Protein Function Defined and Represented?

8.2.1 The Problem of Function Description

The ultimate goal of many present biological studies is to decipher the function of the thousands of genes (or actually of their products) that have been discovered in newly sequenced genomes. But what does *function* exactly mean? Although widely used in biology, the term *gene function* (or *protein function*) is in fact very ambiguous. Textual descriptions such as *tyrosine kinase protein*, *intercellular junction protein* or *axon guidance protein* all refer to protein function descriptions but uncover different functional levels: the first one describes a molecular or biochemical function, the second one a cellular function and the third one a functional role at a tissue level. There is no obvious or implicit link between the above instances, which could be used either for different proteins or also for a single biological entity: for instance, the functional features of the *Drosophila* Abl gene product are correctly depicted with the aforementioned terms. The notion of function therefore appears to be a complex one and the fact that functional characterization of a protein has to be done at several levels in order to obtain an integrated functional view is now largely recognized (see [21] for a more complete discussion).

In order to be able to compare functions, reliable ways of describing functions are needed. If description of function(s) for two proteins are available as two sentences (or a short set of sentences) written in natural language, a human scientist is able to capture subtleties in function description and to make her/his opinion about the relatedness of the functions. However, what is really needed now is an efficient way of describing functions which would be computer-readable, especially if one wants to compare functions between several hundreds of proteins.

A natural way to design such a computer-tractable way of describing a function is language-based and could make use of either keywords, sentences or hierarchical textual descriptions (ontologies). Ideally, a good protein function description system based on textual declarations should be able:

- to describe the function(s) of any protein whatever the biological organism
- to take into account different functional levels (biochemical, cellular, tissue, organismal, ...) for the same protein

- to handle different corresponding standardized controlled vocabularies potentially usable for different species
- to allow more than one functional description within a given functional level (multifunctional proteins)
- to take into account the dynamic or even conflicting nature of functional knowledge. The function of many previously uncharacterized proteins is now known, additional functions can be assessed to proteins of an already known function, a wrongly assigned function for a given protein can be dismissed by newly obtained results and replaced by a new one and finally, there could be potentially incompatible functions (based on conflicting results)
- to allow simple (has A the same function as B?) or elaborated searches for functional similarity (rank n proteins in decreasing functional similarity order)

8.2.2 Attempts Towards Textual Descriptions of Function

Recently, different approaches have been proposed to tackle the problem of a textual description of the term function:

- P. Karp [22] has discussed the concept of biological function diversity and introduced the notions of local function and integrated function that he applied essentially to prokaryotic organisms in the EcoCyc database [23]. In the above example, the biochemical function would be an instance of a local function, whereas the cellular and tissular functions would be instances of an integrated function.
- It has been proposed [21] that a description of the function could be achieved in the context of a hierarchy of structural/organisational levels. There are at least seven natural levels of increasing biological complexity (Fig. 8.1, left column). Five of them correspond to *observable* biological objects (molecules, subcellular organelles, cells, tissues, organisms) but two levels are not directly observable (the molecular network level and the inter-organism level) they need specific representation tools. At each structural/organisational level there is a corresponding functional level (Fig. 8.1, right column). For instance, molecular functions are described in terms of molecular reactions, cellular function as a protein's role in a cellular process, and so on. Although this seven-level description system is not presently implemented in any database, it should be noted that recognizing the importance of function description in such a double hierarchical system has two key advantages: (1) these levels represent different traditional biological disciplines (biochemistry, cellular biology, developmental genetics, physiology, anatomy, ...) for which specific descriptive vocabularies do exist. Ignoring some of these levels or trying to fuse some of them into an artificial single level is likely to produce inconsistencies or errors; (2) we are totally ignorant of the biological laws permitting an inference of knowl-

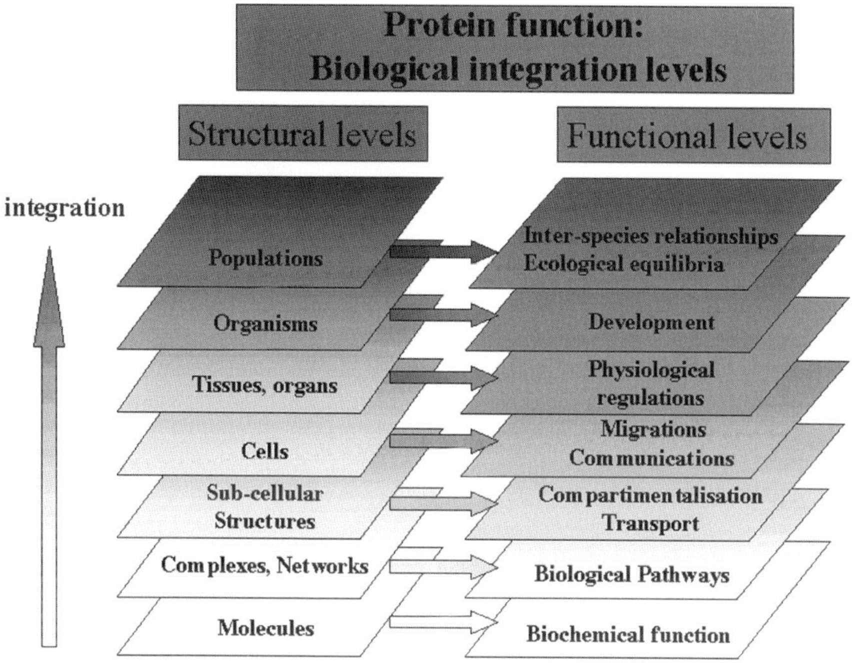

Fig. 8.1. Biological organization and functional levels

edge from a structural or functional level to the upper one: a classic example is to try predicting a cellular or an organismal phenotype from a molecular defect in a gene (filling in the genotype-phenotype gap), which is nearly impossible. In this context, we think that a precise description of each existing level in structural and functional terms could greatly help the understanding of the transition laws between biological levels.

Different hierarchical descriptive representation of functions (ontologies) have been developed, the oldest one probably being the Enzyme Commission's (EC) descriptive numbering system. More recent developments are the MIPS Functional Classification Catalogue [29] and the Gene Ontology (GO) classifications [1]. The GO Consortium aims at 'producing a dynamic controlled vocabulary that can be applied to all organisms, even as knowledge of gene and protein roles in cells is accumulating and changing'. Towards this end, the Consortium has developed three independent ontologies (the *molecular function*, *biological process* and *cellular component* ontologies), based on biological knowledge accumulated for several organisms. Using GO, a protein can be functionally described with a hierarchy of terms at three different levels and the more two proteins share identical descriptions (starting from the root term), the more they are functionally related.

8.2.3 Present Limitations of Functional Descriptions and New Research Directions

Such textual descriptions of functions, even the most elaborate ones like GO, have their own intrinsic limitations. At the moment, the three different ontologies developed by the GO consortium are insufficient in fully describing functions corresponding to the seven existing structural/organisational levels (see above). Also, when proteins are multifunctional (more than one molecular function, involvement in multiple pathways), taking into account several different GO terms in three different hierarchies is not straightforward for comparison purposes. Recent bioinformatic work, such as the development of function grids [25], represents an interesting example of ongoing efforts to integrate many different aspects of protein function with textual descriptions.

There are other aspects of protein function which are simply not considered by existing ontologies. The first one, which is discussed in more detail below, is that proteins do not perform their function in isolation but interact with others, either sequentially during their cellular life or also simultaneously, as part of large protein complexes. A second aspect is that, for a given protein, variations in post-translational modification status (nature of post-translational adducts, number and location of modified residues) can have profound consequences on function. Finally, the spatio-temporal regulation of gene expression is another important functional variable since, for instance, two closely structurally-related proteins such as the products of recently duplicated genes can be expressed in different tissues leading to specific sub-functionalizations.

At this stage, a general comment can be made. A recurrent flaw in all textual approaches described above is that they are totally dependant on human annotations, which could contain a certain amount of subjectivity. It would be extremely interesting to develop new approaches of function description which rely essentially on primary and not secondary, annotated data. A good example of primary data are protein sequences, since the majority of existing sequence analysis and comparison software works on linear sequences of amino-acids represented by letters. However, we have already stressed in the introduction that using sequences for functional purposes has intrinsic limitations. In the following two passages, we will discuss another type of primary datum which seems more appropriate than sequences for functional description and clustering.

8.3 A Protein Network-Based Approach of the Study of Function

8.3.1 Molecular Interactions and Genetic Networks

We discussed above the advantages of describing protein function in the context of a hierarchy of structural and functional levels. In such a view, a new level, that of gene and protein regulatory networks, is increasingly being recognized as essential for understanding gene function: after thirty years of experimental reductionism, where genes and proteins had been mostly considered as individual functioning units, functional genomics approaches (micro-arrays, two-hybrid screens, large-scale isolation of protein complexes...) have shown that genes behave as group sharing a common type of regulation or whose products share common protein interactors. Complex pictures of networks of genes or proteins are emerging and this new conceptual frame is likely to have profound implications in biology in the future. A genetic network can tentatively be defined as a biological unit composed of two entity types: a group of molecular objects (genes and/or proteins, RNAs) and the functional links established between them. The links usually represent direct molecular interactions between biological objects but could also represent indirect ones, like genetic interactions. The latter identify functionally linked genes/proteins located in the same pathway or complex, but which do not necessarily interact physically. Direct physical interactions involving DNA, RNA and proteins, play an essential role in all known biological processes. Until recently, three major types of interactions between these components have accounted for the great majority of known biological macromolecular interactions: the protein-DNA, protein-RNA and protein–protein interactions. Recent reports (see for instance [18, 36] for reviews) have shown that RNA–RNA interactions are also playing an essential role in biological regulation, adding a fourth interaction family to the interaction universe. As will be exemplified below, interaction biology is growing at a rapid pace and the word 'interactome' has been coined to describe the complete set of molecular and genetic interactions occurring within a cell or an organism [34]. Whatever their molecular type, direct interactions between individual molecules then form complex genetic networks which are able to respond to both external stimuli and stresses, as well as to internal changes occurring within the network. A genetic network can be considered as a kind of molecular nervous system since it has a functional role at the level of a cell analogous to that of a nervous system at the level of an organism (although the former is not physically observable, in contrary to a network of neurons). Being able to formally describe interactions and networks, and to query and manipulate them is now largely recognized as essential in the study of gene regulation and function.

8.3.2 Protein–Protein Interaction Data Acquisition, Protein Interaction Databases and Maps

The first step in the study of genetic networks is to obtain experimental data on interactions between specific proteins, genes and RNAs, and then gather and manage lists of functional interactions. Below, we will concentrate on protein–protein interaction data since only these data were further used to develop a method for functional clustering of proteins. Classical biochemical (affinity chromatography, immunoprecipitation, protein–protein cross-linking, two-hybrid, ...) and biophysical (analytical ultracentifugation, spectroscopy, crystallography, ...) methods have long been used to identify and characterize protein–protein interactions. Although very useful, these methods were able to identify only a relatively small number (several hundreds) of specific protein–protein interactions over more than 20 years. In recent years, we have seen the successful development of high-throughput methodologies capable of producing protein–protein interaction data on the proteome scale. Such technologies include high-throughput two-hybrid screens, mass spectrometry analyses of protein complexes and protein microarrays (see [11, 26, 42] for recent reviews). Using these techniques, several thousands of specific interactions have been identified for a few model organisms in less than five years (for review [35, 40]). In order to store and query the rapidly increasing amount of data on protein–protein interactions, several databases have recently been developed throughout the world. Figure 8.2 lists some of the major ones. Bioinformatic softwares also have been developed for visualization and analysis of interaction network data (see for instance [3, 14, 24]).

8.3.3 Protein Networks Studies Allow Us to Revisit the Notion of Function

We are only at the very beginning of *interactomics biology* and major challenges are ahead. One of them is being able to collect large, representative protein–protein interaction lists for several organisms. In addition to the necessity to perform new, proteome-wide interaction screens, an urgent task is also to collect published interactions, this can be difficult since a large proportion of interaction data are not yet present in databases and can be found in scientific literature only. Developing powerful tools to extract specific scientific information from texts (exploring the "textome") will be a helpful strategy in database development and annotations, this is now an active bioinformatic research domain (reviewed in [3]). At the moment, we do not have a precise idea of the number of existing functional interactions either at the level of individual proteins or of the whole proteome. Although the yeast *Saccharomyces cerevisiae* is currently the organism for which the deciphering of the interactome is the most complete [16, 19, 20, 43], a large amount of proteins do

Name	Organism	URL	Description
Molecular Interaction database	All	http://cbm.bio.uniroma2.it/mint/	Interactions between biological molecules (binary interactions, complexes, pathways)
Biomolecular Interaction Network Database	All	http://bind.mshri.on.ca/	Description of interactions, molecular complexes and pathways
Database of Interacting Proteins	All	http://dip.doe-mbi.ucla.edu/	Experimentally determined interactions between proteins
Kyoto Encyclopedia of Genes and Genomes	All	http://www.genome.ad.jp/kegg/	Integration of knowledge on molecular interaction networks involved in biological processes
Protein-Protein Interaction Database	Human, Rat, Mouse	http://www.anc.ed.ac.uk/mscs/PPID/	Binary interactions curated from literature
Munich Information Center for protein sequences	Yeast	http://mips.gsf.de/	Binary interactions, complexes from high-throughput experiments and curated from literature, searchable via the Protein Viewer Pathways
Subtilis Protein Interaction Database	B. Subtilis	http://www-mig.jouy.inra.fr/bdsi/SPiD/	Two-hybrid protein interactions from B. subtilis
Hybrigenics PIMRider	H. Pylori	http://www.hybrigenics.fr/pageb5_cad.htm	Two-hybrid protein interactions from H. pylori
General Repository for Interaction Datasets	Yeast	http://biodata.mshri.on.ca/grid/	Interactions can be viewed and analysed using the Osprey Visualisation System

Fig. 8.2. Main existing interaction databases

not have any known protein interactor(s) yet. From the numerous proteins for which interaction data are available, there seems to exist a great variation in the number of interactors: browsing databases such as DIP or BIND (Fig. 8.2) shows examples of proteins with more than twenty identified partners and many others with only one or two. However, a major difficulty is to assess if this variation is real or partly reflects the incompleteness of our present knowledge, as some proteins have been studied in far more detail than others. At the proteome level, it seems clear that the number of interactions in a cell largely exceeds the number of proteins. Minimal estimations of the number of protein–protein interactions in yeast, based on two-hybrid screens, are in the range of 10,000 to 36,000 for approximately 6,300 proteins [21, 39]. If additional factors which increase the protein diversity (alternative splicing, post-translational modifications) are taken into account, interactions could easily attain the million range in a representative metazoan like *Drosophila*. More realistic evaluations of the size of the protein–protein interactome will have to wait for a correct determination of the number of partners for a set of representative proteins in different functional classes.

Another important challenge, beyond the scope of this chapter, will be to be able to take into account the dynamic nature of interactions and regulatory networks: the development of mathematical models for the simulation and prediction of network behaviour dynamics (see for instance [9] for a review) as well as experimental engineering of small model networks in bacteria and eukaryotes are highly desirable goals.

Finally, it is important to realize that the knowledge of interactions is likely to shed a new light on protein function. The classical view of protein function is defined at the level of the action of a single protein molecule (its binding to another molecule or the catalysis of a given reaction). The present knowledge that the same protein possesses several interaction partners now allows us to 'zoom out' and give an enlarged view of protein function, in which a protein is seen as an element of a network of interacting partners. We will show in the following sections that such a new vision can be exploited in such a way to classify proteins from a functional point of view.

8.4 Functional Clustering of Proteins Based on Interactions

8.4.1 Principle

As soon as the first protein sequences became available, biologists tried to compare them using alignment methods and progressively introduced useful measures for this purpose (identity and similarity percentages, z-scores, Blast scores, ...). Also, at the secondary and tertiary structure level, methods were devised to compare protein structures. Very often, sequence and structural comparisons are used to infer functional relationships between proteins.

Although inferring function from structural comparisons can lead to useful and testable hypotheses, it remains a risky exercise, which could lead to wrong conclusions [10]. More recently, computational methods relying upon genome organization have been developed such as the domain fusion method which establishes that two proteins from a given organism are functionally related when they exist as a single fused polypeptide in another proteome [13, 27]. The fact that genes repeatedly found as neighbors on chromosomes in different organisms may encode functionally related proteins was also used to ground a method [8, 30, 40] as well as the phylogenetic co-inheritance of proteins in several proteomes which may suggest their functional link [31]. Although these methods and their combinations [28] were use to predict the function of a number of proteins, they still suffer from limitations essentially related to the fact that they work better when applied to completely sequenced genomes and they are more appropriate to prokaryotic genome organization compared to the eukaryotic one. In addition to this, they are only valid for a small number of proteins.

In this context, we have recently proposed to use protein–protein interactions data to derive a measure of functional proximity for proteins and to be able to make a functional clustering just as sequence clustering allows to make phylogenetic trees [5, 21]. The central idea in interaction-based functional clustering is not to compare proteins themselves but instead to compare the list of their partners and assume that the more two proteins share interacting partners, the more they should be functionally related. Let us for instance consider three proteins A, B, C, each of them establishing 30 specific interactions (experimentally determined) with other protein partners. If A and C, B and C and A and B have respectively 25, 13 and 2 common interactors, it seems intuitively reasonable to conclude that A and C are highly functionally related, that B and C share at least some functions and that A and B are probably not functionally (or only marginally) related. The feasibility of this idea was first tested on a small set of protein–protein interactions from *Saccharomyces cerevisiae* (14 different proteins and their 47 interactors). The results of two-dimensional matrices (where the 14 proteins were on vertical lines, interactors on horizontal lines and interacting pairs plotted at intersections) indicated that proteins with common interaction partners tend to exhibit a common pattern and further examination of proteins with similar patterns then showed that they correspond to proteins with related functions [21].

These encouraging results prompted us to develop a more general method [4, 5], which allows the calculation of a functional distance. Such a distance should provide a direct measurement of the functional relationships between proteins, independently of their primary sequence similarity. A functional relationship could mean either that proteins are involved in the same biological process or that they share common functional features (molecular or cellular function, belonging to the same functional family, etc.). We first calculate a distance on a set of proteins that share common interactors. Then, using only the distance values, we build classes of proteins. Relevant proteins to be clustered are defined as having at least p interactions ($p > 1$), the other proteins being there only to measure distance values. From all possible pairs of distances, a neighbor-joining algorithm is then applied which finds related proteins and finally the results of the clustering method are displayed as a tree.

8.4.2 Functional Classification of 10% of the Yeast Proteome

We computed 4,794 protein–protein interactions involving 2,139 proteins of *Saccharomyces cerevisiae* (38% of the proteome) with PRODISTIN. The dataset we used was only composed with direct protein–protein interactions extracted from different sources (the MIPS database [29] for two-hybrid excluding high-throughput experiments, in vitro binding, far western, gel retardation, and biochemical experiments, and the core data from Uetz et al. [43] and Ito et al. [20] for high-throughput two-hybrid experiments.

A classification tree containing 602 proteins was built. The grouping of proteins resulting from the tree computation was further analyzed and the clusters found were annotated using the functional keywords from the Yeast Proteome Database [7]. More than 40 clusters of proteins were detected (Fig. 8.3) and it appears that proteins are grouped within the tree according to their cellular function: proteins composing a multiprotein complex, involved in the same pathway or more broadly, in the same biological process, belong to the same cluster. Therefore, the PRODISTIN method is able to functionally classify proteins according to their cellular function solely from interaction data and independently from sequence and structure information. The foundations of the PRODISTIN clustering were further investigated first, by showing with a refined analysis of the keywords distribution within the tree, that the PRODISTIN method cluster proteins according to their cellular function more efficiently than their biochemical function. Secondly, by demonstrating that the clustering of the 602 proteins according to their sequence similarity is drastically different than the PRODISTIN one (Brun et al., submitted). Finally, the robustness of the method was looked into by defining a class robustness index (CRI) based on the topology of the subtrees and by studying the evolution of this parameter in trees built with interactions of decreasing biological confidence (Brun et al., submitted). The correlation observed between the CRI calculated values and the biological confidence of the interactions used endorsed the robustness of PRODISTIN.

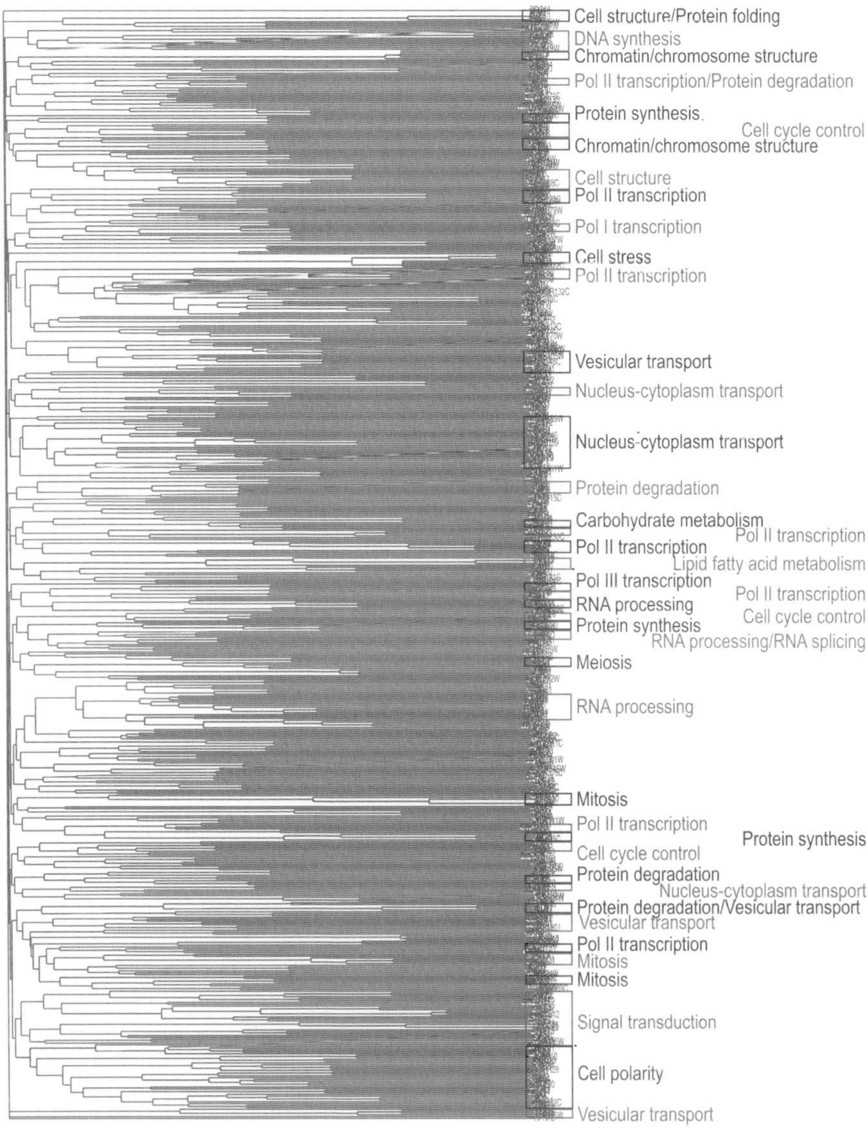

Fig. 8.3. A functional tree for 602 yeast proteins. TreeDyn (http://viradium.mpl.ird.fr/treedyn/) was used to represent the computed classification. Relevant clusters and subtrees are annotated in the margin for their cellular function

8.4.3 The Different Types of Functional Clusters

The detailed analysis of yeast clusters reveals the grouping of proteins known to participate in the same multiprotein complex. For instance, the GIM complex promotes the formation of functional α- and γ-tubulin [17]. All chaperones members of this complex which are present in the tree, form a subtree devoted to cell structure and protein folding (Fig. 8.4a). Similarly, members of pathways are also found clustered in the tree. The MAP kinase cascades regulating mating response and filamentous growth share components [12]. Some of them are found grouped in the same subtree (the MAPKKK STE11, the MAPKK STE7 and the MAPK FUS3 and KSS1) as well as STE5, the scaffold protein upon which the protein kinases can be assembled and be efficiently activated (Fig. 8.4b). In addition, STE50, a regulator of STE11 activity [45] and STE12, a transcription factor activated by FUS3 and acting downstream from the cascades to mediate the mating-pheromone induction of genes involved in the mating response also participate to the cluster. Finally, DIG1 and DIG2, two repressors of the mating and filamentous growth responses acting through a direct inhibition of STE12 [41], and MPT5, a protein involved in recovery from pheromone arrest [46] are also found in this cluster. Interestingly, PST2, a protein uncharacterized so far, belongs to this group of proteins all involved in mating-type responses and differentiation. Consequently, it is tempting to speculate that PST2 also participates in these processes. Therefore, the PRODISTIN classification provides a new means to predict a cellular function for a protein of unknown function, based on the cellular function of the proteins belonging to the same cluster. Based on this idea, we have been able to propose a function for 33/93 unknown proteins present in the tree (Brun et al., submitted).

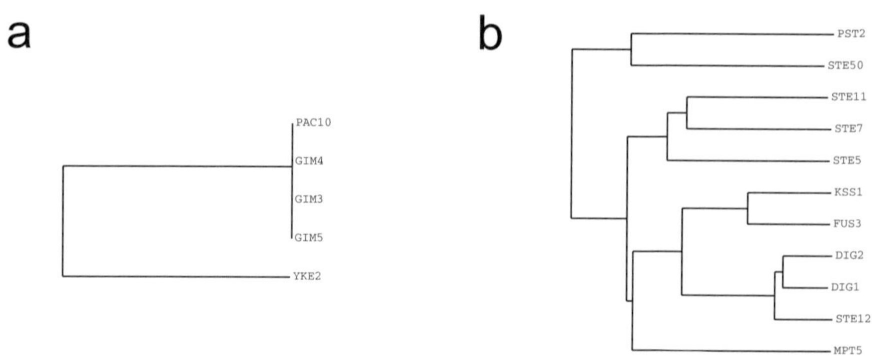

Fig. 8.4. **a** Detail of the cell structure/protein folding subtree. **b** Detail of the signal transduction subtree

8.4.4 Application to Another Proteome: *Helicobacter pylori*

Since the PRODISTIN method relies only on the availability of lists of pairs of protein interactors, it should work for any organism for which such data are available. In order to verify that this was indeed the case, we performed PRODISTIN analyses on another organism, the prokaryote, *Helicobacter pylori*.

When 631 protein–protein interactions involving 518 proteins from *Helicobacter pylori* [33] were processed with PRODISTIN, a tree containing 133 proteins was obtained (Fig. 8.5). Here again, as shown above for the yeast tree, the PRODISTIN classification is able to group proteins known to be involved in the same process and also permits to assign functions to uncharacterized

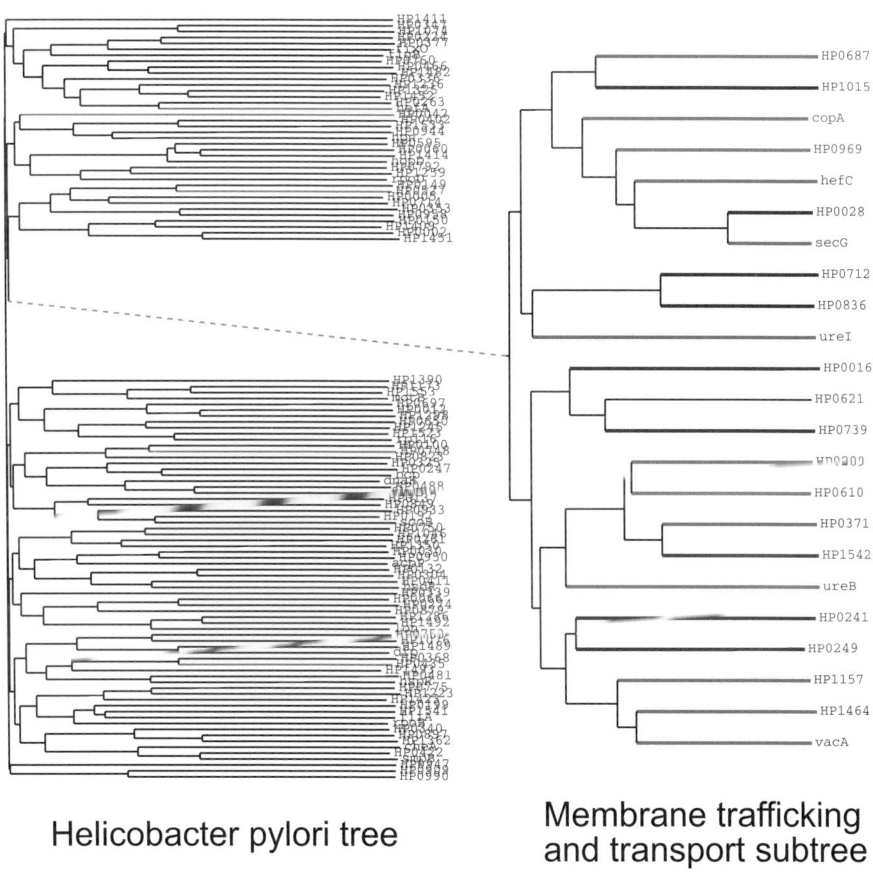

Fig. 8.5. A functional tree for 133 *Helicobacter pylori* proteins and a focus on the membrane trafficking and transport subtree. Branches carrying proteins involved in this process are *grey*. *Black* branches carry proteins of unknown cellular function

proteins. This is particularly well illustrated by the case of the *Helicobacter pylori* proteins known to be involved in various membrane trafficking and transport: the vast majority of the classified proteins involved in these processes (13/17, 76 %) are found grouped in a single subtree containing 23 proteins, further confirming that PRODISTIN clusters have a common cellular function. In addition, the nine remaining proteins of the subtree (except for one) are as yet uncharacterized and may also be involved in membrane trafficking and transport.

In short, we have shown that interaction data with a very simple structure (lists of pairs of interactors) can indeed be used to derive an elaborated functional knowledge, i.e. a functional classification for a significant part of a given proteome. Using the two organisms described above, we have not only been able to propose classifications, which are totally consistent with our present knowledge but also allow us to make functional predictions. Furthermore, additional bioinformatic analyses have also been performed (Brun et al., submitted), allowing us to assess the statistical significance of our results.

8.5 Protein–Protein Interactions and Structural Biology

In the previous parts of this chapter, we mainly insisted on the functional aspects of interactions. Our main point has been that the use of the structure of protein–protein interaction networks now allows one to obtain a functional view of many individual proteins at the cellular level, even though interaction data at the molecular level were used as the sole source of information. We will now consider interactions from a different point of view, which is that high-throughput interaction data can also be used to acquire new knowledge on interaction specificity at the molecular/structural level.

When considered individually, protein three-dimensional structures are generally not informative about the biochemical function(s) performed by proteins. Another level of information is needed, since protein–protein interactions and formation of specific complexes are essential for the functioning of all cell types. When proteins interact to form a complex, the specificity of the interaction depends on the physico-chemical properties of the interface formed by the amino-acids present at the two protein surfaces coming into contact. Protein–protein interaction sites can be very variable in terms of geometry, the chemical nature of the amino-acids at the interface and the number of patches forming the interacting surface [6]. Recent progress has been made with the use of docking methods to predict protein–protein interactions, provided that the 3D structures of partners already known to interact are available [37]. Despite this, it is still impossible to predict whether or not two proteins will specifically interact, even when their three-dimensional structures are known. Clearly, more work has to be done in different directions to reach such a long-term goal:

- The first direction to work in is to determine many morc three-dimensional structures in particular that of more individual proteins from model organisms and more importantly, that of a significant number of protein complexes, since the atomic coordinates of multimolecular assemblies are presently under-represented in PDB, the repository of 3D-macromolecular structures [6]. This would permit a statistically significant analysis of protein interfaces, hopefully leading to a better understanding of protein–protein recognition sites.
- At the post-genomic level, more data pertinent to protein–protein interactions have also to be collected. As already discussed, new screens with powerful methods such as high-throughput two-hybrids are needed to obtain more comprehensive lists of interaction pairs in several model organisms. Also, an increased use of recent methods such as TAP-TAG [32] will permit purification and characterize several hundreds of large protein complexes.
- Making better use of these two data types (structures, post-genomic data, either individually or in combination) in new structural bioinformatic approaches can be valuable. Interacting pairs could be systematically used in docking approaches, provided that structures of partners or those of homologues are available [37]. Another potentially interesting approach is provided by the search for correlated sequence signatures in interacting protein pairs [38]: it has been shown in the yeast *Saccharomyces cerevisiae* by using a statistical analysis performed on all possible combinations of sequence-signature pairs that some are over-represented in a database of yeast interacting proteins and that this could represent a basis for interaction prediction.

Finally, our PRODISTIN method can potentially be used for structural purposes. For instance, it is tempting to propose that a classification of protein domains and their combinations according to cellular functions can be derived from the functional classification of proteins. Indeed, a preliminary analysis showed that proteins containing a RRM motif, a probable diagnostic of RNA binding proteins, all localize in the same part of the tree (Fig. 8.6a). When the cellular function of these proteins is investigated, they are essentially involved in RNA processing/modification (Fig. 8.6b). Furthermore, it appears that proteins containing an RRM motif are involved in RNA splicing (Fig. 8.6 c) but excluded from the RNA decay process (Fig. 8.6d). Therefore, such analysis may provide a new tool to investigate the distribution of protein domains and their combination in the light of cellular function rather than the usual biochemical aspect.

a b c d

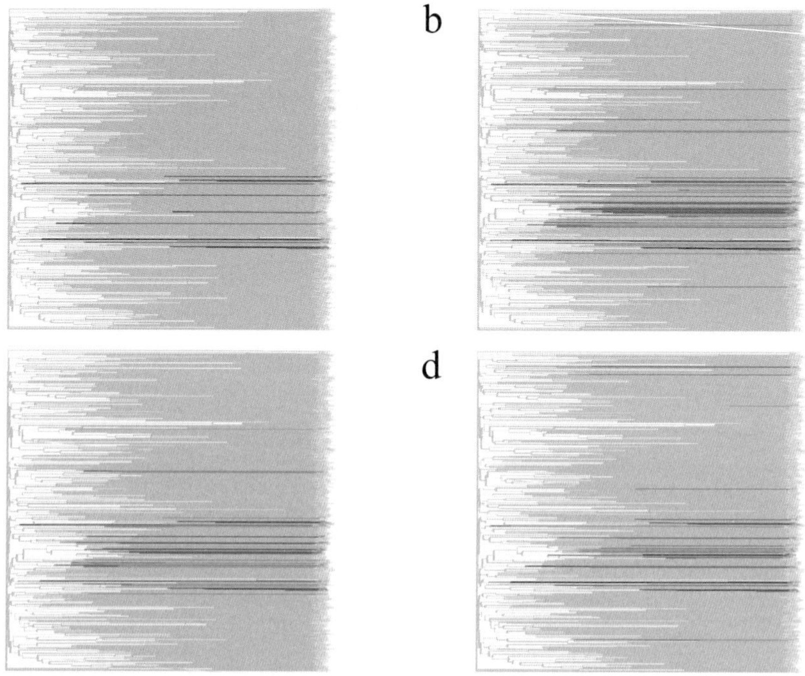

Fig. 8.6. **a** Branches carrying a protein containing a RRM domain are in *grey*. **b** *Black* branches carry proteins involved in RNA processing/modification, **c** RNA splicing; **d** RNA turnover

8.6 Conclusion

Whatever the living organism under consideration is (prokaryote or eukaryote), molecular biologists have worked until recently using essentially a gene-by-gene (or protein-by-protein) approach. Although many different experimental results were often obtained on the same organisms, the same tissue and with the same experimental conditions, they were quite difficult to put together, as if different people worked on isolated pieces of a giant jigsaw puzzle. We advocated that studying the protein function within the new conceptual frame represented by networks of interacting molecules could help to put the pieces together and obtain an integrated view of the biological phenomenon under scrutiny.

In this context, we developed the PRODISTIN method which uses protein–protein interaction data (lists of interacting pairs) to propose functional classifications of proteins. Although our results are still preliminary, the method appears promising in revealing functional resemblances which cannot be detected by sequence comparison programs (Brun et al., submitted). It is presently limited by the amount of interaction data available, but will be more and more useful as data from systematic interaction screens (such as large-scale two-hybrid screens) will be obtained. The method is generic since it could be applied to any type of protein in any organism, as soon as corresponding experimental interaction data are available. Furthermore, it is not limited to a given type of interaction, and although we only applied it to protein–protein interactions so far, it could potentially be used with other interaction types such as protein-DNA and genetic interactions. Moreover, its use for functional evolutionary comparisons between different organisms can be foreseen, provided that lists of orthologous proteins and their interacting partners become available. In this respect, we recently initiated a study of the *Drosophila* and the human proteome using PRODISTIN (C. Brun, A. Baudot and B. Jacq, unpubl.). Finally, we also discussed the potential use of different types of interaction data for structural biological purposes, including the long-term goal of interaction predictions from individual protein 3D structures.

Genome projects have now shown that there is no striking differences in the number of genes between organisms with very different organizational complexities: *Drosophila* has only two to three times more genes than the unicellular yeast and seems to have less genes than the nematode, although its anatomy and behavior are far more complex. Clearly, the absolute number of genes does not seem to be an essential determinant of biological complexity. Rather, it could be that the number of interactions between genes and the structure of the regulatory network that they establish plays a more important role. Studying functional relationships at the entire proteome level (interactomics) is a new frontier in the post-genome era. These studies should assist in making important progress in the understanding of protein function throughout evolution.

Acknowledgements. A preliminary version of this work was presented during the 2002 MPSA conference in Valencia (Spain). The research of B. Jacq's group has been supported by CNRS and by the inter-EPST bioinformatics programme. C. Brun is supported by a Fondation pour la Recherche Médicale postdoctoral fellowship.

References

1. Ashburner M, Ball CA, Blake JA, Botstein D, Butler H, Cherry JM, Davis AP, Dolinski K, Dwight SS, Eppig JT, Harris MA, Hill DP, Issel-Tarver L, Kasarskis A, Lewis S, Matese JC, Richardson JE, Ringwald M, Rubin GM, Sherlock G (2000) Gene ontology: tool for the unification of biology. The Gene Ontology Consortium. Nat Genet 25:25–29
2. Blaschke C, Hirschman L, Valencia A (2002) Information extraction in molecular biology. Brief Bioinform 3:154–165
3. Breitkreutz BJ, Stark C, Tyers M (2003) Osprey: a network visualization system. Genome Biol 4:R22

4. Brun C, Guénoche A, Jacq B (2003) Approach of the functional evolution of duplicated genes in *Saccharomyces cerevisiae* using a new classification method based on protein–protein interaction data. Journal of Structural and Functional Genomics 3:213–224
5. Brun C, Wojcik J, Guénoche A, Jacq B (2002) Bioinformatic study of interaction networks: PRODISTIN, a new method for a functionnal classification of proteins. In: Nicolas J, Thermes C (eds) Journées Ouvertes Biologie Informatique Mathématiques (JOBIM'2002). Saint Malo, France, p 171–182
6. Chakrabarti P, Janin J (2002) Dissecting protein–protein recognition sites. Proteins 47:334–343
7. Costanzo MC, Crawford ME, Hirschman JE, Kranz JE, Olsen P, Robertson LS, Skrzypek MS, Braun BR, Hopkins KL, Kondu P, Lengieza C, Lew-Smith JE, Tillberg M, Garrels JI (2001) YPD, Pombe PD and Worm PD: model organism volumes of the BioKnowledge library, an integrated resource for protein information. Nucleic Acids Res 29:75–79
8. Dandekar T, Snel B, Huynen M, Bork P (1998) Conservation of gene order: a fingerprint of proteins that physically interact. Trends Biochem Sci 23:324–328
9. de Jong H (2002) Modeling and simulation of genetic regulatory systems: a literature review. J Comput Biol 9:67–103
10. Devos D, Valencia A (2000) Practical limits of function prediction. Proteins 41:98–107
11. Drewes G, Bouwmeester T (2003) Global approaches to protein–protein interactions. Curr Opin Cell Biol 15:199–205
12. Elion EA (2001) The Ste5p scaffold. J Cell Sci 114:3967–3978
13. Enright AJ, Iliopoulos I, Kyrpides NC, Ouzounis CA (1999) Protein interaction maps for complete genomes based on gene fusion events. Nature 402:86–90
14. Enright AJ, Ouzounis CA (2001) BioLayout–an automatic graph layout algorithm for similarity visualization. Bioinformatics 17:853–854
15. Galperin MY, Koonin EV (2000) Who's your neighbor? New computational approaches for functional genomics. Nat Biotechnol 18:609–613
16. Gavin AC, Bosche M, Krause R, Grandi P, Marzioch M, Bauer A, Schultz J, Rick JM, Michon AM, Cruciat CM, Remor M, Hofert C, Schelder M, Brajenovic M, Ruffner H, Merino A, Klein K, Hudak M, Dickson D, Rudi T, Gnau V, Bauch A, Bastuck S, Huhse B, Leutwein C, Heurtier MA, Copley RR, Edelmann A, Querfurth E, Rybin V, Drewes G, Raida M, Bouwmeester T, Bork P, Seraphin B, Kuster B, Neubauer G, Superti-Furga G (2002) Functional organization of the yeast proteome by systematic analysis of protein complexes. Nature 415:141–147
17. Geissler S, Siegers K, Schiebel E (1998) A novel protein complex promoting formation of functional alpha- and gamma-tubulin. Embo J 17:952–966
18. Hannon GJ (2002) RNA interference. Nature 418:244–251
19. Ho Y, Gruhler A, Heilbut A, Bader GD, Moore L, Adams SL, Millar A, Taylor P, Bennett K, Boutilier K, Yang L, Wolting C, Donaldson I, Schandorff S, Shewnarane J, Vo M, Taggart J, Goudreault M, Muskat B, Alfarano C, Dewar D, Lin Z, Michalickova K, Willems AR, Sassi H, Nielsen PA, Rasmussen KJ, Andersen JR, Johansen LE, Hansen LH, Jespersen H, Podtelejnikov A, Nielsen E, Crawford J, Poulsen V, Sorensen BD, Matthiesen J, Hendrickson RC, Gleeson F, Pawson T, Moran MF, Durocher D, Mann M, Hogue CW, Figeys D, Tyers M (2002) Systematic identification of protein complexes in Saccharomyces cerevisiae by mass spectrometry. Nature 415:180–183
20. Ito T, Chiba T, Ozawa R, Yoshida M, Hattori M, Sakaki Y (2001) A comprehensive two-hybrid analysis to explore the yeast protein interactome. Proc Natl Acad Sci U S A 98:4569–4574

21. Jacq B (2001) Protein function from the perspective of molecular interactions and genetic networks. Brief Bioinform 2:38–50
22. Karp PD (2000) An ontology for biological function based on molecular interactions. Bioinformatics 16:269–285
23. Karp PD, Riley M, Paley SM, Pellegrini-Toole A, Krummenacker M (1999) Eco Cyc: encyclopedia of *Escherichia coli* genes and metabolism. Nucleic Acids Res 27:55–58
24. Kolchanov NA, Nedosekina EA, Ananko EA, Likhoshvai VA, Podkolodny NL, Ratushny AV, Stepanenko IL, Podkolodnaya OA, Ignatieva EV, Matushkin YG (2002) GeneNet database: description and modeling of gene networks. In Silico Biol 2:97–110
25. Lan N, Montelione GT, Gerstein M (2003) Ontologies for proteomics: towards a systematic definition of structure and function that scales to the genome level. Curr Opin Chem Biol 7:44–54
26. Legrain P, Wojcik J, Gauthier JM (2001) Protein–protein interaction maps: a lead towards cellular functions. Trends Genet 17:346–352
27. Marcotte EM, Pellegrini M, Ng HL, Rice DW, Yeates TO, Eisenberg D (1999) Detecting protein function and protein–protein interactions from genome sequences. Science 285:751–753
28. Marcotte EM, Pellegrini M, Thompson MJ, Yeates TO, Eisenberg D (1999) A combined algorithm for genome-wide prediction of protein function. Nature 402:83–86
29. Mewes HW, Frishman D, Guldener U, Mannhaupt G, Mayer K, Mokrejs M, Morgenstern B, Munsterkotter M, Rudd S, Weil B (2002) MIPS: a database for genomes and protein sequences. Nucleic Acids Res 30:31–34
30. Overbeek R, Fonstein M, D'Souza M, Pusch GD, Maltsev N (1999) The use of gene clusters to infer functional coupling. Proc Natl Acad Sci U S A 96:2896–2901
31. Pellegrini M, Marcotte EM, Thompson MJ, Eisenberg D, Yeates TO (1999) Assigning protein functions by comparative genome analysis: protein phylogenetic profiles. Proc Natl Acad Sci U S A 96:4285–4288
32. Puig O, Caspary F, Rigaut G, Rutz B, Bouveret E, Bragado-Nilsson E, Wilm M, Seraphin B (2001) The tandem affinity purification (TAP) method: a general procedure of protein complex purification. Methods 24:218–229
33. Rain JC, Selig L, De Reuse H, Battaglia V, Reverdy C, Simon S, Lenzen G, Petel F, Wojcik J, Schachter V, Chemama Y, Labigne A, Legrain P (2001) The protein–protein interaction map of Helicobacter pylori. Nature 409:211–215
34. Sanchez C, Lachaize C, Janody F, Bellon B, Roder L, Euzenat J, Rechenmann F, Jacq B (1999) Grasping at molecular interactions and genetic networks in *Drosophila melanogaster* using FlyNets, an Internet database. Nucleic Acids Res 27:89–94
35. Schachter V (2002) Protein-interaction networks: from experiments to analysis. Drug Discov Today 7:S48–54
36. Shi Y (2003) Mammalian RNAi for the masses. Trends Genet 19:9–12
37. Smith GR, Sternberg MJ (2002) Prediction of protein–protein interactions by docking methods. Curr Opin Struct Biol 12:28–35
38. Sprinzak E, Margalit H (2001) Correlated sequence-signatures as markers of protein–protein interaction. J Mol Biol 311:681–692
39. Sprinzak E, Sattath S, Margalit H (2003) How Reliable are Experimental Protein–protein Interaction Data? J Mol Biol 327:919–923
40. Tamames J, Casari G, Ouzounis C, Valencia A (1997) Conserved clusters of functionally related genes in two bacterial genomes. J Mol Evol 44:66-73
41. Tedford K, Kim S, Sa D, Stevens K, Tyers M (1997) Regulation of the mating pheromone and invasive growth responses in yeast by two MAP kinase substrates. Curr Biol 7:228–238
42. Tyers M, Mann M (2003) From genomics to proteomics. Nature 422:193–197

43. Uetz P, Giot L, Cagney G, Mansfield TA, Judson RS, Knight JR, Lockshon D, Narayan V, Srinivasan M, Pochart P, Qureshi-Emili A, Li Y, Godwin B, Conover D, Kalbfleisch T, Vijayadamodar G, Yang M, Johnston M, Fields S, Rothberg JM (2000) A comprehensive analysis of protein–protein interactions in Saccharomyces cerevisiae. Nature 403:623–627
44. von Mering C, Krause R, Snel B, Cornell M, Oliver SG, Fields S, Bork P (2002) Comparative assessment of large-scale data sets of protein–protein interactions. Nature 417:399–403
45. Wu C, Leberer E, Thomas DY, Whiteway M (1999) Functional characterization of the interaction of Ste50p with Ste11p MAPKKK in *Saccharomyces cerevisiae*. Mol Biol Cell 10:2425–2440
46. Xu BE, Skowronek KR, Kurjan J (2001) The N terminus of Saccharomyces cerevisiae Sst2p plays an RGS-domain-independent, Mpt5p-dependent role in recovery from pheromone arrest. Genetics 159:1559–1571

9 Probing Ribosomal Proteins Capable of Interacting with Polyamines

Dimitrios L. Kalpaxis, Maria A. Xaplanteri, Ioannis Amarantos, Fotini Leontiadou and Theodora Choli-Papadopoulou

9.1 Introduction

The ribosome decodes the genetic information, controls the fidelity of codon–anticodon interactions, and catalyzes the peptide bond formation. However, none of these functional properties can be detected in free rRNAs, because ribosomal proteins and ions are required for the attainment of the rRNA proper tertiary structure (Burma et al. 1985). Despite the fact that the binding sites of ribosomal proteins in rRNA are well characterized, our understanding of the ionic environment role in protein-rRNA interactions is still rudimentary, and few exceptions have emerged regarding the role of monovalent and divalent ions (Batey and Williamson 1998, Agalarov et al. 2000, Xing and Draper 1995, Conn et al. 1999, Drygin and Zimmermann 2000).

Polyamines constitute an essential component of the ribosomal ionic environment. Although there has been a significant effort made to understand the polyamine action on tRNA positioning to ribosomes and on the regulation of translation process (Agrawal et al. 1999, Cohen 1998), little information is available concerning the contacts between polyamines and ribosomal constituents. Ribosomal proteins, capable of interacting with polyamines, were first characterized by a fixation of polyamines to ribosomes or ribosomal subunits with homobifunctional cross-linkers (Stevens and Pascoe 1972, Bernabeu et al. 1978, Kakegawa et al. 1986). More recently, photoreactive spermine analogues have been used for labelling proteins in *Escherichia coli* functional ribosomal complexes (Amarantos et al. 2001). The following sections provide a guide to various difficulties and pitfalls related to such experiments and, also, to the type of structural and functional information that can be obtained from them.

Principles and Practice
Methods in Proteome and Protein Analysis
R.M. Kamp, J. J. Calvete, T. Choli-Papadopoulou (Eds.)

9.2 Fixation of Polyamines to Ribosomal Proteins with Homobifunctional Cross-Linkers

To date, two methods have been applied to fix polyamines to proteins in *E. coli* ribosomes, both employing homobifunctional cross-linkers: (1) cross-linking of [^{14}C]spermine to ribosomes or ribosomal subunits with 1,5-difluoro-2,4-dinitrobenzene (Stevens and Pascoe 1972, Bernabeu et al. 1978, Kakegawa et al. 1986); (2) fixation of [^{14}C]spermine to ribosomal subunits with dimethyl suberimidate (Kakegawa et al. 1986). Table 9.1 lists the ribosomal proteins predominantly cross-linked to spermine. Although the extent of spermidine cross-linking was less than that of spermine, the species of ribosomal proteins attached to spermidine were more numerous than those reacting with spermine (Kakegawa et al. 1986).

It should be emphasized that the results reported by Bernabeu et al. (1978) differ from those obtained by Kakegawa et al. (1986), although both research groups used the same cross-linker. This discrepancy has been attributed to differences in the ionic conditions followed during the original binding and cross-linking step, in the concentration of cross-linker, and in the experimen-

Table 9.1 Ribosomal proteins from *Escherichia coli* cells, interacting with polyamines

Method	Primary labelled proteins in the small ribosomal unit	Primary labelled proteins in the large ribosomal subunit
Cross-linking of (^{14}C)spermine to ribosomes with 1,5 difluoro-2,4-dinitrobenzene (Bernabeau et al. 1978)	S4, S5, S9, S18, S19, S20	L2, L6, L13, L14, L16, L17, L18, L19, L22, L27
Cross-linking of (^{14}C)spermine to ribosomal subunits with 1,5-difluoro-2,4-2,4-dinitrobenzene (Kagegawa et al. 1978)	S3, S8, S9	L1, L2, L3, L5, L6, L13, L18, L24, L27
Cross-linking of (^{14}C)spermine to ribosomal subunits with dimethyl-suberimidate (Kagegawa et al. 1978)	S1, S3, S4, S5, S7, S8, S9, S15	L1, L2, L3, L6, L18, L24
Photolabeling of complex C with ABA-(^{14}C)spermine, upon stimulatory conditions (Amarantos et al. 2001)	S3, S4	L2, L3, L4, L6, L15, L17, L18
Photolabelling of complex C with ABA-(^{14}C)spermine, upon inhibitory conditions (Amarantos et al. 2001)	S3, S4, S5, S9, S18	L19, L22, L27

For each method, an appropriate literature reference is given in parentheses. Abbreviations: complex C, AcPhe-tRNA•poly(U)•70S-ribosome; ABA-(^{14}C)spermine, *N*^{1}-azidobenzamidino(^{14}C)spermine tal protocol used for the identification of labelled proteins (Kakegawa et al. 1986). Nevertheless, several limitations appear in both approaches and may lead to artificial results. For instance, the use of homobifunctional cross-linkers can cause protein-protein or protein-rRNA cross-linking, which may consequently lead to loss or inaccurate characterization of the labelled proteins. Moreover, both cross-linkers are amine-reactive reagents, and the pattern of protein labelling depends on whether amino-groups are available in the neighborhood of the polyamine binding site.

9.3 Labelling of Ribosomal Proteins with Photoreactive Spermine Analogues

The application of photoaffinity labelling techniques has been successfully used to map polyamine binding sites in several target molecules or cellular organelles, such as membrane transporters (Felschow et al. 1997), nucleosomes (Clark et al. 1991), and enzymes (Leroy et al. 1995). Recently, we synthesized two photoreactive analogues of spermine (Fig. 9.1), azidobenzamidino (ABA)-spermine and azidonitrobenzoyl (ANB)-spermine, by linking an arylazido group at one of the terminal amino groups of spermine, and we used them for mapping polyamine binding sites in AcPhe-tRNA (free or bound at the P-site of *E. coli* poly(U)-programmed ribosomes) (Amarantos et al. 2000), in 16S rRNA (free or complexed with ribosomal proteins) (Amarantos et al. 2002), and in ribosomal proteins from *E. coli* cells (Amarantos et al. 2001). In the ANB-derivative, one of the spermine charges has been removed, resulting in a compound resembling N^1-acetylspermine or spermidine. In contrast, ABA-spermine preserves a charge in the vicinity of the nearest amino group, more closely resembling the parent compound, spermine. Despite the modification of the spermine molecule by the arylazido group, several tests demonstrated that both photoprobes retain almost all biochemical properties of polyamines (Amarantos et al. 2001). Due to this structural and functional similarity, each of these analogues binds specifically and reversibly to the ribosome in the dark, just like the parent compound. When the photoreactive ligand meets its partner, a further treatment by irradiation with mild UV light induces the formation of constant covalent bonds between the azido group of the analogue and whatever lies adjacently to the target molecule.

The use of ABA- or ANB-spermine for labelling ribosomal proteins appears to have several advantages, compared with the fixation of polyamines to ribosomes with homobifunctional cross-linkers. Firstly, the azido group, after activation by irradiation, reacts easily with a variety of chemical groups in proteins or rRNA (Peters and Richards 1977). Secondly, the photolabelling procedure ensures the specificity of polyamine binding and, importantly, does not require vigorous conditions that may destroy the functional confor-

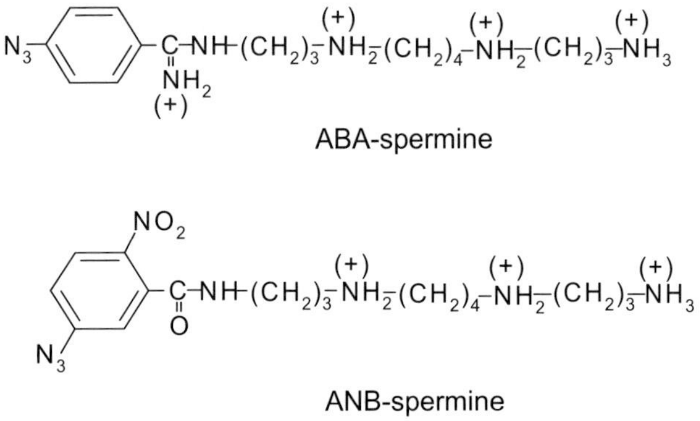

Fig. 9.1. Chemical structures of ABA-spermine and ANB-spermine

mation of the target molecule. It is relevant to mention that the incorporation of photoprobes into ribosome, is inhibited when the spermine present is in excess during photolabelling (Amarantos et al. 2001). Thirdly, the monofunctional character of ABA- and ANB-spermine does not promote protein-protein or protein-rRNA cross-linking, which may lead to loss or ambiguities in protein identification. However, more importantly of all is the preservation of ribosomal functional properties after attachment of photoprobes to ribosomes. The studies using homobifunctional cross-linkers do not provide evidence that the biological activity of ribosome remains unaffected by the covalent binding of labelling. Experimental procedures concerning the synthesis of photoprobes and the photolabelling of ribosomes are described by Amarantos et al. (2000, 2001, 2002), together with a series of control tests and methods for purification of the photolabelled product. Representative results concerning the photolabelling of an initiation ternary ribosomal complex (AcPhe-tRNA·poly(U) ·70S-ribosome, complex C) by ABA-spermine, are given in Table 9.1.

9.4 Functional Implications and Perspectives

By comparing the pattern of ribosomal proteins cross-linked to [^{14}C]spermine with homobifunctional reagents or labelled by ABA-spermine (Table 9.1), it is obvious that in general the patterns do not resemble each other. This is likely due to differences in the experimental approach for ligand incorporation. In addition, differences in the nature of the target molecule (ribosomal subunits vs. ribosomal complexes) or in the ionic conditions used during the original binding and the subsequent cross-linking of ligand to the target molecule (10 mM Mg^{2+} vs. 6 mM Mg^{2+}), may have influenced the results. For instance, two interface proteins (Clemons et al. 1999, Schluenzen et al. 2000, Ban et al. 2000, Gabashvili et al. 2000), S7 and L5, are labelled when ribosomal subunits, not complex C, are used. The association of ribosomal subunits in complex C probably lowers the accessibility of these proteins. Protein S9, is also labelled by ABA-spermine only under inhibitory conditions for the peptidyltransferase activity, but it is always present in the pattern of labelled proteins found by Bernabeu et al. (1978) or by Kakegawa et al. (1986). However, this discrepancy is consistent with the observation that the ionic conditions and spermine concentrations used by the latter investigators favor the inhibitory effect of spermine (Kalpaxis and Drainas 1993). Nevertheless, S3 and S4 are the most strongly labelled proteins in the small ribosomal subunit, independent of whether incorporation of polyamines into the target molecule is carried out under stimulatory or inhibitory conditions for the peptidyltransferase activity. These proteins are implicated in important ribosomal functions (Cohen, 1998), such as aminoacyl-tRNA binding to ribosomes and translocation (S3 and S4), ribosomal accuracy (S4), and peptide chain termination (S4). It is also very interesting that upon conditions promoting high activity on peptidyltransferase, the predominantly labelled proteins in the large ribosomal subunit are those topographically adjacent to the catalytic center of ribosome (Hampl et al. 1981, Nissen et al. 2000, Bischof et al. 1995). This observation suggests that a preferential binding of polyamines to specific ribosomal proteins around the catalytic center has a beneficial effect on the catalytic properties of ribosomal complex. Bound polyamines probably affect the conformation of these proteins, which in turn influences the conformation of rRNA residues involved in catalysis. Alternatively, bound polyamines may provide the active center with an amino group of neutral pK that participates in the mechanism of peptide bond formation. To further investigate this finding, Glu56 of protein L4, a candidate for the binding of spermine, was replaced by aspartic acid or alanine. The distance between this amino-acid residue and the catalytic center is about 21.8 Å (Nissen et al. 2000), which is equal to the length of spermine chain. We found that replacement of Glu56 by aspartic acid marginally affects the peptidyltransferase activity, while ribosomes that possess this mutated protein are susceptible to modulation by spermine. In contrast, replacement of Glu56 by alanine results in a dramatic decrease of peptidyltransferase activity, not capable of improvement by the addition of spermine in the in vitro protein-synthesizing system (unpubl. results).

The pace, at which the structures of ribosomal proteins have been solved, has picked up dramatically. The use of high-speed computers along with the application of crystallography and NMR analysis, has enabled fast determination of the three-dimensional structures (Arnez and Gavarelli 1997, De Guzman et al. 1998, Ramakrishnan et al. 1995). However, much remains to be clarified, concerning the determinants of protein-rRNA interactions. Recent studies revealed that the binding of ribosomal proteins to rRNA occurs via a diverse subset of interactions (Draper and Reynaldo 1999, Zimmermann et al. 2000, Nakashima et al. 2001). In some cases, it has been suggested that these interactions may be directly mediated by divalent cations and polyamines (Batey and Williamson 1998, Agalarov et al., 2000, Xing and Draper 1995, Conn et al., 1999, Drygin and Zimmermann, 2000). On the other hand, the ionic environment can influence the overall shape or fold of the interacting macromolecules and, therefore, may modulate the conformation of the dimmer interface. Further investigation of the role of the ionic environment on the protein-rRNA recognition will likely lead to the identification of new rRNA binding strategies for ribosomal proteins, and this should offer an interpretation of the polyamine effects on protein synthesis. Obviously, the application of photolabelling techniques in combination with sophisticated structural analysis will be a useful tool in this effort.

Acknowledgements. This work was supported in part by a grant (99ED605) from the General Secretariat of Research and Technology, Ministry of Greece, and the European Social Fund.

References

Agalarov SC, Prasad G, Funke PM, Stout CD and Williamson JR (2000) Structure of the S15, S6, S18-rRNA complex: assembly of the 30S ribosomal central domain. Science 288:107–112

Agrawal RK, Penczek P, Grassucci RA, Burkhardt N, Nierhaus KH and Frank J (1999) Effect of buffer conditions on the position of tRNA on the 70S ribosomes as visualized by cryoelectron microscopy. J. Biol. Chem. 274:8723–8729

Amarantos I and Kalpaxis DL (2000) Photoaffinity polyamines: interactions with AcPhe-tRNA free in solution or bound at the P-site of *Escherichia coli* ribosomes. Nucleic Acids Res. 28:3733–3742

Amarantos I, Xaplanteri MA, Choli-Papadopoulou T and Kalpaxis DL (2001) Effects of two photoreactive spermine analogues on peptide bond formation and their application for labelling proteins in *Escherichia coli* functional ribosomal complexes. Biochemistry 40:7641–7650

Amarantos I, Zarkadis IK and Kalpaxis DL (2002) The identification of spermine binding sites in 16S rRNA allows interpretation of the spermine effect on ribosomal 30S subunits functions. Nucleic Acids Res. 30:2832–2843

Arnez JG and Cavarelli J (1997) Structures of RNA-binding proteins. Q. Rev. Biophys. 30:195–240

Ban N, Nissen P, Hansen J, Moore PB and Steitz TA (2000) The complete atomic structure of a large ribosomal subunit at 2.4 Å resolution. Science 289:905–920

Batey RT and Williamson JR (1998) Effects of polyvalent cations on the folding of an rRNA three-way junction and binding of ribosomal protein S15. RNA 4:984–997

Bernabeu C, Vazquez D and Ballesta JPG (1978) Proteins associated with rRNA in the *Escherichia coli* ribosome. Biochem. Biophys. Acta 518:290–297

Bischof O, Urlaub H, Kruft V and Wittmann-Liebold B (1995) Peptide environment of the peptidyl transferase center from *Escherichia coli* 70S ribosomes as determined by thermoaffinity labeling with dihydrospiramycin. J. Biol. Chem. 270:23060–23064

Burma DP, Tewari DS and Srivastava AK (1985) Ribosomal activity of the 16S·23S RNA complex. Arch. Biochem. Biophys. 239:427–435

Clark E, Swank RA, Morgan JE, Basu H and Matthews HR (1991) Two new photoaffinity polyamines appear to alter the helical twist of DNA in nucleosome core particles. Biochemistry 30:4009–4020

Clemons WM, May JLC, Wimberly BT, McCutcheon JP, Capel MS and Ramakrishnan V (1999) Structure of a bacterial 30S ribosomal subunit at 5.5 Å resolution. Nature 400: 833–840

Cohen SS (1998) A Guide to the Polyamines. Oxford University Press, New York

Conn, GL, Draper DE, Lattman EE and Gittis AG (1999) Crystal structure of a conserved ribosomal protein-RNA complex. Science 284:1171–1174

De Guzman RN, Turner RB and Summers MF (1998) Protein-RNA recognition. Biopolymers 48:181–195

Draper DE and Reynaldo LP (1999) RNA binding strategies of ribosomal proteins. Nucleic Acids Res. 27:381–388

Drygin D and Zimmermann RA (2000) Magnesium ions mediate contacts between phosphoryl oxygens at positions 2122 and 2176 of the 23S rRNA and ribosomal protein L1. RNA 6:1714–1726

Felschow DM, Mi Z, Stanek J, Frei, J and Porter, CW (1997) Selective labelling of cell-surface polyamine-binding proteins on leukaemic and solid-tumour cell types using a new polyamine photoprobe. Biochem. J. 328:889–895

Gabashvili IS, Agrawal RK, Spahn CM, Grassucci RA, Svergun DI, Frank J and Penczek P (2000) Solution structure of the *E. coli* 70S ribosome at 11.5 Å resolution. Cell 100:537–549

Hampl H, Schulze H and Nierhaus KH (1981) Ribosomal components from *Escherichia coli* 50S subunits involved in the reconstitution of peptidyltransferase activity. J. Biol. Chem. 256:2284–2288

Kakegawa T, Sato E, Hirose S and Igarashi K (1986) Polyamine binding sites on *Escherichia coli* ribosomes. Arch. Biochem. Biophys. 251:413–420

Kalpaxis DL and Drainas D (1993) Inhibitory effect of spermine on ribosomal peptidyltransferase. Arch. Biochem. Biophys. 300:629–634

Leroy D, Schmid N, Behr J-P, Filhol O, Pares S, Garin J, Bourgarit J-J, Chambaz EM and Cochet C (1995) Direct identification of a polyamine binding domain on the regulation subunit of the protein kinase casein kinase 2 by photoaffinity labeling. J. Biol. Chem. 270:17400–17406

Nakashima T, Yao M, Kawamura S, Iwasaki K, Kimura M and Tanaka I (2001) Ribosomal protein L5 has a highly twisted concave surface and flexible arms responsible for rRNA binding. RNA 7:692–701

Nissen P, Hansen J, Ban N, Moore PB and Steitz TA (2000) The structural basis of ribosome activity in peptide bond synthesis. Science 289:920–930

Peters K and Richards FM (1977) Chemical cross-linking: reagents and problems in studies of membrane structure. Annu. Rev. Biochem. 46:523–551

Ramakrishnan V, Davies C, Gerchman SE, Golden BL, Hoffmann DW, Jaishree TN, Kyila JH., Porter S and White SW (1995) Structures of prokaryotic ribosomal proteins: implications for RNA binding and evolution. Biochem. Cell Biol. 73:979- 986

Schluenzen F, Tocilj A, Zarivach R, Harms J, Gluehmann M, Janell D, Bashan A, Bartels H, Agmon I, Franceschi F and Yonath A (2000) Structure of functionally activated small ribosomal subunit at 3.3 Å resolution. Cell 102:615–623

Stevens L and Pascoe G (1972) The location of spermine in bacterial ribosomes as indicated by 1,5-difluoro-2,4-dinitrobenzene and by ethidium bromide. Biochem. J. 128:279–289

Xing Y and Draper DE (1995) Stabilization of a ribosomal RNA tertiary structure by ribosomal protein L11. J. Mol. Biol. 249:319–331

Zimmermann RA, Alimov I, Uma K, Wu H, Wower I, Nikonowicz EP, Drygin D, Dong P and Jiang L (2000) How ribosomal proteins and rRNA recognize one another. In Garrett RA, Douthwaite SR, Liljas A, Matheson AT, Moore PB and Noller HF (eds) The Ribosome: Structure, Function, Antibiotics, and Cellular Interactions. ASM Press, Washington, D. C., pp. 93–104

10 Applications of Optical Biosensors to Structure-Function Studies on the EGF/EGF Receptor System

Edouard C. Nice, Bruno Catimel, Julie A. Rothacker, Nathan Hall, Antony W. Burgess, Thomas P. J. Garrett, Neil M. McKern and Colin W. Ward

10.1 Introduction

Many biological signalling pathways are regulated by specific protein-protein interactions. The ability to measure such interactions in real time with high sensitivity using instrumental optical biosensors has resulted in the rapid expansion in the use of these technologies for characterising the physico-chemical parameters in many biological signalling pathways (Nice and Catimel 1999, Rich and Myszka 2000).

In biosensor studies one of the interactants is immobilised onto the sensor surface and potential binding partners are applied in solution over the surface using either flow or cuvette based hydraulics. It is possible to analyse binding partners in a diverse range of biological fluids (eg tissue extracts, diluted serum or chromatographic fractions), but homogeneous and well-characterised proteins are required for accurate kinetic analysis. By careful control of the immobilisation chemistry (Johnsson et al. 1995), it is possible to attach a wide range of compounds onto the biosensor surface. No labelling is required with the optical biosensors, as detection is based on evanescent wave technology in which changes in refractive index, caused by changes in mass upon binding at, or near to, the sensor surface modulate the signal. Typically, since the detectors are mass sensitive, the smaller molecular weight binding partner is immobilised to maximise sensitivity of detection. Optical biosensors have been used to study interactions involving a wide range of compounds, including drugs, small chemical entities, peptides, proteins, lipids, carbohydrates, oligonucleotides, bacteria, viral particles, phage and even cells. Typical applications have included, for example, antibody–antigen and receptor–ligand interactions, epitope mapping, analysis of components of signal transduction pathways, and studies on adhesion molecules and nuclear receptors (Rich and Myszka 2000a,b, 2001, 2002). More recently, biosensor

Principles and Practice
Methods in Proteome and Protein Analysis
R.M. Kamp, J. J. Calvete, T. Choli-Papadopoulou (Eds.)

surfaces have also been investigated as microaffinity purification platforms (reviewed in Catimel et al. 2001), in which the surfaces have either been used directly for MALDI analysis (Nelson and Krone 1997, Nedelkov and Nelson 2001a,b), or else samples have been recovered for downstream proteomics analysis (Sonksen et al. 1998, Catimel et al. 2000, Williams 2000).

The general applicability of the technique renders it ideally suited to structure-function studies, where detailed kinetic analysis of different mutants can identify key binding regions; this helps to reveal the biological mechanism of the interactions and facilitate the development of agonists and antagonists with potential therapeutic activity (Van Regenmortel 2001, Gardsvoll et al. 1999, van der Plas et al. 2000). Examples of systems where biosensor studies have yielded valuable structure-function information include growth hormone (Cunningham and Wells 1993), interleukin 1 (Chrunyk et al. 2000), interleukin 5 (Plugariu et al. 2000, Bennet et al. 1995), interleukin 11 (Schleinkofer et al. 2001), insulin like growth factor (Wong et al. 1999, Forbes et al. 2002, Dubaquie and Lowman 1999, Song et al. 2000, Marinaro et al. 1999, Shand et al. 2003) and the epidermal growth factor (EGF)/EGF receptor (EGFR) family (Zhou et al. 1993, Domagala et al. 2000, De Cresenzo et al. 2000, Lenferink et al. 2000, Elleman et al. 2001). In this paper we review a range of biosensor applications and illustrate the power and flexibility of this approach by reference to our recent data on the EGF/EGFR system.

10.2 The EGF/EGFR Family

The EGFR family (EGFR (ErbB1), ErbB2 (HER 2) ErbB3 and ErbB4) are single transmembrane spanning proteins of the receptor tyrosine kinase family. They are monomeric glycoproteins of approximately 1,200 amino acids with an extracellular domain consisting of approximately 620 amino acids, a single transmembrane region and a cytoplasmic tyrosine kinase domain flanked by non-catalytic regulatory regions (Fig. 10.1). The extracellular domain can be divided into four sub-domains based on sequence homology (domains I-IV (Lax et al. 1988), also known as L1, CR1, L2 and CR2 (C.W. Ward et al. 1995) or L1, S1, L2, S2 (Bajaj et al. 1987).

These receptors can be activated by up to a dozen potential ligands which, together with their potential to form homo- or heterodimers, gives rise to a complex signalling network (Groenen et al. 1994, Yarden and Sliwkowski, 2001). This signalling network regulates a wide range of cellular processes including proliferation, differentiation, cell migration and apoptosis (Alroy and Yarden 1997, Riese and Stern 1998, Yarden and Sliwkowski 2001). Increased, or constitutive EGFR signalling has been observed in approximately one third of all human tumours including brain, head and neck, colon, lung and pancreas due to mutations of the receptor, gene amplification, receptor over expression or autocrine stimulation by specific ligands (Sizeland and

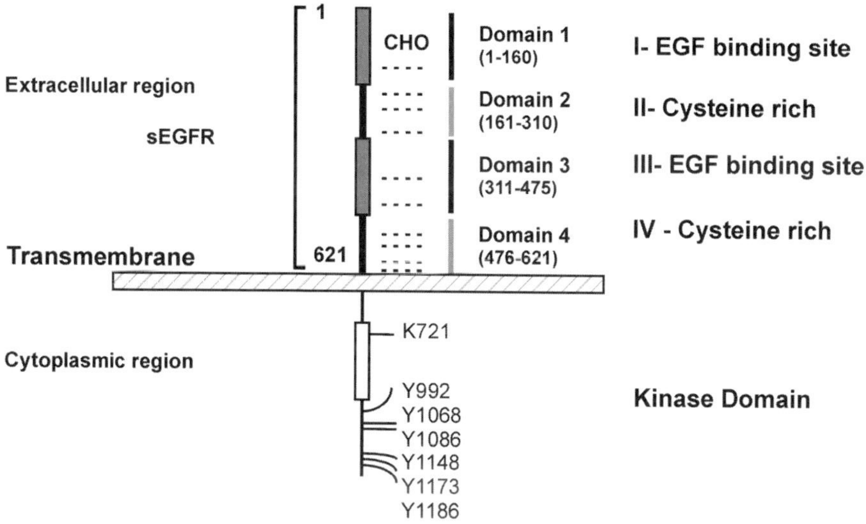

Fig. 10.1. The domain structure of the EGFR. The extracellular region can be divided into four domains (*I–IV*) based on sequence homology (Lax et al. 1991). The presence of potential glycosylation sites is indicated (*dashes*). The cytoplasmic region contains the kinase domain. *K721* is the ATP anchor. The location of tyrosine residues involved in phosphorylation and the recruitment of downstream signalling partners are shown

Burgess 1992, Hynes and Stern 1994, Todd and Wong 1999). Such aberrant signalling is typically associated with poor clinical prognosis including non-responsiveness to chemotherapy and decreased survival (Slamon et al. 1987, Slamon et al. 1989, Mendehlson and Baselga 2000). The EGFR is therefore a significant therapeutic target. Indeed, it has been estimated that the potential world market is in excess of US$6 billion (Steinberg 2002).

The ligands, which are characterised by a distinctive cysteine motif which forms three conserved disulphide bonds, includes EGF, and the transforming growth factor alpha (TGFα), amphiregulin, betacellulin, epiregulin, heparin-binding EGF, epigen and the neuregulins (Groenen et al. 1994; Fig. 10.2). ErbB2 cannot bind ligands (Garrett et al. 2003), but seems to be the preferred partner for heterodimerisation with other EGFR family members (Sundaresan et al., 1998). ErbB3 has an inactive kinase domain, and hence also signals through ErbB family heterodimers (Riese and Stern, 1998). It is now thought (Moriki et al. 2001) that ligand binding stabilises oligomerisation of the receptor and causes re-orientation of the kinase domains resulting in activation and transphosphorylation. Downstream signalling molecules are then recruited to the phosphorylated C-terminal region.

Binding studies on whole cells have shown the presence of two affinity forms of the EGF receptor: a small number of high affinity receptors

Human EGF	1	NSDSECPLSHDGYCLHDGVCMYIEALDKYACNCVVGYIGERCQYRDLKWWELR	53
Mouse EGF	1	NSYPGCPSSYDGYCLNGGVCMHIESLDSYTCNCVIGYSGDRCQTRDLRWWELR	53
Human TGF	1	VVSHFNDCPDSHTQFCFH-GTCRFLVQEDKPACVCHSGYVGARCEHADLLA	50
Amphiregulin	41	KKKNPCNAEFQNFCIH-GECKYIEHLEAVTCKCQQEYFGERCGEK	84
Betacellulin	32	KTHFSRCPKQYKHYCIH-GRCRFVVDEQTPSCICEKGYFGARCERVDLFY	80
Epiregulin	2	AQVSITKCSSDMNGYCLH-GQCIYLVDMSQNYCRCEVGYTGVRCEHFFL	49
HB-EGF	41	KKRDPCLRKYKDFCIH-GECKYVKELRAPSCICHPGYHGERCHGLSLPVE	86
Epigen	53	LKFSHPCLEDHNSYCIN-GACAFHHEL-QAICRCFTGYTGQRCEHLTLTSYA	103
Neuregulin-1	141	AQVSITKCSSDMNGYCLH-GQCIYLVDMSQNYCRCEVGYTGVRCEHFFLTV	222

Fig. 10.2. The EGF family of ligands. The one-letter code for amino acids has been used. The conserved cysteine residues which form the motif characteristic of the EGF family are *highlighted*. The sequences were obtained from entries in the NCBI website except for epiregulin which is from Strachan et al. (2001). The numbering and termini of the sequences were derived from the "mature chain" section of each entry

(K_D=1–20pM) and a large number of low affinity forms (K_D=1–2 nM) (Schlessinger et al. 1983, Yarden and Schlessinger 1987a,b). It is believed that the high affinity form mediates downstream signalling, following ligand activation (King and Cuatrecasas 1982, Kawamoto et al. 1983, Bellot et al. 1990). The binding affinity of purified membranes is influenced by the method of preparation: while Triton X-100 solubilised membranes display low binding affinity (K_D=40–200 nM) for EGF (Yarden and Schlessinger 1987a,b, Yarden et al. 1985), membranes prepared in a carefully designed cytosolic buffer retain the high affinity form of the receptor (Walker and Burgess 1991).

Chemical cross-linking studies have shown that the ligand binding sites in sEGFR are comprised of regions from both domains 1 and 3 (Summerfield et al. 1996, Woltjer et al., 1992). Interestingly, a 40 kDa, proteolytically derived, sEGFR fragment comprising residues 302–503, which would contain only one of the proposed binding sites, is capable of binding EGF with a K_D of approximately 500 nM–1 μM (Kohda et al. 1993).

10.3 Biosensor Analysis

10.3.1 Instrumentation

Biosensors are analytical devices that detect molecular interactions with high selectivity based on molecular recognition. The main feature is a selective active surface consisting of a biological species coupled to an optically or electronically active medium (Cass 1995). Bio-recognition, ideally in a concentration-dependent manner, will alter the properties of the sensor surface enabling suitable transducers to convert biochemical interactions to electronic information that can be amplified and translated into a quantitative signal (Cass 1995). Many of the available instrumental biosensors are multifunction optical biosensors that are suitable for a range of research applications. Six instrumental optical biosensors are now commercially available (Baird and Myszka 2001). The BIAcore range from BIAcore, Uppsala, Sweden (http://www.biacore.com) and the IAsys systems from Affinity Sensors, Cambridge, UK (http://www.affinity-sensors.com) are currently the most widely used (Baird and Myszka 2001). Both instruments use detection principles based on optical evanescence: the flow-based BIAcore systems use surface plasmon resonance detection (Malmqvist 1983) while the cuvette-based IAsys uses a waveguide technique called a prism coupler or resonant mirror (Davies et al. 1994). The BIAcore sensor surface consists of a glass slide coated with a thin (50 nm) gold film to which is attached, by an inert (alkanethiol) linker layer, a chemical matrix onto which one of the binding partners can be immobilised using a number of well-defined chemistries (Nice and Catimel 1999, Catimel et al. 1999; Table 10.1). The sensor chip is interfaced with an integrated fluidic cartridge that consists of a series of flow channels, controlled by micro-diaphragm pneumatic valves, encased in a hardened plastic housing. This geometry forms a flow cell of a 60-nl volume that has four parallel sensor surfaces, each with a surface area of approximately 1.5 mm^2. The IAsys Auto+ is a cuvette-based sensor, with dual microcuvettes (10–80 µl capacity) which have integral sensor surfaces (in either analytical or preparative format, with surface areas of 4 or 12 mm^2 respectively) which are optically coupled to the resonance mirror detector (Catimel et al. 1999). The glass sensor surfaces can be derivatised to give similar functionalities to those available from BIAcore (Table 10.1). Both these instruments continuously monitor the resonance angle and thus can detect changes in refractive index caused by changes in sensor surface mass when a ligand binds to, or dissociates from, its immobilised binding partner. Data are presented as sensorgrams that display the change in resonance units (RU, BIAcore) or angle (arc seconds, IAsys) versus time.

These biosensors display a high sensitivity, which is a prerequisite for any technology which has widespread applicability in biological analysis. In the case of the BIAcore, a signal of 1,000 RU is equivalent to a surface concentration of 1 ng/mm^2 for proteins (Nice and Catimel 1999). Using the BIAcore 3000 the minimum detectable surface concentration of protein is estimated to be approximately 0.5 pg/mm^2, assuming a short-term noise of approximately 0.1 RU and a signal/noise ratio of 5. For the IAsys, a signal of 163 arc seconds for proteins is equivalent to a surface concentration of 1 ng/mm^2 using the CMD surface (Catimel et al. 1999, 2001). The short-term noise of this instrument is approximately 0.2 arc seconds.

This sensitivity typically allows the detection of interactions at low ng/ml concentration, corresponding to the levels of ligand frequently present in biological samples (Nice et al. 1997). However, the concentration of ligand required for a detectable signal for any particular interaction depends on many factors including the relative masses of the interactants, the immobilisation level, association and dissociation rate constants and injection volume.

Table 10.1. BIAcore and IAsys immobilisation chemistries

Surface	Chemistry	Applications
Carboxymethyldextran (CM5 BIAcore, CMD IAsys)	Coupling via primary amines using NHS/EDC chemistry	Compatible with a wide range of protein and peptide applications: Immobilisation of thiol reactive reagents for thiol coupling Immobilisation of hydrazine for coupling of oxydised carbohydrate Immobilisation of reagents for affinity capture
CMD Select Carboxymethyldextran (IAsys)	Large coupling capacity (16-mm^2 surface)	Ideal for ligand fishing
Dextran matrix with low degree of carboxylation(B1 BIAcore)	Coupling via primary amines using NHS/EDC chemistry	Kinetic applications requiring a low density of immobilised ligand
Short carboxymethyldextran (F1 BIAcore) carboxyl group attached directly to the surface (C1 BIAcore) Carboxylate planar surface (IAsys)	Coupling via primary amines using NHS/EDC chemistry	High molecular weight analytes and cell binding studies
Amino planar surface containing primary amine) (IAsys)	Coupling directly to primary amines on the sensor surface using a homobifunctional cross-linker or glutaraldehyde	Coupling of ligands with low pH High molecular weight analytes and cell binding studies
Biotin (IAsys)	Coupling of biotinylated ligand following streptavidin immobilisation	Immobilisation of biotinylated ligands or membranes (eg proteins, oligonuclotides
Streptavidin (SA BIAcore)	Pre-immobilized streptavidin onto CM5 surface	
N-(5-amino-1-carboxypentyl)iminoacetic acid surface (NTA BIAcore)	Metal chelation	Immobilisation of histidine tagged recombinant proteins
Untreated gold surface (J1 BIAcore)	Partial hydrophobic character	To design customised surface chemistry, e.g. self-assembled monolayers, thin polymeric films
Hydrophobic (IAsys)	Hydrophobic binding of biomolecules, such as lipid monolayers	Lipids, liposomes, self-assembled monolayers
Long-chain alkanethiol (HPA BIAcore)		Membrane receptor analyte interactions
Lipophilic modified surface (L1 BIAcore)	Formation of lipid bilayer	Immobilisation of cell membranes, liposomes

10.3.2 Generation of an Active Biosensor Surface

The ability to immobilise one of the binding partners on the sensor surface is fundamental to successful biosensor analysis. Ideally this should be in a defined, biologically active and stable orientation mimicking that found in vivo. Targets for immobilisation may be purified from biological samples, or produced by recombinant or synthetic techniques. Since the biological specificity resides in this surface, the compounds for immobilisation should be homogeneous and well characterised.

A number of suitable chemistries and sensor surfaces are available for the immobilisation of proteins, carbohydrates, nucleic acids and lipids (Table 10.1). With the exception of very acidic species (pI <3.5), proteins and peptides can generally be immobilised on carboxymethylated dextran surfaces via the amino-terminus and ε-amino group of lysine residues using conventional amine coupling chemistry (*N*-ethyl-*N'*-dimethylaminopropyl-carbodiimide (EDC) and *N*-hydroxysuccinimide (NHS)). However, this chemistry frequently gives rise to a heterogeneous surface due to the presence of multiple lysine residues in most proteins, and may result in inactivation due to the involvement of lysine residues in, or near to, the binding site. However, for non-kinetic applications (e.g. screening, ligand searching) amine coupling frequently results in a surface which shows some reactivity since it is likely that some of the molecules will be coupled in an active orientation.

Other immobilisation strategies have included aldehyde and thiol coupling, as well as affinity capture using the biotin/streptavidin interaction, specific binding to protein A, protein G or the Fc domains of antibodies, metal chelation, coiled-coil interactions and antibodies directed against specific tags (e.g. GST, FLAG, Myc, poly-histidine). These methods have been described in detail in several reviews (Nice and Catimel 1999, Catimel et al. 2001).

10.3.3 Kinetic Analysis

The potential to obtain the individual kinetic rate constants (k_a and k_d) by analysis of the association and dissociation phases of the interaction is a major advantage of biosensor technology. Kinetic analysis requires the injection of varying concentrations of analyte over the immobilised ligand to generate a set of binding curves for detailed analysis. However, perhaps contrary to original conceptions, analysis of such data is frequently not trivial. The biosensor binding curves (sensorgrams) can be analysed to determine both association and dissociation rate constants as well as equilibrium binding constants (Nice and Catimel 1999). This can be achieved using a number of different approaches to mathematically describe the biosensor data including linearisation of the primary data, non-linear least squares analysis and, more recently, global fitting. We have found it useful to combine a number of alternative approaches to each data set to verify the appropriateness of the models used (Domagala et al. 2000). Additionally, if equilibrium binding can be obtained, the dissociation constants (K_D or K_A) can be determined by Scatchard Analysis (Req/nC versus Req where Req is the equilibrium response observed on the sensorgarm, C is the concentration and n is the valency) (Chaiken et al. 1992, Nice and Catimel 1999).

10.3.4 Solution Competition Analysis Using Biosensors

Whilst biosensor analysis has generally been used for direct kinetic or thermodynamic analysis, for which one of the interactants is immobilised on the sensor surface, it is also possible to use the technology to determine the dissociation constant (K_A or K_D) from solution competition experiments (Ward et al. 1995, Domagala et al. 2000). In this case, a standard curve is first generated by passing varying concentrations of the binding partner over the sensor surface. Competition experiments are then set up, and after allowing equilibrium to be reached, the concentration of residual free binding partner in solution is measured by injecting the mixture over the sensor surface and determining the concentration directly from the calibration curve (bound material will not be recognised). The data can then be analysed in Scatchard format ([Bound]/[Free] versus [Bound]) where the K_A ($=1/K_D$) can be determined from the slope of the fit. Interestingly, this data can also be transformed (Zeder-Lutz et al. 1993) by dividing both axes by the concentration of the competitor, to give an intercept on the x-axis which will correspond to the molar binding ratio. Another useful transformation of the Scatchard data [Hill Plot, Log ([Bound]/[Free]) versus –Log [Free]) (Hill 1910) enables cooperativity or the presence of multiple binding sites to be investigated (Domagala et al. 2000).

10.4 Biosensor Analysis of the Interactions Between EGF and the EGFR

10.4.1 Immobilisation Strategies for EGF

A number of alternative coupling chemistries have been used for the immobilisation of EGF, or related family members, onto biosensor surfaces. Amine coupling chemistry has been successfully used to immobilise human EGF (hEGF), TGFα and neuregulin 1ß onto biosensor surfaces (Zhou et al. 1993, Domagala et al. 2000, Elleman et al. 2001, Ferguson et al. 2003). Mature human EGF is a 53 amino acid polypeptide that contains only a single lysine (Lys28; Fig. 10.2), which theoretically allows three possible orientations onto the sensor surface (involving Lys28, amino terminal NH2 or both). However for

TGFα, another member of the EGF family, it has been shown that coupling via its equivalent unique lysine (Lys 29; Fig. 10.2) leads to an inactive surface (Domagala et al. 2000, DeCrescenzo et al. 2000). Hence, when using amine coupling chemistry, a pseudo-homogeneous surface is generated in which only one of the possible forms is in an active orientation. In murine EGF, an Asn replaces the single Lys (Fig. 10.2), leaving only the N-terminus available for amine coupling. Interestingly, amino coupling of mEGF in a defined orientation via the N-terminus results in a totally inactive surface, presumably due to steric hindrance (Domagala et al. 2000). It is interesting to note that, while the optimum conditions for immobilisation were being developed, we noticed that mEGF remained strongly retained, onto the carboxymethylated dextran surface, presumably by a charge-based mechanism, prior to activation with NHS/EDC which could result in the binding sites being occluded.

It has been shown previously that for small peptides the addition of a biotin tag can improve surface reactivity, presumably by reducing steric hindrance (Panayotou et al. 1993). Indeed, general biotinylation (which, like amine coupling will react with both the amino terminus and lysines), followed by RP-HPLC purification to purify the individual biotinylated species, has been used to generate an active TGFα surface (De Crescenzo et al. 2000). We have therefore investigated the possibility of generating an active mEGF surface by specific biotinylation at the N-terminus. Murine EGF isolated from mice salivary glands can be resolved by RP-HPLC into two major forms: alpha EGF (Asn1–53) and beta EGF (Ser2–53, des-asparaginyl form) (Burgess et al. 1982, DiAugustine et al. 1985). Both forms have similar receptor binding and mitogenic activities (Burgess et al. 1982).

In an attempt to generate an active, homogeneous mEGF surface that can be immobilised in a specific orientation, we have developed (Wade et al. 2002) a new immobilisation technique that specifically tags N-terminal serine or threonine residues with biotin, which would be generally applicable to other proteins with these residues at the N-terminus (Fig. 10.3). This method uses the potential of N-terminal serine and threonine residues to undergo rapid conversion to their aldehydic form by treatment with periodate. Biotinylation of the aldehyde can then be achieved using simple conjugation with a novel water-soluble dipeptide that contains a lysine residue bearing an Nε-cysteine-derived 1,2-aminothiol and an Nα-biotin moiety. This method was used to selectively biotinylate the amino terminal serine residue of native mEGF2–53. This derivative was then immobilised onto a streptavidin biosensor chip to yield an active surface (K_D=292 nM) (Wade et al. 2002). It should be noted that this reaction is specific for N-terminal residues: serines or threonines elsewhere in the protein chain are unreactive.

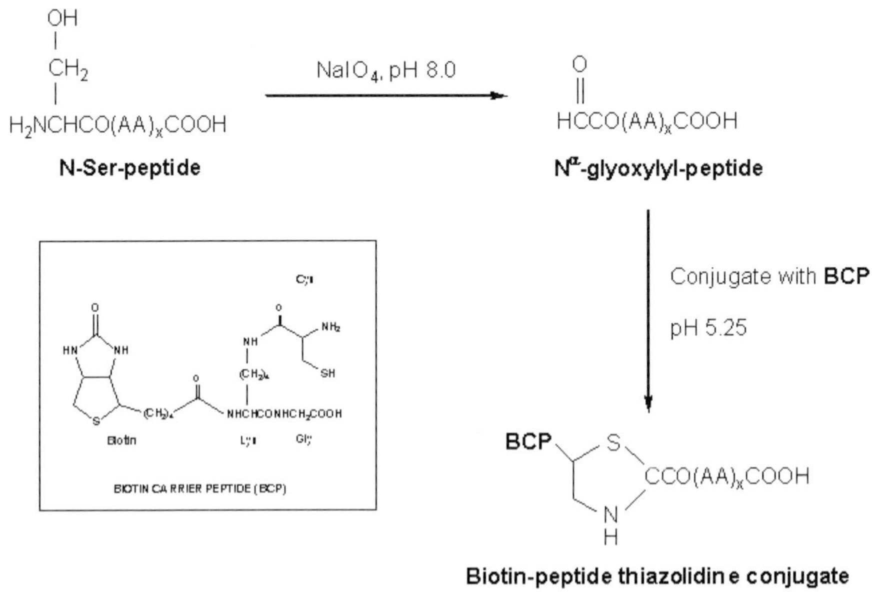

Fig. 10.3. Strategy for the immobilisation of peptides or proteins with an N-terminal serine or threonine residue using thiazolidine conjugation. Details of the biotin carrier peptide are shown in the *inset*

10.4.2 Immobilisation Strategies for sEGFR

Active biosensor surfaces can be generated by immobilising soluble forms of the complete extracellular region of the EGF receptor (sEGFR 1–621) or the high-affinity, truncated form (sEGFR 1–501) using NHS/EDC chemistry (Elleman et al. 2001). This allows kinetic data to be obtained by passing the smaller molecular weight ligands over the receptor surface, and allows the effect of alternative immobilisation conditions to be investigated. sEGFR 1–621 contains 36 lysine residues whilst sEGFR 1–501 has 32. The presence of this large number of lysine residues, while almost certainly giving rise to sensor surface heterogeneity, seems to result in a surface which recognises the ligands, as has been observed previously for antibodies and other large molecular weight compounds. Presumably a large number of the alternative surfaces are attached by lysine residues that are not involved in the binding sites and are therefore in a fully active conformation. Lenferink et al. (2000) have used a similar strategy to immobilise sEGFR 1–621 to analyse EGF-like ligands whilst Berezov et al. (2002) have immobilised recombinant purified ectodomains of EGFR, ErbB2 and ErbB3 for studies on homo- and heterodimerisation and the characterisation of EGFR peptidomimetics.

10.4.3 Kinetic Analysis of the Interaction Between hEGF and the Soluble Extracellular Domain of the EGF Receptor (sEGFR 1–621)

In our early studies on the synergy between micropreparative HPLC and optical biosensors (Nice et al. 1994), we demonstrated the use of the biosensor to monitor binding partners in chromatographic fractions. As part of this study we used immobilised hEGF to monitor the purification of a soluble form of the EGFR produced by A431 tumour cells, which consists of residues 1–615 fused to an additional 18 amino acids (Ullrich et al. 1984). Visual inspection of the biosensor curves obtained with the purified material indicated both rapid on and off rates. Kinetic analysis of the dissociation rate constant (k_d), which is concentration independent, gave values of $1.5–3.1 \times 10^{-2}$ s^{-1}. Since we did not have an accurate measure of the sEGFR concentration in these fractions, we were unable to get a precise value for the k_a.

Detailed binding studies between the EGF receptor and its ligand(s) have been performed using purified recombinant extracellular domain (residues 1–621) of the EGFR receptor (sEGFR) produced in CHO cells, which is fully glycosylated (Zhou et al. 1993, Domagala et al. 2000, De Crescenzo et al. 2000,

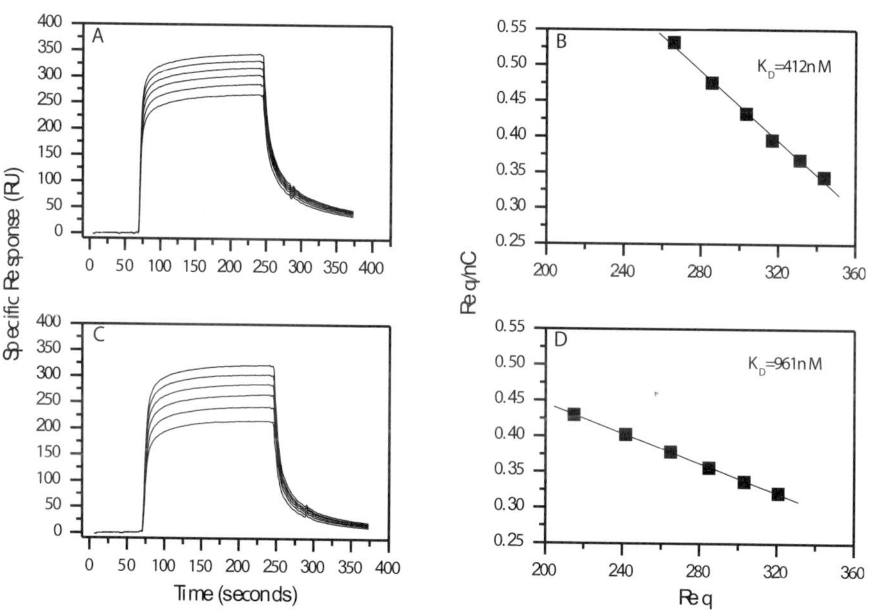

Fig. 10.4. Biosensor analysis of the interaction between sEGFR 1–621 and immobilised ligand. **A** hEGF immobilised by amine coupling using NHS/EDC chemistry. **B** Scatchard analysis using the equilibrium binding responses (Req) observed in **A**. The K_D determined from this analysis is shown. **C** hTGFα immobilised using amine coupling chemistry. **D** The corresponding Scatchard analysis of the data shown in **C**

Elleman et al. 2001). Varying concentrations of receptor were passed over an immobilised ligand to generate data sets for analysis. Amine coupling chemistry was used in the studies of Zhou et al. (1993), Domagala et al. (2000) and Elleman et al. (2001), whereas De Crescenzo et al. (2000) used TGFα biotinylated at the N-terminus coupled to a streptavidin surface. As we had seen for the soluble receptor obtained from A431 cells, the binding curves comprised a fast on-rate, they rapidly reached equilibrium, and then, when the buffer alone was flowing over the sensor surface, showed rapid off-rates (Fig. 10.4). All these studies indicated that complex kinetics were operating since the binding curves could not be fully described by a simple 1:1 Langmuirian model. Thermodynamic analysis of the equilibrium binding data between EGF and EGFR 1–621 gave K_D values of 370 (Domagala et al. 2000) to 560 nm (Zhou et al. 1993) which is in good agreement with the affinities determined using a number of other physicochemical techniques (Greenfield et al. 1989, Gunther et al. 1990, Brown et al. 1994, Lax et al. 1991, Hurwitz et al. 1991). The molar binding ratio indicated from the analysis of the equilibrium binding data, as proposed by Zeder-Lutz et al. (1993) was 1:1. The corresponding analysis of the interaction between TGFα and sEGFR 1–621 gave a K_D of 960 (Elleman et al. 2001) to 1040 nM (De Crescenzo et al. 2000).

Global analysis of the interaction between EGF or TGFα and sEGFR 1–621 (Domagala et al. 2000, De Crescenzo et al. 2000; Fig. 10.5) has been used to test a number of potential interaction models. In the study of Domagala et al., use of the 1:1 model gave a K_D of 347 nM (k_a=1.15×10^5 M^{-1} s^{-1}, k_d=0.04 s^{-1}), although there was significant deviation of the fitted curves from the actual data (Fig. 10.5). In particular the dissociation phase showed non-random residuals (data not shown). This could have been due to mass transport limited kinetics. Refitting using a similar model but including mass transport calculations gave no significant improvement to the fit. However, it was not possible to readily distinguish between other possible models such as surface heterogeneity (K_D1=342 nM, K_D2=247 nM,) receptor heterogeneity (K_D1=493 nM, K_D2=60 nM) or conformational change induced by ligand binding (K_D1= 421 nM, K_D2=6.1 µM). Surface heterogeneity was considered an inappropriate model since, as we have discussed above, only hEGF immobilised in a single orientation via the N-terminus and leads to an active surface. Interestingly, all these models indicated a major component with k_a and k_d values similar to those determined using the Langmurian model.

De Crescenzo et al. (2000) analysed their data using models to describe mass transport limitations, surface heterogeneity, conformational change, dimerisation (assuming a 1:2 or 2:2 stoichiometry) or the presence of two, non-cooperative sites. They concluded that a conformational change model gave the best fit to their data: however they did not attempt to model receptor heterogeneity.

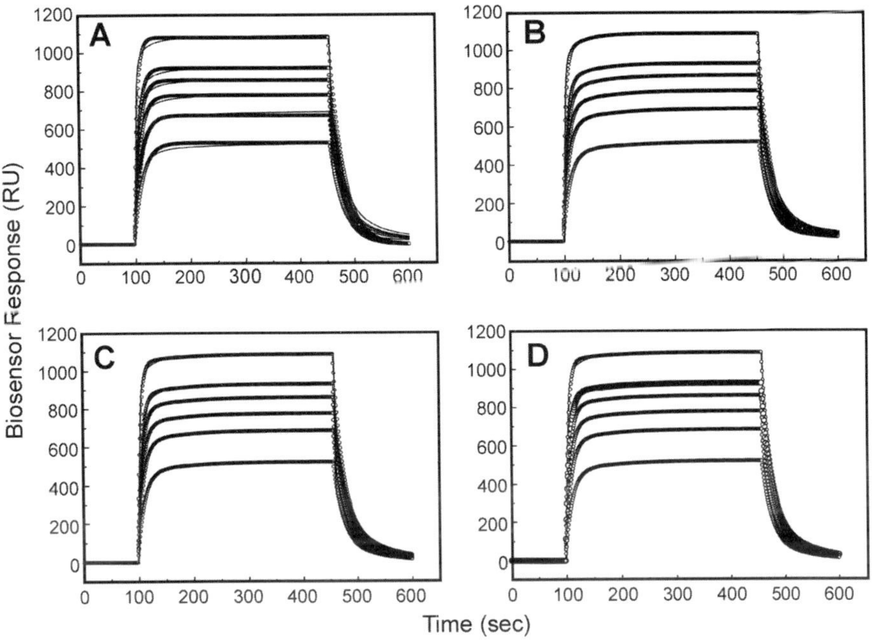

Fig. 10.5. Global analysis of the interaction between immobilised hEGF and sEGFR 1–621. Sensorgrams obtained by passing increasing concentrations of sEGFR 1 – 621 over immobilised hEGF were analysed by global analysis using equations defining alternative binding situations. The fitted data are indicated by *open circles*. **A** 1:1 Langmuirian binding (A+B⇔AB). **B** Surface heterogeneity (A+B⇔AB/A+B*⇔AB*). **C** Sample heterogeneity (A+B⇔AB/A*+B⇔A*B). These data were fitted assuming that there was 95% of one sEGFR species and 5% of the other. **D** Conformational change [A+B⇔AB⇔(AB)*]

10.4.4 Confirmation of the Binding Model

It must be remembered that global analysis merely provides a mathematical description of a particular data set, and that multiple solutions are possible. The biological relevance of the biosensor models needs to be established. We performed solution competition experiments using the biosensor as well as fluorescence anisotropy measurements to further delineate the interaction mechanism. The Scatchard analysis of the solution binding data (Fig. 10.6), generated by biosensor analysis of the concentration of free EGFR in the competition mixtures, gave a biphasic plot that was best fitted by a two-site model with calculated K_Ds of 20 and 550 nM. The intercept on the x-axis suggested that the high affinity site represented 10–15% of the population. The Hill plot gave a coefficient of 0.42, indicative of either multiple binding sites or negative cooperativity.

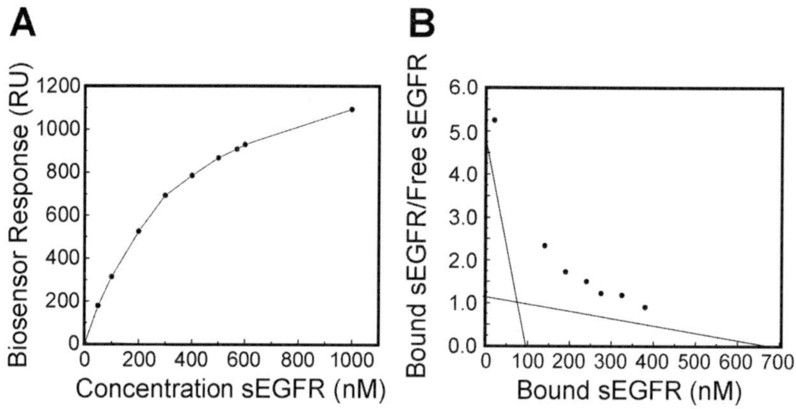

Fig. 10.6. Competition analysis of the interaction between sEGFR 1–621 and EGF derived using the BIAcore. **A** A standard curve of biosensor response versus sEGFR 1–6211 concentration was constructed over the range 20–1000 nM. **B** Scatchard analysis. Solution competition assays were performed, the concentration of free sEGFR being quantitated using the BIAcore and the standard curve (**A**). The concentration of bound receptor can then be obtained directly, without assumption of a specific stoichiometry, and the data plotted in Scatchard format. The biphasic curve was fitted using LIGAND. The K_D values obtained are shown

Fluorescence anisotropy studies indicated a molar binding ratio of 1:1 between EGF and sEFGR. Non-linear least squares and global analyses of the anisotropy data suggested a K_D of approximately 300 nM. However, analysis of the data in Scatchard format again yielded a biphasic plot with K_Ds of 2 and 400 nM. The intercept on the *x*-axis indicated the high affinity site represented approximately 20 % of the population (Domagala et al. 2000). Again, the Hill coefficient was less than 1 (0.67) (data not shown).

Analytical ultracentrifugation was also used to characterise the interacting species in the sEGFR/EGF reaction in solution (Domagala et al. 2000). The average molecular weight of the complex determined using this technique suggested that the dimer is a 2EGF/2sEGFR complex. This was in good agreement with the 1:1 molar binding ratio calculated from the biosensor data using the Van Regenmortel approach (Zeder-Lutz et al. 1993).

Taken together, these studies indicate that there are two populations of sEGFR in solution: one which represents approximately 15– 20 % of the population and binds EGF with a K_D of 2–20 nM and the other with a K_D of 400–550 nM. EGF and sEGFR interact with a molar binding ratio of 1:1, forming a dimer consisting of 2EGF/2sEGFR.

10.4.5 Identification of a Truncated High Affinity Form of thc Soluble Extracellular Domain of the EGF Receptor

One of the major goals of our project, aside from the physical chemical characterisation of the receptor, was to produce sufficient material for crytallisation studies. Crystals could be generated from the glycosylated sEGFR 1–621, but these only diffracted to about 7 Å, and were not suitable for structural analysis. We therefore decided to produce an alternative construct. Using information obtained from the 3D structure of the first three domains of the extracellular region of the structurally related IGF-I receptor (Garrett et al. 1998) we produced a truncated form of the EGFR (sEGFR 1–501) (Elleman et al. 2001). A myc tag was included at the C-Terminus of the sEGFR 1–501 to facilitate purification (McKern et al. 1997).

The sEGFR 1–501 analogue consists of the first three L1/CR1/L2 domains plus the first module of the second Cys-rich region CR2. Its design was based on the following considerations. Firstly, a shorter construct (sEGFR 1–476), comprising the L1/CR1/L2 domains only, did not bind ligand, suggesting that additional regions of the receptor were required. Secondly, the structure of the L1/CR1/L2 fragment of the related IGF-1R showed the presence of a conserved tryptophan residue, Trp176, in the first module of the Cys-rich (CR1) region, which is inserted between the last two rungs of the L1 domain (Garrett et al. 1998). An equivalent tryptophan residue, Trp492, and corresponding hydrophobic pocket are preserved in the first module of the second Cys-rich (CR2) region of EGFR and the L2 domain of EGFR respectively. This interaction is not found in the L2 domains of IGF-1R and IR, which lacks a second Cys-rich region (Garrett et al. 1998). Thirdly, the ligand binding and signalling properties of an 83 amino acid deletion mutant of EGFR (sEGFR 1 – 538) from a human glioma (Humphrey et al. 1991) have been reported to be normal, with a high ligand binding affinity leading to increased EGFR kinase activity, suggesting that much of the CR2 domain (domain IV) can be removed without detriment. Finally, the 40 kDa, proteolytically derived, EGFR fragment (residues 302–503), which is capable of binding EGF with a K_D of 500 nM–1 μM, also includes the first module of CR2 (Kohda et al. 1993). Taken together, these observations suggest that interactions between the first Cys-rich module of CR2 and the L2 domain may be important in maintaining the structural integrity of the EGFR.

The sEGFR 1–501 was produced in stably transfected Lec 8 cells, which are glycosylation defective (Stanley 1989), at levels of ~1.8 mg L^{-1}. We have shown previously using BIAcore analysis (Nice et al. 1996) that removal of carbohydrates using PNGase does not affect the binding of sEGFR 1–621 to the immobilised ligand, in agreement with the concept that glycosylation is required for correct processing but not for biological activity (Bishayee 2000). It was our hope that the reduced glycosylation would be advantageous for the crystallography studies. sEGFR 1–501 was purified using a Mab9E10 anti-c-myc peptide affinity column using peptide elution. The purified protein showed a single symmetrical peak on size exclusion chromatography (apparent molecular mass of ~80 kDa) and migrated as a single band of ~70 kDa on SDS-PAGE under reducing conditions (Elleman et al. 2001). sEGFR 1–501 gave the anticipated N-terminal sequence, LEEKKVXQGT (Ullrich et al. 1984), the X at cycle 7 being due to the presence of a disulphide-bonded cysteine residue at that position. The apparent molecular mass of 70 kDa on SDS-PAGE is due to the residual glycosylation reported for the glycosylation defective Lec 8 cells (Stanley 1989) the calculated mass of human sEGFR 1–501 apo-protein being ~57.5 kDa.

10.4.6 Kinetic Analysis of the Interaction Between EGF and sEGFR 1–501

The kinetics of the interaction between EGFR 1–501 and hEGF and TGFα (Fig. 10.7) was studied using biosensor analysis and compared with the binding characteristics of sEGFR 1–621. Scatchard analysis of the equilibrium

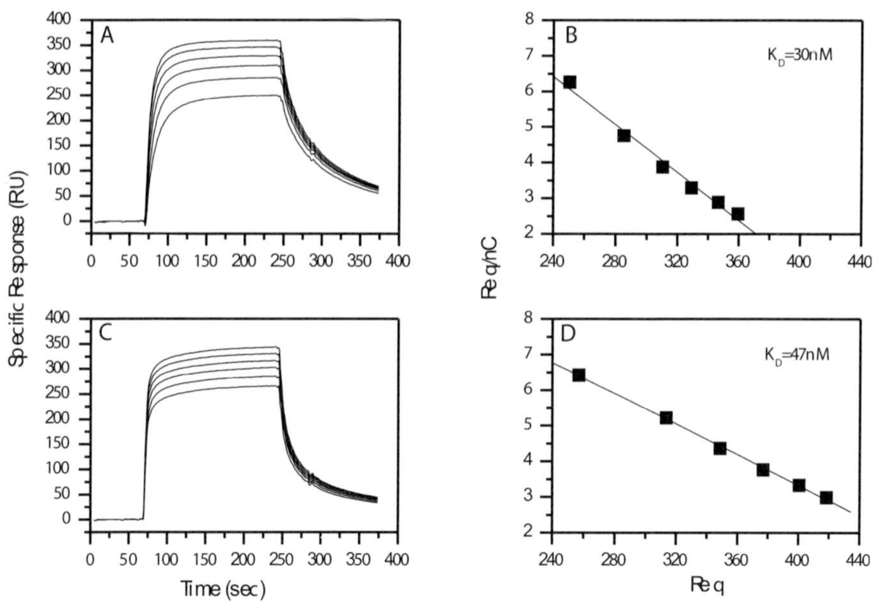

Fig. 10.7. Biosensor analysis of the interaction between sEGFR 1–501 and immobilised ligand. **A** hEGF immobilised by amine coupling using NHS/EDC chemistry. **B** Scatchard analysis using the equilibrium binding responses (Req) observed in **A**. The K_D determined from this analysis is shown. **C** hTGFα immobilised using amine coupling chemistry. **D** The corresponding Scatchard analysis of the data shown in **C**. Note the higher affinity for these interactions compared with those observed for the full-length extracellular domain (Fig. 10.4)

binding data showed that sEGFR 1–501 displayed a higher affinity to hEGF (K_D ~30 nM) compared to sEGFR 1–601 (K_D ~400 nM). The equivalent analyses on the TGFα surface gave K_Ds of ~50 and 1000 nM respectively. The increase in affinity appears to be mainly due to a reduced off-rate (cf. Figs. 10.4 and 10.7).

10.4.7 Analysis of the Receptor/Ligand Interaction Using Immobilised Receptor

In order to verify the kinetic data that we had generated using the immobilised ligand surfaces, it was decided to repeat the analysis in the opposite orientation (i.e. immobilised soluble extracellular domains of the receptor; Fig. 10.8). It can be seen that in this format, since the smaller molecular weight species is causing the change in mass at the sensor surface, that the signals are lower (Figs. 10.4, 10.7 and 10.8). However, the curves showed the same characteristics as we had seen previously (Figs. 10.4, 10.7) in terms of the fast on-rate, rapid equilibrium and the fast off-rate. Scatchard analysis indicated a K_D

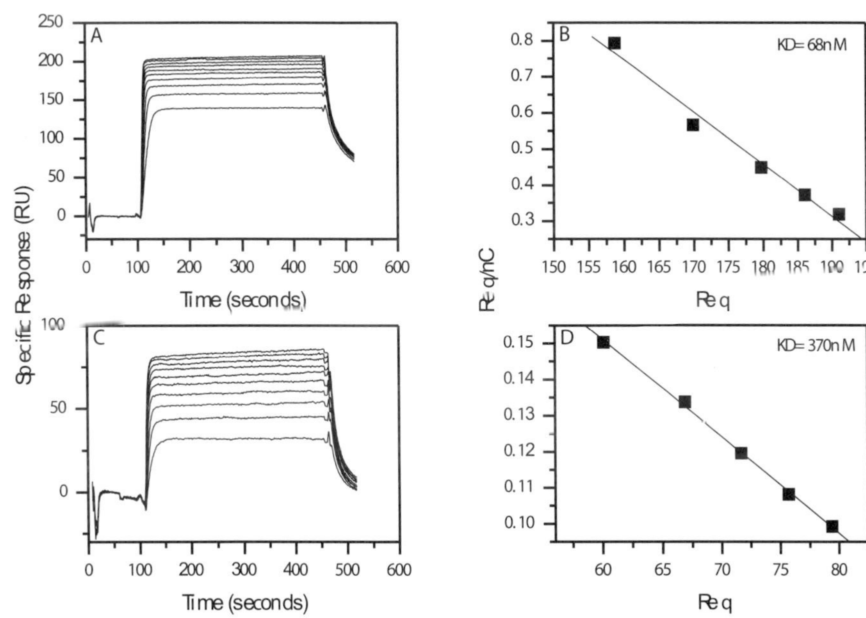

Fig. 10.8. Biosensor analysis of the interaction between immobilised sEGFR 1–501 and 1–621 and EGF. **A** sEGFR 1–501 or **B** sEGFR 1–621 was immobilised on the sensor surface using amine coupling chemistry. Increasing concentrations of EGF were flowed over the surfaces. The corresponding Scatchard analyses, obtained using the equilibrium binding responses, are shown alongside (**B, D**)

of 68 nM for the interaction with sEGFR 1–501 and 370 nM for the interaction with sEGFR 1–621, which agreed with our calculations obtained using alternative immobilisation chemistries.

10.4.8 sEGFR 1–501 and sEGFR 1–621 Are Competitive Inhibitors of EGF-Induced Mitogenesis

The inhibitor capacity of sEGFR 1–621 and sEGFR 1–501 were tested in a mitogenic cell based-assay using the pre-B cell line BaF/3 transfected with the EGFR (Walker et al. 1998). In this assay, EGF induces mitogenic proliferation with an EC50 of approximately 30pM. To test for inhibitors, the cells were stimulated with a constant concentration of mEGF (207pM), inducing maximal stimulation. Varying levels of sEGFR 1–621 and sEGFR 1–501 (0.45 nM–0.5 μM) were used in a competition assay. A neutralising anti-EGF-receptor monoclonal antibody (Mab 528) were used as a positive control. This competition assay (Fig. 10.9) showed that sEGFR 1–501 was almost tenfold more potent as an inhibitor of EGF/EGFR signalling (IC50=0.02 μM) than sEGFR 1–621 (IC50=0.15 μM) and threefold more potent than Mab 528 (IC50=0.06 μM) which is a precursor of the anti-EGFR antibody C225 currently in clinical trials (Normanno et al. 2003, Mendelsohn and Baselga 2000).

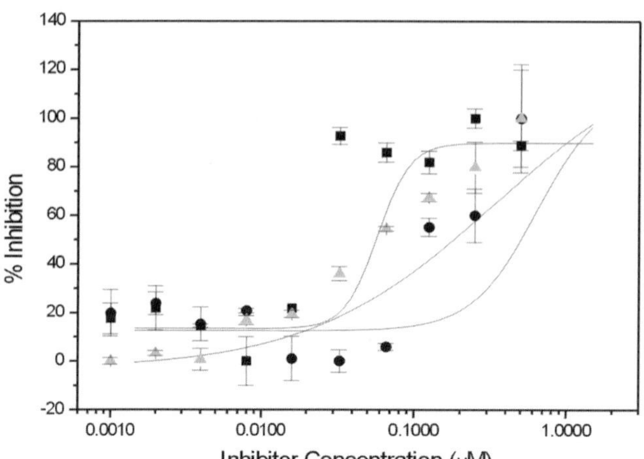

Fig. 10.9. Inhibition of EGF induced mitogenesis by sEGFR 1–501. Inhibition of the mitogenic response using BaFcells transfected with the EGFR and stimulated with mEGF (207 pM) by: sEGFR 1–501 (*filled squares*), sEGFR 1–621 (*filled circles*) or anti-EGFR antibody Mab528 (*filled triangles*). Each point was assayed in triplicate. *Error bars* are shown

10.4.9 Identification of a Determinant of EGF Receptor Ligand Binding Specificity (Chickenising the Human EGF Receptor)

hEGF, mEGF and TGFα bind to human and murine EGF receptors with similar high affinity despite the considerable differences in their amino acid sequences (Fig. 10.2). In contrast, chicken EGF receptor (cEGFR) binds to hEGF and mEGF with an approximately 100-fold lower affinity than to TGFα. Arginine 45 in human EGF was previously identified as the key residue for the reduced binding to cEGFR (van de Poll et al. 1995). Since it has been shown by chemical cross-linking (Summerfield et al. 1996) that Arg 45 binds to the L2 region of the receptor, it was assumed that the region responsible for the differential binding to cEGFR was located within Lys301 – Asp484. Based on the results of van de Poll et al. (1995), we hypothesised that the binding site of cEGFR would contain a basic residue that is neutral or acidic in hEGFR and mEGFR. Six residues in the region Lys301-Asp484 correspond to these criteria in hEGFR (Ser340, Glu367, Asn420, Gly441, Glu472, Gly479). Among them, three residues were selected for mutation studies (lysine substitution): Glu367, which is situated near the epitope of the ligand-competitive LA22

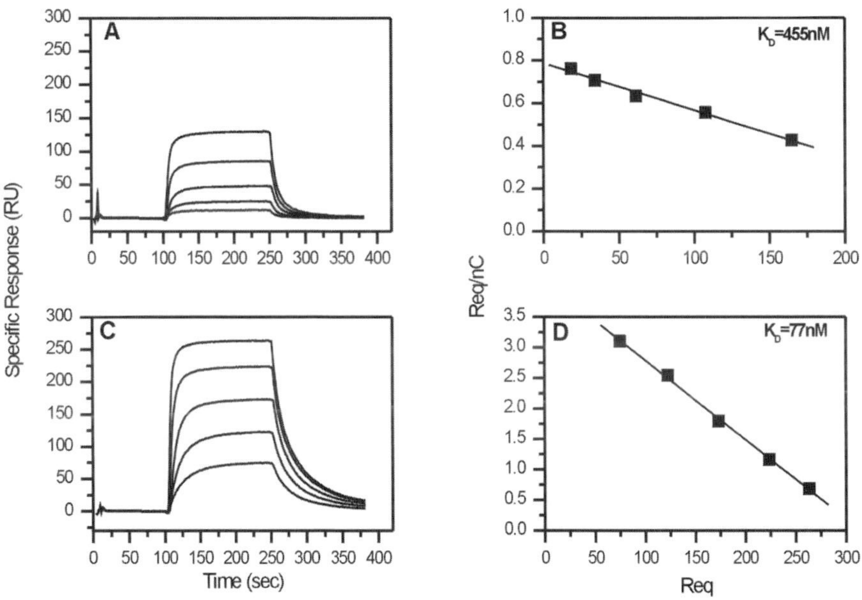

Fig. 10.10. BIAcore analysis of the binding of the Gly^{441}Lys sEGFR 1–501 mutant to immobilised hEGF and hTGFα. Purified Gly^{441}LysEGFR 1–501 (24–385 nM) was passed over immobilised hTGF-a (**A**) or hEGF (**C**). The corresponding Scatchard analysis, using the equilibrium binding values obtained from these sensorgrams, is shown alongside (**B**, **D**)

monoclonal antibody (Wu et al. 1989), Gly441, which is located near the binding site identified by chemical cross-linking (Lys465) and Glu472. The other residues were deemed unlikely to be involved in EGF binding: based on our model of the EGFR extracellular domain (Jorissen et al. 2002): Ser340 is at the back of the L2 domain, Asn420 is a predicted N-linked glycosylation site (Tsuda et al. 2000) and Gly479 is distant from the ligand cross-linking site Lys465 (Jorissen et al. 2002).

Soluble EGFR 1–501 and related mutants (Gly441Lys, Glu367Lys and Glu472Lys) were produced as transient transfectants in human 293T fibroblasts, or as stable transfectants in Lec 8 cells (Stanley, 1989) and injected over immobilised hEGF and TGFα. The Glu367Lys and Glu472Lys mutants showed similar binding characteristics to the wild-type sEGFR 1–501. On the contrary, the Gly441Lys mutant displayed increased binding on the TGFα surface compared with the parallel hEGF channel (Fig. 10.10). This binding selectivity is similar to that observed with chicken EGFR. More detailed kinetic analyses were therefore performed using the Gly441Lys mutant: Equilibrium binding analysis showed that Gly441Lys sEGFR 1–501 had a high affinity interaction with TGFα (K_D of 77 nM) but a low affinity interaction with human EGF (K_D of 455 nM) (Elleman et al. 2001).

10.5 Structural Studies on the EGF Receptor Family

Initially an understanding of the molecular basis of many of the biophysical or cell-based studies on the EGFR family was hindered by a lack of structural and mutational data on the receptors. However, over the last six months several publications have appeared from the USA, Japan and our own laboratories describing the three-dimensional structures for significant portions of the extracellular domains of three of the four EGF receptor family members: EGFR (Garrett et al. 2002; Ogiso et al. 2002, Ferguson et al. 2003), ErbB2 (Garrett et al. 2003; Cho et al. 2003) and Erb B3 (Cho and Leahy 2002).

Interestingly, the structures on the EGFR/ligand dimers (Garrett et al., 2002; Ogiso et al., 2002) reveal a novel mechanism for dimerisation (Fig. 10.11) compared with those seen in other growth factor/receptor structures that have been solved to date (eg NGF/NGFR (Wiesman et al. 1999) VEGF/VEGFR (Wiesman et al. 1997), IL6/gp130 (Chow et al. 2001), hGH/hGHR (de Vos et al. 1992), IL10/IL10R (Josephson et al. 2001). Crystals of TGFα/sEGFR 1–501 or EGF/sEGFR 1–619 contain two molecules of each polypeptide in the asymmetric unit. Dimerisation occurs via homophilic interactions involving a specific loop projecting from each of the first cysteine rich (CR1) regions (back to back dimer). Each ligand is clamped between the first (L1) and third (L2) domains from the same sEGFR 1–501 molecule, and makes contact with only one receptor molecule in the dimer: the two ligands are located on opposite sides of the complex and are more than 70 Å apart.

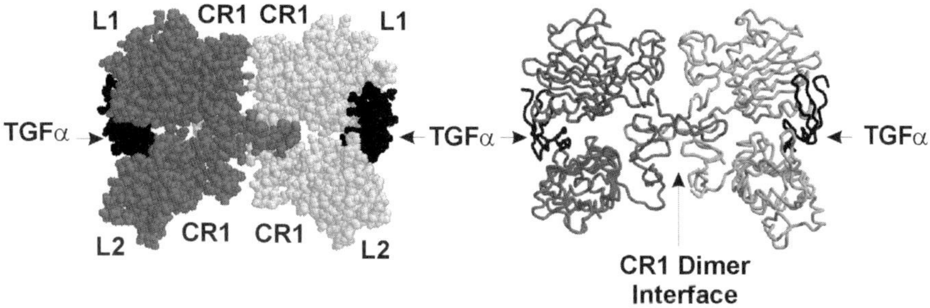

Fig. 10.11. Crystal structure of the EGFR 1–501/TGFα complex showing a side view of the back-to-back dimer. *Left* Space-filled representation of the 2:2 complex; the two sEGFR501 molecules are *shaded light and dark grey*, the two TGFα molecules are shown in *black*. *Right* Backbone representation of the 2:2 complex: the dimer interface via the two CR1 loops is indicated by an *arrow*

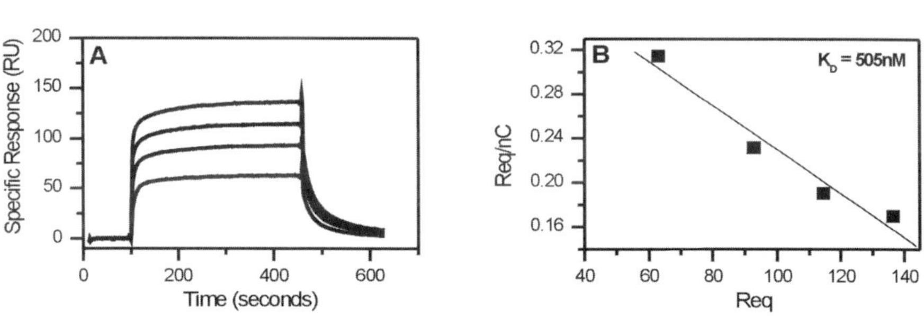

Fig. 10.12. Biosensor analysis of a sEGFR CR1 loop mutant

The biological relevance of the back to back dimer has been confirmed by transfecting BaF cells (a pre-B cell line which lacks endogenous EGFR) with a mutant in which the CR1 loop has been deleted (Garrett et al. 2002), and by biosensor studies (Fig. 10.12) using sEGFR 1–501 in which key residues implicated from the crystal structure in the dimer interaction had been mutated (246YNPTTYAQM253 to 246DAPDTEAD253). Scatchard analysis of the equilibrium binding data showed that the CR1 loop mutant had a considerably (>tenfold) reduced affinity compared to the native protein. Interestingly the affinity observed in the CR1 loop mutant (K_D=505 nM) was now similar to that we had observed with sEGFR 1–621.

10.5.1 Inactivated EGFR Adopts an Autoinhibited Configuration

Our own crystallographic data is further complemented by two related structures that have also been solved recently, namely Erb B3 in the absence of ligand (Cho and Leahy 2002) and EGFR at low pH in which ligand is only bound to L1 (Ferguson et al. 2003). Both these structures reveal potential "autoinhibited" configurations, in which the CR1 dimerisation interface that we have identified is completely occluded by an intramolecular interaction with CR2. In this state, the EGFR (Ferguson et al. 2003) and ErbB3 (Cho and Leahy 2002) ectodomains are folded over in a tethered configuration where the CR1 and CR2 regions are in an antiparallel register that is stabilised by interaction between the CR1 loop and the fifth and sixth modules of CR2. In this configuration the L1 and L2 ligand binding-sites are both exposed but held apart. Ligand binding destabilises this conformation resulting in a dramatic and global change in the configuration of the ectodomain. The L1/CR1 unit rotates and moves away from the CR2 domain to adopt an extended configuration where the ligand becomes clamped between the L1 and L2 domains of a single receptor molecule (Garrett et al. 2002; Ogiso et al. 2002). In this activated state the CR1 and CR2 loops of ErbB receptors are both now exposed and positioned to interact with a second activated partner to form a 2:2 back-to-back complex.

There is also recent biochemical evidence implicating the loop regions of CR2 in mediating inter-receptor dimers. The CR2 region is the site of action for a set of inhibitory peptides originally designed to mimic the CDR3 loop of Herceptin (Park et al. 2000) and shown to compete with Herceptin for binding to the CR2 domain of ErbB2 (Berezov et al. 2001). A subsequent set of inhibitory peptides have been designed which mimic sequences in modules 6 and 7 of CR2 (Berezov et al. 2002), a region shown to contribute to ErbB2 heterodimer formation (Kumagai et al. 2001). Finally, multiple mutagenesis in CR2 modules 3 and 5 of EGFR can lead to dramatic reductions in ligand-induced tyrosine phosphorylation (Saxon and Lee 1999). Interestingly when the CR2 domain of ErbB3 is fitted to the sEGFR501 complex by molecular modelling (not shown) the ends of the CR2 domains (modules 6 and 7) overlap. These interacting CR2 modules 6 and 7 are the same modules targeted by the ErbB2 inhibitory peptides (Berezov et al. 2002).

10.6 Regulation of Homo- and Heterodimerisation

In spite of extensive experimentation, no ligand has ever been identified for Erb B2 (Schlessinger 2002) yet ErbB2 appears to be the major signalling partner for other ErbB receptors. The crystal structures of sErbB2 1–509 (Garrett et al. 2003) or sErbB2 1–631 (Cho et al. 2003) reveal an activated conformation similar to that of the EGFR when complexed with a ligand and very different from that seen in the inactivated forms of ErbB3 or EGFR. The Erb B2 structure (Fig. 10.13) contains only one molecule of the truncated ErbB2 ectodomain in the asymmetric unit. The inability of ErbB2 to bind known ligands is caused by sequence differences and the close juxtaposition of the two L domains, which occludes much of the potential ligand-binding region. While the large loop from the ErbB2 CR1 domain is exposed and clearly available for heterodimerisation, electrostatic calculations suggest that an ErbB2 homodimer would be unlikely and dimers were not observed in either crystal form of the extracellular region. Full-length ErbB2 does not normally homodimerise when expressed in cells, but can do so after high overexpression or mutation within the transmembrane domain. This homodimerisation is mediated in part by a hydrophobic region involving residues 966–968 in the cytoplasmic tail of the tyrosine kinase domain (Penuel et al. 2002).

The conformations of the EGFR and ErbB2 CR1 loops are remarkably similar and we have suggested (Garrett et al. 2003) that the Erb B2 ectodomain is in an active conformation, poised to form heterodimers with other members of the EGFR family. It has been shown (Graus-Potra et al. 1997, Klapper et al. 1999) that Erb B2 is the preferred co-receptor for hetero-oligomerisation.

Biosensor studies have been performed (Ferguson et al. 2000) on EGF and neuregulin induced homo- and hetero-oligomerisation of soluble extracellular domains of the EGFR family. Efficient homodimerisation was observed with EGFR and ErbB4 extracellular domains with EGF and neuregulin respectively. In contrast, ligand induced ErbB receptor extracellular domain hetrodimers were only observed for sErbB2-sErbB4. Fitzpatrick et al. (1998),

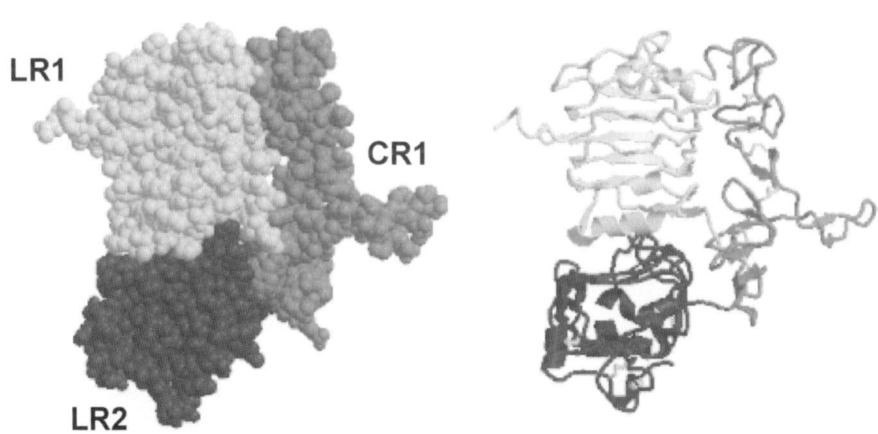

Fig. 10.13. Crystal structure of sErbB2 1 – 509 showing the CR1loop in an active configuration. *Left* Space-filled representation. *Right* Backbone representation. The individual domains are *shaded*

using Scatchard analysis with a radiolabelled ligand, also observed that heterodimers composed of either ErbB3 or ErbB4 with ErbB2 showed high affinity binding with heregulin. To date, in agreement with the conclusions from the structural studies on the truncated form of the ErbB2 extracellular domain (sErbB2 1–509, Garrett et al. 2003) no formation of sErbB2 homodimers has been observed.

Interestingly, Berezov et al. (2002), have used biosensor analysis to investigate the potential of peptides derived from the L2 domain of Erb B2 to inhibit receptor self-association. The peptides have been shown to selectively bind to the EGFR family ectodomains and the isolated L2 domain of ErbB2, and to inhibit heregulin –induced interactions of ErbB3 with other members of the EGFR family.

10.7 Rationalisation of the Structural and Biosensor Data

The availability of the detailed structural data on the EGF/EGFR family members has allowed us to rationalise some of the questions raised by the biosensor studies that we (Nice et al. 1996, Domagala et al. 2000, Elleman et al. 2001), and others (Zhou et al. 1993, De Crescenzo et al. 2000, Ferguson et al. 2003), had undertaken. Firstly, why did EGFR 1–501 bind with higher affinity than the full-length extracellular domain? We had originally hypothesised (Elleman et al. 2001) that domain 4 might be inhibitory and that only single site binding was occuring, but it was not clear what the mechanism of this inhibition might be (one possibility we had considered was that domain 4 might be blocking access of the ligand to one of the potential binding sites, but if this was the case, it was not clear whether this was an artefact of investigating the interaction in solution rather than with the receptor attached to the cell membrane). The binding affinity of the Kohda fragment (500 nm–1 μM), which only contains the L2 domain binding site supported the single site binding hypothesis (Kohda et al. 1993). It is now clear from the crystal structures that domain 4 is indeed involved in the regulation of the binding affinity, since the tether between the CR1 and CR2 loops would only permit binding to a single site without a major conformational change.

The high affinity (20-30 nM) binding site observed with sEGFR 1–501 would then probably fit with binding to both L1 and L2 as observed in the EGFR crystal structure. However, this appears to necessitate dimerisation since the CR1 loop mutant (Fig. 10.12), which would have both sites available since there is no CR1-CR2 loop tether possible in this construct, and which was shown by chemical crosslinking studies to be unable to dimerise, displays the reduced affinity of 500 nM conserved with sEGFR 1–621. Additionally, cross-linked sEGFR dimers (Zhou et al. 1993, Nice et al. 1996), showed similar affinity to that which we had observed with sEGFR 1–501 (K_Ds of 10–25 nM) and also displayed the reduced off-rate (0.002–0.003 s^{-1}).

The binding affinity of TGFα to either sEGFR 1–501 or 1–621 determined using biosensor analysis has reproducibly been lower than that observed for EGF (De Crescenzo et al. 2000, Elleman et al. 2001). It could be that the additional C-terminal residues (WELR) help stabilise the binding of EGF. Interestingly, the role of Gly441, which we had suggested to be important in selectivity in TGFα/EGF binding (Elleman et al. 2001), may also involve interactions with the C-terminus. It is possible that the reduction in binding seen with the Gly441Lys mutant is a function of steric hindrance between the lysine side chain and Glu 51 of the EGF: since TFGα terminates at residue 49 this would not be a problem.

10.8 Conclusion

The use of biosensor analysis for detailed investigation of the interactions of the soluble extracellular domains of the EGFR family with their ligands has revealed a lot of useful information on the structure-function relationships of this important family of signalling proteins. In particular it has shown the presence of both high and low affinity forms of the full-length extracellular domain of the receptor (sEGFR 1–621) in solution (Domagala et al. 2000, De Crescenzo et al. 2000). Surprisingly, biosensor analysis of a truncated form of the receptor (sEGFR 1–501) showed a higher affinity binding, corresponding to the high affinity form observed in sEGFR 1–621. The availability of crystallographic information on the EGFR, ErbB2 and ErbB3 (Garrett et al. 2002; Ogiso et al. 2002, Ferguson et al. 2003, Garrett et al. 2003, Cho and Leahy 2002, Cho et al. 2003) has explained this result. These structures have yielded many surprises, including a clear and dramatic conformational transition on ligand binding (Ferguson et al. 2003), an extra-ordinary back-to-back dimer configuration (Garrett et al. 2002, Ogiso et al. 2002) and the unexpected pre-activation state for the ErbB2 receptor monomer (Garrett et al. 2003). Taken together they suggest a novel mechanism for receptor dimerisation and activation (Fig. 10.14).

These structures will allow this new model of receptor mediated EGFR activation and signalling to be tested rationally by designing appropriate ligand and receptor mutants. These structures also clearly show why previous attempts to generate ligand-based inhibitors have failed, and suggest a number of novel strategies for designing new Erb B antagonists with therapeutic potential. Biosensor studies, similar to the ones we have described in this chapter, will clearly pay a pivotal role in many of these studies.

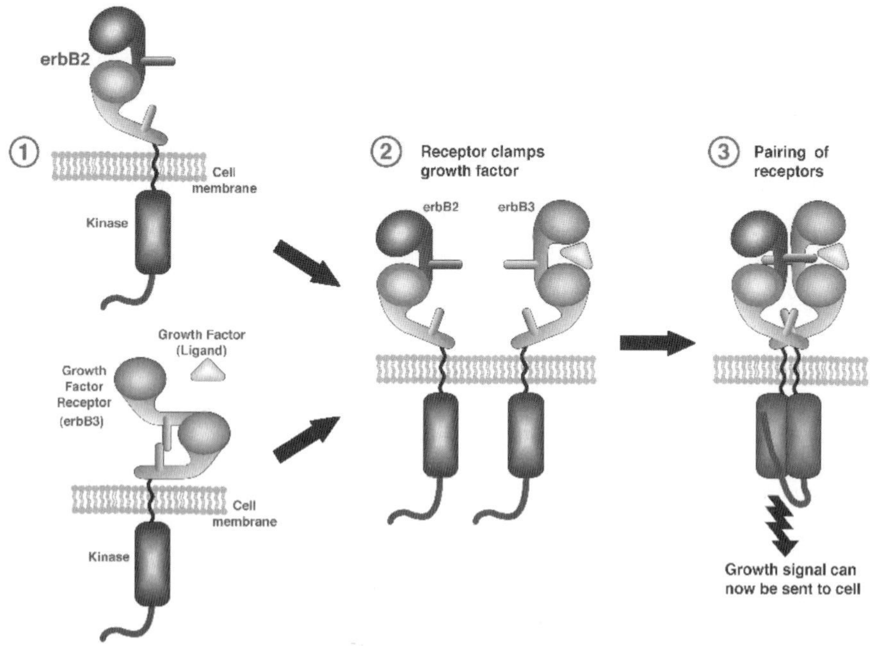

Fig. 10.14. The mechanism of EGFR activation and dimerisation. The unoccupied receptor (**A**) is in an autoinhibited state caused by a tether between the loops in the CR1 and CR2 regions. The receptor remains in the autoinhibited (inactivated) state (**B**) upon ligand binding to only the L1 domain. Simultaneous binding of ligand to both the L1 and L2 domains disrupts the intracellular domain tether between the loops in the CR1 and CR2 domains and involves a large conformational change about a pivot between the CR1-L2 interface (**C**). Rotation of the model by 90° (**D**) highlights the new orientation of the CR1 and CR2 loops. Active dimers are stabilised by the interaction of the CR1 loops (**E**). *1* L1 domain, *2* CR1 domain, *3* L2 domain, *4* CR2 domain, *L* ligand, *M* membrane

References

Alroy I, Yarden Y (1997) The ErbB signalling network in embryogenesis and oncogenesis: signal diversification through combinatorial ligand-receptor interactions. FEBS letters 410:83–86

Baird CL, Myszka DG (2001) Current and emerging commercial optical biosensors. J Mol Recogn 14:261–268

Bajaj M, Waterfield MD, Schlessinger J, Taylor WR, Blundell T (1987) On the tertiary structure of the extracellular domains of the epidermal growth factor and insulin receptors. Biochim. Biophys. Acta 916:220–226

Bellot F, Moolenaar W, Kris R, Mirakhur B, Verlaan I, Ullrich A, Schlessinger J, Felder S (1990) High-affinity epidermal growth factor binding is specifically reduced by a monoclonal antibody, and appears necessary for early responses. J Cell Biol 110:491–502

Bennett D, Morton T, Breen A, Hertzberg R, Cusimano D, Appelbaum E, McDonnell P, Young P, Matico R, Chaiken I. (1995) Kinetic characterization of the interaction of biotinylated human interleukin 5 with an Fc chimera of its receptor alpha subunit and development of an ELISA screening assay using real-time interaction biosensor analysis. J Mol Recognit 8:52–58

Berezov A, Chen J, Liu Q, Zhang HT, Greene MI, Murali R (2002) Disabling receptor ensembles with rationally designed interface peptidomimetics. J Biol Chem 277: 28330–28339

Berezov A, Zhang HT, Greene MI, Murali R. (2001) Disabling ErbB receptors with rationally designed exocyclic mimetics of antibodies: structure-function analysis. J Med Chem 44:2565–2574

Bishayee S (2000) Role of conformational alteration in the epidermal growth factor receptor (EGFR) function. Biochem. Pharmacol 60:1217–1223

Brown PM, Debanne MT, Grothe S, Bergsma D, Caron M, Kay C, O'Connor-McCourt MD (1994) The extracellular domain of the epidermal growth factor receptor. Eur J Biochem 225:223–233

Burgess AW, Knesel J, Sparrow LG, Nicola NA, Nice EC (1982) Two forms of murine epidermal growth factor: rapid separation by using reverse-phase HPLC. Proc Natl Acad Sci USA 79:5753–5757

Cass AEG (1995) Biosensors in Molecular Biology and Biotechnology. RA Meyers editor, VCH publishers, pp110–113.

Catimel B, Rothacker J, Nice EC (2001) The use of biosensors for micro-affinity purification: an integrated approach to proteomics. J Biochem Biophys Methods 49:289–312

Catimel B, Weinstock J, Nerrie M, Domagala T, Nice EC (2000) Micropreparative ligand fishing with a cuvette-based optical mirror resonnance biosensor. J Chrom A 869: 261–273

Catimel B, Domagala T, Nerrie M, Weinstock J, Abud H, Heath JK, Nice EC (1999) Recent applications of instrumental biosensors for protein and peptide structure-function studies. Protein and Peptide Letters 6:319–340

Chaiken I, Rose S, Karlsson R. (1992) Analysis of macromolecular interactions using immobilized ligands. Anal Biochem 201:197–210

Cho HS, Mason K, Ramyar KX, Stanley AM, Gabelli SB, Denney DW Jr, Leahy DJ. (2003) Structure of the extracellular region of HER2 alone and in complex with the Herceptin Fab. Nature 421:756–60

Cho HS, Leahy DJ. (2002) Structure of the extracellular region of HER3 reveals an interdomain tether. Science 297:1330–1333

Chow D, He X, Snow AL, Rose-John S, Garcia KC. (2001) Structure of an extracellular gp130 cytokine receptor signaling complex. Science 291, 2150–2155

Chrunyk BA, Rosner MH, Cong Y, McColl AS, Otterness IG, Daumy GO (2000) Inhibiting Protein-Protein Interactions: A Model for Antagonist Design. Biochemistry 39:7092–7099

Cunningham BC, Wells JA (1993) Comparison of a structural and functional epitope. J Mol Biol 234:554–63

Davies RD, Edwards PR, Watts HJ, Lowe CR, Buckle PE, Yeung D, Kinning TM, Pollard-Knight DV (1994) The resonant mirror: a tool for the study of biomolecular interactions. Techniques in Protein Chemistry V 285–292

De Crescenzo G, Grothe S, Lortie R, Debanne MT, O'Connor-McCourt MD (2000) Real-time kinetic studies on the interaction of transforming growth factor alpha with the epidermal growth factor receptor extracellular domain reveal a conformational change model. Biochemistry 39:9466–9476

de Vos AM, Ultsch M, Kossiakoff AA (1992) Human growth hormone and extracellular domain of its receptor: crystal structure of the complex. Science 255:306–312

DiAugustine RP, Walker MP, Klapper DG, Grove RI, Willis WD, Harvan DJ, Hernandez O (1985) Beta-epidermal growth factor is the des-asparaginyl form of the polypeptide. J Biol Chem 260:2807–2811

Domagala t, Konstantopoulos N, Smyth F, Jorissen RN, Fabri L, Geleick D, Lax I, Schlessinger J, Sawyer W, Howlett GJ, Burgess AW, Nice EC (2000) Stoichiometry, kinetic and binding analysis of the interaction between epidermal growth factor (EGF) and the extracellular domain of the EGF receptor. Growth Factors 18:11–29

Dubaquie Y, Lowman HB (1999) . Biochemistry 38:6386–96

Elleman TC, Domagala T, McKern NM, Nerrie M, Lönnqvist B, Adams TE, Lovrecz GO, Hoyne PA, Richards KM, Howlett GJ, Rothacker J, Jorissen RN, Lou M, Garrett TPJ, Burgess AW, Nice EC, Ward CW (2001) Identification of a determinant of epidermal growth factor receptor ligand-binding specificity using a truncated, high-affinity form of the ectodomain. Biochemistry 40:8930–8939

Ferguson KM, Berger MB, Mendrola JM, Cho HS, Leahy DJ, Lemmon MA (2003) EGF Activates Its Receptor by Removing Interactions that Autoinhibit Ectodomain Dimerization. Mol Cell 11:507–517

Ferguson KM, Darling PJ, Mohan MJ, Macatee TL, Lemmon MA (2000) Extracellular domains drive homo- but not hetero-dimerization of erbB receptors. EMBO J 19:4632–4643

Fitzpatrick VD, Pisacane PI, Vandlen RL, Sliwkowski MX. (1998) Formation of a high affinity heregulin binding site using the soluble extracellular domains of ErbB2 with ErbB3 or ErbB4. FEBS Lett 43:102–106

Forbes BE, Hartfield PJ, McNeil KA, Surinya KH, Milner SJ, Cosgrove LJ, Wallace JC (2002) Characteristics of binding of insulin-like growth factor (IGF)-I and IGF-II analogues to the type 1 IGF receptor determined by BIAcore analysis. Eur J Biochem 269:961–968

Gardsvoll H, Dano K, Ploug M (1999) J Biol Chem 274:37995–8003

Garrett TPJ, McKern NM, Lou M, Elleman TC, Adams TE, Lovrecz GO, Kofler M, Jorissen RN, Nice EC, Burgess AW, Ward CW (2003) The Crystal Structure of a Truncated ErbB2 Ectodomain Reveals an Active Conformation, Poised to Interact with Other ErbB Receptors. Mol Cell 11:495–505

Garrett TPJ, McKern NM, Lou M, Elleman TC, Adams TE, Lovrecz GO, Zhu HJ, Walker F, Frenkel MJ, Hoyne PA, Jorissen RN, Nice EC, Burgess AW, Ward CW (2002) Crystal structure of a truncated epidermal growth factor receptor extracellular domain bound to transforming growth factor alpha. Cell 110:763–773

Garrett TPJ, McKern NM, Lou M, Frenkel MJ, Bentley JD, Lovrecz GO, Elleman TC, Cosgrove L J, Ward CW (1998) Crystal structure of the first three domains of the type-1 insulin-like growth factor receptor. Nature 394:395–399

Graus-Porta D, Beerli RR, Daly JM, Hynes NE. (1997) ErbB-2, the preferred heterodimerization partner of all ErbB receptors, is a mediator of lateral signaling. EMBO J; 16(7): 1647–55.

Greenfield C, Hiles I., Waterfield MD, Federwisc M, Wollmer A, Blundell TL, McDonald N (1989) Epidermal growth factor binding induces a conformational change in the external domain of its receptor. EMBO J 8:4115–4123

Groenen LC, Nice EC, Burgess AW (1994) Structure-function relationships for the EGF/TGFalpha family of mitogens. Growth Factors 11:235–257

Gunther N, Betzel C, Weber W (1990) The secreted form of the epidermal growth factor receptor. J Biol Chem 265:22082–22085

Hill AV (1910) The possible effects of the aggregation of the molecules of haemoglobin on its dissociation curves. J Physiol 40 iv-vii

Humphrey PA, Gangarosa LM, Wong AJ, Archer GE, Lund-Johansen M, Bjerkvig R, Laerum OD, Friedman HS, Bigner DD. (1991) Deletion-mutant epidermal growth factor receptor in human gliomas: effects of type II mutation on receptor function. Biochem Biophys Res Commun. 178:1413–1420

Hurwitz DR. Emanuel SL, Nathan MH, Sarver N, Ullrich A, Felder S, Lax I, Schlessinger J (1991) EGF induces increased ligand binding affinity and dimerization of soluble epidermal growth factor (EGF) receptor. Extracellular domain. J Biol Chem 266:22035–22043

Hynes NE, Stern DF (1994) The biology of ErbB-2/neu/HER-2 and its role in cancer. Biochim Biophys Acta 1198:165–184

Johnsson B, Lofas S, Lindquist G, Edstrom A, Muller Hillgren RM, Hansson A (1995) Comparison of methods for immobilization to carboxymethyl dextran sensor surfaces by analysis of the specific activity of monoclonal antibodies. J. Mol Recognit 8: 125–131

Jorissen RN, Treutlein HR, Epa VC, Burgess AW (2002) Modeling the epidermal growth factor-epidermal growth factor receptor L2 domain interaction: implications for the ligand binding process. J Biomol Struct Dyn 19:961–972

Josephson K, Logsdon NJ, Walter MR. (2001) Crystal structure of the IL-10/IL-10R1 complex reveals a shared receptor binding site. Immunity 15:35 – 46

Kawamoto T, Sato JD, Le A, Polikoff J, Sato GH, Mendelson J (1983) Growth stimulation of A431 cells by epidermal growth factor: identification of high–affinity receptors for epidermal growth factor by an anti–receptor monoclonal antibody. Prot Natl Acad Sci U.S.A. 80:1337–1341

King AC, Cuatrecasas P (1982) Resolution of high and low affinity epidermal growth factor receptor. Inhibition of high affinity component by low temperature, cycloheximide and phorbol esters. J Biol Chem 257:3053–3060

Klapper LN, Glathe S, Vaisman N, Hynes NE, Andrews GC, Sela M, Yarden Y. (1999) The ErbB-2/HER2 oncoprotein of human carcinomas may function solely as a shared coreceptor for multiple stroma-derived growth factors. Proc Natl Acad Sci U S A 96(9):4995–5000

Kohda D, Odaka M, Lax I, Kawasaki H, Suzuki K, Ullrich A, Schlessinger J Inagaki F (1993) A 40-kDa epidermal growth factor/ transforming growth factor a – binding domain produced by limited proteolysis of the extracellular domain of the epidermal growth factor receptor. J Biol Chem 268:1976–1981

Kumagai T, Davis JG, Horie T, O'Rourke DM, Greene MI (2001) The role of distinct p185neu extracellular subdomains for dimerization with the epidermal growth factor (EGF) receptor and EGF-mediated signaling. Proc Natl Acad Sci U S A 98:5526–5531

Lax I, Fischer R, Ng C, Serge J, Ullrich A, Givol D, Schlessinger J (1991) Noncontiguous regions in the extracellular domain of EGF receptor define ligand–binding specificity. Cell Regul 2:337–345

Lax I, Johnson A, Howk R, Sap J, Bellot F, Winkler M, Ullrich A, Vennstrom B, Schlessinger J, Givol D (1988) Chicken epidermal growth factor (EGF) receptor: cDNA cloning, expression in mouse cells, and differential binding of EGF and transforming growth factor alpha. Molec Cellul Biol 8:1970–1978

Lenferink AE, van Zoelen EJ, van Vugt MJ, Grothe S, van Rotterdam W, van De Poll ML, O'Connor-McCourt MD (2000) . J Biol Chem 275:26748–26753

Malmqvist M (1983) Biospecific interaction analysis using biosensor technology. Nature 361:186–187

Marinaro JA, Jamieson, G.P, Hogarth PM, Bach LA (1999) . FEBS 450:240–244

McKern NM, Lou M, Frenkel MJ, Verkuylen A, Bentley JD, Lovrecz GO, Ivancic N, Elleman TC, Garrett TP, Cosgrove LJ, Ward CW (1997) Crystallization of the first three domains of the human insulin-like growth factor-1 receptor. Protein Sci 6:2663–2666

Mendelsohn J, Baselga J (2000) The EGF receptor family as targets for cancer therapy. Oncogene 19(56):6550–6565

Moriki T, Maruyama H, Maruyama IN (2001) Activation of preformed EGF receptor dimers by ligand-induced rotation of the transmembrane domain. J Mol Biol 311:1011–1026

Nedelkov D, Nelson RW (2001a) Analysis of native proteins from biological fluids by biomolecular interaction analysis mass spectrometry (BIA/MS): exploring the limit of detection, identification of non-specific binding and detection of multi-protein complexes. Biosens Bioelectron 16:1071–1078

Nedelkov D, Nelson RW (2001b) Delineation of in vivo assembled multiprotein complexes via biomolecular interaction analysis mass spectrometry. Proteomics 1:1441–1446

Nelson RW, Krone JR (1997) Surface plasmon resonance biomolecular interaction analysis mass spectrometry. 1. Chip based analysis. Anal Chem 69:4363–4368

Nice EC, Catimel B (1999) Instumental biosensors: new perspectives for the analysis of biomolecular interaction. Bioessays 21:339–352

Nice EC, Catimel B, Lackmann M, Stacker S, Runting A, Wilks A, Nicola N, Burgess AW (1997) Stategies for the identification and purification of ligands for orphan biomolecules. Letters in Peptide Science 4:107–120

Nice EC, Smyth F, Domagala T, Fabri L, Catimel B, Burgess AW (1996) Synergies between micropreparative HPLC and an optical biodensor: Application to structure-function studies. Science tools 1:3–5

Nice EC, Lackmann M, Smyth F, Fabri L, Burgess AW (1994) Synergies between micropreparative high-performance liquid chromatography and an instrumental optical biosensor. J Chromatogr 660:169–185

Normanno N, Bianco C, De Luca A, Maiello MR, Salomon DS (2003) Target-based agents against ErbB receptors and their ligands: a novel approach to cancer treatment. Endocr Relat Cancer Mar 10(1):1–21

Ogiso H, Ishitani R, Nureki O, Fukai S, Yamanaka M, Kim JH, Saito K, Sakamoto A, Inoue M, Shirouzu M, Yokoyama S (2002) Crystal structure of the complex of human epidermal growth factor and receptor extracellular domains. Cell 110:775–787

Panayotou G, Gish G, End P, Truong O, Gout I, Dhand R, Fry MJ, Hiles I., Pawson, Waterfield MD (1993) Interactions between SH2 domains and tyrosine-phosphorylated platelet-derived growth factor beta-receptor sequences: analysis of kinetic parameters by a novel biosensor-based approach. Mol Cell Biol 13:3567–3576

Park BW, Zhang HT, Wu C, Berezov A, Zhang X, Dua R, Wang Q, Kao G, O'Rourke DM, Greene MI, Murali R (2000) Rationally designed anti-HER2/neu peptide mimetic disables P185HER2/neu tyrosine kinases in vitro and in vivo. Nat Biotechnol 18:194–198

Penuel E, Akita RW, Sliwkowski MX (2002) Identification of a region within the ErbB2/HER2 intracellular domain that is necessary for ligand-independent association. J Biol Chem 277:28468–28473

Plugariu CG, Wu SJ, Zhang W, Chaiken I (2000) Multisite mutagenesis of interleukin 5 differentiates sites for receptor recognition and receptor activation. Biochemistry 39:14939–49

Rich R, Myszka DG (2000a) Advances in surface plasmon resonance biosensor analysis. Current Opinion in Biotechnology 11:54–61

Rich RL, Myszka DG (2000b) Survey of the 1999 surface plasmon resonance biosensor literature. J Mol Recognit 13:388–407

Rich RL, Myszka DG (2001) Survey of the year 2000 commercial optical biosensor literature. J Mol Recognit 14:273–94

Rich RL, Myszka DG (2002) Survey of the year 2001 commercial optical biosensor literature. J Mol Recognit 15:352–76

Riese DJ, Stern DF (1998) Specificity within the EGF family/ErbB receptor family signaling network. BioEssays 20:41–48

Saxon ML, Lee DC (1999) Mutagenesis reveals a role for epidermal growth factor receptor extracellular subdomain IV in ligand binding. J Biol Chem 274:28356–28362.

Schleinkofer K, Dingley A, Tacken I, Federwisch M, Muller-Newen G, Heinrich PC, Vusio P, Jacques Y, Grotzinger J (2001) Identification of the domain in the human interleukin-11 receptor that mediates ligand binding. J Mol Biol 306(2):263–74

Schlessinger J (2002) Ligand-induced, receptor-mediated dimerization and activation of EGF receptor. Cell 110:669–672

Schlessinger J, Schreiber AB, Levi A, Lax I, Libermann T, Yarden Y (1983) Regulation of cell proliferation by epidermal growth factor. CRC Crit Rev Biochem. 14:93–111

Shand JH, Beattie J, Song H, Phillips K, Kelly SM, Flint DJ, Allan GJ (2003) Specific amino acid substitutions determine the differential contribution of the amino- and carboxyl-terminal domains of insulin-like growth factor binding protein (IGFBP)-5 in binding IGF-I. J Biol Chem in press

Sizeland AM, Burgess AW (1992) Anti-sense transforming growth factor alpha oligonucleotides inhibit autocrine stimulated proliferation of a colon carcinoma cell line. Mol Biol Cell 3:1235–1243

Slamon DJ, Clark GM, Wong SG, Levin WJ, Ullrich A, McGuire WL (1987) Human breast cancer: correlation of relapse and survival with amplification of the HER-2/neu oncogene. Science 235:177–182

Slamon DJ, Godolphin W, Jones LA, Holt JA, Wong SG, Keith DE, Levin WJ, Stuart SG, Udove J, Ullrich A, et al. (1989) Studies of the HER-2/neu proto-oncogene in human breast and ovarian cancer. Science 244:707–712

Song H, Beattie J, Campbell I, Allan W (2000) . J. Mol Endocrinol 24:43–51

Stanley P (1989) Chinese hamster ovary cell mutants with multiple glycosylation defects for production of glycoproteins with minimal carbohydrate heterogeneity. Mol Cell Biol 9:377–383

Sönksen CP, Nordhoff E, Jansson Ö, Malmqvist M, Roepstorff P (1998) Combining MALDI mass spectrometry and biomolecular interaction analysis using a biomolecular interaction analysis instrument. Anal Chem 70:2731–2736

Steinberg D (2002) Closing in on multiple cancer targets. The Scientist, 16(7), 29

Strachan L, Murison JG, Prestidge RL, Sleeman MA, Watson JD, Kumble KD (2001) Cloning and biological activity of epigen, a novel member of the epidermal growth factor superfamily. J Biol Chem 276:18265–18271

Summerfield AE, Hudnall AK, Lukas TJ, Guyer CA, Staros JV (1996) Identification of residues of the epidermal growth factor receptor proximal to residue 45 of bound epidermal growth factor. J Biol Chem. 271:19656–19659

Sundarasen S, Roberts PE, King KL, Sliwkowski, M.X, Mather JP (1998) Biological response to ErbB ligands in nontransformed cell lines correlates with a specific pattern of receptor expression. Endocrinol. 139:4756–4764

Todd R, Wong DT (1999) Oncogenes. Anticancer Res 19:4729–46

Tsuda T, Ikeda Y, Taniguchi N (2000) The Asn-420-linked sugar chain in human epidermal growth factor receptor suppresses ligand-independent spontaneous oligomerization. Possible role of a specific sugar chain in controllable receptor activation. J Biol Chem 275:21988–21994

Ullrich A, Coussens L, Hayflick S, Dull TJ, Gray A, Tam AJ, Yarden Y, Libermann TA, Schlessinger J, Downward J, Mayes ELV, Whittle N, Waterfield MD, Seeburg PH (1984) Human epidermal growth factor receptor cDNA sequence and aberrant expression of the amplified gene in A431 epidermoid carcinoma cells. Nature 309:418–425

van de Poll ML, Lenferink AE, van Vugt MJ, Jacobs JJ, Janssen JW, Joldersma M, van Zoelen EJ (1995) A single amino acid exchange, Arg-45 to Ala, generates an epidermal growth factor (EGF) mutant with high affinity for the chicken EGF receptor. J Biol Chem 270:22337–22343 van der Plas RM, Gomes L, Marquart JA, Vink T, Meijers JC, de Groot PG, Sixma JJ, Huizinga EG (2000) . Thromb Haemost 84:1005–1011

van Regenmortel M (2001) . Cell Mol Life Sci 58:794–800

Wade .JD, Domagala T, Rothacker J, Catimel B, Nice EC (2002) Use of thiazoline-mediated ligation for site specific biotinylation of mouse EGF for biosensor immobilisation. Letters in Peptide Science 58:493–503

Walker F, Hibbs ML, Zhang HH, Gonez LJ, Burgess AW (1998) Biochemical characterization of mutant EGF receptors expressed in the hemopoietic cell line BaF/3. Growth Factors 16:53–67

Walker F, Burgess AW (1991) Reconstitution of the high affinity epidermal growth factor receptor on cell – free membranes after transmodulation by platelet–derived growth factor. J Biol Chem 266:2746–2752

Ward CW, Hoyne PA, Flegg RH (1995) Insulin and epidermal growth factor receptors contain the cysteine repeat motif found in the tumor necrosis factor receptor. Proteins: Struct Funct Genet 22:141–153

Ward LD, Howlett GJ, Hammacher A, Weinstock J, Yasukawa K, Simpson RJ, Winzor DJ (1995) Use of a biosensor with surface plasmon resonance detection for the determination of binding constants: measurement of interleukin-6 binding to the soluble interleukin-6 receptor. Biochemistry 34:2901–2907

Wiesmann C, Fuh G, Christinger HW, Eigenbrot C, Wells JA, de Vos AM. (1997)

Crystal structure at 1.7 A resolution of VEGF in complex with domain 2 of the Flt-1 receptor. Cell, 91, 695–704.

Wiesmann C, Ultsch MH, Bass SH, de Vos AM (1999) Crystal structure of nerve growth factor in complex with the ligand-binding domain of the TrkA receptor. Nature, 401:184 - 188

Williams C (2000) Biotechnology match making: screening orphan ligands and receptors. Cur Opin Biotechnol 11:42–46

Woltjer RL, Lukas TJ, Staros JV (1992) Direct identification of residues of the epidermal growth factor receptor in close proximity to the amino terminus of bound epidermal growth factor. Proc Natl Acad Sci U S A 89:7801–7805

Wong MS, Fong CC, Yang M (1999) Biosensor measurement of the interaction kinetics between insulin-like growth factors and their binding proteins. Biochim Biophys Acta 1432:293–301

Wu DG, Wang LH, Sato GH, West KA, Harris WR, Crabb JW, Sato JD (1989) Human epidermal growth factor (EGF) receptor sequence recognized by EGF competitive monoclonal antibodies. Evidence for the localization of the EGF-binding site. J Biol Chem 264:17469–17475

Yarden Y, Sliwkowski MX (2001) Untangling the ErbB signalling network. Nat Rev Mol Cell Biol 2:127–137

Yarden Y, Schlessinger J (1987a) Epidermal growth factor induces rapid, reversible aggregation of the purified epidermal growth factor receptor. Biochemistry 26:1443–1451

Yarden Y, Schlessinger J (1987b) Self-phosphorylation of epidermal growth factor receptor: evidence for a model of intermolecular allosteric activation. Biochemistry 26: 1434–1442

Yarden Y, Harari I, Schlessinger J (1985) Purification of and active EGF receptor kinase with monoclonal antireceptor antibodies. J Bio Chem 260:315–319

Zeder - Lutz G, Wenger R, Van Regenmortel, MHV and Altschuh D (1993) Interaction of cyclosporin A with an Fab fragment or cyclophilin. FEBS 326, 153 - 157

Zhou M, Felder S, Rubinstein M, Hurwitz DR, Ullrich A, Lax I, Schlessinger J (1993) Real-time measurements of kinetics of EGF binding to soluble EGF receptor monomers and dimers support the dimerization model for receptor activation. Biochemistry 32:8193–8198

11 The Functional Interaction Trap: A Novel Strategy to Study Specific Protein-Protein Interactions

ALOK SHARMA, SUSUMU ANTOKU and BRUCE J. MAYER

11.1 Protein–Protein Interactions in Cellular Systems

Protein–protein interactions play a central role in almost all aspects of the living cell, and understanding these interactions holds the key to understanding a host of cellular processes. Whether it is an enzyme modifying its substrate, the assembly of subunits of a multiprotein complex, the recognition and binding of a specific ligand, or the polymerization of monomeric subunits such as those of actin, protein–protein interactions are essential for regulating and organizing virtually all physiological responses. Much recent research has focused on understanding these interactions in detail, and while progress has been rapid in many areas, it is clear that new tools to study specific protein-protein interactions are sorely needed. For example, our ability to identify actual or potential protein-protein interactions has outpaced our ability to validate the functional significance of those interactions, or of post-translational modifications that might result from those interactions. In this article we will discuss the Functional Interaction Trap (FIT) approach [13, 44], a novel proteomic tool designed to elucidate the physiological outputs that are mediated by specific protein-protein interactions or post-translational modifications in cellular signaling. We will also discuss in detail a specific application of the FIT approach, where it is used to dissect the functional consequences of tyrosine phosphorylation of specific substrates in the cell.

11.2 Signal Transduction

Cellular signaling utilizes an extensive and complex array of protein-protein interactions to transduce and interpret extracellular signals [16, 36]. In multicellular organisms, signaling is critical for normal development and daily functioning of the cell and the organism. This is particularly true of complex

Principles and Practice
Methods in Proteome and Protein Analysis
R.M. Kamp, J. J. Calvete, T. Choli-Papadopoulou (Eds.)

functions of the cell such as adhesion, migration, differentiation, and survival, which depend entirely on the proper functioning of the signaling machinery. A large number of protein-protein interactions are required for a signal to be transferred from the outside of the cell to the cytoplasm, from the cytoplasm to the nucleus, from the nucleus back to the cytoplasm, and in some cases from the cytoplasm back to the outside of the cell to be carried away to other cells. In general, signal transduction results from an input signal which modifies the activity of a one or more cell-surface proteins (by post-translational modification and/or physical association with other proteins), which in turn modifies the activity of additional proteins downstream. One protein can alter the activity of multiple proteins, and the process of successive post-translational modifications (or physical associations) of proteins leads to the formation of a signaling cascade. Each of the protein interactions or modifications produces effects that are integrated together to produce a physiological response.

An important aspect of signaling is that it generally does not involve simple linear pathways, but instead is dependent on highly interconnected networks of interactions. An analogy can be drawn between the signaling process and a computer program. The input signal is the *execute* or *run* command that starts to decipher the program, which has been split up into smaller modules. Each module consists of several commands that work together to produce an output. The output from each module is then "compiled" together to produce the desired final output. In a complex environment, the same modules could be used for running different parts of different programs. Similarly, cellular signaling takes advantage of the same modules (such as small G proteins or MAP kinase cascades) to perform many different tasks.

The resulting complexity of signaling networks leads to a profound experimental problem: understanding the output produced by a specific interaction between two proteins, (for example a kinase and a substrate) is very difficult, because the kinase is likely to modify multiple substrate proteins and a particular substrate may be modified by multiple kinases. Thus, in order to study the effects produced by the interaction between two specific proteins, we need to develop methods that will promote an interaction *exclusively* between the two proteins of interest, thus permitting analysis of the downstream effects mediated by that specific pairwise interaction. As will be subsequently described, the FIT approach that we have developed [13, 44] was designed to allow the analysis of the functional consequences of specific protein-protein interactions or post-translational modifications in the cell.

11.3 Tyrosine Phosphorylation and the Identification of Physiologically Relevant Substrates

The tyrosine kinase activity of the v-Src oncogene product was discovered in 1980 [17, 45]. Soon thereafter other viral oncogenes such as v-Abl were also shown to possess tyrosine kinase activity, implying a central role in cancer [39]. Of course the endogenous cellular homologues of the viral oncogenes (such as c-Src and c-Abl) were found to be present in normal cells and were assumed to play an important role in normal signaling [40, 46, 52]. Subsequently, it was shown that transmembrane receptors for some mitogenic growth factors also possessed tyrosine kinase activity, leading to the classification of tyrosine kinases as either receptor tyrosine kinases (such as the EGF, PDGF, and insulin receptors) or non-receptor tyrosine kinases (such as Src or Abl). With the discovery of more tyrosine kinases came the realization that diseases other than cancer, for example disorders of the immune system, could also result from improper tyrosine kinase signaling. The most recent analyses have shown that the human genome encodes at least 90 tyrosine kinases [29], which participate in a wide variety of processes in the cell.

Tyrosine kinases such as Src phosphorylate multiple substrates, thereby modulating diverse cellular functions such as proliferation, adhesion, migration, differentiation and survival [5, 34, 37, 49]. How and why then do v-Src and v-Abl cause cancer? Or on the other hand, what are the substrates that mediate the normal physiological roles of c-Src and c-Abl? Clearly, one of the keys to understanding the profound biological activities of tyrosine kinases is identifying the substrates whose phosphorylation is responsible for physiological responses. It is logical to expect that relatively few substrates must be responsible for each of the responses, normal or abnormal. While a large number of substrates of tyrosine kinases such as Src and Abl have been identified [28], the question of which substrate phosphorylation leads to which particular response has remained largely unexplored mostly due to the lack of specific experimental tools to address this issue.

Identifying the substrates responsible for biological outputs is not merely of academic interest, as those substrates may represent novel targets for drug discovery. In order to treat cancer or other disorders resulting from defective tyrosine kinase signaling, it would obviously be preferable to target the improper functioning unit of the signaling cascade rather than trying to manipulate the physiological manifestations of the disease itself. The tyrosine kinases themselves are excellent drug targets, as recently demonstrated by the spectacular success of Gleevec, which specifically targets BCR-Abl in human chronic myelogenous leukemia (CML) [9, 10]. However, in some cases, inhibiting the kinase may have unacceptable side effects, or (as has been seen in CML) the kinase may mutate to become resistant to the inhibitor [15]. In other cases, it may be desirable to inhibit only one of several downstream outputs from a kinase. In such instances, identification of the substrates responsible for the output to be inhibited is an obvious first step in devising inhibitors of those pathways.

A closer look at the structure of these tyrosine kinases and how they recognize their substrates will shed some light on the difficulty of the task, and on a potential strategy which would address this problem. Tyrosine kinases consist largely of modular domains, each conferring a specific function. In the case of nonreceptor tyrosine kinases such as Src, these include an N-terminal myristoylation signal, followed by a unique region, followed by modular protein interaction domains termed Src homology 3 and 2 (SH3 and SH2) domains, and finally the catalytic domain [5, 34, 37, 49]. Considerable data has shown that both the SH3 and SH2 domains play an important role in binding to putative substrates, allowing the relatively weak and nonspecific catalytic domain to efficiently phosphorylate the bound substrate, often on multiple sites [32, 37, 38, 47].

The key to substrate phosphorylation then lies in the specificity of the SH3 and SH2 domains. SH3 domains recognize and bind to proteins containing a left-handed polyproline helix with a core consensus of PxxP, while SH2 domains bind specifically to tyrosine-phosphorylated peptides [22, 31, 36]. Although both SH3 and SH2 domains have some inherent specificity (that is, different SH3 domains will bind best to different spectra of proline-rich binding sites, and different SH2 domains bind best to different phosphotyrosine-containing sites), there is a considerable overlap in this binding specificity; therefore, the same substrate may bind to and be phosphorylated by multiple tyrosine kinases such as Src or Abl. Furthermore, many different potential substrates may bind with similar affinity to a particular kinase. Thus, an activated tyrosine kinase may phosphorylate a multitude of proteins, while different kinases phosphorylate overlapping sets of substrates. How then can we begin to dissect which responses are being mediated by which substrates?

One approach to identifying key substrates is to individually knock out the expression of candidate substrates, either by homologous recombination or more recently by RNAi. While this may indicate whether a substrate is *necessary* for a particular output, it cannot address whether phosphorylation of that substrate is *sufficient* for that output. Furthermore, such an analysis would be compromised if the substrate has some other important cellular function that does not depend on its phosphorylation. While in principle this problem could be addressed by knocking in mutant genes lacking specific phosphorylation sites, this approach is in most cases prohibitively slow, costly, and labor-intensive for vertebrate cells, and it presupposes a great deal of information about potential substrates and phosphorylation sites. Therefore it is necessary for us to to develop an alternative method with which we could experimentally induce the phosphorylation of a single substrate of choice in the cell.

11.4 The Functional Interaction Trap as a Novel Strategy to Promote Specific Protein-Protein Interactions and Post-Translational Modifications

How can we achieve a pairwise interaction between a kinase and a single substrate, when the activation of tyrosine kinases generally leads to interaction with multiple substrates? One approach is to modify the kinase and the substrate of interest in such a manner that the modified kinase recognizes, binds to, and phosphorylates only the modified substrate, while failing to recognize, bind to or phosphorylate any normal endogenous substrates. For this approach to be successful two criteria must be met. Firstly, it is important to eliminate the ability of the modified kinase to phosphorylate endogenous substrates. In the case of the nonreceptor tyrosine kinases, it was already known that mutants lacking the SH3 and SH2 substrate binding domains had low activity in vivo, although the catalytic domain was intact [33, 50]. Secondly, this approach requires that the modified substrate and kinase can be forced to interact. In the FIT strategy, interaction between two proteins of choice is mediated via an engineered, highly specific binding interface. This experimental strategy is depicted schematically in Fig. 11.1.

We first investigated whether single-chain antibodies and their peptide epitopes could function as the artificial binding interface between the Abl tyrosine kinase and potential substrates [13]. While the results were promising, we found that binding mediated by single-chain antibodies was rather weak in vivo, and their relative instability in the intracellular environment posed additional experimental problems. In a search for an alternate artificial binding interface, we next turned to "leucine zipper" coiled-coil domains [27]. These short amphipathic alpha helical segments are relatively small and can be selected for a propensity to heterodimerize as opposed to homodimerize. Specifically, we took advantage of the WinZipA and WinZipB (hereafter ZipA and ZipB) coiled-coil segments first selected by Plückthun and colleagues for efficient heterodimerization in vivo [1]. We expected that the addition of a small coiled-coil segment would only minimally disrupt the normal secondary and tertiary structure of the modified proteins. This is important for the FIT approach, as any significant disruption of the structures of the proteins could alter their biological activities, or prevent substrate phosphorylation due to steric hindrances.

The coiled-coil mediated interaction between modified Src and substrates was robust and specific, leading to a specific pairwise interaction as measured both by co-immunuprecipitation, and by the specific phosphorylation of the modified substrates [44]. We could demonstrate that, in cells expressing Src kinase lacking the SH3 and SH2 domains and bearing a ZipA coiled-coil segment, highly efficient and specific tyrosine phosphorylation of a substrate of choice was observed when that substrate was fused with the complementary ZipB segment. As expected, phosphorylation was dependent on the presence

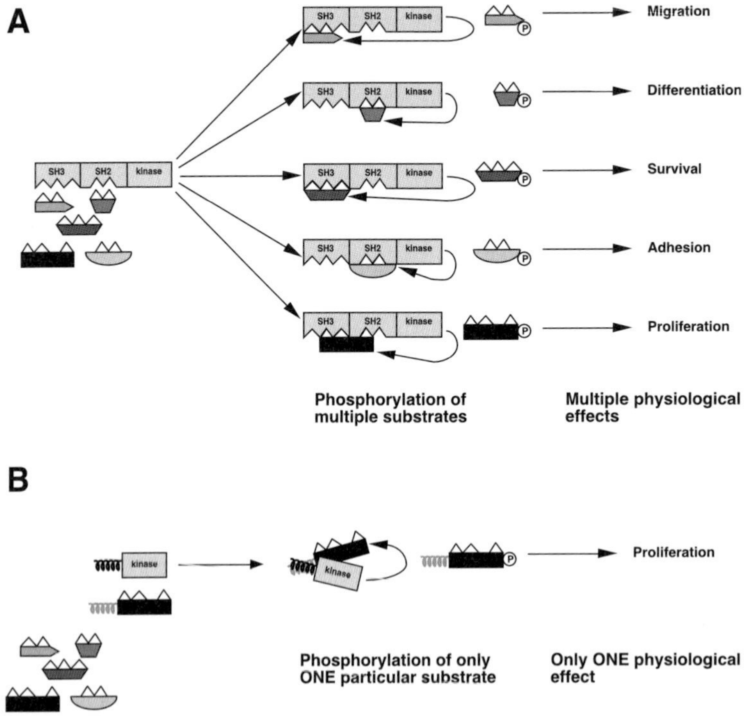

Fig. 11.1. Schematic representation of the FIT strategy to analyze functional consequences of tyrosine phosphorylation. **A** Upon kinase activation, the binding of a large number of substrates to the SH3 and SH2 domains of the kinase results in the phosphorylation of multiple substrates, each leading to a specific physiological output. **B** Replacing the SH3 and SH2 domains of the kinase with one partner of an artificial binding interface (such as coiled coils), which recognizes the second partner fused to the substrate, will enable a specific protein-protein interaction between the kinase and a single substrate, enabling specific phosphorylation of that particular substrate without phosphorylation of other endogenous substrates. Any biological output produced as a result can thus be attributed to the specific phosphorylation of that particular substrate

of ZipA in the kinase and ZipB in the substrate, and could be seen under conditions where no apparent increase in tyrosine phosphorylation of other proteins in the cell was observed.

11.4.1 Coiled-Coil Segments Can Act as a Specific Artificial Binding Interface Between the Abl Tyrosine Kinase and Substrates

We similarly designed FIT-compatible mutants of the Abl kinase and its substrates by fusing them to coiled-coil domains, and investigated whether we could demonstrate specific association of the modified proteins and phosphorylation of the modified substrate. Our strategy involved replacing the SH3 and SH2 substrate binding domains of Abl with a coiled-coil segment (Zip A), and adding the second, complementary coiled-coil segment (ZipB) to the substrate. Plasmids expressing the two modified proteins were then transfected together in low amounts into tissue culture cells. In these Abl constructs the location of the ZipA segment was different from that in Src constructs, where the ZipA segment was fused to the extreme amino or carboxyl terminus [44]. In this case the ZipA segment was precisely inserted into Abl in the position normally occupied by the SH3 and SH2 domains, thus the N-terminal myristoylation site and the C-terminal actin binding domains of Abl [48] were unaffected in the resulting chimeric protein (diagrammed in Fig. 11.2).

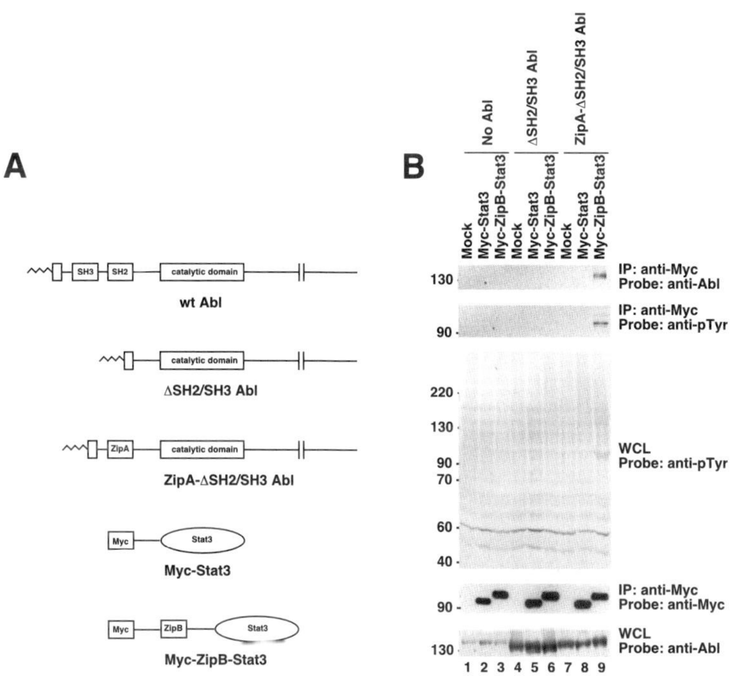

Fig. 11.2. Coiled-coil domains promote specific interaction between Abl kinase and Stat3. **A** The structures of Abl and Stat3 constructs are depicted diagrammatically. Approximate positions of myristoylation site (*wavy line*), SH3, SH2, and tyrosine kinase catalytic domains are indicated, as are ZipA and ZipB coiled coil segments and Myc epitope tag. **B** 293T cells were transfected as indicated and harvested 24 h later. Equal amounts of cell lysates were separated by SDS-polyacrylamide gel electrophoresis and transferred to membranes. Immunoblots were probed with antibodies as indicated on the *right*. *Anti-pTyr* Phosphotyrosine-specific monoclonal antibody. *WCL* Whole cell lysate. Approximate positions of molecular weight markers are depicted on the *left* (in kDa)

Figure 11.2 illustrates the typical results obtained using FIT to promote a pairwise interaction between modified Abl and a substrate, in this case the Stat3 transcription factor [51]. When plasmids encoding Abl lacking its SH3 and SH2 domains, or Abl lacking its SH3 and SH2 domains but fused to Zip A, were transfected along with plasmids encoding Stat3 fused to either the Myc tag alone or Myc tag with ZipB, co-immunoprecipitation of Abl and Myc-Stat3 was detected only when complementary coiled coils were present on both the kinase and the substrate (Fig. 11.2B, top panel, lane 9). This implies that Abl lacking its substrate binding domains cannot recognize and bind to endogenous substrates (shown by the failure of Stat3 or the modified ZipB-tagged Stat3 to co-immunoprecipitate with the Abl ΔSH3/SH2 mutant); however, the presence of ZipA on Abl enabled it to recognize and bind to Stat3 containing ZipB, while still failing to bind to endogenous substrates or Stat3 lacking ZipB. Thus, a pairwise interaction between a kinase and a substrate has been achieved through the coiled-coil domains.

More importantly, this coiled-coiled interaction is not only physical, but also has functional consequences: Abl fused to Zip A specifically phosphorylated tyrosine residues on Stat3 fused to ZipB, but not Stat3 lacking ZipB (Fig. 11.2B). It is important to note that the increased tyrosine phosphorylation of ZipB-Stat3 was observed in the apparent absence of tyrosine phosphorylation of other endogenous cellular proteins (Fig. 11.2B). The expression of Stat3, ZipB-Stat3 or Abl was not significantly different between the different transfected cells. Thus, the coiled-coiled interaction is specific, promoting a pairwise interaction between the modified kinase and substrate that leads to the tyrosine phosphorylation of the substrate. Similar results were obtained with other Abl substrates, including HS-1, Dok1, and $p130^{Cas}$ (data not shown).

In previous studies with Src, we also examined whether the position of the coiled-coil domain within the kinase affected the level of tyrosine phosphorylation of the substrate. When the ZipB coiled coil was positioned at the N-terminus of substrates such as $p130^{Cas}$, no difference was observed in FIT-induced phosphorylation by Src constructs containing the ZipA coiled coil either at the N- or C-terminus [44]. Thus, the addition of coiled coils to the kinase in either location does not provide any steric hindrance to the effective association between the kinase and the substrate. In the case of the Abl mutants described above, the ZipA segment is located internally, between the N-terminal unique domain and the catalytic domain of Abl. Thus, it appears that the FIT approach is quite flexible and robust, and that the ability of FIT-modified kinases to phosphorylate FIT-modified substrates is not highly sensitive to the location of the ZipA and ZipB segments in the proteins of interest.

11.4.2 Coiled-Coil Segments can Activate Physiological Downstream Signaling Events

Although coiled-coil mediated interaction can lead to specific tyrosine phosphorylation of substrates, this in itself may not mimic physiological phosphorylation of that substrate during normal signaling. For example, FIT-mediated phosphorylation might occur on inappropriate or irrelevant sites, and thus may not be sufficient to induce a physiological downstream response. In addition to phosphorylation of inappropriate sites, the FIT strategy might introduce other problems. For example, ZipA-ZipB binding is relatively strong (K_d=24 nM [1]), thus the possibility exists that the substrate may not be able to freely dissociate from the kinase to allow its relocalization or downstream interaction with other proteins. Alternatively, fusion of the substrate to coiled-coil segments could prevent its normal function or result in aberrant and/or constitutive activity. We therefore tested whether FIT-mediated phosphorylation of Stat3 led to the expected downstream response.

Under physiological conditions, phosphorylation of Stat3 by JAK family kinases leads to the formation of Stat3 homodimers, which translocate to the nucleus thereby inducing Stat-dependent gene transcription [8]. To test whether Stat3 homodimers can still form, translocate, and activate gene transcription, we used a Stat3-dependent transcriptional reporter system. When Stat3 with or without ZipB was transfected along with Abl lacking SH3 and SH2 domains with or without Zip A, together with a Stat3-dependent reporter construct driving luciferase expression (pGL2Basic [23]), luciferase activity was significantly increased only in the cells expressing Abl and Stat3 fused to the complementary coiled-coil segments (Fig. 11.3). This indicates that Stat3 molecules whose phosphorylation was mediated via coiled coils were still able to form phosphorylated Stat3 dimers and subsequently activate gene transcription. The failure of ZipA-modified Abl to stimulate transcription when Stat3 lacked ZipB demonstrates that the addition of coiled coils to both the kinase and the substrate was required. Furthermore, the dependence of the ability of ZipB-Stat3 to stimulate transcription in the presence of ZipA-Abl demonstrates that the addition of the coiled-coil segment to Stat3 neither hampered its ability to promote gene transcription, nor did it activate it constitutively. Thus FIT-mediated phosphorylation using coiled coils is compatible with downstream signaling and can lead to a physiological response. Similar results were observed when Src was used instead of Abl to induce Stat3 phosphorylation by FIT [44].

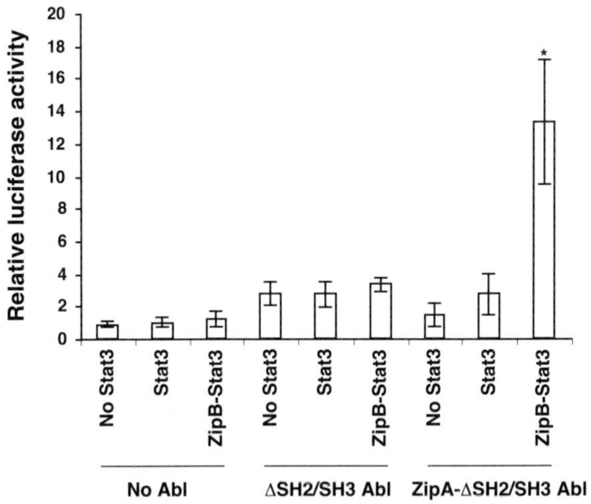

Fig. 11.3. Coiled-coil interaction between Abl kinase and Stat3 can promote gene transcription and hence downstream events. 293T cells were transfected with plasmids expressing Abl and Stat3 constructs as indicated, along with a Stat3-dependent reporter construct expressing luciferase, and harvested in reporter lysis buffer 24 h later. An aliquot of the lysate was mixed with luciferase assay substrate and the emitted light measured in a luminometer. Another aliquot of the lysate was used for protein estimation for normalization. Relative luciferase activity was calculated by normalizing the no Abl, no Stat3 group to a value of 1. Data expressed as mean ±SD of three observations and analyzed by two-way ANOVA. *$p<0.05$ as compared to respective control groups. The luciferase assay kit was obtained from Promega

11.4.3 Implications of FIT for Analysis of the Functional Consequences of Specific Tyrosine Phosphorylation

We have shown that the FIT approach provides a novel tool for probing the functional consequences of tyrosine phosphorylation. It is worth considering here some of implications and potential uses of this approach. The most straightforward application of the method, of course, will be in identifying and validating those substrates whose phosphorylation is sufficient to induce biological phenotypes of interest. For example, a more detailed and systematic study of the many known or suspected substrates of Src or Abl should identify those that play an important role in cellular processes such as cellular transformation. It is remarkable how little progress has been made on this front despite more than 20 years of intensive effort highlighting the potential usefulness of novel approaches like FIT. Such studies will be spurred on by the expectation that essential downstream targets of tyrosine kinases are likely to include novel targets for drug discovery, given the demonstrated importance of tyrosine kinases in human disease.

Dominant versus recessive effects: It should be pointed out that while it might be relatively easy to use FIT to study dominant effects such as transformation, recessive effects might be more difficult to attribute to the phosphorylation of a particular substrate. In other words, the presence of normal, wild-type kinases and substrates in the cell makes it problematic to use FIT to identify substrates involved in normal physiological roles. In the cases of Src and Abl, however, this drawback can be overcome by taking advantage of already available genetic knock-out cells lines. For example, SYF cells (mouse embryonic fibroblasts lacking the Src, Yes and Fyn tyrosine kinases, the only Src family kinases detectably expressed in fibroblasts) are known to have defects in cellular processes such as migration [19]. Expression in these cells of FIT-compatible Src and candidate substrates should permit the identification of those substrates whose phosphorylation is required for normal cell migration. Similarly, mouse fibroblasts that lack Abl and its close relative, Arg are available [20]. In principle, any recessive process could be studied by FIT, as long as cells that lacked one of the two candidate proteins (either the kinase or candidate substrate) were available. Of course, in yeast, the ability to easily replace endogenous genes by homologous recombination means that FIT could be applied to the study of recessive processes more easily.

Validating cellular effects previously attributed to phosphorylation of specific substrates: The phosphorylation of an enormous number of substrates has been implicated in a variety of cellular responses, and FIT can be used to validate the results obtained by more indirect methods. For example, Src-induced tyrosine phosphorylation of $p130^{Cas}$ and paxillin-α have been suggested to exert opposing effects on cell migration and contact inhibition of growth [53]. FIT could be used to confirm this model: varying the relative amounts of ZipB-tagged $p130^{Cas}$ and ZipB-tagged paxillin in cells expressing ZipA-tagged Src or Abl should result in a predictable, graded range of cell migration or contact inhibition responses. In another example, increased tyrosine phosphorylation of $p130^{Cas}$ in the FG-M human pancreatic carcinoma cell line has been correlated with increased migratory activity on vitronectin, whereas increased tyrosine phosphorylation of $p130^{Cas}$ is not seen in FG cells, which do not show enhanced cell migration on vitronectin [18]. If the tyrosine phosphorylation of $p130^{Cas}$ is directly responsible for the ability of FG-M cells to migrate on vitronectin, FIT-induced phosphorylation of $p130^{Cas}$ in FG cells should stimulate cell migration on vitronectin to the level seen in FG-M cells.

Addressing the role of individual phosphorylation and binding sites: Many substrates contain multiple tyrosine phosphorylation sites; for example, $p130^{Cas}$ contains more than ten such sites [3, 34]. The biological consequences of phosphorylation of each site may be different, serving as docking sites for different SH2 domain-containing partners, for example, or inducing specific allosteric changes. The individual roles of these phosphorylation sites can be addressed by FIT by using specific tyrosine→phenylalanine mutants, in which one or more sites are mutated while the remaining sites retain their ability to be phosphorylated by the kinase. At present, the functional in vivo role of individual substrate phosphorylation sites can only be addressed definitively by time-consuming genetic knock-in experiments.

Validating cellular signaling models: FIT can readily be used to test and validate critical aspects of signal transduction models which are currently based on indirect evidence. Only one of many possible examples is given here. Integrin signaling is believed to occur as follows: upon activation of integrin receptors, the FAK tyrosine kinase is recruited to the receptor along with Src, paxillin and p130^{Cas}, which become tyrosine phosphorylated and therefore bind to the Crk SH2 domain. Crk via its SH3 domains may then recruit proline-rich effector molecules such as C3G and Sos, which subsequently mediate Rap1 and MAPK activation [7, 11].However, it has not been firmly established whether FAK phosphorylation is sufficient to initiate this process. If this were the case, then FIT-induced tyrosine phosphorylation of a kinase-inactive FAK would be sufficient to induce focal adhesion complexes to form on the membrane and result in activation of the Rap1 and MAPK pathways, even in the absence of integrin receptor activation.

11.5 Broader Uses of the FIT Strategy

Functional consequences of other post-translational modifications: We have described the use of FIT to study the specific interaction between tyrosine kinases and their substrates, but the FIT strategy was originally envisioned as a more general means to induce the specific, pairwise association of signaling proteins [13]. Given our success with tyrosine kinases, it is likely that FIT could also be used to study the functional consequences of post-translational modification by other enzymes with relatively weak catalytic activity, especially those with distinct substrate binding domains. These would include, but would not be limited to, serine/threonine kinases such as MAP kinases and cyclin-dependent kinases, proline isomerases, and enzymes involved in histone acetylation or methylation. In each of these cases, FIT has the potential to generate a wealth of information about the functional contribution of specific substrates to physiological responses of interest. The only requirement for this approach is the availability of mutants of the enzymes of interest, which are catalytically active, yet functionally inert in vivo because their interaction with normal substrates is compromised.

Analysis of downstream events dependent on protein-protein interactions: Many signaling proteins contain modular protein interaction domains such as the SH3, SH2, PTB, WW, PDZ, and EVH1 domains. These domains function to mediate the assembly of multiprotein complexes essential for the transmission of signals [35, 36]. Although these domains generally have considerable binding specificity, that specificity is by no means absolute. From an experimental standpoint, this means that many candidate proteins are likely to bind to a protein of interest with similar affinity, making it difficult or impossible to tease out which of these binding partners is important for downstream biological effects. Using the FIT approach, it will be possible to modify a protein of interest by inactivating or deleting a protein interaction domain, and then testing the effects of pairwise interaction of the modified protein with a panel of candidate interaction partners. Thus the functional consequences of each specific interaction can be elucidated, and those responsible for a biological output identified.

Physiological responses elicited by cell-surface complexes regulated by dimerization or oligomerization: Receptor tyrosine kinases such as ephrin receptors are known to homo- and heterodimerize, with specific dimers eliciting specific downstream physiological effects [6, 21]. FIT could be used to experimentally induce the formation of specific homo- and heterodimers of these receptors in vivo. Similarly, G-protein coupled receptors such as the dopamine receptors are believed to undergo homo- and heterodimerization resulting in various permutations and combinations, each of which may induce distinct downstream signals [12, 25, 41, 43]. FIT has the potential to promote the formation of specific combinations of receptors and thus this aspect of receptor dimerization can also be investigated in detail. Ion channel receptors such as the nicotinic acetylcholine receptors, $GABA_A$ receptors and NMDA receptors are composed of subunits, and the role of each of these and their various combinations could also be studied by forcing specific combinations to associate in vivo via FIT. Recently, it has also been shown that G-protein coupled receptors can directly interact with ion channels via direct protein-protein interactions [24, 26]. As these and other interactions are uncovered, FIT could be a valuable tool to study and validate their functional importance.

Screening libraries for protein-protein interactions important for physiological responses: One of the most powerful potential uses of the FIT methodology will be to screen large libraries of candidate partners to identify those whose interaction with or modification by a target protein leads to a biological activity of interest. With the advent of efficient recombination-based subcloning methods and the deciphering of the complete human genome, it is now possible to create representative libraries of full-length cDNAs as fusion proteins [4]. Thus, for example a retroviral expression library of ~30,000 human cDNA clones could be constructed as ZipB fusion proteins, representing virtually the entire protein coding capacity of the human genome. This library could be introduced into cells expressing a target protein, for example, a tyrosine kinase, fused to ZipA. Thus in each infected cell, the target protein will be forced to interact with a single ZipB-tagged protein. By screening several tens of thousands of such infected cells, each expressing a different ZipB fusion, it will be possible to essentially identify all proteins in the proteome whose interaction with the target protein leads to a biological activity of interest. High-throughput cell-based screening methods could easily be devised to identify interaction partners that result in alterations in morphology, motility, proliferation, and a host of other biological properties. Such an approach will highlight only those proteins whose interaction with or modification by the target protein leads to *functionally important* biological outputs. One of the strengths of such a screen is that it does not presuppose any knowledge about the identity of potential partners before the actual experiment is performed.

11.6 Advantages and Disadvantages of FIT

As we have described, the advantages of using FIT as a means to study the physiological responses mediated upon protein phosphorylation are many. A specific pairwise interaction is achieved and thus the responses produced can be confidently attributed to the phosphorylation of the particular substrate of interest by the specific kinase. Appropriate controls can easily be performed, including those in which either the kinase or the substrate lacks the coiled-coil interaction domain, to control for any biological outputs that are not directly due to the interaction between the tagged kinase and tagged substrate. The kinase and the substrate can be expressed in very low amounts, thus ensuring minimal disturbance of the cellular machinery and thus the problem of compensatory responses and responses due to overexpression of a particular protein are avoided or kept at a minimum. And of course the FIT approach is not as labor-intensive as alternatives such as genetic knock-out and knock-in approaches, and is capable of providing complementary information.

In addition, the FIT approach allows the possibility of multiplexing. The complexity and richness of an FIT experiment can be increased by simultaneously introducing multiple ZipA- or ZipB-tagged interaction partners, thus allowing multiple interactions to occur in the same cell. For example, in the case of tyrosine phosphorylation, the ability to induce the phosphorylation of multiple substrates of choice by FIT will allow us to study responses where phosphorylation of a single substrate might be insufficient to elicit a physiological response. One can imagine an analysis using a variety of combinations of known or suspected substrates of Src to decipher which *combinations*, when phosphorylated, are sufficient to induce various parameters of malignant transformation such as proliferation in low serum, anchorage-independent growth, and so on.

One potential disadvantage is that in the present state, FIT can only be used to study long-term responses to constitutive interaction, as opposed to immediate downstream events. Many phosphorylation events, for example, lead to rapid responses (within seconds), followed by densensitization and/or feedback inhibition to attenuate the signal. FIT depends on the introduction of expression constructs, and the synthesis and accumulation of Zip-tagged proteins, a relatively slow process. To address this, efforts are underway to establish an inducible FIT system, where association between a kinase and a substrate can be induced rapidly in the cell. Various systems have been described in which the heterodimerization of two proteins can be rapidly induced by cell-permeable small-molecule chemical inducers of dimerization [2, 14], but we have not yet demonstrated that such systems are compatible with FIT-induced substrate phosphorylation. When available, such an inducible FIT system would provide us with a tool to study the time-dependence of specific responses to substrate phosphorylation. For example, the duration of the activation of the ERK family of MAP kinases after stimulation seems to be a critical factor in determining the physiological response elicited in the cell [30, 42].

One aspect of the FIT strategy that should be considered is that positive results obtained by this approach will be much more informative than negative results. For example, if it is observed that the specific phosphorylation of cortactin leads to increased cell migration in SYF cells, it can be inferred that tyrosine phosphorylation of cortactin is important for cell migration. If, on the other hand, specific phosphorylation of $p130^{Cas}$ does not lead to increased cell migration, it cannot be concluded that its phosphorylation does not play a role; it is entirely possible that the specific phosphorylation of $p130^{Cas}$ alone may be insufficient, and that the concomitant phosphorylation of other substrates such as paxillin or scaffolding proteins such as FAK may be required. This of course could be addressed directly by co-transfecting multiple ZipB-tagged substrates. However, a more general concern is the possibility that fusion of a particular substrate to the coiled-coil domain may compromise its function in some way, or that FIT may not promote phosphorylation of the relevant sites due to steric effects or other reasons. In short, there are many potential reasons for false negative results with FIT. On the other hand, from a practical standpoint, false negatives are far less troublesome than false positives, each of which might take considerable time and resources to track down.

11.7 Concluding Remarks

In conclusion, the FIT approach can be used to promote a specific pairwise interaction between two proteins of interest, leading to physiological responses (such as phosphorylation) which are compatible with downstream signaling (such as activation of gene transcription). The determination of which protein-protein interactions and post-translational modifications are important for mediating normal and abnormal physiological responses might lead to identification and validation of new targets for treatment of human disorders resulting from disturbances of signaling cascades. The flood of proteomic information that is now being generated regarding protein-protein interactions and post-translational modifications will only be useful insofar as we can validate its functional significance in the cell; FIT is a powerful new tool with which to begin this all-important validation process. Future improvements and refinements to this approach, including new artificial interaction interfaces with a range of affinities in vivo as well as inducible interaction interfaces, will allow this strategy to be applied to an even wider range of biological questions.

References

1. Arndt KM, Pelletier JN, Müller KM, Alber T, Michnick SW, Plückthun A (2000) A heterodimeric coiled-coil peptide pair selected in vivo from a designed library-versus-library ensemble. J. Mol. Biol. 295, 627–639
2. Belshaw PJ, Ho SN, Crabtree GR, Schreiber SL (1996) Controlling protein association and subcellular localization with a synthetic ligand that induces heterodimerization of proteins. Proc. Nat. Acad. Sci. USA 93, 4604–4607
3. Bouton AH, Riggins RB, Bruce-Staskal PJ (2001) Functions of the adapter protein Cas: signal convergence and the determination of cellular responses. Oncogene 20, 6448–6458
4. Brizuela L, Braun P, LaBaer J (2001) FLEXGene repository: from sequenced genomes to gene repositories for high-throughput functional biology and proteomics. Mol. Biochem. Parasitol. 118, 155–65
5. Brown JL, Stowers L, Baer M, Trejo J, Coughlin S, Chant J (1996) Human Ste20 homologue hPAK1 links GTPases to the JNK MAP kinase pathway. Curr. Biol. 6, 598–605
6. Bruckner K, Klein R (1998) Signaling by Eph receptors and their ephrin ligands. Curr Opin Neurobiol 8, 375–382
7. Buday L (1999) Membrane-targeting of signaling molecules by SH2/SH3 domain-containing adaptor proteins. Biochim Biophys Acta 1422, 187–204
8. Chen X, Vinkemeier U, Zhao Y, Jeruzalmi D, Darnell JEJ, Kuriyan J (1998) Crystal structure of a tyrosine phosphorylated STAT-1 dimer bound to DNA. Cell 93, 827–839
9. Druker BJ (2002) Inhibition of the Bcr-Abl tyrosine kinase as a therapeutic strategy for CML. Oncogene 21, 8541–8546
10. Druker BJ, Talpaz M, Resta DJ, Peng B, Buchdunger E, Ford JM, Lydon NB, Kantarjian H, Capdeville R, Ohno-Jones S, Sawyers CL (2001) Efficacy and safety of a specific inhibitor of the BCR-ABL tyrosine kinase in chronic myeloid leukemia. N. Engl. J. Med. 344, 1084–1086
11. Feller SM (2001) Crk family adaptors-signalling complex formation and biological roles. Oncogene 20, 6348–6371
12. Franco R, Ferre S, Agnati L, Torvinen M, Gines S, Hillion J, Casado V, Lledo P, Zoli M, Lluis C, Fuxe K (2000) Evidence for adenosine/dopamine receptor interactions: indications for heteromerization. Neuropsychopharmacology 23(4 Suppl), S50–59
13. Fujiwara K, Poikonen K, Aleman L, Valtavaara M, Saksela K, Mayer BJ (2002) A single-chain antibody / epitope system for functional analysis of protein-protein interactions. Biochemistry 41, 12729–12738
14. Ho SN, Biggar SR, Spencer DM, Schreiber SL, Crabtree GR (1996) Dimeric ligands define a role for transcriptional activation domains in reinitiation. Nature 382, 822–826

15. Hochhaus A, Kreil S, Corbin AS, La Rosee P, Muller MC, Lahaye T, Hanfstein B, Schoch C, Cross NC, Berger U, Gschaidmeier H, Druker BJ, Hehlmann R (2002) Molecular and chromosomal mechanisms of resistance to imatinib (STI571) therapy. Leukemia 16, 2190–2196
16. Hunter T (2000) Signaling-2000 and beyond. Cell 100, 113–127
17. Hunter T, Sefton BM (1980) Transforming gene product of Rous sarcoma virus phosphorylates tyrosine. Proc Natl Acad Sci, USA 77, 1311–1315
18. Klemke RL, Leng J, Molander R, Brooks PC, Vuori K, Cheresh DA (1998) CAS/Crk coupling serves as a "molecular switch" for induction of cell migration. J. Cell Biol. 140, 961–972
19. Klinghoffer RA, Sachsenmaier C, Cooper JA, Soriano P (1999) Src family kinases are required for integrin but not PDGFR signal transduction. EMBO J 18, 2459–2471
20. Koleske AJ, Gifford AM, Scott ML, Nee M, Bronson RT, Miczek KA, Baltimore D (1998) Essential roles for the Abl and Arg tyrosine kinases in neurulation. Neuron 21, 1259–1272
21. Kullander K, Klein, R. (2002) Mechanisms and functions of Eph and ephrin signalling. Nat Rev Mol Cell Biol 3, 475–486
22. Kuriyan J, Cowburn D (1997) Modular peptide recognition domains in eukaryotic signaling. Annu. Rev. Biophys. Biomol. Struct. 26, 259–288
23. Leaman DW, Pisharody S, Flickinger TW, Commane MA, Schlessinger J, Kerr IM, Levy DE, Stark GR (1996) Roles of JAKs in activation of STATs and stimulation of c-*fos* gene expression by epidermal growth factor. Mol. Cell. Biol. 16, 369–375
24. Lee FJS, Xue S, Pei L, Vukusic B, Chery N, Wang Y, Wang YT, Niznik HB, Liu F (2002) Dual regulation of NMDA receptor functions by direct protein-protein interactions with the dopamine D1 receptor. Cell 111, 219–230
25. Lee SP, O'Dowd BF, Ng GY, Varghese G, Akil H, Mansour A, Nguyen T, George SR (2000) Inhibition of cell surface expression by mutant receptors demonstrates that D2 dopamine receptors exist as oligomers in the cell. Mol Pharmacol 58, 120–128
26. Liu F, Wan Q, Pristupa ZB, Yu XM, Wang YT, Niznik HB (2000) Direct protein-protein coupling enables cross-talk between dopamine D5 and gamma-aminobutyric acid A receptors. Nature 403, 274–280
27. Lupas A (1996) Coiled coils: new structures and new functions. Trends Biochem Sci 21, 375–382
28. Manning BD, Cantley LC (2002) Hitting the target: emerging technologies in the search for kinase substrates. Sci STKE 2002(162), PE49
29. Manning G, Whyte DB, Martinez R, Hunter T, Sudarsanam S (2002) The protein kinase complement of the human genome. Science 298, 1912–1934
30. Marshall CJ (1995) Specificity of receptor tyrosine kinase signaling: transient versus sustained extracellular signal-regulated kinase activation. Cell 80, 179–185
31. Mayer BJ (2001) SH3 domains: complexity in moderation. J. Cell Sci. 114, 1253–1263
32. Mayer BJ, Hirai H, Sakai R (1995) Evidence that SH2 domains promote processive phosphorylation by protein-tyrosine kinases. Curr. Biol. 5, 296–305
33. Mayer BJ, Jackson PK, Van Etten RA, Baltimore D (1992) Point mutations in the *abl* SH2 domain coordinately impair phosphotyrosine binding in vitro and transforming activity in vivo. Mol. Cell. Biol. 12, 609–618
34. Panetti TS (2002) Tyrosine phosphorylation of paxillin, FAK, and p130Cas: effects on cell spreading and migration. Front. Biosci. 7, d143–150
35. Pawson T, Nash P (2000) Protein-protein interactions define specificity in signal transduction. Genes Dev. 14, 1027–1047
36. Pawson T, Scott JD (1997) Signaling through scaffold, anchoring, and adaptor proteins. Science 278, 2075–2080

37. Pellicena P, Miller WT (2002) Coupling kinase activation to substrate recognition in Src-family tyrosine kinases. Front. Biosci. 7, d256–267
38. Pellicena P, Miller WT (2001) Processive phosphorylation of p130Cas by Src depends on SH3-polyproline interactions. J. Biol. Chem. 276, 28190–28196
39. Ponticelli AS, Whitlock CA, Rosenberg N, Witte ON (1982) In vivo tyrosine phosphorylations of the abelson virus transforming protein are absent in its normal cellular homolog. Cell 29, 953–960
40. Reddy EP, Smith MJ, Srinivasan A (1983) Nucleotide sequence of Abelson murine leukemia virus genome: Structural similarity of its transforming gene product to other *onc* gene products with tyrosine-specific kinase activity. Proc. Natl. Acad. Sci. USA 80, 3623–3627
41. Rocheville M, Lange DC, Kumar U, Patel SC, Patel RC, Patel YC (2000) Receptors for dopamine and somatostatin: formation of hetero-oligomers with enhanced functional activity. Science 288, 154–157
42. Roovers K, Assoian RK (2000) Integrating the MAP kinase signal into the G1 phase cell cycle machinery. Bioessays 22, 818–826
43. Scarselli M, Novi F, Schallmach E, Lin R, Baragli A, Colzi A, Griffon N, Corsini GU, Sokoloff P, Levenson R, Vogel Z, Maggio R (2001) D2/D3 dopamine receptor heterodimers exhibit unique functional properties. J Biol Chem 276, 30308–30314
44. Sharma A, Antoku S, Fujiwara K, Mayer BJ (2003) Functional interaction trap: A strategy for validating the functional consequences of tyrosine phosphorylation of specific substrates in vivo. Mol Cell Proteomics (in press), published online on 29th September, 2003
45. Sefton BM, Hunter T, Beemon K, and Eckhart W (1980) Evidence that the phosphorylation of tyrosine is essential for cellular transformation by Rous sarcoma virus. Cell 20, 807–816
46. Sefton BM, Hunter, T., and Beemon, K. (1980) Relationship of polypeptide products of the transforming gene of Rous sarcoma virus and the homologous gene of vertebrates. Proc Natl Acad Sci, USA 77, 2059–2063
47. Shokat KM (1995) Tyrosine kinases: modular signaling enzymes with tunable specificities. Chem. Biol. 2, 509–514
48. Smith JM, Mayer BJ (2002) Abl: mechanisms of regulation and activation. Front. Biosci. 7, d31–42
49. Thomas SM, Brugge J (1997) Cellular functions regulated by Src family kinases. Ann. Rev. Cell Dev. Biol. 13, 513–609
50. Tian M, Martin GS (1997) The role of the Src homology domains in morphological transformation by v-src. Mol. Biol. Cell 8, 1183–1193
51. Turkson J, Jove R (2000) STAT proteins: novel molecular targets for cancer drug discovery. Oncogene 19, 6613–6626
52. Wang JYJ, Baltimore D (1983) Cellular RNA homologous to the abelson murine leukemia virus transforming gene: expression and relationship to the viral sequence. Mol. Cell. Biol. 3, 773–779
53. Yano H, Uchida H, Iwasaki T, Mukai M, Akedo H, Nakamura K, Hashimoto S, Sabe H (2000) Paxillin alpha and Crk-associated substrate exert opposing effects on cell migration and contact inhibition of growth through tyrosine phosphorylation. Proc. Natl. Acad. Sci., USA 97, 9076–9081

12 Analysis of Protein–Protein Interactions in Complex Biological Samples by MALDI TOF MS. Feasibility and Use of the Intensity-Fading (IF-) Approach

JOSEP VILLANUEVA, OSCAR YANES, ENRIQUE QUEROL, LUIS SERRANO and FRANCESC X. AVILÉS

12.1 Introduction

12.1.1 Mass Spectrometry as a Modern Approach to Study Protein–Protein and Protein–Ligand Interactions

Non-covalent macromolecular interactions is nowadays one of the most challenging fields of postgenomic biology. Genome projects have revealed that the true complexity of cellular biology exists at the level of proteins rather than at the level of genes (Pandey and Mann 2000). From the raw genetic sequences many important properties of proteins, such as three-dimensional structure, function, cellular or tissular localization or expression levels, cannot be predicted. These properties are been actively studied nowadays by proteomics approaches. Other approaches are addressing the analysis of protein-protein and protein-ligand interactions, and the derivation of protein networks (Gavin et al., 2002). The understanding of how proteins interact with each other, and with their main ligands, is an issue of fundamental interest for the comprehension of cellular functions (Alberts, 1998), as well as for the efficient application of most biomedical or biotechnological strategies based on them.

Several experimental methodologies have been used to study non-covalent interactions involving proteins. Among them, optical spectroscopy (UV absorption, fluorescence, circular dichroism), nuclear magnetic resonance (NMR), light scattering, ultracentrifugation, titration and differential scanning calorimetry, Biacore and, in the last few years, mass spectrometry (Hensley and Myszka 2000). Mass spectrometry (MS) has recently emerged as an effective tool for the study of macromolecules, due to its high sensitivity and accuracy, fast analysis and low sample consumption (Mann et al. 2001). The introduction of two soft ionization methods, electrospray ionization (ESI)

Principles and Practice
Methods in Proteome and Protein Analysis
R.M. Kamp, J. J. Calvete, T. Choli-Papadopoulou (Eds.)

and matrix-assisted laser desorption/ionization (MALDI), has led to a revolution in the application of mass spectrometry to large biomolecules, including the analysis of their interactions (Mann et al. 2001). In ESI the liquid sample is sprayed using a high voltage, forming droplets that, after evaporation of the solvent, produce multiply charged ions (Fenn et al. 1989; Fenn et al. 1990). MALDI is based on the co-crystallization of the sample with an excess of a chemical matrix, generally on a metal surface. By means of short laser pulsed irradiation, the matrix sublimates carrying the sample molecules to the gas phase, mainly as monocharged ions (Hillenkamp et al. 1991; Knochenmus et al. 2003; Karas et al. 2003). MALDI ionization has been frequently used coupled to time-of-the flight analytical procedures (MALDI-TOF) for the separation and mass measurements of biomolecules (Mann et al. 2001).

12.1.2 Characterization of Non-Covalent Interactions Using ESI

Up to now, due to the softness of this ionization procedure, ESI has been the most common mass spectrometric approach chosen to study non-covalent interactions. Some of the complexes detected with ESI-MS involve protein-small ligand, protein-DNA and protein-protein non-covalent interactions (Ogorzalek et al. 1993; Veenstra et al. 1997; Loo 1997). Furthermore, the evolution of ESI into nanoflow-ESI, allowing lower sample consumption and more native-like ionization, together with the coupling of ESI with time of flight (tof) analyzers, has made ESI an excellent method used to study intact large non-covalent complexes (Verentchikov et al. 1994; Wilm and Mann 1996). Using this approach, macromolecular complexes of more than 1MDa have been detected, stoichiometry of protein complexes has been derived and the assembly of multimeric proteins has been studied in real time (Rostom et al. 2000; Sobott et al. 2002).

12.1.3 Characterization of Non-Covalent Interactions Using MALDI

The study of non-covalent interactions by MALDI MS can be divided in two strategies (Fig. 12.1).

12.1.3.1 MALDI-Based Indirect Methods

When applied to the study of non-covalent interactions, MALDI MS has been generally used as an indirect method, in approaches consisting of different methodologies coupled to MS detection. Thus, MALDI MS coupled to hydrogen exchange determines the number and location of hydrogens involved in an interaction by using different hydrogen isotopes (Mandell et al. 1998). Limited proteolysis followed by MALDI MS has been applied to study DNA-pro-

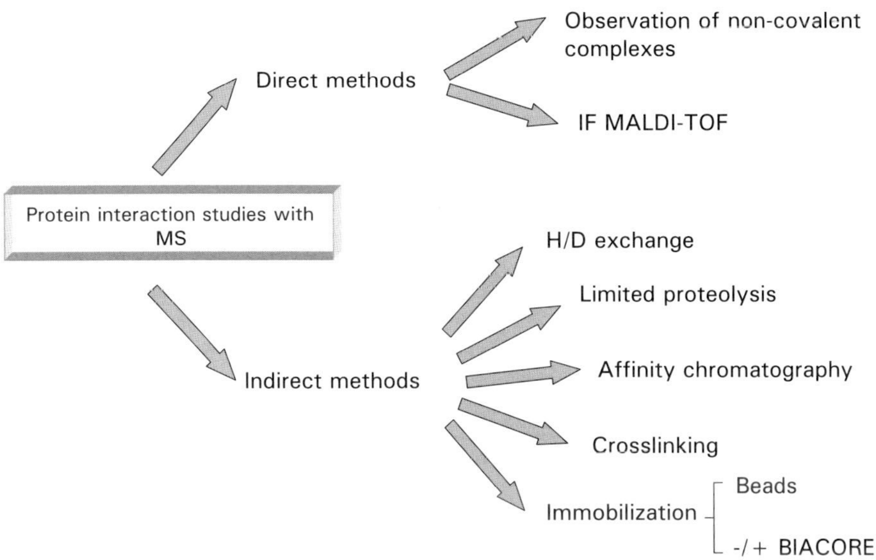

Fig. 12.1. Schematic representation of the different strategies for the study of protein interactions by mass spectrometry (*MS*)

tein and protein-protein complexes, allowing the identification of regions protected against degradation (Cohen et al. 1995; Kriwacki et al. 2000). Chemical cross-linking analyzed by MALDI-TOF has been used to determine the spatial organization of multicomponent complexes and stoichiometry of protein oligomers (Rappsilber et al. 2000; Bennett et al. 2000). Immobilization of proteins on magnetic beads and on Biacore chips has also been applied to detect complexes and determine affinity constants between biomolecules (Nelson et al. 1999). Also, affinity-chromatography followed by MALDI-TOF mass spectrometry takes advantage of a selection step previous to the identification of the interacting molecules (Rigaut et al. 1999; Gavin et al. 2002).

12.1.3.2 MALDI-Based Direct Methods

MALDI-TOF mass spectrometry has been rarely used as a direct method to study non-covalent interactions. Several steps of the MALDI process, such as sample preparation and laser desorption, generated a concern about its capabilities to detect non-covalent complexes. MALDI sample preparation is generally carried out by co-crystallization of the analytes with a molar excess of an organic compound, called matrix, which facilitates energy absorption (from a laser light) and analyte ionization. This matrix is prepared in the presence of an organic solvent at an acidic pH, which may give rise to a denaturing environment for biomolecules. Despite this logical concern, a growing number of reports have demonstrated that MALDI, in certain conditions, can be used to detect non-covalent complexes (Farmer and Caprioli 1998). These reports include the determination of stoichiometry of complexes, the study of enzyme-substrate complexes and the characterization of leucine zippers dimers (Woods et al. 1995; Rosinke et al. 1995; Glocker et al. 1996).

Different alternatives have been adopted to overcome the above-mentioned MALDI limitations. Milder conditions for sample preparation, raising the pH and ionic strength of the matrix, has been shown to be useful to maintain non-covalent complexes in the gas phase (Woods et al. 1995). Also, the so-called "first shot" method, based in the use of spectral signals generated by the first laser shots on parts of the sample not previously irradiated (Rosinke et al. 1995), has been reported to be more respectful for the integrity of non-covalent complexes. However, in general, the approaches currently used do not prevent the spectral signals of the complexes having a low intensity and low reproducibility compared to those of the individual components.

12.1.3.3 The Intensity-Fading (IF) MALDI-T of Approach

In this work we describe a new direct method to detect the formation of non-covalent complexes involving any kind of biomolecules, in particular proteins, in solution, by MALDI-TOF MS. This method, called Intensity Fading MALDI-TOF MS (IF MALDI-TOF MS), allows the detection of non-covalent complexes through the decrease (fading) of the molecular-ion intensities of the partners as directly compared to the MALDI mass spectrum of the mixture (problem and control molecules) following the addition of the target molecule. To show the capability and feasibility of this method, examples of model interactions, involving enzymes and small protein inhibitors with different affinity for them have been tested. Furthermore, the approach has been applied to highly complex samples consisting of extracts from invertebrate organisms. Following this, several high affinity protein ligands for trypsin-like serine proteases (SPs) and metallo-carboxypeptidases (CPs) from a sea anemone and from a leech have been identified, in a fast and simple way.

The present work will mainly focus on the description of this procedure and of the results obtained with such model systems.

12.2 Experimental Procedures

12.2.1 Biomolecule Interaction Experiments

12.2.1.1 General Sample Preparation

Lyophilized samples of PCI and Δ3PCI were dissolved in 10 mM Tris-HCl (pH 7.5) at a concentration of 16 μM. Non-binding control (Insulin) were dissolved in 10 mM Tris-HCl (pH 7.5) at the concentration which gave similar

MALDI MS ion intensities to the ligands assayed in parallel (similar concentration). Proteases were diluted in deionized water at 10 molar excess concentrations with respect to the counterpart inhibitors. Both the *Stichodactyla helianthus* and the *Hirudo medicinalis* extracts were dissolved in deionized water at a concentration of 20 mg/ml. These solutions were centrifuged at 8,000 *g* for 10 min, the supernatants had been processed by a reverse-phase resin-based protocol to clean and concentrate peptides and small proteins (Villanueva et al. 2001). The reverse-phase resin-bound molecules were eluted with 50 % isopropanol, lyophilized and dissolved in 10 mM Tris-HCl (pH 7.5).

12.2.1.2 Protease–Inhibitor Interaction

Incubation of 0.5 μl of each protease inhibitor (PCI and PCI-desCT), 0.5 μl of the non-binding control and 0.5 μl of 10 mM Tris-HCl (pH 7.5) (control sample), or 0.5 μl of protease, was carried out for 3 min at room temperature. In the case of the *Stichodactyla helianthus* and *Hirudo medicinalis* samples, 1 μl of the dissolved final lyophilized solutions (see above) was mixed with 0.5 μl of 10 mM Tris-HCl (pH 7.5) (control sample) or with 0.5 μl of carboxypeptidase A or trypsin and incubated for 3 min at room temperature. In the case of samples labile to proteolysis, it had been found advantageous to use proteases pre-treated by chemical methods to eliminate their degradative power, whilst keeping their affinity for protein substrates (i.e. anhydrotrypsin).

12.2.2 MALDI-TOF Mass Spectrometry

All mass spectra were acquired on a Bruker Biflex TOF mass spectrometer equipped with a nitrogen laser with an emission wavelength of 337 nm. The spectra were obtained in the linear mode at an accelerating voltage of 19 KV. Deflection of the low-mass ions was used to enhance the target protein signal. Spectra were acquired by averaging 50 to 100 shots.

12.2.2.1 Preparation of Samples for MALDI-TOF Mass Spectrometry

Sample preparation is critical for the success of the methodology, as in other MALDI TOF MS based approaches (Kussmann et al. 1997). An excess of one component in a mixture may suppress the ion signals of the other components. In addition, impurities in a complex mixture may interfere with analyte detection and reduce sensitivity. Similarly, the preparation of the matrix is essential: a MALDI matrix preparation close to neutral pH is frequently very important in obtaining satisfactory results; the MALDI matrices used in this work had a pH of 5.5 to 6 (see below). Experiments performed using a matrix containing 0.1 % trifluoroacetic acid perturbed in many cases the complexes that we studied here. We also noticed that the concentration of organic solvent is essential to maintain the integrity of our complexes: the MALDI matrices used in this work contained 30 % of acetonitrile (which diluted to 22.5 % after mixing with the aliquots to be analyzed).

12.2.2.2 MALDI-TOF Matrix Preparation

After a series of trials, a 30/70 % (v/v) acetonitrile/deionized water solution at pH 5.5–6 was the solvent composition for the matrices selected to perform all of the experiments reported in this work. This allowed keeping the organic solvent concentration low enough when mixed with the biological sample, at 1/3 sample/matrix (v/v) ratio (see below).

The matrix solution contained 10 mg/ml of sinapic acid or 10 mg/ml of a-cyano-4-hydroxycinnamic acid (HCCA) in the above described solvent. The spectral signals of entire complexes (data not shown) were detected by addition of 1 M amonium citrate to the matrix preparation (30/70 %, v/v, acetonitrile/1 M ammonium citrate solution, at pH 6) (Woods et al. 1995).

12.2.2.3 Sample-Matrix Preparation

Aliquots from the protein-ligand interaction model assay, from the assays to detect the entire complexes and from the biological extracts (*Stichodactyla helianthus* and *Hirudo medicinalis*) were successfully studied by using a modification of a reported sample preparation method derived from the standard dried-droplet method (Hillenkamp and Karas 1988; Kusmann et al. 1997). According to it, 0.5 μl of the matrix is deposited onto the target and allowed to dry. After this, 0.5 μ1 of the sample/matrix (1/3, v/v) mixture is added to the preformed matrix layer. Shortly after *crystallization* starts (1–2 min, at room temperature), 2 μl of cold deionized water is added onto the sample/matrix mixture, giving rise to a droplet. After 5–10 s the droplet is pipetted off, 0.5 μl of matrix is added and it is allowed to dry. In our hands, the mixture sample/matrix 1/3 (v/v) gave the best results, in terms of statistics, to analyze the complex mixtures described in this work.

12.3 Results and Discussion

12.3.1 Basis for the Detection of Non-Covalent Complexes by MALDI-TOF MS

The method described here is based on the analysis of the MALDI-TOF MS ion intensities to detect the occurrence of complexes between proteins and biomolecules among model interactions and complex biological mixtures (Fig. 12.2). The detection is based on a statistically significant *fading* of the relative intensities of one or more molecules in a control sample, after the addition of the target molecule, which can be firmly attributed to a non-covalent complex formation between these molecules. The remaining relative ion

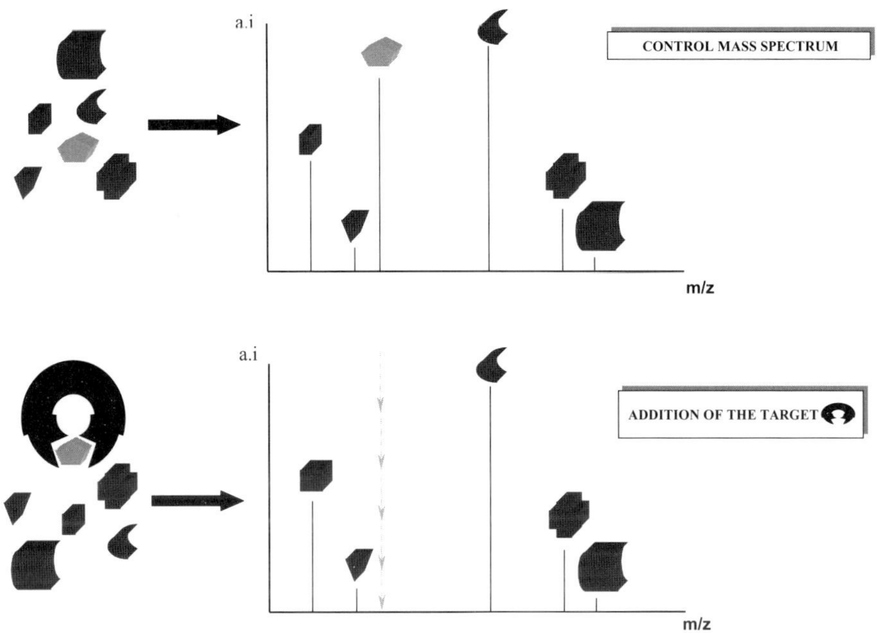

Fig. 12.2. Scheme of the basis for the detection of non-covalent complexes by intensity fading MALDI-TOF MS

signal/s (or non binding molecules) are not statistically affected by the presence of the target. The control molecules selected have no affinity (experimentally demonstrated by other methods referred to in the bibliography, or by previous trials) to the target, and as a consequence of this, their signal intensities are not affected by target addition. The results from analyses of all the ionized species is based on averaged measures (percentages) and the standard deviation of the relative intensities of all the ionized species represented in the MALDI-TOF mass spectrum, with respect to the control, before and after the addition of the target molecule by means of ten independent experiments. The information is represented graphically to examine whether the relative intensities of some of the ionic species in the sample are statistically significant or, on the contrary, it is a method error effect.

12.3.2 Suggested Mechanism for "Intensity Fading" (IF-) in MALDI-MS

Our rationale for developing and promoting the use of *intensity fading* (IF-) in MALDI-TOF MS is that complexes are more difficult to ionize than the individual molecules. It seems that the matrix used does not perturb the stability of the complexes (when acid free), although these ones behave as poorly ionizable species. This hypothesis relies on the fading of the ion signals of the interacting partners after the addition of their targets and on the fact that, very frequently, faint and broad signals from the complexes are detected at the same time, at spectral regions corresponding to larger masses.

There are some reports detecting non-covalent complexes by MALDI-TOF, most of them based on direct visualization of entire complexes by means of modifications of MALDI-TOF sample preparation, particularly by salt addition to the matrix (Woods et al. 1995). In the IF MALDI-TOF approach that we propose, such a procedure of adding salt (i.e. ammonium citrate 1 M) has been discarded because one of our main goals is to develop a screening method capable of processing a high number of possible ligands and complex biological samples. The ionization (and spectral visualization) of non-covalent complexes with saline matrices usually requires trials at different saline conditions for each case and, besides, generally gives rise to a decline in resolution, sensitivity and reproducibility, limiting the potentiality of the approach as a massive screening method.

12.3.3 Semiquantitative Determination of the Affinities Between the Interacting Partners

Given the well-known high sensitivity, easy use and high-throughput capabilities of MALDI-TOF MS, it would be very interesting to tune the IF-approach for the obtention of quantitative or semiquantitative information about affinities between a selected target (different proteases in our case) and various ligands (i.e. protease inhibitors used in this study). Based on our experiences and in the recently referred to bibliography (Bucknall et al. 2002), we have collected evidence that MALDI-TOF mass spectrometry can be a useful tool to obtain such information at the semiquantitative level provided that compounds of similar molecular weight, structural analogs or isotopomers were incorporated into each analysis. Molecules with similar structure and amino acid composition usually show similar ionization behavior. One of the main factors affecting the relative intensity of the ionized molecules during the acquisition process is their concentration in the matrix. Obviously, this is true when a homogeneous crystallization of the sample is obtained. The analyte-matrix co-crystallization is a crucial step for a successful desorption-ionization process. An unsuitable co-crystallization can account for a three-dimensional heterogeneous pattern, with spots rich in matrix or in a particular analyte. On the other hand, proper sample-matrix homogeneity favors the statistical analysis (by diminishing standard deviation) and subsequent quantitation of affinity differences between ligands.

To demonstrate the wild-type form of the carboxypeptidase A, an inhibitor from potatoes (PCI) ($Ki=1.5\times10^{-9}$ M) and a site-directed mutant of this inhibitor (PCI-desCT) ($Ki=40\times10^{-6}$ M, with much lower affinity for the enzyme)(Marino-Buslje et al. 2000), were mixed simultaneously, at the same concentration (16 μM), with the target molecule (carboxypeptidase A, CPA). The reaction conditions were those described in the previous section (see the Experimental Procedures section), but in this case different concentrations of CPA were applied in the assay, in a decreasing trend. At CPA concentration reflecting a tenfold molar excess of protease (165 μM) with respect to the inhibitors (wtPCI and PCI-desCT), the spectral signals from both inhibitors disappear from the mass spectrum as a consequence of the binding to the target (see Fig. 12.3). While gradually diluting CPA, it can be observed a competition process between wtPCI and Δ3PCI for the enzyme, with the release of a higher amount of the latter in solution and therefore a higher relative intensity in the mass spectra. At about 1 μM CPA concentration, the Δ3PCI mutant has completely recovered the same relative intensity as in the control spectrum, while wtPCI still shows that 50 % of its signal has faded (Fig. 12.3). If CPA dilution is continued, the spectral signals from both inhibitors recover the intensities of the control samples and become equivalent.

Further work is required to test whether this approach is able to provide true quantitative data (i.e. to derive association constants), and to tune it with such a purpose. However, by now it gives qualitative or semiquantitative information about affinities between the added enzymes and various ligands in the examined solution. Molecular ion intensity differences between the inhibitors at decreasing concentrations of CPA can be correlated with their different binding affinities, experimentally obtained by classic enzymologic methods (Marino-Buslje et al. 2000). In the case described above, it is clear that PCI is a better ligand than Δ3PCI.

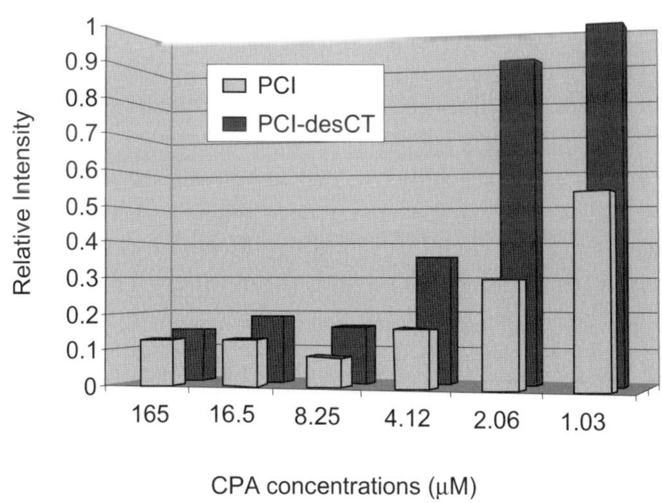

Fig. 12.3. Changes promoted in the relative intensities (*RI*) of the protein molecules wtPCI and PCI-des CT after the addition of CPA at decreasing concentrations

12.3.4 Detection of Protein Ligands in Complex Samples

Once the methodology was set up with model molecules the next step was to apply it to complex samples. With this aim, highly heterogeneous extracts from invertebrate organisms were analyzed by MALDI-TOF MS to detect protease inhibitors, assaying the effect of the addition of different proteases on their mass spectra. Two kinds of invertebrate organisms, which have been revealed as an important source of bioactive molecules, were used for such a purpose: the sea anemone *Stichodactyla helianthus* and the aquatic leech *Hirudo medicinalis*. Protease inhibitors, cytolysins, cardiotoxins, and neurotoxins have recently been isolated from these organisms (Delfin et al. 1996; Lanio et al. 2001; Schoofs and Salzet 2002).

12.3.4.1 Ion Suppression Effects in MALDI-TOF MS, and Sample Preparation for Complex Biological Samples

A remarkable MALDI causative phenomenon of many problems encountered in the direct analysis of complex mixtures by MALDI-TOF MS is the "ion suppression effect". It takes place at both the matrix-analyte (matrix suppression effect, MSE) and analyte-analyte (analyte suppression effect, ASE) reactions competing for analyte ionization associated to MALDI (Knochenmuss et al. 2000; Knochenmuss and Zenobi 2003; Karas and Kruger 2003). We have found that this phenomenon is particularly important when working with very complex samples (composed of many analytes with different properties and proton affinities), the relative intensity ion signals of their components been frequently affected by each other, and by the addition of a perturbing partner, in a different extent. The reasons of such behavior are probably multiple, such as competition among analytes for a proper ionization, and an unequal desorption/ionization process of the sample as a consequence of the three-dimensional variation in the pattern of analyte-matrix co-crystallization, among others (Zhu et al. 1995; Knochenmuss et al. 2000; Knochenmuss and Zenobi 2003; Karas and Kruger 2003). The minimization of these problems is essential, in order to obtain trustable results with the *Intensity-fading* approach .

To reduce these problems we have found that it is essential to carefully prepare the sample, in order to avoid contaminants that might interfere with a good and homogeneous *co-crystallization* with the matrix, with the stability of the complexes to be analyzed, and with the proper ionization of the molecular species. The removal of salts, buffers, detergents and denaturants from the sample, in the first step, is of great help. Also, a careful selection of matrix, and analyte relative concentrations, adjustment of pH, crystallization conditions, use of additives and on-target sample clean-up, are important factors in the second step.

We frequently perform the first step by means of an RP-cleaning protocol for crude extracts to concentrate proteins and peptides and minimize the effects of salts and impurities (Villanueva et al. 2001). However, in order to apply IF MALDI-TOF for screening studies on complex biological samples or proteomes, we recommend the pre-fractionation of the biological extract by simple and fast true chromatographic techniques or a combination of them, to reduce the analyte-analyte suppression effect. As shown in Fig. 12.4, a simple size exclusion chromatography can greatly improve the number of ionized species (proteins) visualized in the leech extract used as a model. Figure 12.5 shows a comparison between the direct MALDI-TOF spectrum of the same sample (upper part) and the composite spectrum obtained by overlapping the spectra from eight fractions from the size exclusion chromatography (lower part). More than twofold ionized species can be observed in the second one.

In the present case, and for the second step, the choice of matrix has to be adapted to the properties of the analytes. Sinapic acid 10 mg/ml in 30% $CH_3CN:H_2O$, acid free (pH 5–6), has been found to be our best matrix for pro-

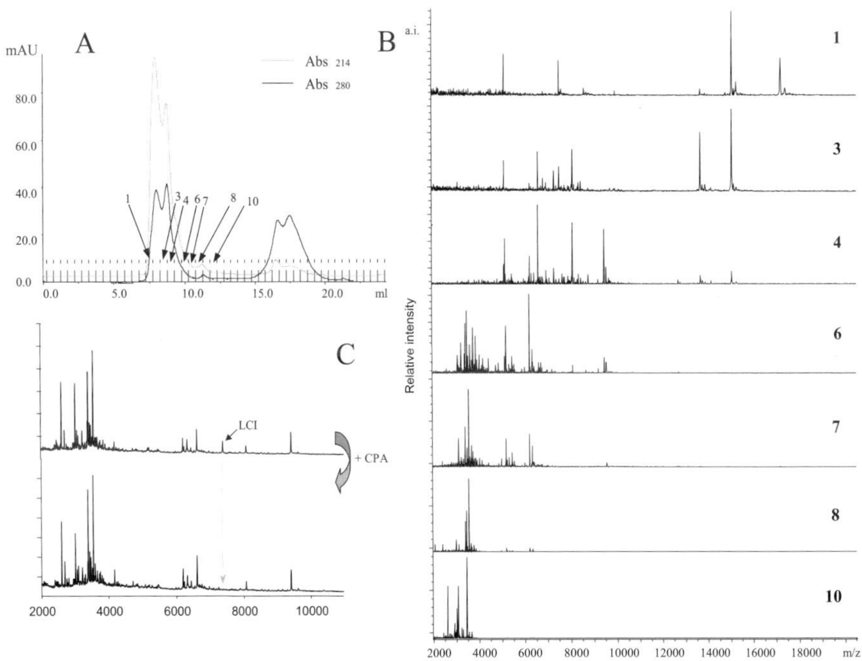

Fig. 12.4. Prefractionation of complex biological samples before MALDI-TOF MS. **A** Size exclusion chromatography in a Superdex Peptide HR 10/30 of 2 mg of *Hirudo medicinalis* saliva extract. Fractions of 0.5 ml were collected. **B** MALDI-TOF MS analysis of seven of the collected fractions during the size exclusion chromatography. **C** MALDI-TOF MS of a *Hirudo medicinalis* extract before (*above*) and after (*below*) mixing with carboxypeptidase A (*CPA*)

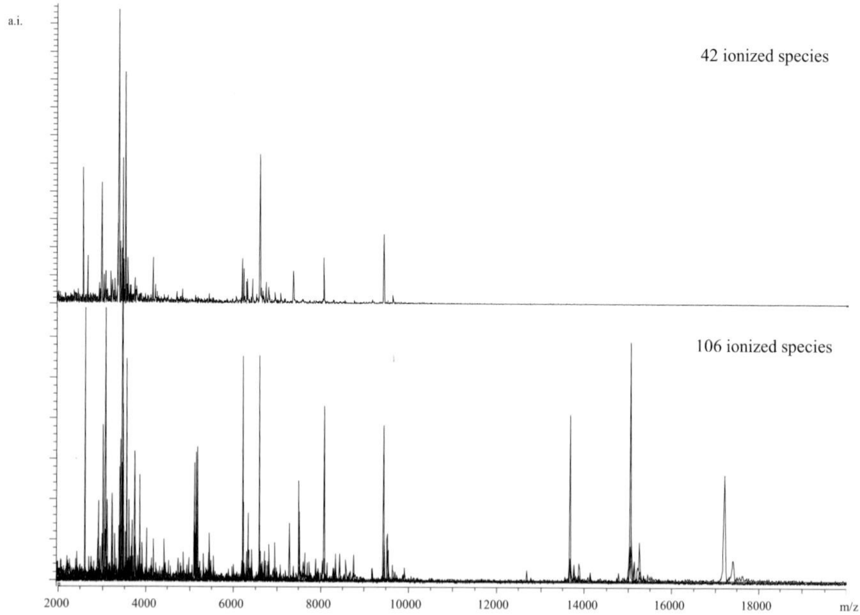

Fig. 12.5. Comparative analysis of the direct MALDI-TOF MS of the *Hirudo medicinalis* extract (*above*) and of the composite spectrum (*below*) generated by overlaying the seven MALDI-TOF spectra of Figure 12.4B

tein models and complex biological samples in IF MALDI-TOF analysis. The application of this matrix at a volumetric ratio of 1:3 (sample:matrix) for complex biological mixtures provides excellent results in terms of homogeneity co-crystallization on the MALDI target, and mass resolution, sensitivity and reproducibility among MALDI-TOF spectra. The addition to the sample of a non-interacting protein, of adequate mass, as an internal reference, is also a useful probe to control the ion suppression effects.

12.3.4.2 Leech Saliva IF MALDI-TOF Analysis

As previously mentioned, heterogeneous extracts from invertebrate animals were used to test the feasibility of applying the "Intensity-fading" (IF-) approach to complex samples. Figure 12.4C shows the MALDI-TOF mass spectra corresponding to an extract from the saliva of the aquatic leech *Hirudo medicinalis* before and after the addition of bovine CPA, the selected target for binding. After the reaction with CPA, one peak of the MALDI-TOF mass spectrum with a m/z of 7326 clearly fades. By mass spectrometry analysis, and also by parallel HPLC fractionation and protein sequence analysis, such a peak has been identified as the Leech Carboxypeptidase inhibitor

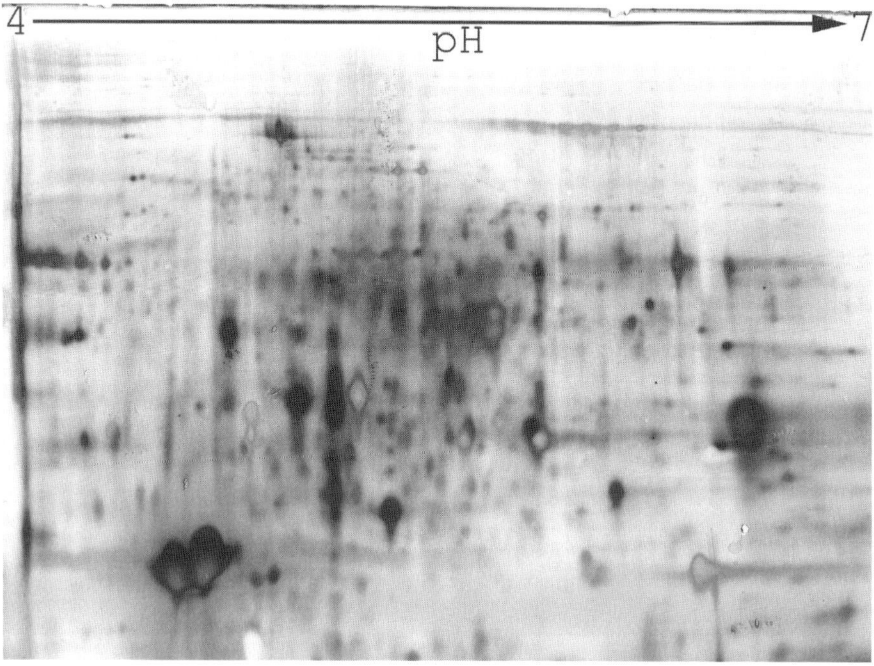

Fig. 12.6. Silver stained 2D-Electrophoresis in acryllamide gels of the standard pattern of *Hirudo medicinalis* saliva. A previous isoelectrofocusing (pH 4–7) was performed

(LCI), a 66 residues protein previously reported by our group (Reverter et al. 1998; Reverter et al. 2000). Figure 12.8A shows a plot of the changes in some of the molecular ions present in the mass spectrum of the same extract after reaction with CPA (data of ten independent experiments were used to draw the plot). The results demonstrate that only the molecular ion with a m/z of 7326 is statistically affected by the addition of CPA.

It should be mentioned that we have focused our analysis only on proteins from the leech saliva with molecular masses below 15.000, because this is the span of greater interest for our present purposes and also because this is the spectral range in which MALDI TOF spectrometer shows its best capabilities. However, the protein content of such a sample is also very rich above that size threshold, as shown by a *standard* 2D-electrophoresis analysis on acrylamide gel, in the pH range 4–7, which is displayed in Fig. 12.6. After scanning and computational analysis, we have estimated that the number of spots on it is above 600. This figure probably would be much higher if the analysis is performed in a more accurate way (i.e. in a wider pH range, and with smaller loadings, giving rise to sharper spots). Therefore, given that this is not a simple sample, we should expect that significant signal suppression effects take place on its mass spectrometry analysis.

12.3.4.3 Sea Anemone Extract IF MALDI-TOF Analysis

12.3.4.3.1 Trypsin as the Target Molecule

Figure 12.7 shows the MALDI-TOF mass spectra corresponding to the whole body extract of the sea anemone *Stichodactyla helianthus* before and after the addition of bovine trypsin. After the reaction with trypsin, one signal of the MALDI-TOF mass spectrum, with a molecular mass of 6110 m/z, has been assigned as the SphI-1 Kunitz protease inhibitor (Delfin et al. 1996). The isolation of this species by HPLC, followed by sequencing of its nine N-terminal residues by Edman degradation, confirmed such identification. Figure 12.8B shows a plot of the relative intensities of some of the molecular ions present in the MALDI-TOF mass spectrum of such an extract after the reaction with trypsin (data of ten independent experiments were used to draw the plot). It is evidenced that only the molecular ion with a m/z of 6110 is statistically affected by the addition of trypsin.

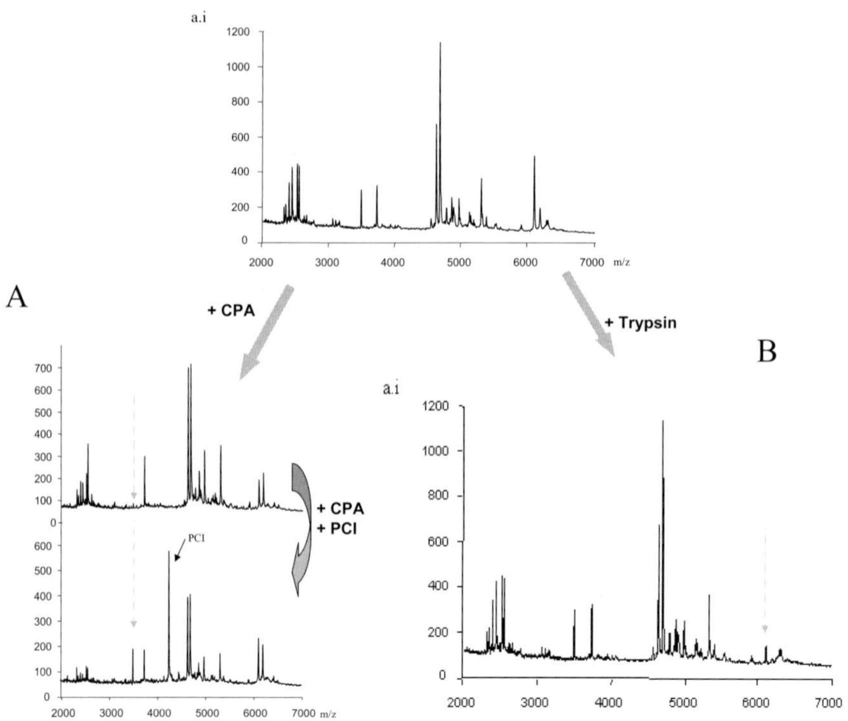

Fig. 12.7. Intensity fading MALDI-TOF MS of a sea anemone extract. **A** MALDI-TOF MS of a *Stichodactyla helianthus* extract before (*above*) and after (*center*) the addition of carboxypeptidase A (*CPA*). The effect of competition with the potato inhibitor of carboxypeptidase A (*PCI*), added in excess, is shown (*below*). **B** MALDI-TOF MS of a *Stichodactyla helianthus* extract before (*above*) and after (*below*) mixing with trypsin

12.3.4.3.2 Carboxypeptidase A as the Target Molecule

Figure 12.7 also shows the addition of bovine carboxypeptidase A instead of trypsin to the extract which gives rise to the fading of another spectral signal with a m/z of 3484. In order to validate a detected interaction with sufficient confidence, we believe that a convenient procedure is to perform a competition assay with a known competitive partner (an exogenous inhibitor in our case, desirably with a similar or stronger affinity). If the fading is *reversed* by the addition, a direct and elegant MALDI-based prove is provided which, at the same time, may confirm the specificity of the interaction. In our selected case, the recovery of the spectral signal of the 3484 m/z compound in a competition assay with PCI (the potato carboxypeptidase inhibitor), added in excess, gives us confidence that such a signal corresponds to a true CPA binding molecule (Fig. 12.7A, lower), probably interacting in a similar site of the enzyme (the active site). A similar IF MALDI-TOF assay could also be carried

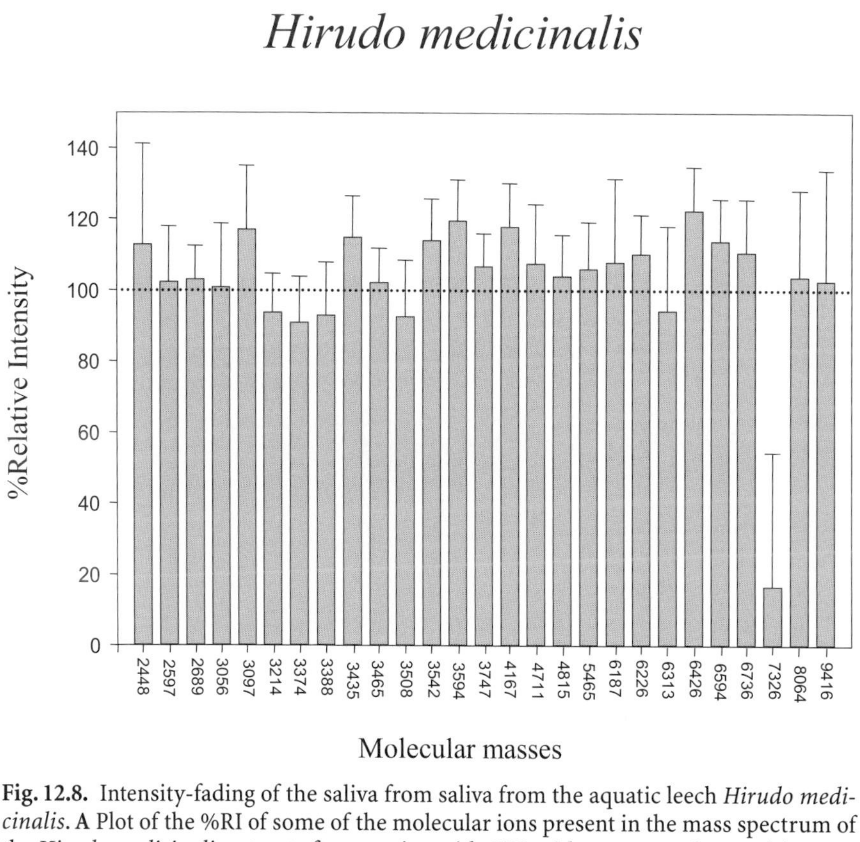

Fig. 12.8. Intensity-fading of the saliva from saliva from the aquatic leech *Hirudo medicinalis*. **A** Plot of the %RI of some of the molecular ions present in the mass spectrum of the *Hirudo medicinalis* extract after reaction with *CPA*, with respect to the RI of the control (before *CPA* addition). The *dotted line* indicates the %RI of the ionized species before *CPA* addition. Ten independent experiments were used to draw the plot.

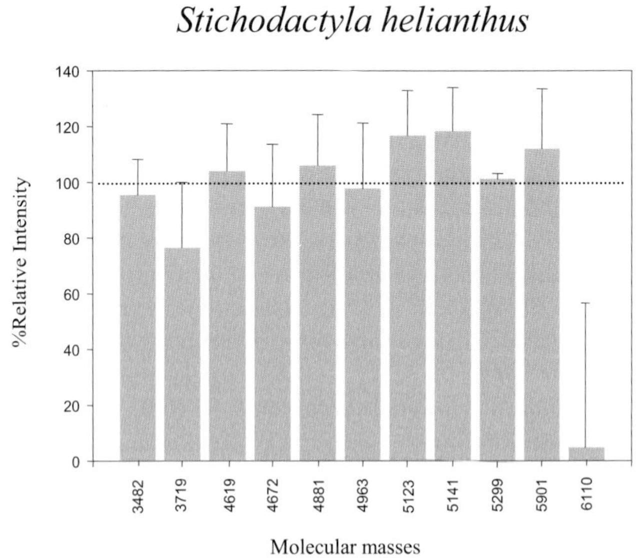

Fig. 12.8. (*Continued*) **B** Plot of the % RI of some of the molecular ions present in the mass spectrum of the *Stichodactyla helianthus* extract after reaction with trypsin, with respect to the RI of the control (before trypsin addition). The *dotted line* indicates the %RI of the ionized species before trypsin addition. Ten independent experiments were used to draw the plot

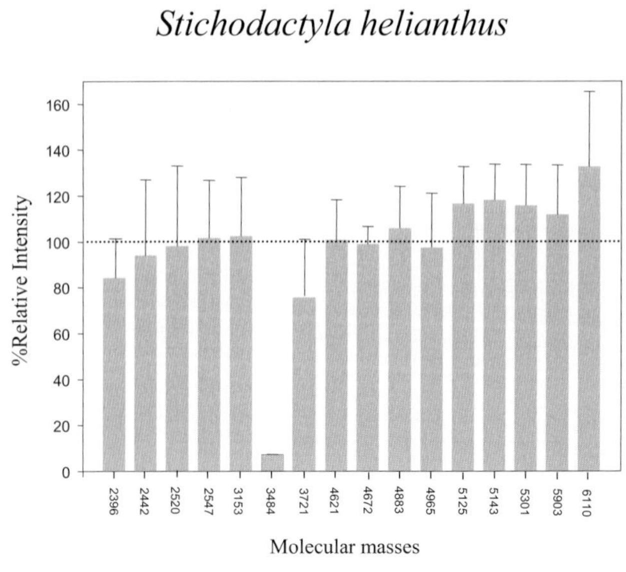

Fig. 12.8. (*Continued*) **C** Plot of the % RI of some of the molecular ions present in the mass spectrum of the *Stichodactyla helianthus* extract after reaction with *CPA*, with respect to the RI of the control (before *CPA* addition). The *dotted line* indicates the %RI of the ionized species before *CPA* addition. Ten independent experiments were used to draw the plot

out at different concentrations of the target molecule and/or at different pH and temperature conditions to verify the interaction. Figure 12.8C represents the results of repeated assays, showing that only the molecular ion with a m/z of 3484 is statistically affected by the addition of CPA.

12.4 General Discussion

One of the great challenges nowadays in molecular and cellular biology is the characterization of the network of interactions established by every protein and its neighbors in each cellular and tissular locations, and the changes suffered by such networks along development (Gavin et al. 2002). Although the methodologies focused on such a purpose are more powerful (i.e., sensitive and capable of massive analysis) and reflecting (in some cases) in vivo states, further technological developments for less demanding (but, nevertheless, quite complex) biological extracts are still required . Here we have described an approach based on the use of MALDI-TOF MS ion intensities to detect the formation of complexes between proteins and other ligands in biological samples. We call it *intensity fading* (IF-) MALDI-TOF MS because it follows the decrease produced in the intensity of the spectral signal of the partners when they interact (Villanueva et al. 2003). The simplicity and rapidity of the approach, together with its high operational sensitivity, should prove a valuable tool for many biological and biotechnological applications.

Although here we have focused our attention on the analysis of protein-protein interactions, either in simple or in complex biological mixtures, we have evidence that it may be used for detecting and characterizing the interaction of proteins with many other ligands (peptides, nucleotides, organic molecules, large macromolecules ...) (Villanueva et al. 2003). The approach should be applicable to study interactions established between any kind of "problem" biomolecules, providing that they are analyzable by MALDI-TOF MS and that their complexes are stable in the conditions used for MALDI-TOF sample preparation.

Our procedure should also facilitate the derivation of binding affinities (by measuring spectral signals at different ratios between the partners). More research should be done on this issue, to explore the best conditions for such quantification. By now, we think that our approach is more trustable at the qualitative and semiquantitative levels.

The strategy described here for characterizing in-depth the detected ligands in complex mixtures has been quite simple, following rather *traditional* approaches, generally through their fractionation in parallel HPLC experiments and subsequent sequencing or spectroscopic analysis of the isolated species. However, nowadays it should be feasible to perform such a characterization in a continuous (and quite automatic) way using additional stages of physical fragmentation of the selected molecular ion of the ligand, followed by analysis of the fragmentation pattern (i.e., MS/MS or MS^n variants). Also, it should be feasible that differential screening of very complex biological samples of various sources (tissues or fluids, along the development, normal and pathogenic, from different organisms) can be performed in a high-throughput way, following the current approaches of proteomics.

It is worth mentioning the tremendous analytical power that could be expected for MALDI-TOF MS, either through the proposed Intensity-fading (IF-) approach, or by the use of more technological demanding approaches (such as MS/MS variants above mentioned), if the *signal suppression effects* (Knochenmuss and Zenobi 2003; Karas and Kruger 2003) are further minimized or eliminated. The great improvement observed in the number of resolved spectral peaks, of molecular species, when simple pre-fractionation procedures are applied (such as gel filtration), as previously shown, confirms such potentiality and the impact of the signal suppression problem. Therefore, this is an issue that clearly deserves further investigation.

No doubt that the simplicity of the operation and interpretation, fast analysis, great sensitivity, tolerance to salts and contaminants of small molecular mass, and other relevant properties of MALDI TOF MS will keep attracting the interest of scientists for further application to research fields at the interfaces between many disciplines around protein science (protein chemistry and biophysics, molecular and cellular biology, proteomics, drug discovery...) that may benefit from such properties. Recent reports on diverse applications for the monitoring of protein expression, post-translational modifications analysis, protein folding and stability studies, and protein conformational analysis, among others, some of them contributed by our groups (Villanueva et al. 2000, 2001, 2002), are examples of such a trend.

Acknowledgements. This work has been supported by grant BIO2001-02046 from MCYT (Ministerio de Ciencia y Tecnologia, Spain), and by the Centre de Referencia en Biotecnologia (CERBA, Generalitat de Catalunya). O. Yanes acknowledges a fellowship from MCYT.

References

Alberts B (1998) The cell as a collection of protein machines: preparing the next generation of molecular biologists. Cell 92:291–294

Bennett KL, Kussmann M, Bjork P, Godzwon M, Mikkelsen M, Sorensen P, Roepstorff P. (2000) Chemical cross-linking with thiol-cleavable reagents combined with differential mass spectrometric peptide mapping-a novel approach to assess intermolecular protein contacts. Protein Sci. 9:1503–1518

Bucknall M, Fung KYC, Duncan MW (2002) Practical quantitative biomedical applications of MALDI-TOF mass spectrometry. J Am Soc Mass Spectrom 13:1015–1027

Cohen SL, Ferre-D'Amare AR, Burley SK, Chait BT (1995) Probing the solution structure of the DNA-binding protein Max by a combination of proteolysis and mass spectrometry. Protein Sci 4:1088–99

Delfin J, Martinez I, Antuch W, Morera V, Gonzalez Y, Rodriguez R, Marquez M, Saroyan A, Larionova N, Diaz J, Padron G, Chavez M (1996) Purification, characterization and immobilization of proteinase inhibitors from Stichodactyla helianthus. Toxicon 34:1367–1376

Farmer TB, Caprioli RM (1998) Assessing the multimeric states of proteins: studies using laser desorption mass spectrometry. Biol Mass Spectrom 20:796–800

Fenn JB, Mann M, Meng CK, Wong SF, Whitehouse CM (1989) Electrospray Ionization for Mass Spectrometry of Large Biomolecules. Science 246:64–71

Fenn JB, Mann M, Meng CK, Wong SF, Whitehouse CM (1990) Electrospray ionization-principles and practice. Mass Spectrom Rev 9:37–70

Gavin AC et al (38 authors overall) (2002) Functional organization of the yeast proteome by systematic analysis of protein complexes. Nature 415:141–147

Glocker MO, Bauer SH, Kast J, Volz J, Przybylski M (1996) Characterization of specific noncovalent protein complexes by UV matrix-assisted laser desorption ionization mass spectrometry. J Mass Spectrom 31:1221–1227

Hensley P, Myszka G (2000) Analytical biotechnology: sorting needles and haystacks. Curr Opin Biotecnol 11:9–12

Hillenkamp F, Karas M, Beavis RC, Chait BT (1991) Matrix-assisted laser desorption/ionization mass spectrometry of biopolymers. Anal Chem 63:1193A-1203A

Karas M, Kruger R (2003) Ion Formation in MALDI: The Cluster Ionization Mechanism. Chem Rev. 103:427–440

Knochenmuss R, Stortelder A, Breuker K, Zenobi R (2000) Secondary ion-molecule reactions in matrix-assisted laser desorption/ionization. J Mass Spectrom 35:1237–1245

Knochenmuss R, Zenobi R (2003) MALDI Ionization: The Role of In-Plume Processes. Chem Rev 103:441–452

Kriwacki RW, Siuzdak G (2000) Probing protein-protein interactions with mass spectrometry. Methods Mol Biol 146:223–238

Kussmann M, Nordhoff E, Rahbek-Nielsen H, Haebel S, Rossel-Larsen M, Jakobsen L, Gobom J, Mirgorodskaya E, Kroll-Kristensen A, Palm L, Roepstorff P (1997) Matrix-assisted laser desorption/ionization mass spectrometry sample preparation techniques designed for various peptide and protein analytes. J Mass Spectrom 32:593–601

Lanio ME, Morera V, Alvarez C, Tejuca M, Gomez T, Pazos F, Besada V, Martinez D, Huerta V, Padron G, de los Angeles Chavez M (2001) Purification and characterization of two hemolysins from Stichodactyla helianthus. Toxicon 39:187–194

Loo JA (1997) Studying Noncovalent Protein Complexes by Electrospray Ionization Mass Spectrometry. Mass Spectrom Rev 16:1–23

Mann M, Hendrickson RC, Pandey A (2001) Analysis of proteins and proteomes by mass spectrometry. Annu Rev Biochem 70:437–473

Marino-Buslje C, Venhudova G, Molina MA, Oliva B, Jorba X, Canals F, Aviles FX, Querol E (2000) Contribution of the C-tail residues of potato carboxy-peptidase inhibitor in the binding to carboxypeptidase A. Eur J Biochem 267:1502–1509

Mandell JG, Falick AM, Komives EA (1998) Measurement of amide hydrogen exchange by MALDI-TOF mass spectrometry. Anal Chem 70:3987–3995

Nelson RW, Krone JR (1999) Advances in surface plasmon resonance biomolecular interaction analysis mass spectrometry (BIA/MS). J Mol Recognit 12:77–93

Ogorzalek Loo RR, Goodlett DR, Smith RD, and Loo JA (1993) Observation of Noncovalent Ribonuclease S-Protein/S-Peptide Complex by Electrospray Ionization Mass Spectrometry. J Am Chem Soc 115:4391–4392

Pandey A, Mann M (2000) Proteomics to study genes and genomes. Nature 405:837–846

Rappsilber J, Siniossoglou S, Hurt EC, Mann M (2000) A generic strategy to analyze the spatial organization of multi-protein complexes by cross-linking and mass spectrometry. Anal Chem 72:267–275

Reverter D, Vendrell J, Canals F, Horstmann J, Aviles FX, Fritz H, Sommerhoff CP (1998) A carboxypeptidase inhibitor from the medical leech Hirudo medicinalis. Isolation, sequence analysis, cDNA cloning, recombinant expression and characterization. J. Biol. Chem. 273:32927–32933

Reverter D, Fernandez-Catalan C, Baumgartner R, Pfander R, Huber R, Bode W, Vendrell J, Holak T, Aviles FX (2000) Three-dimensional structure of a novel leech carboxypeptidase inhibitor (LCI) determined free in solution and in complex with human carboxypeptidase A2. Nature Struct Biol 7:322–328

Rigaut G, Shevchenko A, Rutz B, Wilm M, Mann M, Seraphin B (1999) A generic protein purification method for protein complex characterization and proteome exploration. Nat Biotechnol 17:1030–1032

Rosinke, B., Strupat, K., Hillenkamp, F., Rosenbusch, J.P., Dencher, N., Krüger, U. Galla, H-J (1995) Matrix-assisted laser desorption/ionization mass spectrometry (MALDI-MS) of membrane proteins and non-covalent complexes. J. Mass Spectrom 30:1462–1468

Rostom AA, Fucini P, Benjamin DR, Juenemann R, Nierhaus KH, Hartl FU, Dobson CM, Robinson CV (2000) Detection and selective dissociation of intact ribosomes in a mass spectrometer. Proc Natl Acad Sci USA 97:5185–5190

Schoofs L, Salzet M (2002) Trypsin and chymotrypsin inhibitors in insects and gut leeches. Curr Pharm Des 8:125–133 & 483–491

Sobott F, Benesch JL, Vierling E, Robinson CV (2002) Subunit exchange of multimeric protein complexes. Real-time monitoring of subunit exchange between small heat shock proteins by using electrospray mass spectrometry. J Biol Chem 277:38921–38929

Veenstra TD, Johnson KL, Tomlinson AJ, Naylor S, Kumar R (1997) Determination of calcium-binding sites in rat brain calbindin D28 K by electrospray ionization mass spectrometry. Biochemistry 36:3535–3542

Verentchikov AN, Ens W, Standing KG (1994) Reflecting time-of-flight mass spectrometer with an electrospray ion source and orthogonal extraction. Anal Chem 66:126–133

Villanueva J, Canals F, Villegas V, Querol E & Aviles FX (2000) Hydrogen exchange monitored by MALDI-TOF mass spectrometry for rapid characterization of the stability and conformation of proteins. FEBS Lett. 472:27–33

Villanueva J, Canals F, Querol E, Aviles FX (2001) Monitoring the expression and purification of recombinant proteins by MALDI-TOF mass spectrometry. Enzyme. Microb Technol. 29:99–103

Villanueva J, Villegas V, Querol E, Aviles FX, Serrano L (2002) Protein secondary structure and stability determined by combining exoproteolysis and matrix-assisted laser desorption/ionization time-of-flight mass spectrometry. J Mass Spectrom 37:974–984

Villanueva J, Yanes O, Querol E, Serrano L & Aviles FX (2003) Identification of protein ligands in complex biological samples using Intensity-fading (IF-) MALDI-TOF Mass Spectrometry. Anal Chem 75(14):3053–63

Wilm M, Mann M (1996) Analytical properties of the nanoelectrospray ion source. Anal Chem 68:1–8

Woods AS, Buchsbaum JC, Worrall TA, Berg JM, Cotter RJ (1995) Matrix-Assisted Laser Desorption/Ionization of noncovalently bound compounds. Anal. Chem 67:4462–4468

Zhu YF, Lee KL, Tang K, Allman SL, Taranencko NI, Chen CH (1995) Revisit of MALDI for small proteins. Rapid Commun Mass Spectrom 9:1315–1320

13 Accelerator Mass Spectrometry in Protein Analysis

John S. Vogel, Darren J. Hillegonds, Magnus Palmblad, Patrick G. Grant and Graham Bench

13.1 Introduction

Tools for protein quantitation have advanced through several phases. Common proteins in readily studied sources, albumin and globulins in plasma, were discovered in the nineteenth century and quantified by gravimetric mass determination of chemical precipitates. Finer fractionations and scales led to quantitation of increasingly less common proteins, but the method has obvious limits. Some proteins, enzymes, are quantified by their *activity*, measured as the increasing chemical product or a shrinking amount of substrate. No absolute quantitation of enzyme concentration is available without having a well-known standard for each enzyme expressing a particular action. Immunochemical methods successfully quantitate proteins, as long as antibodies are selective for the protein under study. Standard curves are required for quantification, since binding affinities are sensitive to assay conditions. Non-specific interactions further confound quantitation. Electrophoretic and chromatographic separations allowed analysis of even smaller concentrations, with quantitation accomplished through various optical properties, such as absorbance, fluorescence, or chemiluminescence. Not all proteins are responsive to the latter effects, and optical density measurements often have large uncertainties and poor sensitivity.

The application of 2D gel electrophoresis as a primary tool for proteome analysis led to a range of sensitive stains that make optical (including UV and X-ray) densitometry more quantitative (Rabilloud et al. 1994, Tuma et al. 1999). These stains still suffer from limitations in a dynamic range and variations among proteins, depending on protein structure and composition. Mass spectrometry is rapidly becoming the tool of choice for exploring proteome diversity at a very low copy number, because it has exquisite resolution of protein and peptide molecular mass, and hence identity. Variations in ionization, fragmentation, and detection response across the protein mass range lead to

Principles and Practice
Methods in Proteome and Protein Analysis
R.M. Kamp, J. J. Calvete, T. Choli-Papadopoulou (Eds.)

complexity in quantitation (Arnott et al. 1993, Giometti et al. 2002). This is partially solved by incorporation of isotopically distinct internal standards at known or control concentrations that are resolvable and comparable to the protein being quantified. The results are only comparative, however, and not absolute. Stable isotopes are used in this tagging (Gygi et al. 2002, Keller et al. 2002) which quantitates differences in identical proteins among two or more separate experiments. Radioisotope labels on proteins are linearly quantified over large dynamic ranges with low backgrounds in gel separations (Lees and Richards 1999) but suffer from high radiation dangers, especially for short-lived isotopes that provide the highest sensitivities. This is inherent in the detection of radioisotopes using their decay products. Here, we discuss a mass spectrometry quantitation of radioisotopic labels that provides quantitation over a wide range with a low background at very low radiation levels, often below the levels defined for radioactive waste.

Radioactivity, measured in decays per minute (dpm), is an inefficient measure of the quantity of the atoms present for isotopes with half-lives greater than about 10 years. Radiocarbon (^{14}C) is a good example of an isotope that is frequently decay counted for biochemical applications as well as for carbon dating of previously live materials, including humans. The half-life of ^{14}C is 5730 years, to give a mean life [$=t_{1/2}/\ln(2)$] of 8270 years or 4.35×10^9 min. The radioactivity of a sample is just the number of ^{14}C in the sample divided by the mean life. Thus, 1 dpm requires 4.35×10^9 ^{14}C atoms in the sample. Quantifying 1% of the ^{14}C in a sample requires continuous counting for 1% of the mean life, or 82.7 years. The inefficiency is less striking for tritium, but 1 dpm still represents 9.6 million tritium atoms. The inefficiency is much worse for long-lived isotopes such as ^{26}Al ($t_{1/2}$=750,000 years), ^{36}Cl ($t_{1/2}$=350,000 years), or ^{41}Ca ($t_{1/2}$=104,000 years)(Vogel et al. 1995). This problem led to the development 25 years ago, of high energy mass spectrometers whose function was to count these atoms directly, without regard to their decay properties(Muller 1977, Nelson et al. 1977). Carbon dating was the primary impetus for development of this accelerator– based mass spectrometry (AMS), but the technique has been applied to biochemical studies for over a decade (Turteltaub et al. 1990).

The sensitivity of AMS is demonstrated by the archaeological use: plants and animals that are in equilibrium with the biosphere (that is, alive) have natural ^{14}C in them from atmospheric sources at about 100 attomoles (amol) per mg of carbon, an amount that is called *Modern*. Agriculture rose shortly after the end of the last large ice age, about 10,000 years ago. The collagen from a bone of a person who has been dead for 10,000 years contains 30 amol ^{14}C per mg carbon (30% Modern). A sample of this collagen would have to contain 240 g of carbon to be decay counted at 10 dpm, taking 17 h to get 10,000 counts. An ancient bone or artifact often had to be destroyed to be carbon dated using decay counting (Nelson et al. 1986). Fifty thousand year old carbon contains only 0.25 amol ^{14}C per mg C (0.25% Modern) and was rarely dated prior to AMS, although this time period includes the displacement of Neanderthals by Cro-Magnons in Europe and the beginning of migration of peoples toward the Americas. AMS dates the ^{14}C in a mg of carbon back to around 50,000 years, quantifying a few hundred zeptomoles (zmol) of ^{14}C to 10 % precision or better. AMS count rates are several hundred per second for Modern material. AMS precision exceeds 1 % for Modern materials, again providing zmol sensitivity for isotopic labels above the natural level of ^{14}C. Biochemical tracing is not limited by the natural concentration of ^{14}C, but AMS does have operational upper limits at concentrations of about 1000 Modern, or 100 fmol ^{14}C per mg carbon, providing a linear dynamic range of 5 orders of magnitude from zmol to fmol. This range corresponds to 10^{-6} to 0.2 Becquerel (Bq) or 30×10^{-6} to 0.5 picoCuries (pCi), emphasizing that radioisotopes are used at low radiation exposure.

The dynamic range of protein concentrations range over 10 or 11 orders of magnitude, so even AMS cannot cover the entire complement, but AMS has precise quantitation at amol levels that is not available by other means. AMS is particularly suitable for tracing isotope labeled compounds in humans for kinetic and dynamic studies, because both the radiation dose and the chemical dose can both be very small. Adult humans contain about 100 nCi (3.7 kBq) of ^{14}C, which is not the dominant radioisotope naturally found in people. Nutrition research with AMS typically uses 100 nCi of labeled nutrients (Buchholz et al. 1999, Dueker et al. 2000) that expose the volunteer to a smaller radiation risk than is obtained in a short airplane flight. Here, we have applied AMS, and other technologies based on energetic ions, to the study of protein binding in vivo and to high sensitivity amino acid sequencing to determine protein identity. The technique is becoming more available in the research community.

13.2 Accelerator Mass Spectrometry

AMS obtains high isotope counting efficiency using the method outlined in Fig. 13.1. Atoms from about 1 mg of carbon are ionized with an added electron and accelerated to a moderate energy before a mass analysis at mass 14. The vast majority of these ions are molecular hydrides of the lighter carbon isotopes: $^{12}CH_2$, ^{12}CD, and ^{13}CH. Nitrogen, the dominant mass 14 isotope, does not form a negative ion and does not appear in the initial ion beam. As in MS-MS, molecules are broken by collisions in a thin foil or gas cell. Macromolecules dissociate at kilo-electronVolt (keV) energies, but an acceleration through several million Volts is necessary to destroy hydrides in a single collision. Insulation for millions of volts meant that the initial spectrometers, based on nuclear physics accelerators, were building-sized instruments. More recently, multiple collisions at sub-megaVolt potentials have been used to destroy these molecules, ushering in a series of smaller and more affordable

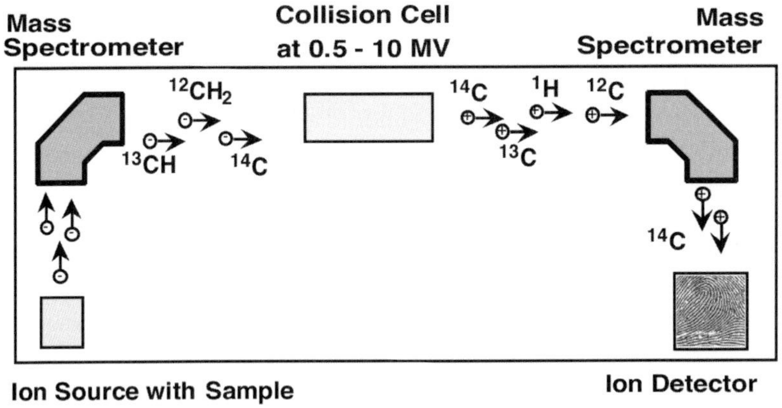

Fig. 13.1. Schematic of the basic ideas behind AMS: tandem mass spectrometry with a charge-changing collision cell at 1 million V potentials between the spectrometer elements. Each ion is *fingerprinted* by how it loses energy in the final detector

spectrometers (Hughey et al. 2000, Synal et al. 2000, Ognibene et al. 2002). Molecular dissociation produces positive ions that accelerate away from the high positive potential back to ground potential. Multiple magnetic and electrostatic elements follow the accelerations and molecular breakup to select the desired high energy ^{14}C ions from the molecular debris composed of H, D, ^{7}Li, ^{12}C, and ^{13}C. This selection is effective to parts per billion or better and reduces the filtered ion stream to hundreds of particles per second. Transport through the high energy spectrometer ends in a detector that uniquely identifies each ion by quantifying how quickly it loses energy in a gas cell.

The specificity of AMS lies in the two processes that require high energies provided by inclusion of an accelerator in the mass spectrometer: destruction of molecular isobars in energetic collisions, and identification of high energy ions through quantification of energy loss. These two processes are present in AMS detection of all isotopes and are straightforward for ^{3}H and ^{14}C, because their nuclear isobars (isotopes of other elements having the same mass as the counted ion) do not form negative ions. Other isotopes use specific chemical and ionization *tricks* to reduce nuclear isobars to levels that can be handled by the identifying detector. AMS is used in isotope ratio mode by momentarily pulsing the mass 13 negative ion beam through the spectrometer, with the accelerated ^{13}C ions being counted in a Faraday cup in the analysis stage. Samples of standard materials are quantified intermittently among the samples to provide normalization to recognized ^{14}C concentrations (Vogel et al. 2002).

Samples are introduced into the cesium bombardment ion source as elemental carbon. All samples are first combusted to a well mixed homogeneous form, CO_2, that is then reduced to elemental carbon on an iron-group catalyst

(Vogel 1992). The process requires from 0.1 to 1 mg of carbon, much of which can be below ^{14}C carrier carbon in the case of very small biological or chemical isolates. A 3 mm diameter plug from a 1-mm-thick 15 % acrylamide gel contains about 0.5 mg carbon, making an excellent low background carrier for any protein and the ^{14}C isolated with it.

13.3 Biomolecular Targets of Labeled Compounds

Environmental toxins are activated during metabolism to chemical forms that readily bind to proteins and DNA. Quantifying such binding in animal hosts and human volunteers has been a common application of AMS over the past decade (Dingley et al. 1999, Mauthe et al. 1999). We have recently used [^{14}C]-diisopropylflurophosphate (DFP) as a reporter compound to quantitate the effects of co-administered chemicals in related pesticide families on the protein retention of DFP in mouse plasma, brain, and other tissues (Vogel et al. 2002). The samples were taken 48 h after dosing, well after all unbound DFP was metabolized and excreted. We now use these tissues to identify the in vivo target proteins of DFP. Previous studies of target proteins in chickens and rats have involved adding isotope labeled DFP to homogenized brain tissue (Carrington and Abou-Donia 1985, Richards et al. 1999). Since DFP binds covalently to a wide class of serine hydrolases, including esterases and proteases, the biological importance of the relative protein binding in tissue homogenates is poorly defined. We used very small doses of DFP (1 $\mu g\ kg^{-1}$) administered in the food of unstressed mice to discover the relative in vivo binding to targets without inducing toxicities or behavioral changes.

Complete protein target specification uses 2D gel electrophoresis, with spot localization by stain followed by excision. However, we wish to avoid ambiguities of staining, lest a low copy protein that has a strong affinity for the labeled compound be missed. This is unlikely with the DFP case in point, but the approach needs to be general. Dicing the entire 2D gel for AMS quantitation at 100×100 pixels would produce 10,000 samples. This creates too many samples for routine use or funding, representing about 1 million US$ worth of measurements. Instead, we make virtual 2D gels by 1–1.5 mm sectioning of both a 1D isoelectric focusing (IEF) gel strip and an SDS PAGE molecular weight (MW) lane of the same protein sample. The high sensitivity of AMS means that the gels need not be overloaded to enhance detection, leading also to better gel reproducibility.

Representative ^{14}C-"density" plots are shown in Fig. 13.2 (MW gel) and Fig. 13.3 (IEF gel) for plasma proteins from a mouse dosed with 30 ng (2 nCi) of ^{14}C-DFP. The optical density of the Coomassie stained MW gel is shown in Fig. 13.2 as a shaded area. Albumin was not removed prior to analysis and caused an excessive bulge out of the lane around 60 kDa. The highest fraction of protein binding also occurs in this mass range, but is not related to non-

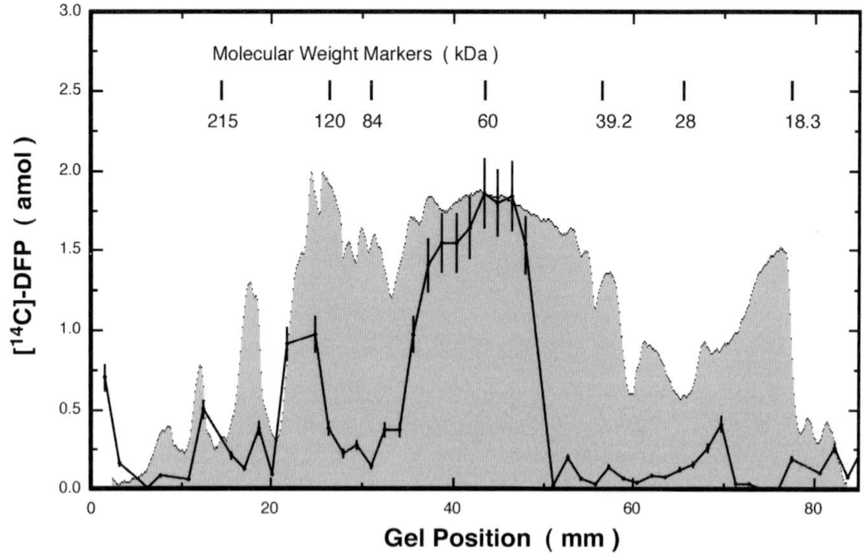

Fig. 13.2. AMS quantitation of bound [^{14}C]-DFP is shown for an SDS-PAGE gel lane of mouse plasma proteins. Optical density of the Coomassie stained lane is shown as the *shaded area* (arbitrary scale), and the location of molecular weight markers are shown in another *lane*. Zeptomoles of bound DFP were quantified from the AMS isotope ratio using the known amount of carbon in each piece of excised gel

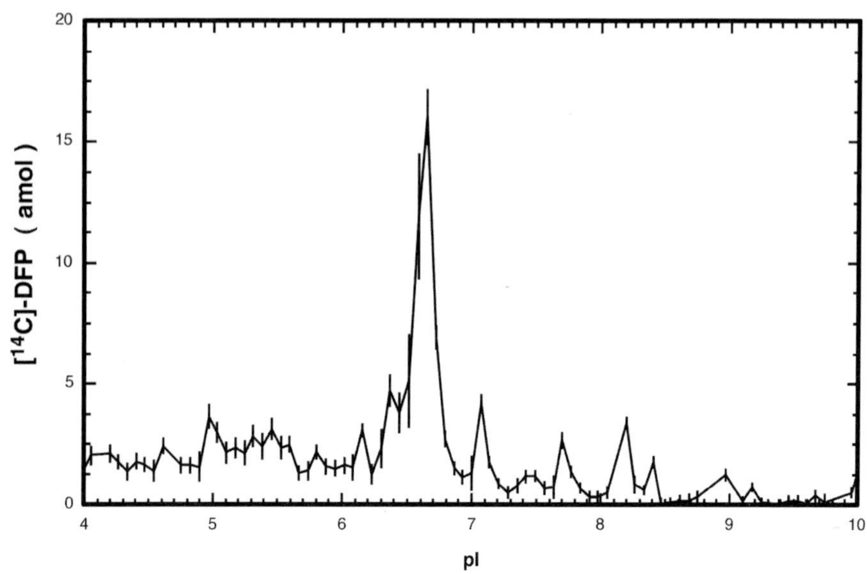

Fig. 13.3. AMS quantitation of bound [^{14}C]-DFP is shown for an IEF gel lane of mouse plasma proteins

specific albumin binding, as shown in Fig. 13.3. The greatest ^{14}C content of the IEF strip occurs at a pI of 6.5, and relatively little signal is found at the albumin pI of 5.5. An expected plasma target of DFP, butyrl-cholinesterase has a MW of 65 kDa and a pI of 6.54, matching the primary binding locus of the ^{14}C in our virtual 2D gel. About eight of the IEF peaks correspond to what can be likely seen as serine hydrolase proteins. This assignment of target is not conclusive, of course. Even a tryptic digest and mass spectrometric analysis of the protein(s) in the narrow IEF peak cannot confirm that the binding is to the cholinesterase. The technique can best quantitate differential effects. Perturbations to this and other significant peaks quantify changes in plasma binding patterns of similarly dosed mice that are undergoing stress, co-administration of chemicals, or other stimulations. The differential effects are quantitated not only between differently treated animals, but also among the proteins from each animal. The quantitative level of the data is emphasized by the vertical scales and error bars of the two figures. The bound DFP, which has on average more than 2 ^{14}C's per molecule, is quantitated at tens to hundreds of zmol.

13.4 Specific Binding Affinity

The sensitivity of AMS for quantitating bound compounds is not matched by the quantitation of the amount of protein to which the labeled compound is bound. Thus, uncertainty in specific affinity of proteins for natural or anthropogenic substrates is dominated by difficulties in measuring protein amounts. Common methods, including those aimed at improved sensitivity and dynamic range, make use of optical properties of proteins or stains bound to them. These depend on chemical structures and are not independent of the nature of the protein. We sought a method that would be independent of chemical structure. Energetic ions lose energy while passing through materials in well understood interactions with atomic electrons (Berger et al. 2000). These interactions do not depend on chemical bonds and are linearly proportional to the total number of electrons in the ions' paths. Thus, the average energy loss of an accelerated proton, alpha particle, or other atomic ion is directly proportional to the amount of material through which it passes (Lewis 1968). We use this property to measure protein masses by passing energetic alpha particles (5.3 MeV) from the radioactive decay of ^{210}Po through a thin (150 nm) silicon nitride (SiN) window on which a purified protein sample has been deposited by pipette or electrospray.

Figure 13.4 shows the energies of alpha particles that passed through various parts of a 1.5-μg protein sample placed on an SiN window. Most of the ions pass through parts of the window that have no protein, forming a peak at the same energy as seen on the blank window alone (correcting for any electronics drift using the pulser's peak). A number of ions passed through

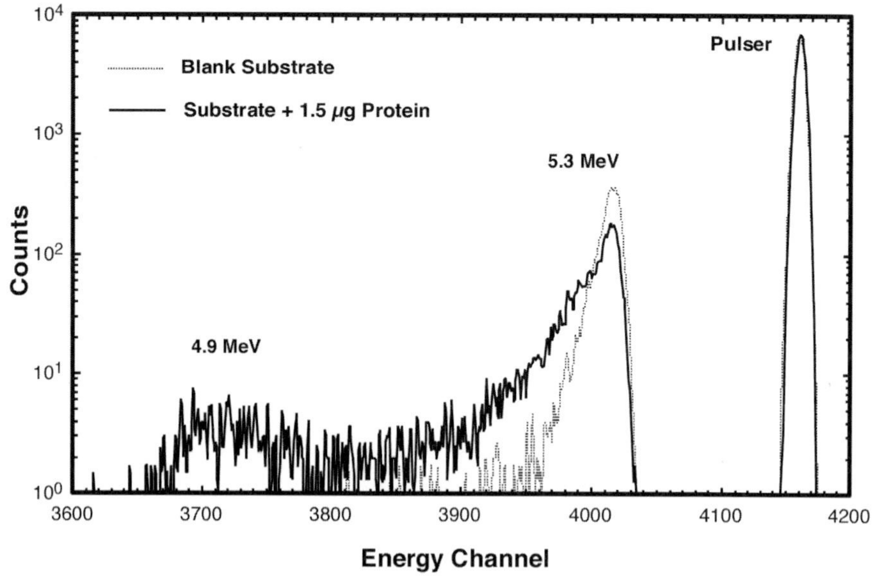

Fig. 13.4. Alpha particle energy spectra from ^{210}Po taken through a 150-nm-thick *SiN* window, both before and after pipetting of 1.5 μg of BSA onto the window. Particles passing through areas containing protein lose more energy than the ones passing through only the *SiN*. The average energy loss of all particles is used to quantitate the total amount of protein on the window. An electronic pulser allows corrections for electronic drift

enough protein to lose about 400 keV, forming a broad peak to the left of the blank window's peak. Thinner parts of the protein spot remove less energy from the transmitted alpha particles, producing the counts added to the low energy side of the blank peak. An algorithm is used to calculate the average energy loss and hence the average protein density on the exposed SiN window. This number provides a measure of the total protein mass. Measurements of six BSA samples pipetted onto separate windows at a nominal 1.5 μg apiece in a 1 μl of water yielded a mean and standard error in the mean of 1.554±0.073 μg, a 4.7 % relative standard deviation.

The precision of AMS quantitation was used to check the linearity of the energy loss measurements over a range of 50 ng to 5 μg of [^{14}C]-BSA on the SiN substrates. Protein aliquots were pipetted onto windows that were made to fit within our usual AMS combustion tubes, quantified by energy loss of alpha particles, converted to an AMS ^{14}C sample, and analyzed for their ^{14}C content. Note that the same aliquot of protein was used both in the mass determination and in the AMS quantification of ^{14}C. This removes the uncertainty in protein binding that arises from using separate aliquots in different analytical modes. The BSA specific ^{14}C activity was then used to also derive

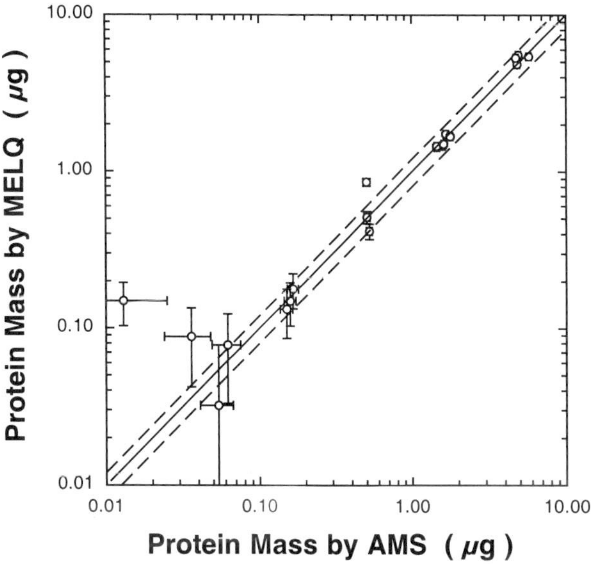

Fig. 13.5. The mass of [^{14}C]-BSA spotted onto *SiN* windows as measured by alpha particle energy loss is given versus that derived from AMS measurement of the same samples using the specific activity of the protein. *Dashed lines* represent ±10 % deviations from concordance. The method is quantitative to about 100 ng. Deposition of higher masses requires spray deposition that is not limited by the solubility of protein in pure water

the masses of spotted protein. Figure 13.5 shows the correlation of the mass by energy loss quantitation (MELQ) with the mass determined by isotope counting with AMS. The solid line shows perfect concordance, and the dotted lines show ±10 % deviations. The present system quantifies to ±20 % down to about 100 ng, showing linearity to 5 μg. The solubility of protein in pure water limited these experiments to this range, but electrospraying the protein onto the window will deliver more massive samples without this limitation. The system should show perfect linearity to hundreds of micrograms for these aerosol-deposited samples.

13.5 Attomole Edman Sequencing

Gas phase Edman sequencing using HPLC identification is a standard and commonly used method for obtaining de novo amino acid (AA) sequences of small amounts of unknown proteins. Capillary column HPLC readout of the derivatized amino acids has pushed Edman's sensitivity to femtomole levels, but signal to noise ratios have proven hard to maintain at lower quantities. Radioisotopes have extremely high signal to noise ratios if detected efficiently by AMS, and an AMS-Edman sequence technique has been demonstrated (Miyashita 2002). Retention of signal, rather than background noise, is the limiting factor in processing attomoles of an isotope-labeled moiety through such a complex system. This was solved by expressing protein in cells growing on a ^{14}C-containing substrate. Labeled protein was diluted to the fmol range and sequenced in the presence of a large excess of an unlabeled protein (25 pmol beta-lactoglobulin) in a 477 automated sequencer with a 120A HPLC (PE Biosystems). The phenylthiohydantoin (PTH) AA derivatives from the lactoglobulin passivated the surfaces of the reaction chamber, the transport tubing, and the HPLC medium. The PTH-AA peaks were collected as fractions and each assayed for ^{14}C by AMS for each Edman cycle. Full peak collection was assisted by the presence of equimolar amounts of all possible PTH-AA, provided by peptide-coated beads that were also in the reaction chamber. These beads produce equal amounts of all PTH-AA's on each cycle. These passivation and peak defining methods are the keys to having sufficient sample recovery for amol samples of isotope labeled protein. These methods can only be used, however, if the protein being sequenced contains a chemically invisible label that provides distinction from the passivating and marker proteins, such as an isotope.

Figure 13.6 shows the first 6 PTH-AA peaks of 1.76 fmol of [^{14}C]-glutathione sulfur transferase (GST) sequenced in this manner. Note the single AA peak per cycle in the AMS response, showing a very clean transport and isolation of each PTH-AA through the HPLC. The fourth AA is proline, which does not fully derivatize during its cycle, and the following HPLC scans contain successively more, but easily distinguished, peaks. There is insufficient carbon in the HPLC eluents after evaporation of the eluting solvents, so a well-

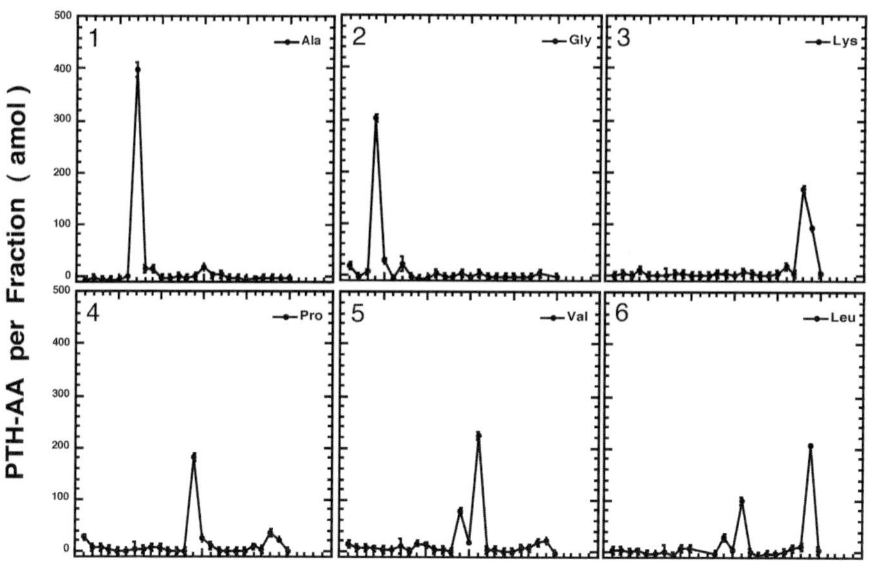

Fig. 13.6. The first six HPLC spectra are shown for an Edman degradation analysis of 1.8 fmol of [^{14}C]-GST. Carryover of lysine from the third cycle is seen, but multiple peaks are prominent only after the poorly derivatized proline in cycle 4

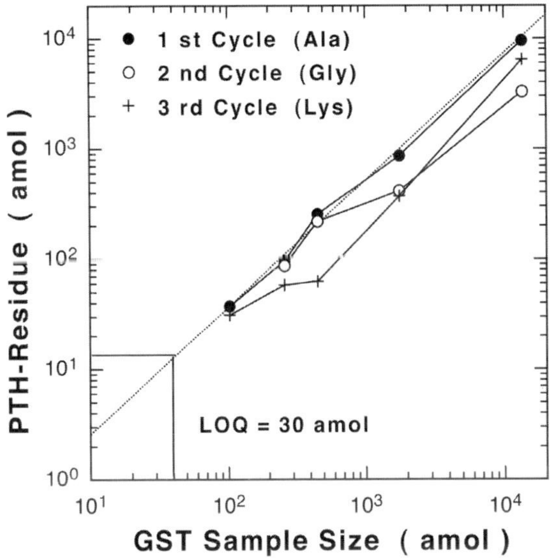

Fig. 13.7. The quantitative yield of the first three cycles of the Edman sequence of [^{14}C]-GST is plotted against the amount of sample placed in the reaction chamber. A limit of quantitation was set at 30 amol of sample, equal to five times the standard deviation of HPLC background levels

defined amount (1.19 mg) of carbon is added to each fraction to enable absolute quantitation of the amount of PTH-AA from the measured isotope ratio. This carrier compound contained 10 amol of ^{14}C per mg carbon, which was subtracted from the plotted data.

The ultimate sensitivity of the sequencing was determined by obtaining the first three PTH-AA's of successively smaller samples. The amounts of measured PTH-AA for the three cycles are shown as functions of the applied protein quantity in Fig. 13.7. The concentration of the applied samples was also quantified by AMS and showed the expected non-linearity of serial dilutions at such low concentrations. The limit of quantitation (LOQ) was set at five times the standard deviation of the non-peak HPLC fractions for each cycle. The LOQ for the first cycle was 30 amol, and nearly the same for the other two cycles. The overall efficiency of the method (protein to PTH-AA) varied from 55 % for the first cycle to 20 % for the third cycle. Repetitive yield over ten cycles was 88 % per cycle at 1.8 fmol. The method is admittedly cumbersome in that 20 samples are measured for each Edman cycle, but the quantitation of each cycle is absolute, and the 200 samples required for ten cycles are easily measured in a single day of AMS. The general lesson of surface passivation and marker AA incorporation may be applicable to other labeling schemes. For example, the isotope could be attached to the derivatizing agent, as long as total and efficient elimination of the non-bound label were possible through evaporation or other processes.

13.6 Conclusion

The sensitive quantitation of AMS has been linked to several protein analysis technologies, including gels, capillary electrophoresis, size exclusion chromatography, and molecular weight filters. Zeptomoles of expressed proteins can be quantified. Attomoles of labeled proteins can be AA sequenced. Protein binding of labeled chemicals and biomolecules is quantitated in the amol to a pmol range by AMS. Another energetic ion technology, MELQ, combines a more precise and universal method to quantitate protein sample mass with the isotope measurements of AMS to determine even low protein specific affinities for labeled compounds to a high accuracy.

AMS spectrometers are becoming smaller and more affordable for eventual placement in advanced quantitative laboratories. At present, there are several AMS facilities serving the biomedical community, including the NIH National Research Resource for Biomedical AMS at Livermore National Laboratory.

Acknowledgments. This work was performed in part under the auspices of the US Department of Energy by University of California Lawrence Livermore National Laboratory under Contract No. W-7405-Eng-48. Funding was provided by LLNL Laboratory Directed Research and Development 01-ERI-006 and National Institutes of Health RR-13461 and ES-004699.

References

Arnott D, Shabanowitz J, Hunt DF (1993). Mass Spectrometry of Proteins and Peptides–Sensitive and Accurate Mass Measurement and Sequence Analysis. Clinical Chemistry 39(9): 2005–2010

Berger MJ, Coursey JS, Zucker MA (2000). Stopping-Power and Range Tables for Electrons, Protons, and Helium Ions. NIST Information resource 4999: http://physics.nist.gov/PhysRefData/Star/Text/contents.html

Buchholz BA, Arjomand A, Dueker SR, Schneider PD, Clifford AJ, Vogel JS (1999). Intrinsic erythrocyte labeling and attomole pharmacokinetic tracing of ^{14}C-labeled folic acid with accelerator mass spectrometry. Analytical Biochemistry 269(2): 348–52

Carrington CD, Abou-Donia MB (1985). Characterization of [^{3}H]di-isopropyl phosphorofluoridate-binding proteins in hen brain. Rates of phosphorylation and sensitivity to neurotoxic and non-neurotoxic organophosphorus compounds. Biochem J 228(3): 537–44

Dingley KH, Curtis KD, Nowell S, Felton JS, Lang NP, Turteltaub KW (1999). DNA and protein adduct formation in the colon and blood of humans after exposure to a dietary-relevant dose of 2-amino-1-methyl-6- phenylimidazo[4,5-b]pyridine. Cancer Epidemiol Biomarkers Prev 8(6): 507–12

Dueker S, R., Lin Y, Buchholz BA, Schneider PD, Lame MW, Segall HJ, Vogel JS, Clifford AJ (2000). Long-term kinetic study of beta-carotene, using accelerator mass spectrometry in an adult volunteer. Journal of Lipid Research 41(11): 1790–1800

Giometti CS, Reich C, Tollaksen S, Babnigg G, Lim HJ, Zhu WH, Yates J, Olsen G (2002). Global analysis of a "simple" proteome: *Methanococcus jannaschii*. Journal of Chromatography B: Analytical Technologies in the Biomedical & Life Sciences 782(1–2): 227–243

Gygi SP, Rist B, Griffin TJ, Eng J, Aebersold R (2002). Proteome analysis of low-abundance proteins using multidimensional chromatography and isotope-coded affinity tags. Journal of Proteome Research 1(1): 47–54

Hughey BJ, Skipper PL, Klinkowstein RE, Shefer RE, Wishnok JS, Tannenbaum SR (2000). Low-energy biomedical GC-AMS system for C-14 and H-3 detection. Nuclear Instruments & Methods in Physics Research Section B-Beam Interactions with Materials & Atoms 172:40–46

Keller A, Nesvizhskii AI, Kolker E, Aebersold R (2002). Empirical statistical model to estimate the accuracy of peptide identifications made by MS/MS and database search. Analytical Chemistry 74(20): 5383–5392

Lees JE, Richards PG (1999). Rapid, high-sensitivity imaging of radiolabeled gels with microchannel plate detectors. Electrophoresis 20(10): 2139–2143

Lewis VE (1968). An Alpha-Particle Thickness Guage. Nuclear Instruments & Methods in Physics Research 64:293–296

Mauthe RJ, Dingley KH, Leveson SH, Freeman S, Turesky RJ, Garner RC, Turteltaub KW (1999). Comparison of DNA-adduct and tissue-available dose levels of MeIQx in human and rodent colon following administration of a very low dose. International Journal of Cancer 80(4): 539–545

Miyashita, M, Presley, JM, Buchholz, BA, Lam, KS, Lee, YM, Vogel, JS, and Hammock, BD (2001) Attomole level protein sequencing by Edman degradation coupled with accelerator mass spectrometry. Proc. Nat. Acad. Sci. 98(8): 4403–4408

Muller RA (1977). Radioisotope Dating With a Cyclotron. Science 196:489–494

Nelson DE, Korteling RG, Stott WR (1977). Carbon-14: direct detection at natural concentrations. Science 198:507–8

Nelson DE, Morlan RE, Vogel JS, Southon JR, Harrington CR (1986). New dates on northern Yukon artifacts: Holocene not upper Pleistocene. Science 232:749–751

Ognibene TJ, Bench G, Brown TA, Peaslee GF, Vogel JS (2002). A new accelerator mass spectrometry system for C-14-quantification of biochemical samples. International Journal of Mass Spectrometry 218(3): 255–264

Rabilloud T, Vuillard L, Gilly C, Lawrence JJ (1994). Silver-Staining of Proteins in Polyacrylamide Gels– a General Overview. Cellular & Molecular Biology 40(1): 57–75

Richards P, Johnson M, Ray D, Walker C (1999). Novel protein targets for organophosphorus compounds. Chem Biol Interact 119–120:503–11

Synal HA, Jacob S, Suter M (2000). The PSI/ETH small radiocarbon dating system. Nuclear Instruments & Methods in Physics Research Section B-Beam Interactions with Materials & Atoms 172:1–7

Tuma RS, Beaudet MP, Jin XK, Jones LJ, Cheung CY, Yue S, Singer VL (1999). Characterization of SYBR gold nucleic acid gel stain: A dye optimized for use with 300 nm ultraviolet transilluminators. Analytical Biochemistry 268(2): 278–288

Turteltaub KW, Felton JS, Gledhill BL, Vogel JS, Southon JR, Caffee MW, Finkel RC, Nelson DE, Proctor ID, Davis JC (1990). Accelerator mass spectrometry in biomedical dosimetry: relationship between low-level exposure and covalent binding of heterocyclic amine carcinogens to DNA. Proceedings of the National Academy of Sciences of the United States of America 87(14): 5288–5292

Vogel JS (1992). Rapid production of graphite without contamination for biomedical AMS. Radiocarbon 34:344–350

Vogel JS, Turteltaub KW, Finkel R, and Nelson DE . (1995). Accelerator Mass Spectrometry: Isotope Quantification at Attomole Sensitivity. Analytical Chemistry 67:353A-359A

Vogel JS, Keating GA, Buchholz BA (2002). Protein Binding of Isofluorophate in Vivo after Coexposure to Multiple Chemicals. Environmental Health Perspectives 110(Supplement 6): 1031–1036

14 The Use of Microcalorimetric Techniques to Study the Structure and Function of the Transferrin Receptor from *Neisseria meningitidis*

TINO KRELL and GENEVIÈVE RENAULD-MONGÉNIE

14.1 Introduction

Meningococcal disease continues to be a worldwide health problem and can lead to death within several hours if untreated (Begg et al. 1999). There is currently no vaccine to prevent serogroup B meningococcal disease. The proteins which form the transferrin receptor of *Neisseria meningitidis* are promising candidates for inclusion in such a vaccine (Gorringe et al. 1995). The receptor consists of two types of subunits, transferrin-binding protein A and B (TbpA and B), which both have the capacity to independently bind their ligand, human transferrin (htf) (Cornelissen and Sparling 1996). TbpA (100 kDa) is thought to be a porin-like integral membrane protein, which is proposed to serve as a channel for the transport of iron across the outer membrane. TbpB (65–85 kDa) is considered to be an outer membrane protein, which is anchored to the membrane via the lipidated N-terminal part of the protein and an interaction between TbpA and TbpB has been demonstrated (Fuller et al. 1998).

The immunological properties of both proteins have been studied extensively. The receptor (TbpA+B) was shown to induce bactericidal antibodies in laboratory animals and is protective in a mouse infection model (Danve et al. 1993). The immunogenic potential of individual TbpA and TbpB has been studied extensively (Rokbi et al. 2000, West et al. 2001).

In contrast to the detailed knowledge concerning the immunogenicity of the receptor, a number of fundamental questions about the structure and function of the transferrin receptor remain to be answered. Firstly, it is generally accepted that the receptor consists of TbpA and TbpB. However, the architecture of the receptor is unclear and the stoichiometry of TbpB:TbpA in the receptor complex has been described in various ways. Secondly, the study of meningococcal and gonococcal mutants lacking the gene for either TbpA or B has shown that both proteins are required for the optimal uptake of htf iron

Principles and Practice
Methods in Proteome and Protein Analysis
R.M. Kamp, J. J. Calvete, T. Choli-Papadopoulou (Eds.)

(Irwin et al. 1993, Pintor et al. 1998). In contrast to TbpA, which forms a transmembrane pore permitting iron internalization, the role of TbpB in iron uptake remains unclear. Thirdly, it needs to be elucidated whether htf-binding by TbpA and TbpB occurs in an independent or sequential fashion. Both modes of ligand binding are supported by previous reports (Boulton et al. 1999) and reports differ in the number of htf-binding sites per receptor. Fourthly, little is known about the affinities with which the iron-free and iron-loaded forms of htf are bound by the bacterial receptor and how these affinities compare to the human transferrin receptor.

We have decided to address these questions using microcalorimetric techniques, which have become increasingly popular for the study of protein-protein interactions. Two basic microcalorimetric approaches can be distinguished which were termed isothermal titration calorimetry (ITC) and differential scanning calorimetry (DSC).

ITC is a universally applicable technique to determine the thermodynamic parameters for the binding of two ligands. It is a *classic* titration (see Fig. 14.1) of two ligands where the heat changes caused by binding events are recorded. A single ITC experiment permits the determination of the association constant (K_a), stoichiometry (n), free energy (ΔG), enthalpy (ΔH) and entropy (ΔS). The main advantage of ITC over alternative techniques is that both ligands are unmodified and in solution. The fact that no surface immobilization of ligands is necessary was found to be especially important for the study of membrane proteins, which due to their hydrophobicity, tend to bind in a non-specific manner.

DSC is used to determine the thermodynamic parameters of protein unfolding, which is an endothermic process. The protein sample and a reference containing the buffer are heated up at a constant rate and the change of heat capacity as a function of temperature is recorded. Raw data are analysed by curve-fitting procedures to different models using generally either equations for two state transitions (native and denatured) or multiple state transitions. This permits the determination of the thermodynamic parameters of

Fig. 14.1. Isothermal titration calorimetry (ITC) data for the binding of human holo- and apo-transferrin *(htf)* to individual TbpA, TbpB and the receptor complex (TbpA+TbpB). All experiments were carried out at 25 °C. **A** Binding of holo-and apo-htf to TbpA. **B** Binding of holo-htf to TbpB, apo-htf does not bind to TbpB. **C** Binding of holo-htf to the entire receptor (TbpA+TbpB). **D** Binding of apo-htf to the entire receptor. The ITC raw data are shown in the *upper panels* and integrated peak areas in the *lower panels*. From the curve fit, the parameters ΔH (reaction enthalpy), K_D (dissociation constant) and n (reaction stoichiometry) can be determined directly. From the values K_D and ΔH, the change in free energy (ΔG) and entropy change (ΔS) can be calculated using the equation: $\Delta G = -RT\ln(1/K_D) = \Delta H - T\Delta S$, where R is the universal molar gas constant and T the absolute temperature. Thermodynamic parameters are listed in Table 14.1

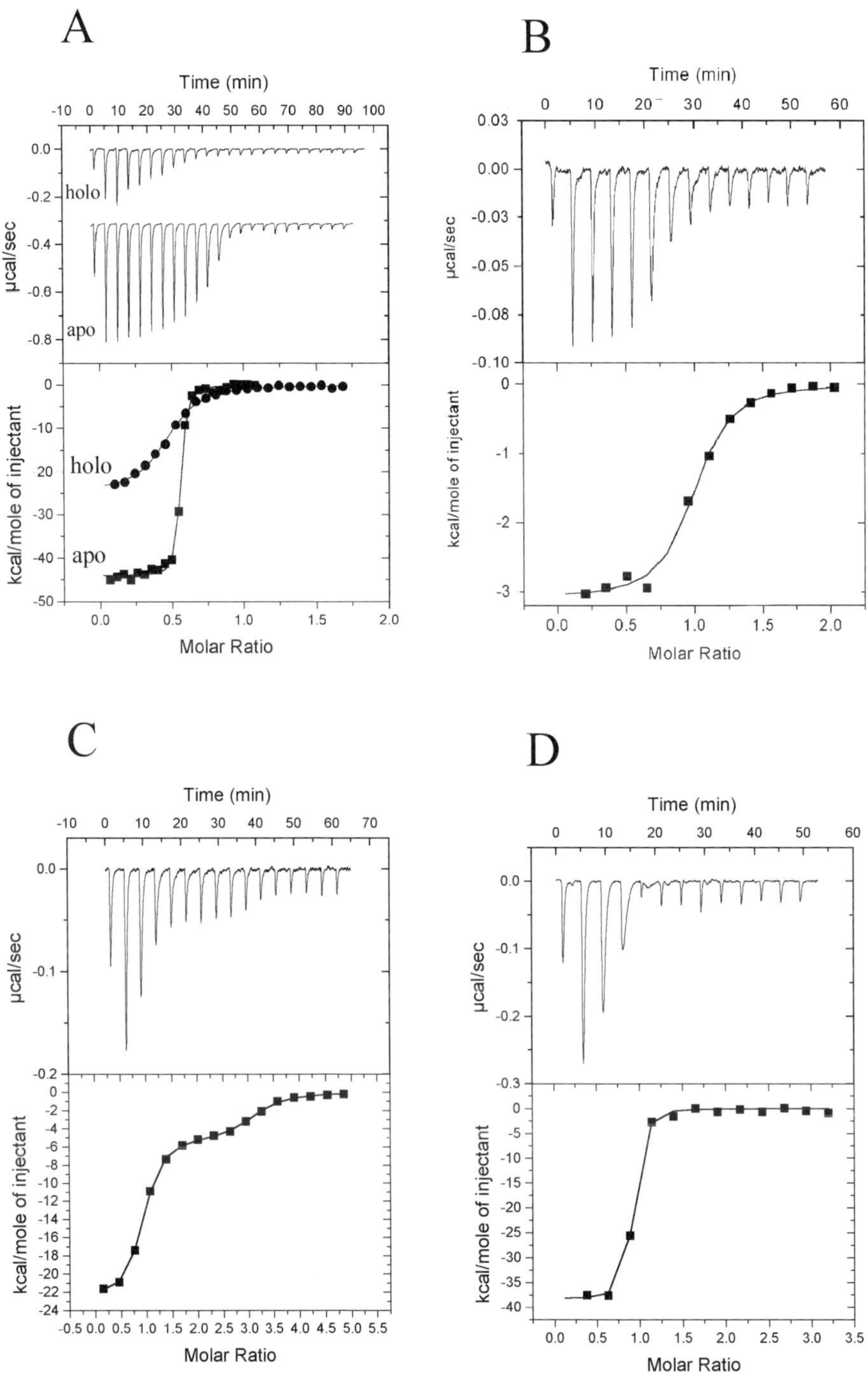
A
Time (min)
µcal/sec
holo
apo
kcal/mole of injectant
holo
apo
Molar Ratio
B
Time (min)
µcal/sec
kcal/mole of injectant
Molar Ratio
C
Time (min)
µcal/sec
kcal/mole of injectant
Molar Ratio
D
Time (min)
µcal/sec
kcal/mole of injectant
Molar Ratio

the thermal unfolding, such as the midpoint of protein unfolding transition (T_m), the enthalpy change on protein denaturation (ΔH_{cal}) which corresponds to the integrated peak area and the change in van't Hoff enthalpy (ΔH_{vH}), expressed by the peak width, which compared to ΔH_{cal} provides information on the content of the cooperatively folding unit. For further information on both techniques see the reviews of Holdgate (2001) and Jelesarov and Bosshard (1999).

14.2 Microcalorimetric Titrations of Individual TbpA, TbpB and the Meningococcal Receptor Complex with Human Iron-Free (apo) and Iron-Loaded (holo) Transferrin

Human transferrin (htf) was shown to bind to both individual proteins of the transferrin receptor and our studies were aimed at assessing the contribution of each protein in the receptor function. Therefore, initial ITC htf-binding studies were carried out using recombinant TbpA (strain K454) and TbpB (strain M982) (Oakhill et al. 2002, Krell et al. 2002). In a second step, binding experiments were repeated with the receptor complex (TbpA+TbpB) purified from *Neisseria menigitidis* M982 (Ala'Aldeen et al. 1994).

14.2.1 Binding of Transferrin to TbpA

Calorimetric titrations of TbpA with apo- and holo-htf are shown in Fig. 14.1A and derived thermodynamic parameters are listed in Table 14.1. The binding of both apo-htf and holo-htf to TbpA was driven by a large enthalpy change and opposed by an unfavourable entropy change. The ΔH for apo-htf binding was substantially larger than the value obtained for the binding of holo-htf, which might indicate that the complex between apo-htf and TbpA is maintained by more or stronger interactions as compared to the complex between holo-htf and TbpA. This difference in ΔH is also reflected in differences in the binding constants. TbpA has an affinity approximately 20 times greater for apo-htf than holo-htf. This observation, in combination with the fact that in vivo the iron-free state is the predominant form of htf, indicates that in the absence of TbpB, iron acquisition from htf is slowed down by a quasi-saturation of TbpA with apo-htf. These data explain previous reports of TbpB-defective bacterial mutants which showed either slower growth on htf-containing medium or reduced iron uptake capacity (Irwin et al. 1993, Pintor et al. 1998). The large unfavourable entropy change observed for the binding of both htf forms by TbpA is most likely to be the consequence of a loss of conformational flexibility upon htf-binding which might also be related to the conformational changes in TbpA following ligand-binding as demonstrated in *N. gonorrhoeae* by Cornelissen et al. (1997b).

Table 14.1. Thermodynamic parameters derived from the microcalorimetric titration of individual TbpA, TbpB and the receptor complex (TbpA+TbpB) from isotype II strains with human holo- and apo-transferrin (see Fig. 14.1). Data are means of at least two independent experiments

Receptor protein[a]	Transferrin form	n	K_D (nM)	K_A (M^{-1})	ΔH (kcal/mol)	$-T\Delta S$ (kcal/mol)	ΔG (kcal/mol)
TbpA	holo	0.48±0.03	68.9±21	$(14.5±4.2)×10^6$	−26.6±2.7	16.8±2.7	−9.8±0.17
TbpA	apo	0.48±0.07	3.7±0.02	$(27.1±0.2)×10^7$	−41.8±3.0	30.4±3.0	−11.4±0.01
TbpB	holo	0.82±0.10	78.1±2.5	$(12.8±0.4)×10^6$	−4.0±0.8	−5.6±0.8	−9.6±0. 02
TbpB	apo	Very weak signal, no analysis possible					
Receptor (TbpA+TbpB):	holo						
Site 1 (binding at TbpA)		0.74±0.13	0.71±0.2	$(14.0±4.2)×10^8$	−23.6±1.3	11.1±1.3	−12.5±0.18
Site 2 (binding at TbpB)		2.20±0.07	22.2±10.4	$(45.0±19.3)×10^6$	−4.6±0.6	−5.8±0.7	−10.4±0.27
Receptor (TbpA+B): binding at TbpA	apo	0.81±0.06	2.0±0.3	$(49.0±8.8)×10^7$	−38.2±1.3	26.4±1.3	−11.8±0.11

[a] TbpA, transferrin-binding protein A; TbpB, transferrin-binding protein B

14.2.2 Binding of Transferrin to TbpB

In contrast to TbpA, the htf-binding to TbpB is driven by favourable enthalpy and entropy changes (Fig. 14.1B, Table 14.1). Around 60 % of the binding energy can be attributed to entropy changes and around 40 % to enthalpy changes (Table 14.1). The observed ΔH of –4.0 kcal/mol is below the average for a protein-protein interaction. The binding constants of holo-htf to individual TbpA and TbpB are similar. In agreement with previous studies (Renauld-Mongénie et al. 1998) the interaction of TbpB with apo-htf was negligible.

14.2.3 Binding of Transferrin to the Receptor Complex (TbpA+TbpB)

The transferrin receptor complex was purified from *N. meningitidis* strain M982 using htf-Sepharose affinity chromatography. The resulting protein was analysed by SDS-PAGE electrophoresis and was found to be very pure, with the two bands identified as TbpA and TbpB.

In contrast to experiments with individual TbpA or TbpB, two different binding events can be distinguished for the binding of holo-htf to the receptor complex (Fig. 14.1C and Table 14.1). Data analysis was carried out using the experimentally determined molecular weight of 300 kDa per receptor complex (Boulton et al. 1998). The data fitted well to the "two independent binding sites" model provided by the ORIGIN software (Microcal, Northampton, USA) and the thermodynamic parameters obtained are listed in Table 14.1. A high-affinity binding site (K_D=0.71 nM) can be distinguished from a low-affinity site (K_D=22.2 nM), and both sites appear to be independent. Most interestingly, the data show around one high-affinity site and around two low-affinity sites (Table 14.1, parameter n), indicating that in total three molecules of holo-htf are needed to saturate the receptor.

Cornelissen and Sparling (1996) have studied the htf-binding to whole cells of *N. gonorrhoeae* using solid-phase and liquid-phase binding assays. The authors conclude that there are two independent htf-binding sites with K_D values of 0.8 and 16 nM. The calorimetric titration of purified receptor with holo-htf provides evidence for there being two independent binding sites with a K_D of around 0.7 and 22 nM (Table 14.1). Considering the technique used by Cornelissen and Sparling is entirely different to our approach, the similarity of both sets of data is remarkable. Furthermore, these values indicate that the neisserial receptor can compete successfully with the human transferrin receptor, which was shown to bind holo-htf with affinities between 5–20 nM. However, the technique employed by Cornelissen and Sparling did not allow the attribution of obtained K_D values to either TbpA or TbpB. This has been achieved by ITC. The iron-free form of htf is known to bind exclusively to TbpA but not to TbpB (Renauld-Mongénie et al. 1998,

Boulton et al. 1998). This implies that the signal observed in the titration of the entire receptor with apo-htf corresponds to binding to TbpA. An ITC titration of the receptor complex with apo-htf at a ligand concentration similar to that used with holo-htf is shown in Fig. 14.1D. The same experiment with a threefold lower concentration of apo-htf was used to determine the binding parameters (not shown). In both cases, there was only around one binding event and the number of binding sites was found to be approximately 1 (n=0.81, Table 14.1). This n-value was close to the corresponding value of the high-affinity binding site seen in the titration with holo-htf (n=0.74, Table 14.1), which identifies the high-affinity site as binding at TbpA. The fact that the low-affinity binding site corresponds to binding to TbpB was also confirmed by the similarity of binding parameters of this low-affinity site present in the receptor and the parameters for the titration of TbpB alone with holo-htf (Table 14.1).

As stated above, the experimentally determined molecular weight of 300 kDa/receptor complex was used for data analysis (Boulton et al. 1998). Under these conditions we provide evidence that the receptor has around one high-affinity site at TbpA and around two low-affinity sites at TbpB. This indirect information on the receptor stoichiometry needs to be validated by a direct analysis of the TbpA-TbpB stoichiometry. The receptor protein (TbpA+TbpB) analysed in this study was purified from *N. menigitidis* M982 grown under iron-limiting conditions. We have studied the expression of tbpB and tbpA genes of *N. menigitidis* M982 grown under iron-limiting conditions using TaqMan real-time quantitative reverse transcriptase-PCR and have observed a ratio of 2:1 for the expression of tbpB and tbpA genes, respectively (Renauld-Mongénie et al., unpubl. data). Using a similar approach a ratio of 2:1 has been reported for the tbpB:tbpA gene expression in *N. gonorrhoeae* (Ronpirin et al. 2001). The data of the tbpB/tbpA gene expression are consistent with the ITC data reported.

14.2.4 Conclusions Concerning the Structure and Function of the Receptor

The model of the transferrin receptor as suggested by Boulton et al. (1999) contains a single htf-binding site. It was proposed that initial htf-binding occurs at TbpB, which then triggers conformational changes in htf and binding by TbpA. This proposition was based on the observation by surface plasmon resonance technology that the affinities of TbpA and TbpB for htf are comparable (Boulton et al. 1999). Here we show that holo-htf binding by the receptor complex occurs with around 30 times higher affinity at TbpA than at TbpB (Table 14.1, K_D of 0.71 vs. 22.2 nM for binding at TbpA and TbpB, respectively), which does not suggest an initial binding event at TbpB. Our data seem to indicate that htf binds directly to the TbpA component of the receptor. In addition, we provide evidence that there are around three binding sites for htf per receptor. TbpA has been identified as the high-affinity site whereas the other two binding sites represent binding at TbpB. Both binding sites appear to be independent and no evidence for cooperative binding was obtained.

We have demonstrated that TbpA (as a component of the receptor) binds holo-htf with an approximately threefold greater affinity than apo-htf (Table 14.1, K_D of 0.71 vs. 2.0 nM for the binding of holo- and apo-htf, respectively). In strong contrast, TbpA alone had a strong ligand preference for apo-htf (Table 14.1, K_D of 68.9 vs. 3.7 nM for the binding of holo- and apo-htf, respectively). We now propose that, within the receptor complex, TbpB shifts the ligand specificity of TbpA towards the iron-loaded form of htf. This proposition is consistent with the data from Cornelissen and Sparling (1996) who demonstrated that the interaction between TbpB and TbpA in *N. gonorrhoeae* is accompanied by conformational changes in both proteins. We also suggest that the specific binding of holo-htf by the TbpB component of the receptor increases the local concentration of free holo-htf in proximity of TbpA. As a consequence, this reduces apo-htf binding by TbpA, increasing the efficiency of iron uptake from htf. Thus, we propose a double function for TbpB: firstly, to shift the ligand specificity of TbpA towards holo-htf and secondly, to increase the local concentration of holo-htf in the proximity of TbpA.

14.3 Generation of Recombinant N- and C-Terminal Domains of TbpB and the Study of Their Interaction

The sequence of the N-terminal half of TbpB can be aligned with its C-terminal half. This observation has led to the hypothesis that TbpB consists of two domains, which are joined by a linker region and it has been suggested that TbpB (isotype II) is a result of a gene duplication event (Mazarin et al. 1995). In order to characterize both domains and to study their implication in binding to htf the two domains were produced as individual histidine-tagged proteins, which were termed N-ter and C-ter.

14.3.1 Isothermal Titration Calorimetry (ITC) Binding Studies

14.3.1.1 Calorimetric Titrations of TbpB, N-ter and C-ter with holo-htf

Previous studies, mainly using solid-phase htf-TbpB binding assays, have indicated that htf-binding occurs primarily at the N-terminal domain of TbpB (Cornelissen et al. 1997a), but the participation of the C-terminal domain in ligand binding has also been demonstrated (Renauld-Mongénie et al. 1997).

We found by isothermal calorimetric studies that C-ter did not bind htf whereas N-ter bound the ligand with a similar affinity as the full-length protein (data not shown).However, the modes of htf-binding by N-ter and His_6-TbpB are very different. Htf-binding by the N-ter was enthalpy driven, with a four times larger ΔH as compared to the full-length protein, and counterbalanced by an unfavourable entropy change. In contrast, htf-binding of full-length TbpB was dominated by a relatively large and favourable entropy change, which compensates for the low enthalpy change. As mentioned above, enthalpy changes are often related to the creation of molecular interactions and one might hypothesize that there is an optimal structural fit between N-ter and htf (large ΔH) resulting in a loss of conformational flexibility (unfavourable entropy change). For the interaction of the full-length TbpB with htf, this optimal fit might not be achieved (maybe because of a steric hindrance caused by the presence of C-ter) and consequently a reduced enthalpy change and a favourable entropy change are observed.

14.3.1.2 Calorimetric Titration of the N-terminal Domain of TbpB with its C-Terminal Domain

Figure 14.2A shows a calorimetric titration of N-ter with C-ter at 25 °C and it is evident that both domains strongly interact with each other. For the titration at 25 °C a K_D of 47 nM [$K_{A=}21.4\times10^6$ M^{-1}, ΔG=–10.0 kcal/mol] was determined. Binding was driven by a large enthalpy change (–41.0 kcal/mol), which was compensated by an also large unfavourable entropy change (–$T\Delta S$= 31.0 kcal/mol), demonstrating a substantial loss in the degree of conformational flexibility upon domain interaction.

This titration was repeated at different temperatures and the enthalpy changes as function of the temperature was plotted (data not shown). The parameter ΔC_p (change of heat capacity) can be calculated from the slope of the linear fit of these data. Negative changes in ΔC_p can be correlated to the burial of apolar surfaces. The ΔC_p for the interaction of both domains of TbpB was found to be –3.16 kcal mol^{-1} K^{-1} (R^2=0.98). The magnitude of the ΔCp value suggests that extensive apolar surface area is buried upon domain interaction. This ΔC_p change was significantly larger than the average value identified for a protein-protein ligand interaction (–0.2 to –0.7 kcal mol^{-1} K^{-1}). To our knowledge, such a large value for ΔCp has not been reported in the literature and might indicate that the nature of protein-protein ligand interaction is rather different to inter-domain interactions. Furthermore, a stoichiometry of 1:1 has been determined (Fig. 14.2A, lower panel) indicating that the recombinant domains are homogeneous samples of correctly folded protein. This domain interaction is accompanied by secondary structure changes as observed by circular dichroism spectroscopy (see below).

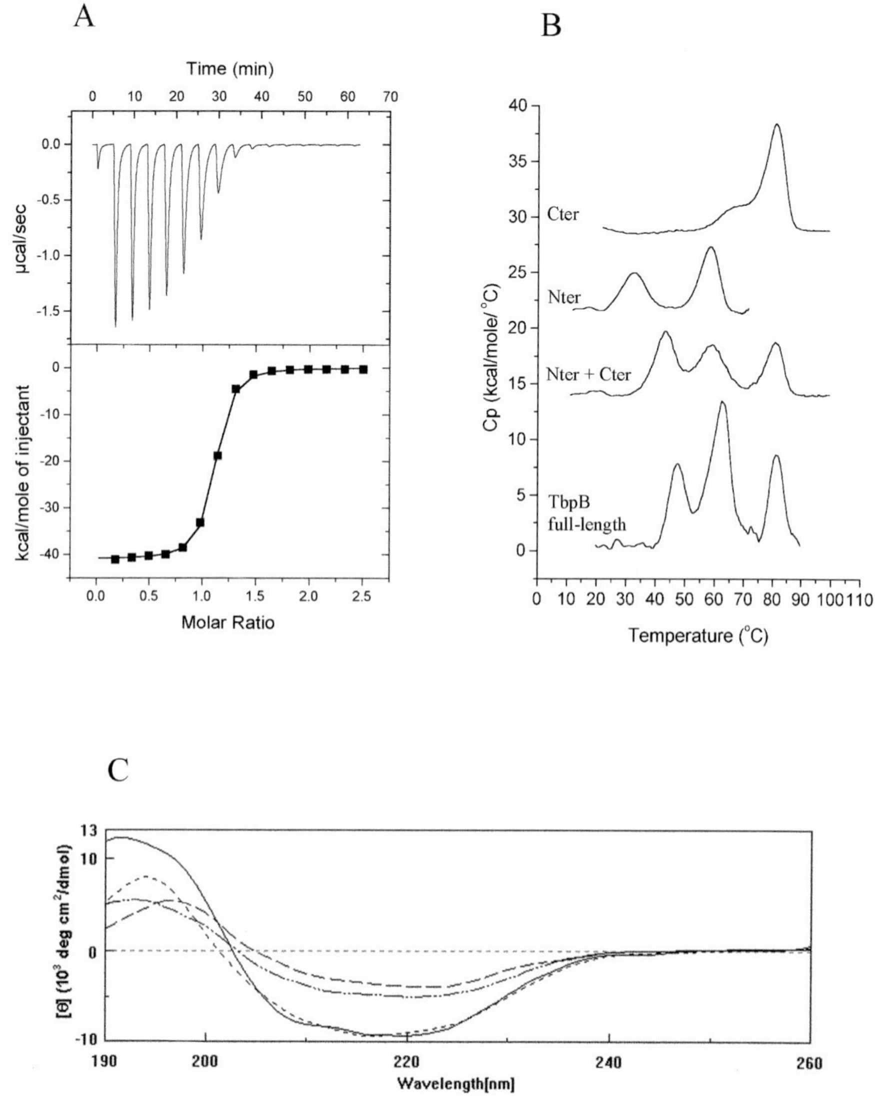

Fig. 14.2. The interaction between the recombinant N- and C-terminal domains of TbpB monitored by isothermal titration calorimetry (ITC), differential scanning calorimetry (DSC) and circular dichroism spectroscopy. **A** ITC data for the injection of recombinant C-terminal domain into a solution of the N-terminal domain of TbpB at 25 °C. **B** DSC analysis of individual recombinant domains of TbpB, a stoichiometric mix of both domains and of the full-length protein. For clarity reasons, traces were moved arbitrarily on the y-axis. Derived thermodynamic parameters are listed in Table 14.2. **C** Far UV circular dichroism spectra of recombinant full-length TbpB and its individual recombinant domains. His_6-TbpB (*solid line*), N-terminal domain (N-ter, *long-dashed line*), C-terminal domain (C-ter, *dashed-dotted line*), a stoichiometric mix of N- and C-terminal domain (*short-dashed line*)

14.3.2 Thermal Denaturation Studies Monitored by Differential Scanning Calorimetry (DSC)

Full-length TbpB and its individual domains have been studied by DSC (Fig. 14.2B, Table 14.2). The denaturation of both the N-ter and C-ter was characterized by two separate unfolding transitions. DSC profiles showing two separate unfolding transitions have been observed in the past for proteins containing two separate domains. It can be proposed that both N-ter and C-ter contain two sub-domains. This is a structural organization also found for the ligand transferrin. The C-terminal domain was found to be fairly thermostable with a T_m beyond 80 °C, which is astonishing considering the relatively low amount of secondary structure determined by cd spectroscopy (see below). N-ter was found to be much less thermostable with transitions centred around 33 and 58 °C. The DSC profile of TbpB was fairly similar to the profile of the stoichiometric mix of both domains (Fig. 14.2B), indicating a close structural resemblance between the full-length protein and the stoichiometric mix of both domains. The first two transitions in both profiles can

Table 14.2. Thermodynamic parameters for the unfolding of the recombinant C- and N-terminal domains of TbpB (C-ter, N-ter), a stoichiometric mix of both domains and of full-length TbpB. (see Figure 14.2B)

Sample	T_m (°C)	ΔH_{cal} (kcal/mol)	ΔH_{vH} (kcal/mol)	$\Delta H_{vH}/\Delta H_{cal}$	Reversibility[a] (%)
C-ter	70.5±0.48	46±2.8	52±2.7	1.13	>80[b]
	80.6±0.04	67±2.3	134±3.2	2.00	81
N-ter	32.6±0.09	35±0.7	71±1.8	2.02	0
	58.0±0.04	49±0.6	111±1.8	2.26	32
N-ter+C-ter	43.2±0.05	53±0.6	82±1.1	1.54	0
	58.9±0.06	56±0.7	58±1.1	1.03	0
	80.3±0.04	38±0.5	124±1.9	3.26	85
TbpB full-length	47.7±0.08	58±1.3	103±2.8	1.77	32[b]
	62.0±0.05	119±1.4	94±1.4	0.79	90[b]
	81.3±0.06	56±1.1	150±3.8	2.68	83

Abbreviations: C-ter, histidine-tagged C-terminal domain of transferrin-binding protein B strain M982; DSC, differential scanning calorimetry; htf, human transferrin; ITC, isothermal titration calorimetry; N-ter, histidine-tagged N-terminal domain of transferrin-binding protein B strain M982; TbpA, transferrin-binding protein A; TbpB, transferrin-binding protein B

[a] Reversibility was defined as: % reversibility=$\Delta H_2/\Delta H_1 \times 100\,\%$ with ΔH_2 being the change of enthalpy from the second up-scan and ΔH_1 the change of enthalpy from the first up-scan of the same protein sample

[b] Interference with the appearance of misfolded protein resulting from unfolding transitions with a higher T_m

be attributed to the N-terminal domain whereas the transition centered at around 80 °C corresponds to the denaturation of the C-terminal domain. The most striking difference between the DSC profiles of the stoichiometric mix of both domains and the profiles of the individual domains was the up-shift of the first transition of N-ter by around 11 °C accompanied by an increase in ΔH from 35 to 53 kcal/mol (Table 14.2). This observed stabilization of one sub-domain of N-ter on interaction with C-ter is thought to account for a large part of the strong exothermic signal observed for the ITC titration of N-ter with C-ter. These data indicate an involvement of one N-terminal sub-domain in inter-domain interaction.

Information obtained from DSC can show insight into the structural arrangement of a protein but can also be used to predict long-term stability of a protein. It is generally accepted that an increased T_m and an elevated degree of the reversibility of protein denaturation indicate favourable long-term stability of proteins, which is a very desirable property of any vaccine component. The thermal denaturation of C-terminal domain either as separate protein, in complex with N-ter or as part of the full-length protein showed a reversibility of 80–85 % (Table 14.2), which is above the average for a protein, whereas the denaturation of the N-terminal domain was much less reversible. The observed thermostability of C-ter and its high degree of reversibility predict increased long-term stability of this domain. These biochemical and thermodynamic properties, together with its immunological properties (Renauld-Mongénie et al. 1997, Rokbi et al. 2000) suggest considerable potential for C-ter as a vaccine antigen.

14.3.3 Circular Dichroism Spectroscopy

The interaction of both recombinant domains has also been monitored by circular dichroism spectroscopy and spectra of both individual TbpB domains, a stoichiometric mix of both domains and of the full-length TbpB are shown in Fig. 14.2C. The molecular ellipticity at 220 nm can be regarded as a measure of the secondary structure content of a protein. Interestingly, both the N-ter and C-ter were found to contain very little secondary structure and a secondary structure content below 10 % was estimated for both recombinant proteins, with C-ter being slightly more structured than N-ter. This is consistent with a large part of both proteins being present in a coil conformation. However, the secondary structure content of a stoichiometric mix of both domains was found superior to that of both domains on their own (Fig. 14.2C). This is evidence that the domain interaction is accompanied by structural rearrangements. The observed conversion of disordered parts of the domains into structured regions is likely to be related to the loss of rotational and translational freedom expressed by the very unfavourable entropy change ($-T\Delta S$=31.0 kcal/mol, see above). In addition, the cd spectra of TbpB was almost superimposable upon the spectrum of the stoichiometric mix of both domains, which confirms the DSC data and indicates that the observed domain interaction corresponds to the interaction of both domains in the full-length TbpB. The role of these structural changes in the function of TbpB remains to be elucidated.

14.3.4 Conclusions Concerning the Structure of TbpB

The data presented confirm that TbpB contains two domains, which can be obtained as individual, active, recombinant proteins. The two recombinant domains were shown to strongly interact with each other. This interaction is accompanied by an increase in secondary structure. ITC studies have shown hat the N-terminal domain of TbpB is involved in htf-binding, whereas the C-terminal does not bind to htf. DSC data suggest that both domains contain two subdomains, a structural organization also found in htf. Furthermore, the recombinant C-terminal domain was found to have thermodynamic parameters, which are favourable for utilization as a vaccine antigen.

References

Ala'Aldeen DAA, Stevenson P, Griffiths E, Gorringe AR, Irons LI, Robinson A, Hyde S, Borriello SP (1994) Immune responses in humans and animals to meningococcal transferrin-binding proteins: implications for vaccine design. Infect. Immun. 62: 2984–2990

Begg N, Cartwright KAV, Cohen J, Kaczmarski EB, Innes JA, Leen CLS, Nathwani D, Singer M, Southgate L, Todd WTA, Welsby PD, Wood MJ (1999) Consensus statement on diagnosis, investigation, treatment and prevention of acute bacterial meningitis in immunocompetent adults. J. Infect. 39:1–15

Boulton IC, Gorringe AR, Allison N, Robinson A, Gorinsky B, Joannou C., Evans RW (1998) Transferrin-binding protein B isolated from *Neisseria menigitidis* discriminates between apo and diferric human transferrin. Biochem. J. 334:269–273

Boulton IC, Gorringe AR., Shergill JK, Joannou CL, Evans RW (1999) A dynamic model of the meningococcal transferrin receptor. J. Theor. Biol. 198:497–505

Cornelissen CN, Anderson JE, Sparling PF (1997a) Characterization of the diversity and the transferrin-binding domain of gonococcal transferrin-binding protein 2. Infect. Immun. 65:822–828

Cornelissen CN, Anderson JE, Sparling PF (1997b) Energy-dependent changes in the gonococcal transferrin receptor. Molec. Microbiol. 26:25–35

Cornelissen CN, Sparling PF (1996) Binding and surface exposure characteristics of the gonococcal transferrin receptor are dependent on both transferrin-binding proteins. J. Bacteriol. 178:1437–1444

Danve B, Lissolo L, Mignon M, Dumas P, Colombani S, Schryvers AB, Quentin-Millet M-J (1993) Transferrin-binding proteins isolated from *Neisseria menigitidis* elicit protective and bactericidal antibodies in laboratory animals. Vaccine 11:1214–1220

Fuller CA, Yu R, Irwin SW, Schryvers AB (1998) Biochemical evidence for a conserved interaction between bacterial transferrin binding protein A and transferrin binding protein B. Microb. Pathog. 24:75–87

Gorringe AR, Borrow R, Fox AJ, Robinson A (1995) Human antibody response to meningococcal transferrin binding proteins: evidence for vaccine potential. Vaccine 13:1207–1212

Holdgate GA (2001) Making cool drugs hot: The use of isothermal titration calorimetry as a tool to study binding energetics. Biotechniques 31:164–184

Irwin SW, Averil N, Cheng CY, Schryvers AB (1993) Preparation and analysis of isogenic mutants in the transferrin receptor protein genes, *tbpA* and *tbpB*, from *Neisseria menigitidis*. Mol. Microbiol. 8:1125–1133

Jelesarov I, Bosshard H (1999) Isothermal titration calorimetry and Differential scanning calorimetry as complementary tools to Investigate the energetics of biomolecular recognition. J. Mol. Recog. 12:3–18

Krell T, Chevalier M, Lissolo L (2002) Affinity-purification of Transferrin-binding protein B under nondenaturing conditions. Protein Express. Purif. 24:323–328

Mazarin V, Rokbi B, Quentin-Millet M-J (1995) Diversity of the transferrin-binding protein Tbp2 of *Neisseria meningitidis*. Gene 158:145–146

Oakhill JS, Joannou CL, Buchanan SK, Gorringe AR, Evans RW (2002) Expression and purification of functional recombinant meningococcal transferrin-binding protein A. Biochem. J. 364:613–616

Pintor M, Gomez JA, Ferron L, Ferreiros CM, Criado MT Analysis of TbpA and TbpB functionality in defective mutants of *Neisseria meningitidis*. (1998) J. Med. Microbiol. 47:757–760

Renauld-Mongénie G, Latour M, Poncet D, Naville S, Quentin-Millet M-J (1998) Both the full-length and the N-terminal domain of the meningococcal transferrin-binding protein B discriminate between human iron-loaded and apo-transferrin. FEMS Microbiol. Lett. 169:171–177

Renauld-Mongenie G, Poncet D, von Olleschik-Elbheim L, Cournez T, Mignon M, Schmidt, MA, Quentin-Millet M-J (1997) Identification of human transferrin-binding sites within meningococcal transferrin-binding protein B. J. Bacteriol. 179:6400–6407

Rokbi B, Renauld-Mongenie G, Mignon M, Danve B, Poncet D, Chabanel C, Caugant D A, Quentin-Millet M-J (2000) Allelic diversity of two transferrin binding protein B gene isotypes among a collection of *Neisseria meningitidis* strains representative of serogroup B disease: Implication for the composition of a recombinant TbpB-based vaccine. Infect. Immun. 68:4938–4947

Ronpirin C, Jerse AE, Cornelissen CN (2001) Gonococcal genes encoding transferrin-binding proteins A and B are arranged in a bicistronic operon but are subject to differential expression. Infect. Immun. 69:6336–6347

West D, Reddin K, Matheson M, Heath R, Funnell S, Hudson M, Robinson A, Gorringe A (2001) Recombinant *Neisseria menigitidis* transferrin binding protein A protects against experimental meningococcal infection. Infect. Immun. 69:1561–1567

15 The Quantitative Advantages of an Internal Standard in Multiplexing 2D Electrophoresis

John Prime, Andrew Alban, Edward Hawkins and Barry Hughes

15.1 Introduction

Two-dimensional electrophoresis (2-DE) is the leading tool in proteomics research today, capable of visualising many components of complex proteomes in a single gel (Gorg et al. 2000). This tool separates proteins by pI (isoelectric point) and molecular weight producing a pattern of spots on an SDS-PAGE (polyacrylamide gel electrophoresis) gel, which can be visualised via a range of staining and labelling systems. Protein spot patterns can be compared between different samples of interest, e.g. normal versus diseased, to determine which protein spots demonstrate changes in abundance. The most common 2-DE work system incorporates the analysis/identification of differentially expressed proteins of interest using Mass Spectrometry. Figure 15.1 illustrates a 2-DE workflow from sample preparation to Mass Spectrometry analysis.

The Ettan DIGE (difference gel electrophoresis) system includes CyDye™ DIGE fluor Cy2™, Cy3 and Cy5 minimal dyes. These are spectrally resolvable mass and charge-matched fluorescent dyes that enable the multiplexing of up to three samples within the same 2-D gel. Approximately 3% of lysine residues present in the sample are labelled at the e-amino group. A protein labelled with any of the CyDye DIGE Fluors will migrate to the same position on a 2-D gel, allowing direct spot volume ratio measurement between samples on the same gel. Figure 15.2 illustrates the workflow for the Ettan DIGE system.

Traditional comparative 2-DE methods involve the separation of samples independently on separate gels. This 'one sample per gel' approach exposes the data to a high level of system variation, for example, the variations in protein uptake into the first dimension strip, second dimension gel running, etc. The system variation is composed of positional variation and volume variation, i.e. the same amount of the same protein run on different gels can migrate to slightly different positions with different spot intensities. This can

Principles and Practice
Methods in Proteome and Protein Analysis
R.M. Kamp, J. J. Calvete, T. Choli-Papadopoulou (Eds.)

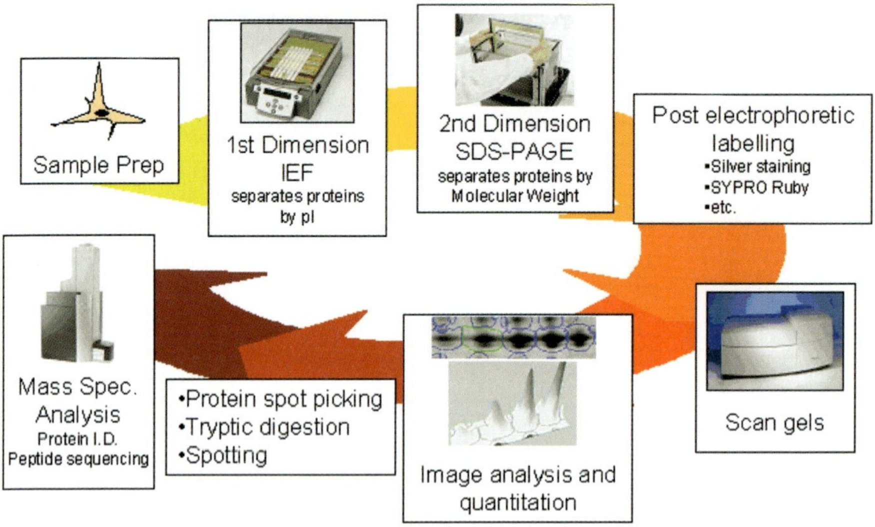

Fig. 15.1. Workflow for a conventional 2-DE system. The basic work process from sample preparation to mass spectrometric analysis are illustrated including: post electrophoretic, protein labelling and the picking of protein spots of interest

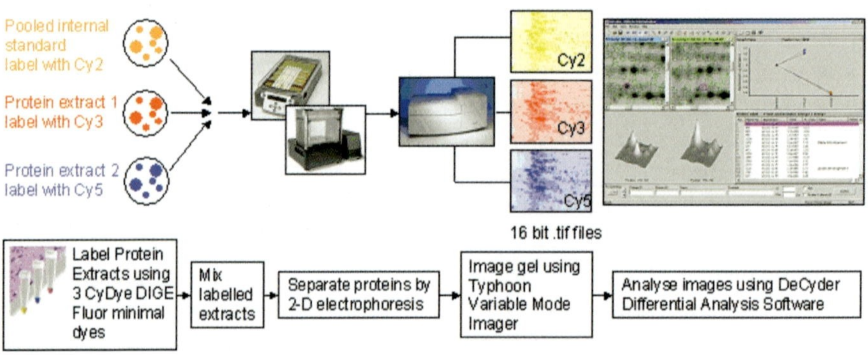

Fig. 15.2. Ettan DIGE system workflow. Different protein samples are labelled with CyDye DIGE Fluor minimal dyes, mixed, undergo 2-D electrophoresis, the resultant gels are scanned and the images analysed with DeCyder to determine significant differences in protein abundance between the different samples

cause problems with accuracy of spot volume quantitation and normalisation across multiple gels for comparison, and also positional gel-to-gel matching of spots (Voss and Haberl 2000, Victor et al. 2002). This high level of system variation can mask the often subtle induced biological changes that the experiment is intended to detect, for example, the differences that are caused by a disease state, drug treatment or life-cycle stage. To compound this problem, it is also necessary to dissect the induced biological changes within an experiment from the inherent biological variations, i.e., the differences between two individual animals, cultures, plants or flies, that are present irrespective of the applied experimental test conditions. To achieve this, multiple sample replicates must be incorporated within each experimental design. This will result in the analysis of a large number of samples, which is a slow and time-consuming process as only one sample can be separated on each individual gel.

The multiplexing properties of the Ettan DIGE system can be exploited to effectively remove system variation by the inclusion of the same pooled standard sample on each gel. Ideally, the pooled standard sample will comprise equal amounts of each sample in an experiment, so that all spots present in an experiment are represented in the standard and therefore will be present on each gel. This experimental design has benefits in terms of system variation and reproducibility, and also gel-to-gel matching. Protein spots are quantified using the novel DeCyder™ Differential Analysis Software. This software has been specifically designed for the Ettan DIGE System incorporating the pooled standard experimental design, by measuring the in-gel volume ratio of a standard:sample in an overlaid co-detected image pair. Differences in spot intensity, for example due to protein loss during entry into the strip, will not affect quantification. As the standard and sample were separated on the same gel, the same relative amount of a protein in each colour channel will have been lost and therefore the measured ratio will be unaffected. The benefits of matching occur as the pooled standard is used to match spot patterns from different gels. The fact that the same sample is matched from different gels rather than different samples, which may have different spot patterns, means that the confidence of gel-to-gel matching is higher.

A range of studies using Ettan DIGE for quantitative proteome analysis have demonstrated the sensitivity and reproducibility of the system. These include the characterisation of protein expression changes in Escherichia coli induced by treatment with benzoic acid (Yan et al. 2002), identification of common protein expression changes caused by different genetic alterations in mice (Skynner et al. 2002), identification of oesophageal cancer specific protein markers (Zhou et al. 2002) and characterisation of proteins differentially expressed in response to the growth factor ErbB-2 in epithelial cells (Gharbi et al. 2002).

A recent study has demonstrated the advantages of using the pooled standard experimental design with the Ettan DIGE system to measure known changes in lysates spiked with known proteins (Alban et al. 2003). The study described here, was designed to compare quantitative proteome analysis by conventional 'one sample per gel' 2-DE with the Ettan DIGE System incorporating the pooled standard experimental design.

15.2 Materials and Methods

C[illegible]Cy3 and Cy5, Pharmalytes (pH 3–10), DryStrip Cover Fluid, PlusOne™ Bi[illegible]ilane and ReadySol acrylamide solution (40 % w/v acrylamide mon[illegible]er solution containing 3 % w/v *N,N*-methylenebisacrylamide) were from Amersham Biosciences UK Ltd. (Buckinghamshire, UK). Conalbumin, bovine serum albumin (BSA), glyceraldehyde-3-phosphate dehydrogenase (GAPDH) and trypsin inhibitor were from Sigma (Dorset, UK). Anhydrous dimethylformamide (DMF) was from Aldrich (Dorset, UK). SYPRO™ Ruby protein gel stain was from Molecular Probes (Oregon, USA).

15.2.1 Sample Preparation and Labelling

E. coli strain ER1647 (Amersham Biosciences, Buckinghamshire, UK) was grown overnight in glucose rich MOPS media at 37 °C followed by harvesting by centrifugation for 10 min at 4 °C at 12,000×*g*. The cell pellet was washed twice with wash buffer (10 mM Tris pH 8.0, 0.5 mM magnesium acetate). Cells were then resuspended in lysis buffer (8 M urea, 4 % w/v CHAPS, 10 mM Tris pH 8.0) and lysed by sonication (3×10-s pulses on ice). The protein concentration of the *E. coli* lysate was determined using the Bio-Rad Dc Protein Assay as described by the manufacturer (Bio-Rad, Hertfordshire, UK). The model proteins were dissolved in a lysis buffer to give stock solutions with final concentrations of 1 mg/ml which were then used to prepare the experimental samples.

The *E. coli* lysate was spiked with the four proteins, at eight different amounts to give eight sample types – simulating time points. Each protein spike was added to the lysate in such a way that the trend of the level of the spike over the eight samples was different for each protein, as shown in Fig. 15.3. Table 15.1 indicates the amount of model protein in each sample sufficient for one gel loading (50 μg protein per CyDye per gel or 50 μg protein per gel for the conventional 'one sample per gel' analysis).

In the case of the samples prepared for the Ettan DIGE system analysis, a pooled standard was prepared by combining equal aliquots from each of the eight samples. Samples for ten gels were prepared and the pooled standard sample was produced by pooling 100 µg protein from each of the samples, prior to labelling.

Samples for the Ettan DIGE system study were combined for electrophoresis as described in Table 15.2. Each sample was run in triplicate for both the conventional 2-DE and the Ettan DIGE system to allow statistical analysis, consequently requiring 24 gels for the conventional 2-DE (see Table 15.3) and 12 gels for the Ettan DIGE system (see Table 15.2).

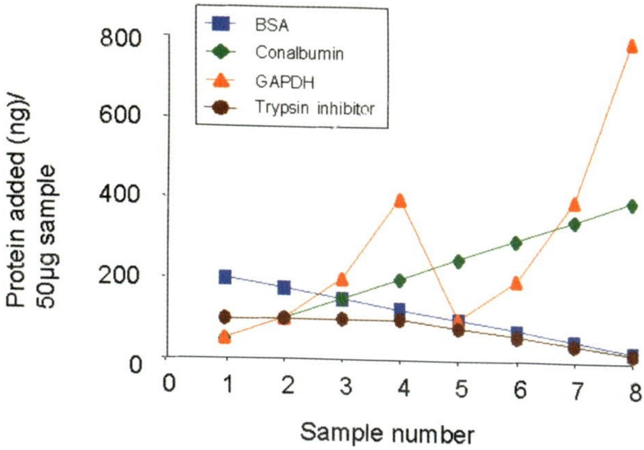

Fig. 15.3. Graph illustrating differing amounts of the protein spikes added to each 'simulated time-point' sample (*1–8*)

Table 15.1. Amounts of model proteins used in each sample

Sample No.	1	2	3	4	5	6	7	8
BSA (ng)	200	175	150	125	100	75	50	25
Conalbumin (ng)	50	100	150	200	250	300	350	400
GAPDH (ng)	50	100	200	400	100	200	400	800
Trypsin Inhibitor (ng)	100	100	100	100	80	60	40	20

Table 15.2. Ettan DIGE experimental design

Gel	Cy2	Cy3	Cy5
Gels 1–3	Pooled standard	$Sample_1$	$Sample_2$
Gels 4–6	Pooled standard	$Sample_3$	$Sample_4$
Gels 7–9	Pooled standard	$Sample_5$	$Sample_6$
Gels 10–12	Pooled standard	$Sample_7$	$Sample_8$

Table 15.3. Conventional 'one sample per gel' experimental design

Gel	Sample (post-stained with SYPRO Ruby)
Gels 1–3	$Sample_1$
Gels 4–6	$Sample_2$
Gels 7–9	$Sample_3$
Gels 10–12	$Sample_4$
Gels 13–15	$Sample_5$
Gels 16–18	$Sample_6$
Gels 19–21	$Sample_7$
Gels 22–24	$Sample_8$

15.2.2 CyDye Pre-Labelling of Protein Samples for the Ettan DIGE System

Cyanine dyes were reconstituted in 99.8 % anhydrous DMF and added to labelling reactions in a ratio of 400 pmol CyDye: 50 µg protein. Protein labelling was achieved by incubation on ice in the dark for 30 min. The reaction was quenched by the addition of 10 mM lysine (1 µl per 400 pmol dye) followed by incubation on ice for a further 10 min.

15.2.3 2-D Gel Electrophoresis

Prior to isoelectric focusing (IEF), the pre-labelled samples for the Ettan DIGE system analysis to be separated in the same gel were mixed and added to an equal volume of 2x sample buffer (8 M urea, 4 % w/v CHAPS, 130 mM DTT, 2 % v/v Pharmalytes 3–10). The individual samples for the conventional 2-DE analysis were also added to an equal volume of 2X sample buffer as above.

Both sets of samples underwent IEF and SDS-PAGE using the following conditions. 2D electrophoresis was performed using Amersham Biosciences 2-D PAGE equipment and PlusOne reagents (Amersham Biosciences, Buckinghamshire, UK). Immobiline™ DryStrip gels (pH 3–10 NL, 24 cm) were rehydrated overnight in 450 µl rehydration buffer (8 M urea, 4 % w/v CHAPS, 1 % Pharmalytes (pH 3–10), 2 mg/ml DTT) were overlaid with 2.5 ml DryStrip Cover Fluid, in an Immobiline DryStrip reswelling tray. Samples were applied to IPG strips via cup loading near the basic end of the strips. Strips were focused using IPGphor cup loading strip holders on the IPGphor isoelectric focusing system for a total of 120 kVh at 20°C. Prior to SDS PAGE, each strip was equilibrated with 10 ml equilibration buffer A (8 M urea, 100 mM Tris-HCl pH 6.8, 30 % v/v glycerol, 1 % w/v SDS, 5 mg/ml DTT) on a rocking table for 10 min, followed by 10 ml equilibration buffer B (8 M urea, 100 mM Tris-HCl pH 6.8, 30 % v/v glycerol, 1 % w/v SDS, 45 mg/ml iodoacetamide) for a further 10 min.

The strips were then loaded and run on 12.5 % acrylamide isocratic Laemmli gels (Laemmli 1970) using the Ettan DALT*twelve* apparatus. The gels to be used for the conventional 2-DE gels were bind silane treated on the back plate (wiped over with 3.5 ml of bind silane solution [0.1 % (v/v) Bind silane, 80 % (v/v) ethanol and 2 % (v/v) acetic acid and left for 1 h]. Gels were run at 5 W per gel constant power at 20 °C until the proteins had entered the resolving gel followed by 10 W per gel constant power at 20 °C until the bromophenol blue dye front had run off the bottom of the gels.

15.2.4 Image Acquisition of Ettan DIGE System Gels

The CyDye fluor labelled proteins in Ettan DIGE system gels were visualised using the Typhoon 9410 imager (Amersham Biosciences, Buckinghamshire, UK). The Cy2 images were scanned using a 488-nm laser and an emission filter of 520 nm BP (band pass) 40. Cy3 images were scanned using a 532-nm laser and an emission filter of 580 nm BP30. Cy5 images were scanned using a 633-nm laser and a 670-nm BP30 emission filter. The narrow BP emission filters ensure that there is negligible cross-talk between fluorescence channels (<1 %). All gels were scanned at 100 μm resolution. Images were cropped to remove areas extraneous to the gel image using ImageQuant Version 5.0 (Amersham Biosciences, Buckinghamshire, UK) prior to analysis with the DeCyder Version 4.0 Differential Analysis software.

15.2.5 SYPRO Ruby Post-Staining of Conventional 2-DE Gels

Upon completion of the SDS-PAGE run, the front plates from these gels were carefully removed (the bind silane ensuring gel remained attached to the back plate). The gels were put in 30 % (v/v) ethanol, 7.5 % (v/v) acetic acid for 1 h. They were then stained overnight with undiluted SYPRO™ Ruby protein gel stain that had been filtered through Whatman™ filter paper No. 1 qualitative 18.5 cm. The stain was then removed and the gels were rinsed in four changes of deionised H_2O for 2 h to remove SYPRO Ruby crystals. Destaining was then performed for 1 h in 10 % (v/v) methanol, 7 % (v/v) acetic acid solution. The glass front plates were replaced on the gels and they were scanned on the Typhoon 9410 imager at 100-μm resolution using a 457-nm laser and an emission filter of 610 nm BP30. The images were cropped as with the Ettan DIGE system gels prior to analysis using Progenesis analysis software.

15.3 Results

An *E. coli* lysate was spiked with varying amounts of four commercially available model proteins as shown in Table 15.1. For the Ettan DIGE system analysis, three samples were run on each gel; a pooled standard labelled with Cy2 and two different samples labelled with Cy3 and Cy5 giving a total of eight data points over the range of four gels. The pooled standard contained an equal (μg protein) aliquot from each of the samples and represented an average of all the samples being compared. For the 'one sample per gel' analysis, the same range of samples were analysed but with only one sample on each gel. These gels were visualised with SYPRO ruby staining. The protein loading for both experiments was 50 μg of each sample per gel, and each sample was run three times in order to obtain valid statistical data. In the Ettan DIGE experiment the eight different samples were analysed and compared in triplicate by running 12 gels (Table 15.2). The same number of samples required 24 gels for this analysis using the conventional single sample techniques (Table 15.3).

15.3.1 Ettan DIGE System Analysis

The gel images from the Ettan DIGE system were analysed using DeCyder Differential Analysis Software, a 2-D analysis platform designed specifically for use with the Ettan DIGE system. Fully automated spot detection and quantitation was performed on overlaid Cy2 (Standard Pool)/Cy3 (Sample_x) and Cy2 (Standard Pool)/Cy5 (Sample_y) image pairs from each gel, followed by automated gel-to-gel matching and statistical analysis (see Fig. 15.4).

An advantage of sample multiplexing in DIGE is that detected spot boundaries from the same gel will overlay perfectly, and therefore gel warping is not required to match the spots within image pairs from the same gel. This allows direct measurement of spot volume ratios of spots present in an image pair. The fundamental benefit of this technique is the ability to co-detect and compare each sample in-gel with a pooled standard. This enables accurate automated gel-to-gel matching, as the same complete spot map is present on each gel. As the Cy3 and Cy5 spot maps were all co-detected with a pooled standard Cy2 image, the Cy3 and Cy5 spot maps from different gels were also simultaneously matched (see Fig. 15.4). This demonstrates a key advantage of the pooled standard experimental design and the Ettan DIGE system, in that the same sample is used for gel-to-gel matching, rather than matching gels that contain different samples and which may have different spot patterns.

For a given spot, the standardised abundance was expressed as a volume ratio between the pooled standard and a co-detected sample from the same gel. When comparing spot abundance between samples across different gels, DeCyder software compares how the abundance ratio measurements of the same protein spot from different samples and gels relate to the standard sample and further normalisation is not necessary. In this way, as each sample spot map is co-detected with a standard spot map, all of the spots are compared in-gel to the same pooled standard. This separates system variation from *real* induced biological changes in protein spot abundance.

Analysis of variance (ANOVA) was applied to matched spots and the data was filtered to retain spots with ANOVA p values of 0.05 or less, and spots that had an appearance in 36 or more of the 48 spot maps (for each gel the standard spot map appears twice in the analysis, Cy2/Cy3 and Cy2/Cy5, giving a total of 48 spot maps). This resulted in the display of 52 spots, 43 of which occurred in a series of four clusters of spots (Fig. 15.5a). The spots that were significant but did not occur in the clusters included artefacts such as streaks.

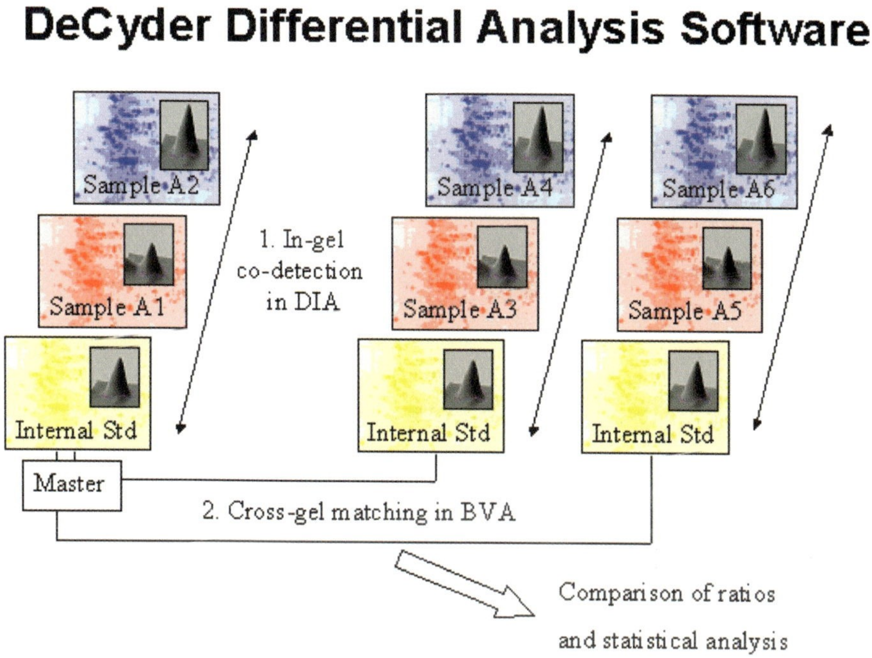

Fig. 15.4. DeCyder matching and analysis. Pictorial description of the DeCyder image analysis (DIA) process. The two samples and internal standard perfectly overlay in the same gel, consequently, this enables image pairs consisting of the internal standard and a sample from the same gel to be co-detected and the spot volume ratios between the sample and standard determined. A master (usually the best gel containing the most spots) is matched to the internal standards of the other gels in the study, allowing comparison of spot volume ratios and the accurate production of abundance ratios between samples on different gels. These data can then undergo statistical analysis. Matching and statistical analysis are performed in the DeCyder-BVA (biological variation analysis) software module

The data were ordered by ANOVA p value and the 37 most significant spots were all situated in the four clusters. Out of these spots the most significant had an ANOVA p value of 2.2×10^{-19} and the least significant a p value of 8.4×10^{-4}. The positions on the gels of the four spot clusters relative to each other are consistent with the predicted pI and molecular weights of the protein spikes. This suggests that each of the four model proteins is present as multiple differently charged isoforms in the gels. The spot clusters are shown in Fig. 15.5b. Cluster 1 is consistent with the pI and molecular weight of BSA, cluster 2 is consistent with conalbumin, cluster 3 with GAPDH and cluster 4 with trypsin inhibitor.

Graphs of standardised abundance for a representative spot from each of the clusters, shown in Fig. 15.6, were generated automatically by the DeCyder

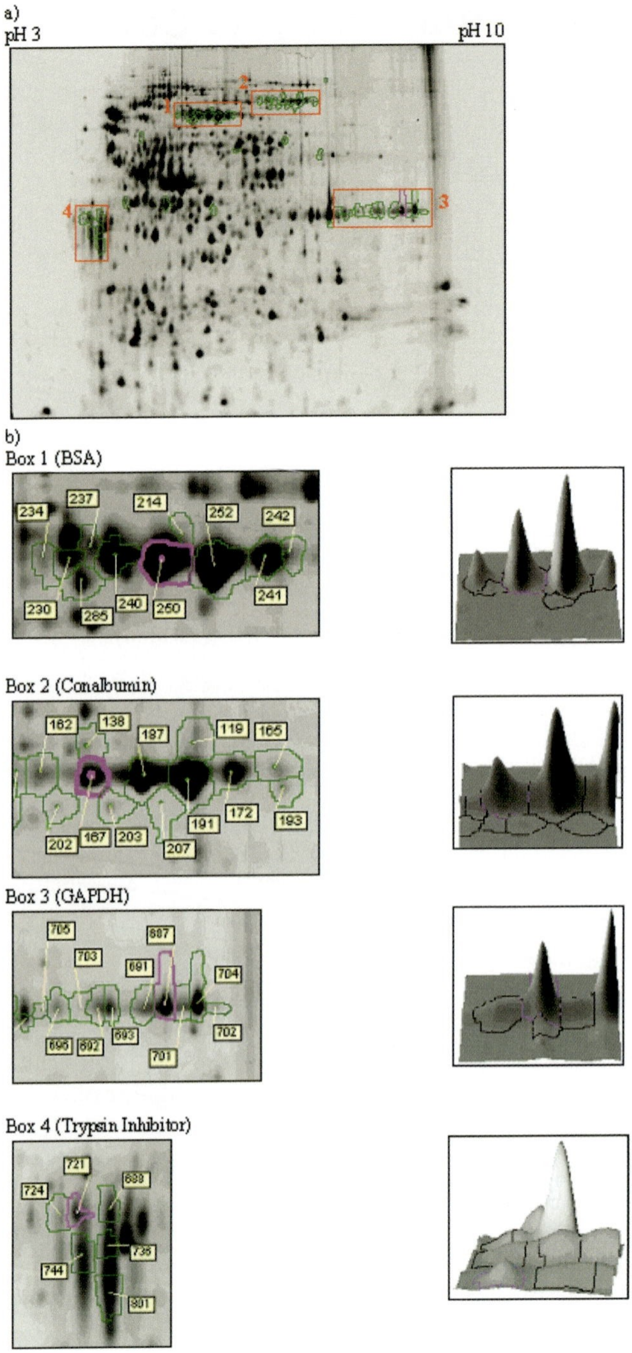

Fig. 15.5. **a** Location of the four spiked protein spot clusters on an Ettan DIGE gel. *Box 1* BSA, *Box 2* conalbumin, *Box 3* GAPDH and *Box 4* trypsin inhibitor. **b** Two- and three-dimensional views of the spiked protein clusters from the DeCyder analysis. Data for the graphs (Fig. 15.6) were taken from the spots highlighted

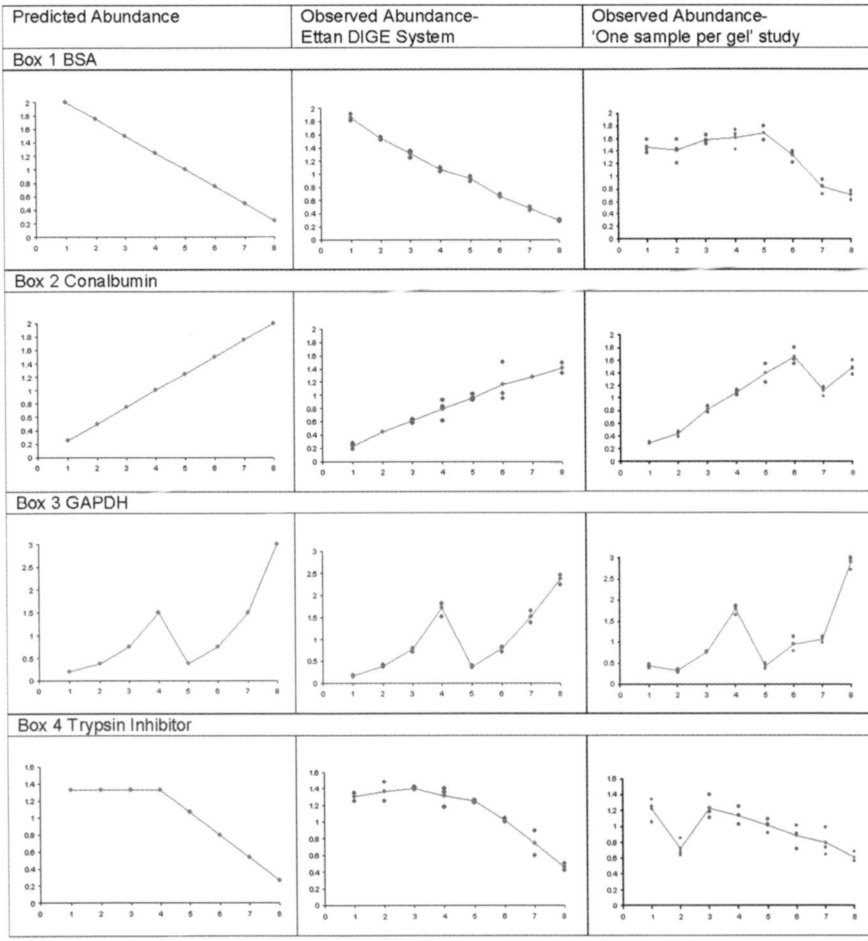

Fig. 15.6. Comparison of predicted abundance of spiked proteins with those observed with the Ettan DIGE system and the 'one sample per gel' study. The *spots* indicate individual data points, and the *line* indicates the trend in abundance for each protein given by plotting it through the mean abundance values. These graphs give the results as normalised volume ratios

software by displaying the standardised abundance values relative to the pooled standard. Standardised protein abundance increases along the *y*-axis and the time points are displayed along the *x*-axis from 1 to 8. Each dot on the graph represents the same matched spot from a different gel. The spread of the dots in the y-axes direction indicates the variation in abundance of the protein spot in the triplicate gels belonging to the same time point following normalisation. Predicted standardised abundance graphs were generated by averaging the protein loadings for each protein in Table 15.1, and then dividing each individual loading value by the average. The range and the overall profile of the standardised abundance graphs closely resemble the predicted abundance graphs of all four proteins.

15.3.2 Image Analysis of Conventional 'One Sample Per Gel' SYPRO Ruby Stained Gels with Progenesis

The conventional 'one sample per gel' SYPRO Ruby stained gel images were all cropped to a similar size to reduce peripheral spot finding. The images were analysed using Progenesis™ Discovery v 2003.01, the latest top of the range version available from NonLinear Dynamics (NLD). Progenesis Discovery is specifically designed for the analysis of conventional 'one sample per gel' images produced from 2-DE studies comparing differing proteome samples.

Multiplexing is not an option within the conventional 'one sample per gel' SYPRO Ruby /Progenesis Discovery system utilised here, and so the overall experiment required running a total of 24 gels in 8 triplicate sets. All 24 gels were included in a single automated Progenesis analysis, set up so that each group of three replicate gels were initially analysed individually. An average (representative) gel is then created by matching and averaging the three gels in each of the eight groups.

The final stage of the automatic analysis is to create an overall reference gel, which is derived from matching and averaging all eight group averaged gels. Following fully automated analysis, every gel was viewed individually, and spot editing carried out as necessary. Following spot editing and background subtraction the images were re-matched using both manual and automated matching methods. The final stage of the analysis was to extract the required spot information and to calculate average values for each spot.

Results for each spiked area were obtained by individual analysis of each of the 24 images; the normalised result was used in each case. The normalised spot value was obtained by determining the percentage ratio of the total spot density for each gel compared to the density of the spot of interest in that gel. The three values for each spot were averaged, except in one or two cases, where results were not available for all three gels in a group, in which case only two values were used. Spots were not easily identifiable for BSA at time point 5 and conalbumin and GAPDH at time points 3 and 5 in one of their respective gels. The same representative spot was again chosen from each of the clusters produced by the spiked proteins as were chosen in the Ettan DIGE system work.

In a number of instances normalised data were also generated by a *group* analysis, in which all three images were processed automatically. In some instances the automatic spot matching worked sufficiently well to allow averaged values to be calculated. On every occasion where all three spots in the three images were fully matched, results obtained were compared to the manually obtained results. In every case, agreement between the manual and the automatically generated data was extremely high, differing only by a few percent.

These protein abundances are given in Fig. 15.6 alongside those determined by the Ettan DIGE system and the predicted abundance values. Individual results for each proteins representative spot at each time point are illustrated along with the average abundance value and trend. Figure 15.7 illustrates a SYPRO Ruby stained gel from the conventional 'one sample per gel' 2-DE analysis with the four clusters of spots for the four spiked proteins highlighted in Fig. 15.7a. Figure 15.7b contains two- and three-dimensional plots from the Progenesis Discovery image analysis software illustrating the representative spot from each cluster during analysis.

A number of drawbacks and problems were observed with the Progenesis analysis. One major problem was that although regions of interest were set for all images (defining the region taken for normalisation) and these were checked to make sure that they were equal in all images, some artefacts may have made a significant difference to the calculated percentages for specific spots. Some streaks can significantly affect overall percentages. Efforts were made to remove such density from images where feasible, but it was not always possible to do this in a realistic timeframe. This problem is not encountered in the Ettan DIGE system as normalisation of a specific spot is performed against the same spot in the pooled standard and therefore is not effected by the presence of artefacts or other deleterious effects present elsewhere on the gel.

The analysis was completely automated, from spot finding through to warping, matching and quantitation. However, spot detection in Progenesis, even using the top of the range Discovery product, was not as accurate and reliable as in DeCyder. As a result, every image had to be manually edited to ensure that spot boundaries were as accurate as possible. Thus the results obtained, which are clearly based on the spot boundaries used, are in effect user dependent and subjective rather than entirely objective. Two different users are extremely likely to generate two different sets of results, depending on interpretation, thus creating user to user variability. This is an inherent weakness with any 2-DE image analysis system that relies upon the operator to edit the spot boundaries and undoubtedly reduces the confidence in the accuracy with which any results could be held.

The GAPDH spots were often poorly resolved in the SYPRO Ruby images. As a result, rather than selecting a specific spot and following its values throughout the analysis, it was necessary to quantitate a group of unresolved spots, and to compare these results. The individual spots could only be clearly identified in a few instances. The images in Fig. 15.8 shows the problem. The image on the left-hand side in Fig. 15.8.a is typical of the spot separation; the right-hand image 8.b was found in only a few of the 24 images. This spot separation could be due to a number of causes, including poor separation of the

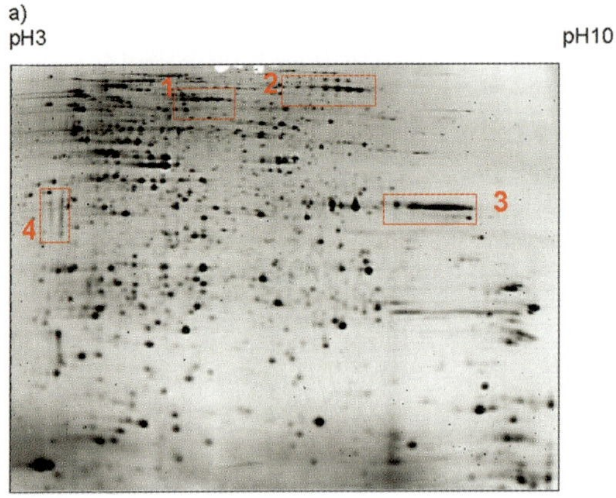

Fig. 15.7. **a** Location of the four spiked protein spot clusters on a SYPRO Ruby stained conventional 2-DE gels. *Box 1* BSA, *Box 2* conalbumin, *Box 3* GAPDH and *Box 4* trypsin inhibitor. **b** Two- and three-dimensional views of the spiked protein clusters from the Progenesis Discovery image analysis software. Data for the graphs (Fig. 15.6) were taken from the spots highlighted

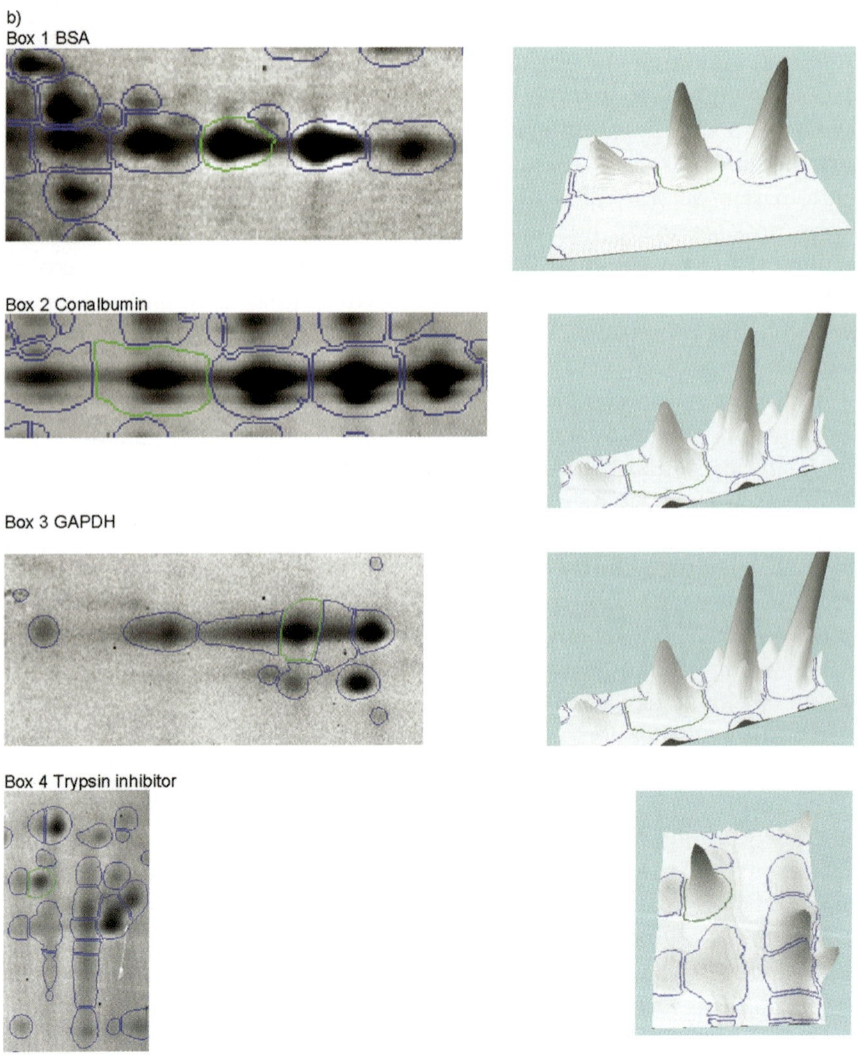

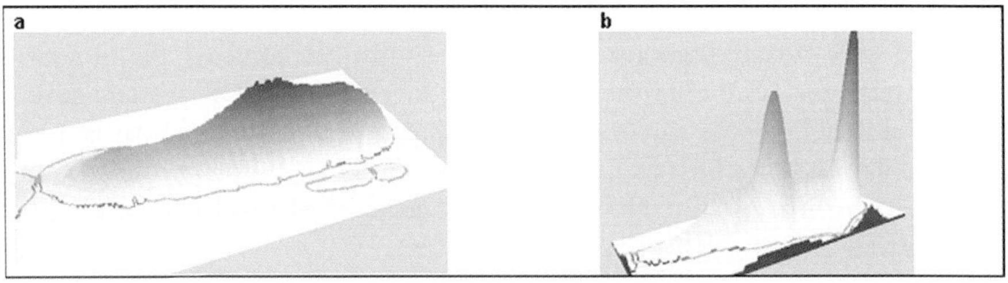

Fig. 15.8. Three-dimensional images from Progenesis illustrating **a** typical poor spot separation found in the GAPDH cluster in 'one sample per gel' 2-DE gels and **b** good quality spot separation evident in only a couple of the same gels

GAPDH protein isoforms during IEF, GAPDH is a basic protein and consequently difficult to resolve. Some problems with GAPDH spot resolution were also encountered with the Ettan DIGE system but the representative spot was still identifiable and sufficiently resolved to be individually analysed on each gel. Consequently, the GAPDH protein abundance graph for the conventional 2-DE system must be taken in the light that the spot abundance data were not generated from the same representative spot as was the case with the Ettan DIGE system. However defined amounts of GAPDH protein were added (spiked) into the different time-point samples from the same source for both the Ettan DIGE system and conventional 2-DE samples, and they therefore should contain the same relative proportions of each GAPDH isoform. It can be postulated that although the actual amounts of GAPDH in the spots could not be compared between the Ettan DIGE system and conventional 2-DE samples, both the total density for the GAPDH cluster or the individual representative spot density were still effective indicators of GAPDH abundance, as long as they were used consistently for the analysis of the conventional 2-DE system. In this case the total density for the GAPDH protein cluster was used for all 24 gels in the 'one sample per gel' study.

15.3.3 Comparison of Quantitative Proteome Analysis Results Between the Two Systems

The predicted and observed protein abundances obtained from the two 2-DE systems can be seen in Fig. 15.6. A comparison of the results between these two systems in relation to the predicted abundance graphs is detailed below.

15.3.3.1 BSA

The Ettan DIGE system gave a protein abundance profile across the eight time points for BSA (Fig. 15.6, box 1), which very closely matched the predicted trend. Additionally the individual abundance points for the replicates grouped very closely together for each time point. In contrast, the observed abundances for BSA obtained with the conventional 2-DE system gave a poorly matched profile in comparison with the predicted trend. In fact, the observed abundance bears a much closer resemblance to the predicted trend for the Trypsin inhibitor. The grouping of individual results for each time point is tighter than for the Ettan DIGE system.

15.3.3.2 Conalbumin

Again, the resultant observed protein abundance trend for conalbumin (Fig. 15.6, box 2) bears a close resemblance to the predicted graph, although the slope of the line is not as steep and reaches ~1.4 at time point 8 rather than the expected time point 2. Individual spot groupings vary from being very tight to a wide spread in the case of time points 4 and 8. In general, these individual spot groupings are similar to those observed by the conventional 2-DE system. However, the conventional 2-DE system graph is not a simple line increasing in protein abundance with increasing timepoint. Time points 1 to 6 do maintain a good correlation with the predicted graph but time point 7 decreases, indicating a consistent result for its three replicate gels and time point 8 increases again, but is not in line with the original trend established by the first six time points.

15.3.3.3 GAPDH

Of the 4 protein spikes the GAPDH (Fig. 15.6, box 3) abundance profile produced by the conventional 2-DE system most closely resembled the predicted graph, despite it being the most complex of the four. The Ettan DIGE system graph also strongly correlated with the predicted abundance graph. Both systems detected the increasing level of GAPDH in the first four time points and clearly delineated the reduction in time point 5. The increase from time points 5 to 8 was again detected by both systems although not as accurately as the first increase in the profile. The Ettan DIGE system detected the second increase with a similarly steepening curve but did not peak as high as the predicted level. Whereas the conventional 2-DE system peaked at a similar level to that in the predicted graph. However, the curve was not as accurate mainly due to the three time point 7 gels underestimating the level of GAPDH. The grouping of individual results at most time points were again close for both systems. One must consider the fact that the 2-DE conventional results for GAPDH will have been affected by the merging of the spots in the cluster necessitating the quantitative analysis of all the spots in most gels rather than the specified representative spot. This could be responsible in part for the accuracy of these results or could have contributed to the reduction in abundance seen at time point 7. If the former is true then this increases the doubt over the capability of the tested conventional 2-DE system to accurately quantitate abundance changes to specific protein spots.

15.3.3.4 *Trypsin Inhibitor*

The Ettan DIGE system detected this spiked protein as a gentle curving reduction from the predicted level down to a level slightly higher than that indicated in the predicted graph. A small reduction is indicated at time point 4 and so the Ettan DIGE system has not determined the plateau in trypsin inhibitor (Fig. 15.6, box 4) abundance to time point 4 with complete accuracy. Nevertheless, the increasing reductions in protein level from time point 5 upwards and the general decreasing slope up to time point 8 have been captured by the Ettan DIGE system. The conventional 2-DE system has also detected the reduction in trypsin inhibitor abundance but a consistently low observation at time point 2 gives the graph a misleading dip for one time point in trypsin inhibitor protein abundance where there should have been a plateau between time points 1 and 4.

The trends for the four spiked protein observed with the Ettan DIGE system closely match the predicted abundance profiles. Although the observed trends are not perfect matches to the predicted values a user performing a quantitative proteome study over a series of time points should be confident that the changes and trends in observed protein abundances for proteins of interest would closely resemble those actually occurring in the system under study. Conclusions could be made with a high degree of confidence about changes in protein abundance over a range of time points utilising this experimental system and incorporating the internal standard experimental design.

In contrast, the conventional 2-DE system did not reliably determine the changes and trends in observed protein abundance. Of the four protein spikes conalbumin and GAPDH bore the closest resemblance to the predicted graphs. The GAPDH results suggest that this conventional 2-DE system is capable of accurately elucidating multifaceted changes in protein abundance from within complex proteome samples, however, the evidence from the other protein spikes tested suggests this system cannot do this reliably. Even in the case of conalbumin and GAPDH some subtle changes in the observed trends were apparent. Furthermore, in the case of BSA and trypsin inhibitor the trends could be easily mistaken for those of other proteins e.g. BSA for trypsin and vice versa. These disparities between the actual and observed protein abundances could easily lead to the misinterpretation of results.

15.4 Conclusions

This study and the work of Alban et al. have demonstrated that a 2-DE experimental design incorporating an internal standard allows accurate quantitative comparison whilst carrying out complex studies like time courses, dose responses etc. (Alban et al. 2003). Inclusion of an internal standard in all the gels in a study, aids gel to gel matching, increasing the confidence in detection and quantitation between different samples on different gels. The internal standard additionally allows the accurate comparison of protein abundance between samples.

The Ettan DIGE system has the added advantage of an extended linear dynamic range coupled to high sensitivity (5 orders of magnitude and >125 pg protein per spot) enhancing the range of protein differences that can be detected to include the subtle smaller differences as evidenced in Fig. 15.6. Conventional 'one sample per gel' 2-DE analysis is able to detect gross changes in protein abundance but even these were with a reduced confidence due to the spurious and potentially misleading variations introduced by this system.

The Ettan DIGE system incorporating an internal standard has an added benefit over the conventional 'one sample per gel' system due to the reduced number of gels required to run the same number of samples, in this case 12 gels compared with 24. Utilising the 2-DE equipment used in this study reduces the number of IEF and SDS-PAGE runs required to perform a study of this size, thus, potentially halving the time taken to perform the first and second dimension sections of the experimental work. This reduction in the number of gels inherently reduces the potential level of gel to gel variability. Furthermore, pre-labelling the samples with the CyDye DIGE Fluor minimal dyes required approximately 2 h to perform compared to the SYPRO Ruby protein staining protocol, utilised in this study to obtain maximum sensitivity, which took an extra day. It should also be noted that the Progenesis image analysis process is significantly more *hands on* than DeCyder, requiring manual spot editing and more manual matching. Other image analysis software packages and protein labelling systems may reduce the time taken with image analysis and protein staining but will still be burdened with the heavier 'one sample per gel' workflow. Thus the Ettan DIGE system with the internal standard offers the experimenter the potential for higher throughput or shorter study times.

If 2-DE is to reach its full potential then it must be able to accurately detect and quantitate differences in protein abundance between samples of interest. These differences must be the result of real biological variations and not due to system/experimental variation. Furthermore, the experimenter needs statistical confidence in these differences if the results of a study are to carry scientific weight. As in some other image analysis packages a statistical confidence level is assigned to all protein abundance changes observed in the Ettan DIGE system, while Progenesis Discovery is not, at present, as sophisticated in terms of its statistical options only offering options for re-sampling and coefficient of variance determination. The inclusion of an internal standard in every gel, which is made possible by the multiplexing nature of the Ettan DIGE system, offers improved experimental accuracy and gives the experimenter greater confidence in their results.

Acknowledgements. The authors gratefully acknowledge the assistance of Judith Pickering, Imogen Horsey and Paul Pashby in the preparation of this manuscript (Amersham Biosciences UK Ltd., 2003. all rights reserved).

References

Alban A, David SO, Bjorkesten L, Andersson C, Sloge E, Lewis S, Currie I, (2003) A novel experimental design for comparative two-dimensional gel analysis: Two-dimensional difference gel electophoresis incorporating a pooled internal standard. Proteomics 3:36–44

Asirvatham VS, Watson BS, Sumner LW, (2002) Analytical and biological variances associated with proteomic studies of *Medicago truncatula* by two-dimensional polyacrylamide gel electrophoresis. Proteomics, 2:960–968

Gharbi S, Gaffney P, Yang A, Zvelebil MJ, Cramer R, Waterfield MD, Timms JF. (2002) Evaluation of Two-dimensional Differential Gel Electrophoresis for Proteomic Expression Analysis of a Model Breast Cancer Cell System, Molecular and Cellular Proteomics 1:91–98

Görg A, Obermaier C, Boguth G, Harder A, Scheibe B, Wildgruber R, Weiss W, (2000) The current state of two-dimensional electrophoresis with immobilised pH gradients. Electrophoresis, 21:1037–1053

Laemmli UK, (1970) Cleavage of structural proteins during the assembly of the head of bacteriophage T4. Nature, 227:680–685

Skynner HA, Rosahl TW, Knowles MR, Salim K, Reid L, Cothliff R, McAllister G, Guest P, (2002) Alterations of stress related proteins in genetically altered mice revealed by two-dimensional differential in-gel electrophoresis analysis. Proteomics, 2:1018–1025

Voss T, Haberl P, (2000) Observations on the reproducibility and matching efficiency of two-dimensional electrophoresis gels: Consequences for comprehensive data analysis. Electrophoresis, 21:3345–3350

Yan JX, Devenish AT, Wait R, Stone T, Lewis S and Fowler S (2002) Fluorescence 2-D difference gel electrophoresis and mass spectrometry based proteomic analysis of *Escherichia coli.* Proteomics, 2:1682–1698

Zhou G, Li H, DeCamp D, Chen S, Shu H, Gong Y, Flaig M, Gillespie JW, Hu N, Taylor PR, Emmert-Buck MR, Liotta LA, Petricoin EF 3rd, Zhao Y (2002) 2D Differential In-gel Electrophoresis for the Identification of Esophageal Scans Cell Cancer-specific Protein Markers. Molecular and Cellular Proteomics 1:117–124

16 Genetic Engineering of Bacterial and Eukaryotic Ribosomal Proteins for Investigation on Elongation Arrest of Nascent Polypeptides and Cell Differentiation

Fotini Leontiadou, Christina Matragkou, Filippos Kottakis, Dimitrios L. Kalpaxis, Ioannis S. Vizirianakis, Sofia Kouidou, Asterios S. Tsiftsoglou and Theodora Choli-Papadopoulou

16.1 Introduction

This paper describes our efforts to investigate the role of specific bacterial and eukaryotic ribosomal proteins in crucial cell functions such as elongation arrest of the nascent polypeptides and cell differentiation. These objectives have been approached by: (1) engineering the L4 bacterial ribosomal protein, which has been shown by crystallographic data to be a candidate molecule for controlling the nascent polypeptides before their emerge from the ribosome, to elucidate specific amino-acids functions in highly conserved regions; and (2) isolating specific eukaryotic genes, like S5 and L35a, that encode ribosomal proteins to study their possible involvement in hematopoietic cell differentiation and apoptosis.

The large ribosomal subunit (50S) in prokaryotes catalyses peptide bond formation and binds initiation, termination and elongation factors. Proteins are abundant everywhere on its surface except in the active site where peptide bond formation occurs and where it contacts the small subunit. The peptide bond formation has been studied for a long time but only recently it is strongly believed to be enlightened by crystallographic data (Nissen et al. 2000). However, the catalytic mechanism of the reaction on the ribosome and the ribosomal groups involved in catalysis is not known and there are still controversies concerning the reaction of the peptide bond formation (Katunin et al. 2002).

Another very important step after peptide bond formation and translocation is the passage of the nascent polypeptide chain through the tunnel, the existence of which is confirmed by crystallographic data (Nissen et al., 2000). It appears very likely from the structure (Tenson and Ehrenberg 2002) that all

Principles and Practice
Methods in Proteome and Protein Analysis
R.M. Kamp, J. J. Calvete, T. Choli-Papadopoulou (Eds.)

nascent polypeptides pass through the exit tunnel before emerging from the ribosome. The length of the tunnel from the site of peptide synthesis to its exit is about 100 Å, which is consistent with the length of nascent polypeptide that is protected from proteolytic cleavage by the ribosome (Picking et al. 1992), and the minimum length required for antibody recognition at the exit. The narrowest part of the tunnel is formed by proteins L22 and L4, which approach the tunnel from opposite sides, forming what appears to be a gated opening (Nissen et al. 2000). This partial elucidated ribosome crystal structure from the large subunit shows that the tunnel walls largely consist of hydrophilic non-charged groups, thereby facilitating the passage of all kinds of peptide sequences. Tunnel dependence on the peptide sequences and current data from several systems show that the exit cannot be sequence-neutral. Nascent peptides in prokaryotes and eukaryotes contain special sequence motifs, and when these effector sequences are situated in the exit tunnel of translating ribosomes, they can dramatically affect both protein elongation and peptide termination (Lovett and Rogers 1996).

Another interesting issue concerning extra functions of ribosomal proteins, is their possible involvement in cell differentiation and apoptosis of hematopoietic and other cell lines. For example, it has been observed that down-regulation of expression of genes encoding ribosomal proteins occurs in differentiating leukemia HL-60 and K562 cells, as well as 3T3-L1 adipocytes (Mailhammer et al. 1992, Lin et al. 1994, Xu et al. 1994). Furthermore, apoptotic PC12 cells exhibited an up-regulated level of rpL4 gene expression before DNA fragmentation, and COS-7 cells transfected with rpL4 cDNA showed DNA fragmentation in an extent proportional to the amount of rpL4 cDNA transfected into these cells (Kajikawa et al. 1998). Moreover, NIH3T3 cells stably transfected with rpS3a cDNA become apoptotic in numbers proportional to the expression level of exogenous rpS3a gene of these cells (Naora et al. 1998).

In the following sections of this paper, we present experimental approaches for investigating the role of prokaryotic and eukaryotic ribosomal proteins in the elongation arrest of nascent peptide chains and hematopoietic cell differentiation process.

16.2 The Involvement of L4 Ribosomal Protein on Ribosome Elongation Arrest

Proteins L4 and L22 are known to be among the early assembly proteins of the large ribosumal subunit (Herold and Nierhaus 1987, Stelzl et al. 2000). These proteins play a scaffolding role, and their modifications could perturb the assembly of the 50S particle, resulting in large-scale deformations. Changes in L4 and L22 proteins – a single amino acid substitution Lys63Glu in L4 and a deletion of three amino acid residues, Met82, Lys83, and Arg84 in L22; (Chittum and Champney 1994) – lead to erythromycin-resistance mutations (Gabashvili et al. 2000) that are due to the alteration of the conformation probably of their tails extending from the globular parts (Unge et al. 1998).

In contrast to chloramphenicol and lincosamides, macrolides of the erythromycin class do not block peptidyl transferase activity (Vazquez 1975). It is likely that erythromycin blocks the tunnel that channels the nascent peptides away from the peptidyl-transferase centre (Nissen et al. 2000, Yonath et al. 1987, Milligan and Unwin 1986). On the light of these erythromycin-related data the following experiments were designed in our laboratory to study the regulatory role of ribosomal proteins in the rate of elongation of different polypeptides. These experiments include: (1) production of mutated recombinant ribosomal proteins; (2) development of mutant bacteria bearing specific mutated ribosomal proteins, and (3) evaluation of the effects that the mutated ribosomal proteins exert in the transcription/translation machinery in a cell-free system. The following is a brief description of these experiments.

1. Mutations of *specific* highly conserved amino acids located within the extended loop of the eubacterial L4 protein from *Thermus thermophilus* which is involved in the tunnel function according to crystallographic data presented elsewhere (Nissen et al. 2000). Isolation and characterization of recombinant plasmids bearing either the wild-type L4 (wtThL4), or the mutated forms of L4 cDNA.
2. The resulting recombinant plasmids carrying either the wtTthL4 or the mutants L4 are used to transform *E. coli* BL21 (DE3) and the transformed cells are grown for the preparation of the different cell-free systems according to the method devised by Zubay (1973). Furthermore, ribosome isolation and protein identification was carried out in order to verify the incorporated mutants.
3. The coupled transcription/translation of different genes from proteins with different molecular masses, sequences and conformations is performed using the different isolated S30 extracts in the presence of [^{35}S]Met (2.6 dpm/pmol) as described by Zubay (1973). Incubation times range from 10 to 45 min in order to investigate the different arising pause-site peptides. The cell-free translation products are analyzed by SDS-PAGE followed by autoradiography using the molecular dynamics phosphor-imager. Representative results derived from these experiments are included in Figs. 16.1 and 16.2. In Fig. 16.1 it is shown that the incorporation of a *Thermus thermophilus* L4 mutant (TthL4), which is within the highly conserved glutamic acid 56, has been substituted by glutamine, which was successfully incorporated in ribosomes. Also, the data of Fig. 16.2 show an example for the different pause-site peptide patterns produced after in vitro transcription/translation of the IF2 initiation factor by *E. coli* ribosomes harboring the wild-type *E. coli* L4 and ribosomes with *Thermus thermophilus* L4 mutant (Glu56 was changed to glutamine). These results confirm the appearance of

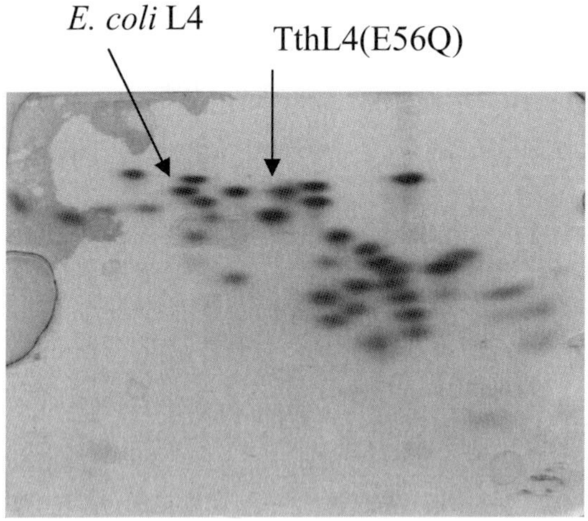

Fig. 16.1. 2D gel electrophoresis of Total Proteins isolated from 70S *E. coli* ribosomes (TP-70). The *arrows* indicate the endogenous *E. coli* L4 and the incorporated *T. thermophilus* L4 mutant (TthL4-gln[56])

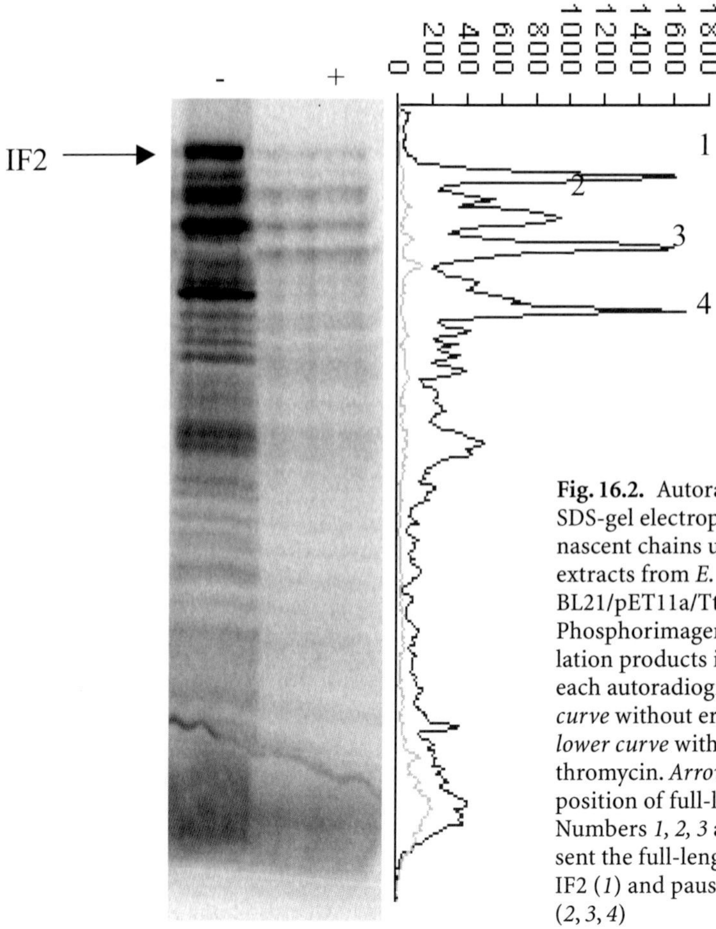

Fig. 16.2. Autoradiogram of SDS-gel electrophoresis of IF2 nascent chains using the S30 extracts from *E. coli* cells BL21/pET11a/TthL4-Gln[56]. Phosphorimager scan of translation products is given next to each autoradiograph: *upper curve* without erythromycin, *lower curve* with erythromycin. *Arrows* indicate the position of full-length product. Numbers *1, 2, 3* and *4* represent the full-length product IF2 (*1*) and pause-site peptides (*2, 3, 4*)

the pause-site peptides, which are produced besides the full-length nascent chain of IF2. The arrow indicates the full-length product and the peaks correspond to the different intensities to the pause-site peptides.

Similar experiments with other highly conserved amino acids would provide information about the different rate of translation on the elongation level of different proteins and therefore important information corroborating the dependence on the translation on the unique amino acid sequence of the nascent polypeptides, as well as, the involvement of these mutated amino acids of the ribosomal protein on the elongation arrest.

16.3 Down-Regulation of rpS5 and rpL35a Gene Expression During Murine Erythroleukemia (MEL) Cell Differentiation: Implications for Cell Differentiation and Apoptosis

The virus transformed murine erythroleukemia, (MEL, or Friend) cells have been widely used as a suitable model system for red blood cell maturation (erythropoiesis). MEL erythroid cell differentiation resembles the differentiation of early erythroid progenitor CFU-E cells into orthochromatophilic normoblasts (Tsiftsoglou and Wong 1985, Marks et al. 1987, Tsiftsoglou et al. 1987, 1992). During the development of red blood cells, several genes are activated transcriptionally (e.g. genes encoding heme biosynthetic enzymes and hemoglobin), while others progressively repress (Marks et al. 1987, Tsiftsoglou et al. 2002). The group of non-globin genes that are inactivated quite early in the differentiation process include genes that encode ribosomal RNA (rRNAs) that are tandemly repeated and transcribed into 28S, 18S and 5.8S rRNAs (Tsiftsoglou et al. 1982). This marked reduction in the transcription of rRNAs genes, taken together with a dramatic decrease in protein synthesis seen in differentiated cells, and with the number of ribosomes decreasing as the cells progress from the erythroblastic stages into reticulocyte-like cells, suggest that both changes in rRNA synthesis and in the genesis of ribosomes occur in a coordinating manner as part of the erythroid differentiation process. Apparently, abnormalities in these processes could lead to some of the reticulocyte disorders like reticulocytosis (Bessis et al. 1983).

Detailed Northern blot hybridization analysis of differentiating MEL cells revealed that induction of differentiation is associated with down-regulation of rpS5 and rpL35a genes that encode discrete ribosomal proteins of small and large ribosomal subunits respectively (Vizirianakis et al. 1999, Pappas et al. 2001). This gradual repression of mouse ribosomal protein genes occurred independently to the inducer used. Our finding, however, that blockade of MEL cell differentiation by N^6-methyladenosine, (N^6mAdo), an inhibitor of commitment and RNA methylation, allowed the rpS5 and rpL35a genes to be actively transcribed, thus indicating that the repression of rpS5 and rpL35a genes is closely related to the initiation of commitment (Vizirianakis et al. 1999, Pappas et al. 2001). These results are consistent with early observations showing that rRNAs genes are inactivated exclusively in terminal differentiated cells (Tsiftsoglou et al. 1982). Our findings are also consistent with the observation that human promyelocytic HL-60 cell differentiation induced by various agents is accompanied by substantial down-regulation of rpL3 gene expression and that human erythroleukemia K-562 cells induced by phorbol ester treatment along to megakaryocytic differentiation pathway exhibit a reduction in the steady-state level of rpL35, rpL31, rpL27 and rpL21 mRNAs (Mailhammer et al. 1992, Lin et al. 1994). Similar changes in ribosomal protein mRNA levels during rat intestinal differentiation have been reported (Maheshwari et al. 1993). However, on other occasions such as during terminal differentiation of myoblasts to myotubes, (where cells become permanently post-mitotic), no change in the expression level of rpL37 gene has been observed (Su and Bird 1995). Overall, since the same data were observed during differentiation of two different types of cells (MEL and neuroectodermal RD/TE-671), it suggests that the suppression of rpS5 and rpL35a gene expression represents a more or less physiologic event associated with the differentiation process itself. The finding that similar down-regulation of expression of genes encoding ribosomal proteins occurs also in differentiating HL-60, K562 cells and 3T3-L1 adipocytes further supports this conclusion (Mailhammer et al. 1992, Lin et al. 1994, Xu et al. 1994).

In addition to the role of ribosomal protein gene expression in cell differentiation, several other studies have implicated these genes in apoptosis (Naora et al. 1998, Kajikawa et al. 1998). In PC12 cells induced to apoptosis by 5-azacytidine the level of rat rpL4 gene expression increased before DNA fragmentation, whereas in transfected COS-7 cells DNA fragmentation occurred to an extent proportional to the amount of rpL4 cDNA transfected into the cells (Kajikawa et al. 1998). Moreover, NIH3T3 cells stably transfected with rpS3a cDNA, the expression level of exogenous rpS3a gene was correlated with apoptosis observed in these cells (Naora et al. 1998). In our studies, however, no suppression of rpS5 and rpL35a genes was observed either in apoptotic or cell cycling MEL cells (Vizirianakis et al. 1999, Pappas et al. 2001). These data are consistent with the observation that rpL37, rpS6, rpS11 and rpS14 genes are constitutively expressed during transition from quiescence to active proliferation in different cell types (Su and Bird 1995, Ferrari et al. 1990).

Eukaryotic cells contain several ribosomal proteins of various structure and function (e.g. 80 ribosomal proteins have been isolated and characterized in the rat) (Wool et al. 1995). We do not know precisely how all these proteins contribute to the formation of the 40S and 60S ribosomal subunits and to what extent contribute to the protein synthesis. In MEL cells for example, it has been proposed that the ribosomal proteins are added sequentially during the formation of the 40S small ribosomal subunit (Todorov et al. 1983). Although the role of these proteins is structural, recent studies have shown that the ribosomal proteins have different structural motives and may have a functional role as well (Wool et al. 1995, Wool 1993). Evidence now exists to indicate that rpS3 can act as a putative receptor for suberoylanilide hydroxamic acid, (SAHA), that promotes erythroid MEL cell differentiation (Webb et al. 1999). RpS5, rpS4 and rpS12 have been implicated in the signaling step for the formation of a peptide bond, (after the binding of tRNA in the ribosome), during the translation process (Purohit and Stern 1994, Schroeder 1994). Moreover, it has been suggested that rpL35 and rpL23a perhaps mediate the targeting of a ribosome-nascent chain complex to the endoplasmic reticulum (Pool et al. 2002). The studies presented above dissect the functions of ribosomal proteins in the ribosome structure and protein biosynthesis from those involved in cell proliferation, differentiation, or apoptosis. In this regard, it will be of great interest to demonstrate whether or not rpS5, rpL35a, and/or other ribosomal proteins are involved in the accumulation of translationally inactive rather than active salt-labile 80S ribosomal complexes seen in differentiating MEL cells, as it has been claimed that this peculiarity is in part due to the lack of the poly(A)-binding protein (PABP) in RNA transcripts bound into the ribosomes (Hensold et al. 1996).

Studying ribosomal protein gene expression patterns during differentiation and apoptosis in leukemia cells will provide valuable information on the role of these genes during normal hematopoiesis.

References

Bessis M, Lessin LS and Beutler E (1983) Morphology of the erythron. In: Williams, WJ, Beutler E, Erslev AJ, Lichtman MA (eds) Hematology. New York: McGraw-Hill, pp 257–279

Chittum HS and Champney WS (1994) Ribosomal protein gene sequence changes in erythromycin-resistance mutants of *Escherichia coli*. J. Bacteriol. 176:6192–6198

Ferrari S, Afredini R, Tagliafico E, Rossi E, Donelli A, Torelli G and Torelli U (1990). Non-coordinated expression of S6, S11, and S14 ribosomal protein genes in leukemic blast cells. Cancer Res. 50:5825–5828

Gabashvili IS, Agrawal RK, Spahn CM, Grassucci RA, Svergun DI, Frank J and Penezek P (2000) Solution structure of the *E. coli* 70S ribosome at 11.5 Å resolution. Cell 100: 537–549

Hensold JO, Barth-Baus D and Stratton CA (1996) Inducers of erythroleukemic differentiation cause messenger RNAs that lack poly(A)-binding protein to accumulate in translationally inactive, salt-labile 80 S ribosomal complexes. J. Biol. Chem. 271, 23246–23254

Herold M and Nierhaus KH (1987) Incorporation of six additional proteins to complete the assemply map of the 50S subunit from *Escherichia coli* ribosomes. J.Biol.Chem. 262:8826–8833

Kajikawa S, Nakayama H, Suzuki M, Takashima A, Murayama O, Nishihara M, Takahashi M and Doi K (1998) Increased expression of rat ribosomal protein L4 mRNA in 5-azacytidine-treated PC12 cells prior to apoptosis. Biochem. Biophys.Res. Commun. 252:220–224

Katunin VI, Muth GW, Strobel SA, Wintermeyer W and Rodnina MV (2002) Important contribution to catalysis of peptide Bond Formation by a single ionizing group within the ribosome. Mol. Cell 10:339–346

Lin CH, Palma JF and Solomon WB (1994) Phorbol ester induction of differentiation and apoptosis in the K562 cell line is accompanied by marked decreases in the stability of globin mRNAs and decreases in the steady state level of mRNAs encoding for ribosomal proteins L35, L31, L27, and L21. Cell. Mol. Biol. Res. 40:13–26

Lovett PS and Rogers EJ (1996). Translation attenuation regulation of chloramphenicol resistance in bacteria–a review. Microbiol. Rev. 60:366–385

Maheshwari Y, Rao M, Sykes DE, Tyner AL and Weiser MM (1993) Changes in ribosomal protein and ribosomal RNA synthesis during rat intestinal differentiation. Cell Growth Differ. 4:745–752

Mailhammer R, Szots H, Bonisch J and Dormer P (1992) Downregulation of messenger RNA levels for ribosomal proteins in differentiating HL-60 cells. Exp. Cell Res. 200: 145–148

Marks PA, Sheffery M, Rifkind RA (1987) Induction of transformed cells to terminal differentiation and the modulation of gene expression. Cancer Res. 47:659–666

Milligan RA and Unwin PN (1986) Location of exit channel for nascent protein in 80S ribosome. Nature 319:693–695

Naora H, Takai I, Adachi M and Naora H (1998) Altered cellular responses by varying expression of a ribosomal protein gene: Sequential coordination of enhancement and suppression of ribosomal protein S3a gene expression induces apoptosis. J. Cell Biol., 141:741–753

Nissen P, Hansen J, Ban N, Moore PB, and Steitz TA (2000) The structural basis of ribosome activity in peptide bond synthesis. Science 289:920–930

Pappas IS, Vizirianakis IS, Tsiftsoglou AS (2001) Cloning, sequencing and expression of a cDNA encoding the mouse L35a ribosomal protein during differentiation of murine erythroleukemia (MEL) cells. Cell Biol. Int. 25:629–634

Picking WD, Picking WL, Odom OW and Hardesty B (1992) Fluorescence characterization of the environment encountered by nascent polyalanine and polyserine as they exit *Escherichia coli* ribosomes during translation. Biochemistry 31: 2368–2375

Pool MR, Stumm J, Fulga TA, Sinning I and Dobberstein B (2002) Distinct modes of signal recognition particle interaction with the ribosome. Science 297:1345–1348

Purohit P and Stern S (1994) Interactions of a small RNA with antibiotic and RNA ligands of the 30S subunit. Nature 370:659–662

Schroeder R (1994) Translation: Dissecting RNA function. Nature 370:597–598

Stelzl U, Spahn CMT and Nierhaus KH (2000) Selecting rRNA binding sites for the ribosomal proteins L4 and L6 from randomly fragmented rRNA: application of a method called SERF. Proc. Natl. Acad. Sci. USA 97:4597–4602

Su S and Bird RC (1995) Cell cycle, differentiation and tissue-independent expression of ribosomal protein L37. Eur. J. Biochem. 232:789–797

Tenson T. and Ehrenberg M. (2002) Regulatory Nascent Peptides in the ribosomal tunnel. Cell 108:591–594

Todorov IT, Noll F and Hadjiolov AA (1983) The sequential addition of ribosomal proteins during the formation of the small ribosomal subunit in Friend erythroleukemia cells. Eur. J. Biochem. 131, 271–275

Tsiftsoglou AS, Wong W, Volloch V, Gusella J, Housman D (1982) Commitment of murine erythroleukemia (MEL) cells to terminal differentiation is associated with coordinated expression of globin and ribosomal genes. Prog. Clin. Biol. Res. 102A:69–79

Tsiftsoglou AS and Wong W (1985) Molecular and cellular mechanisms of leukemic hematopoietic cell differentiation: An analysis of the Friend system. Anticancer Res. 5:81–100

Tsiftsoglou AS, Hensold J, Robinson SH, Wong W (1987) The regulatory role of commitment in gene expression during induction of leukemic cell differentiation. In: Harrap KR, Connors TA (eds) New Avenues in Developmental Cancer Chemotherapy. Academic Press, New York, pp 205–227Nature Rev. Genet. 1:57–64

Tsiftsoglou AS, Robinson SH (1992) Differentiation of murine erythroleukemia cells and human HL-60 cell lines. In: Murphy MJ, Jr (eds) Concise Reviews in Clinical and Experimental Hematology. AlphaMed Press, Dayton, Ohio, pp 295–306

Tsiftsoglou AS, Pappas IS and Vizirianakis IS (2002) The developmental program of murine erythroleukemia cells. Oncol. Res. 13:339–349

Unge J, Aberg A, Al-Kharadaghi S, Nikulin A, Nikonov S, Davydova NL, Nevskaya N, Garber M and Liljas A (1998) The crystal structure of the ribosomal protein L22 from *Thermus thermophilus*: insights into the mechanism of erythromycin resistance. Structure 6:1577–1586

Vazquez D (1975) Inhibitors of Protein Synthesis. Springer, Berlin

Vizirianakis IS, Pappas IS, Gougoumas D, Tsiftsoglou AS (1999) Expression of ribosomal protein S5 cloned gene during differentiation and apoptosis in murine erythroleukemia (MEL) cells. Oncol. Res., 11:409–419

Webb Y, Zhou X, Ngo L, Cornish V, Stahl J, Erdjument-Bromage H, Tempst P, Rifkind RA, Marks PA, Breslow A and Richon VM (1999) Photoaffinity labeling and mass spectrometry identify ribosomal protein S3 as a potential target for hybrid polar cytodifferentiation agents. J. Biol. Chem. 274:14280–14287

Wool IW (1993) The bifunctional nature of ribosomal proteins and speculations on their origins. In Nierhaus KH, Franceschi F, Subramania AR, Erdmann VA and Wittmann-Liebold B (eds) The Translational Apparatus: Structure, Function, Regulation, Evolution. Plenum Press, New York pp. 727–737

Wool IG, Chan YL and Gluck A (1995) Structure and evolution of mammalian ribosomal proteins. Biochem. Cell Biol. 73, 933–947

Xu L, He GP, Li A and Ro HS (1994) Molecular characterization of the mouse ribosomal protein S24 multigene family: a uniquely expressed intron-containing gene with cell-specific expression of three alternatively spliced mRNAs. Nucleic Acid Res. 22:646–655

Yonath A, Leonard KR and Wittmann HG (1987) A tunnel in the large ribosomal subunit revealed by three-dimensional image reconstruction. Science 236:813–816

Zubay G (1973) In vitro synthesis of protein in microbial systems. Annu. Rev. Genet. 7.267–287

17 MALDI-MS Analysis of Peptides Modified with Photolabile Arylazido Groups

William Low, James Kang, Micheal DiGruccio, Dean Kirby, Marilyn Perrin and Wolfgang H. Fischer

Abstract

The ability of MALDI-MS to analyze photolabile arylazido peptide derivatives was investigated. Peptides containing UV-labile p-azidobenzoyl groups were subjected to MALDI-MS analysis in a variety of matrices. As a standard MALDI-MS employs a UV laser (337 nm), we investigated conditions that would allow detection of the intact molecule ions for these modified peptides. When using α-cyano-4-hydroxycinnamic acid (ACHC) or 2,5 dihydroxybenzoic acid (DHB) as the matrix, photoinduced degradation products were prevalent. In contrast, when employing the matrix sinapinic acid, the intact molecule ion corresponding with the azido peptide was the predominant signal. The protection of photolabile azido derivatives correlates with the UV absorbance properties of the matrix employed, i.e. sinapinic acid which exhibits a strong absorbance near 337 nm most efficiently protects the azido derivative from photodegradation.

17.1 Introduction

Photoactivatable reagents are powerful tools in the study of molecular interactions [1, 2]. Arylazido modified peptides can conveniently be prepared by reacting an activated p-azidobenzoic acid derivative with amino groups present either at the N-terminus or at lysyl side chains. Activation of these p-azidobenzoyl peptides is achieved by UV-light irradiation, which results in loss of nitrogen (N_2) and formation of a reactive nitrene intermediate. The reactive species can then form a covalent bond with a nearby protein chain. This can lead to the identification of a binding site, e.g. of a ligand and a receptor or a substrate and an enzyme.

A convenient way of preparing azidobenzoyl peptides involves reacting the peptide with an activated azidobenzoic acid derivative, such as the hydroxy-

Principles and Practice
Methods in Proteome and Protein Analysis
R.M. Kamp, J. J. Calvete, T. Choli-Papadopoulou (Eds.)

succinimidyl ester. These reagents react with the primary amino groups in the peptide, the N-terminal amino group or the ε-amino group in lysyl residues. In cases where several amino groups are present, one or more of these reactive groups can become modified. It is important to determine which amino group carries the modification to reproducibly generate the one that exhibits the highest affinity for its receptor or other binding partner. The full characterization of photolabile components represents a challenge as many analytical procedures involve the use of UV light, e.g. UV detection during HPLC purification and UV laser irradition during MALDI-MS analysis. We have therefore investigated conditions for MALDI-MS analysis that would allow the detection of intact photoactivatable peptides.

The CRF family of neuropeptides is involved in essential responses to stress [3]. Centrally these peptides activate the pituitary adrenal axis. Peripheral effects include an influence on blood pressure as well as immune and inflammatory responses. To date four members of the CRF peptide family have been characterized in mammals [4]. The receptors for the CRF peptides belong to the G-protein coupled seven transmembrane receptor family. Two subtypes have been characterized that exhibit differential affinity for the ligands [5, 6]. We have recently identified the first extracellular domain of the CRF type I receptor as playing a major role in agonist and antagonist binding [7]. In particular the synthetic peptide antagonist Astressin binds the first extracellular portion with high affinity. The present study resulted from our interest in generating an Astressin analog that could be used to photolabel the binding region of the first extracellular domain of the type I CRF receptor.

17.2 Results and Discussion

The corticotropin releasing factor antagonist $DTyr^{12}$-astressin [8] was reacted with N-hydroxysulfosuccinimidyl-4-azidobenzoate. The peptide contains one lysine residue and a free N-terminal amino group. Acylation can thus result in modification of either of the amino groups or of both. Consistent with this fact, three major products were isolated after HPLC separation. Inspection of the sequence revealed that digestion with V8 protease would generate separate fragments containing the N-terminal amino acid and the single lysyl residue. Digestion with V8 protease was carried out in ammonium bicarbonate buffer at pH 8.3 for all three products. An aliquot of the digestion mixture was resolved by reversed phase HPLC. The HPLC fractions as well as the crude digests were subjected to MALDI-MS analysis. The following analyses were carried out on the C-terminal peptide H-Gln-Leu-Ala-Gln-Glu*-Ala-His-Lys*-Asn-Arg-Lys(pAzbz)-Leu-Nle-Glu-OH, where the side chains of Glu* and Lys* are connected by a lactam bridge and Lys(pAzbz) represents the p-azidobenzoyl modified lysine. The N-terminal peptide was also analyzed giving similar results (data not shown). Initial analyses were carried out employ-

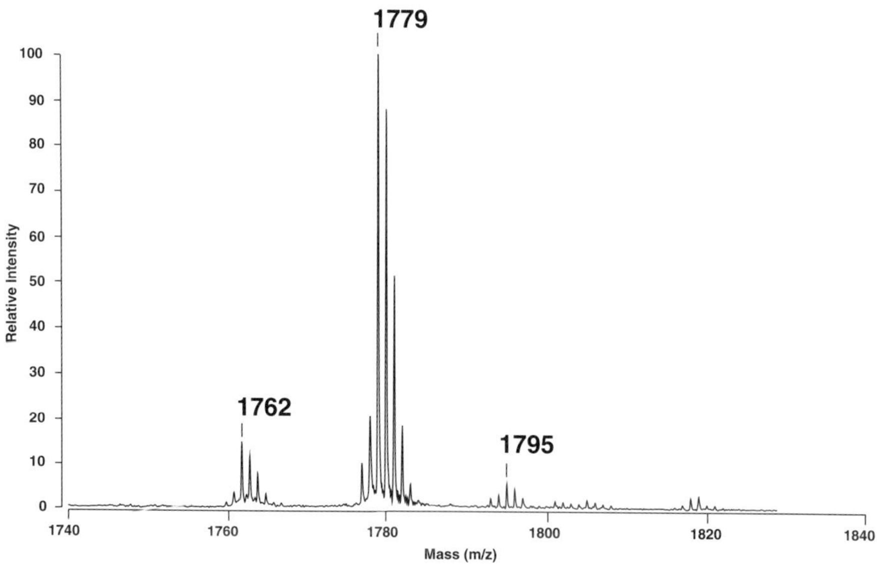

Fig. 17.1. MALDI-MS spectrum of arylazido peptide employing ACHC matrix. The calculated mass for the intact arylazido peptide is $[M+H]^+=1805$ Da

ing α-cyano-4-hydroxycinnamic acid (ACHC) as the matrix. This matrix is commonly used when analyzing peptides with molecular weights <5000 Da. The modified peptides exhibited a strong signal 26 mass units lower than that expected for the benzoylazido peptides (Fig. 17.1). Additional weak signals were observed that were 10 and 43 mass units lower than expected for the $[M+H]^+$ of the modified peptide. No significant signal was observed at m/z=1805, the expected mass for the intact arylazido peptide. The major signal at m/z=1779 is consistent with a loss of nitrogen (N_2) and an addition of 2 hydrogen atoms (H) to the nitrene which could have occurred during storage or purification of the peptide or during analysis in the mass spectrometer. To investigate these possibilities, the modified peptides were analyzed in different matrices by MALDI-MS and, utilized an ionization procedure which did not involve UV-irradiation, by ESI ion trap MS.

Figure 17.2 shows the MALDI-MS spectrum obtained for the N-terminal peptide using 2,5 dihydroxybenzoic acid (DHB) as the matrix. The predominant species is again observed at m/z=1779 with a minor signal present at m/z=1795. Again, no significant signal could be observed for the intact arylazido peptide (calculated $[M+H]^+=1805$).

The signal corresponding to the intact photolabile peptide at m/z=1805 was the major signal when using sinapinic acid as the matrix (Fig. 17.3). In addition, a strong signal at m/z=1779 and a very weak signal at m/z=1762 were observed. When the peptide was analyzed by electrospray ionization ion

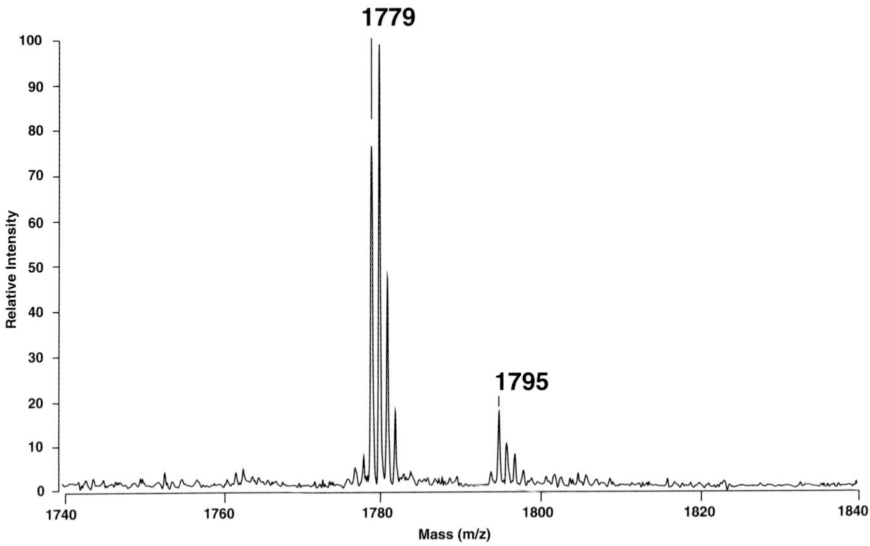

Fig. 17.2. MALDI-MS spectrum of arylazido peptide employing DHB matrix. The calculated mass for the intact arylazido peptide is $[M+H]^+=1805$ Da

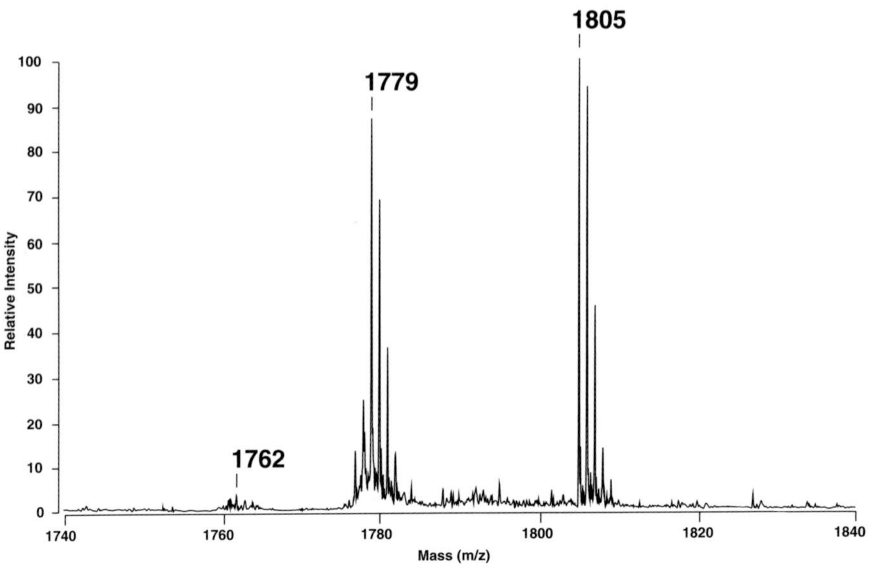

Fig. 17.3. MALDI-MS spectrum of arylazido peptide employing sinapinic acid matrix. The calculated mass for the intact arylazido peptide is $[M+H]^+=1805$ Da

trap mass spectrometry, the predominant signal corresponded with the intact arylazido peptide, none of the degradation products were observed (data not shown). This indicates that the photodegradation products were formed in the MALDI mass spectrometer.

In Fig. 17.4, the traces of mass spectra obtained employing ACHC and sinapinic acid are overlayed. The mass differences to the intact $[M+H]^+$ for the arylazido peptide are shown. The photodegradation products are observed at a signal of $[M+H-10]^+$, $[M+H-26]^+$ and $[M+H-43]^+$, respectively. The chemistry of azido compounds and that of their photoactivation has been studied extensively [1, 2]. The first step of the activation involves elimination of N_2 and formation of the reactive nitrene. The nitrene then reacts by insertion into C–H or C–C bonds. Alternatively, other reactive intermediates such as the seven-membered heterocyclic tropolone can form and in turn react with functional groups in neighboring proteins. The MALDI-MS analysis of ATP azido derivatives has been described by Chen et al. [9]. These authors recorded MALDI spectra in the negative mode and detected mainly $[M-H-26]^-$ and $[M-H-12]^-$ species. The $[M-H-26]^-$ is proposed to arise from elimination of N_2 and addition of 2H, whereas the $[M-H-12]^-$ is likely to result from

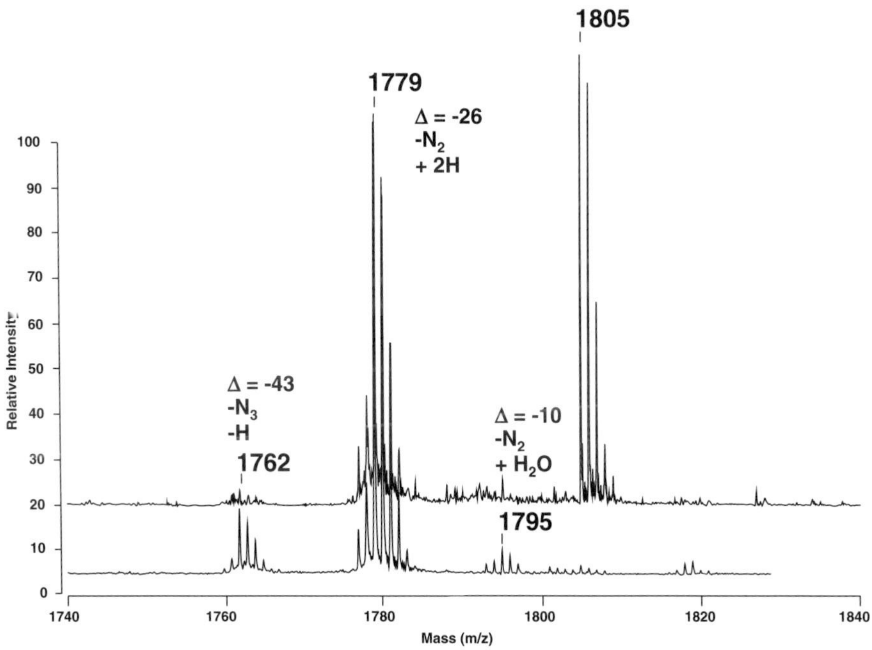

Fig. 17.4. Overlay of spectra recorded with sinapinic acid (*top*) and ACHC matrix (*bottom*). The differences of observed masses to the the $[M+H]^+$ of the intact arylazido peptide are shown

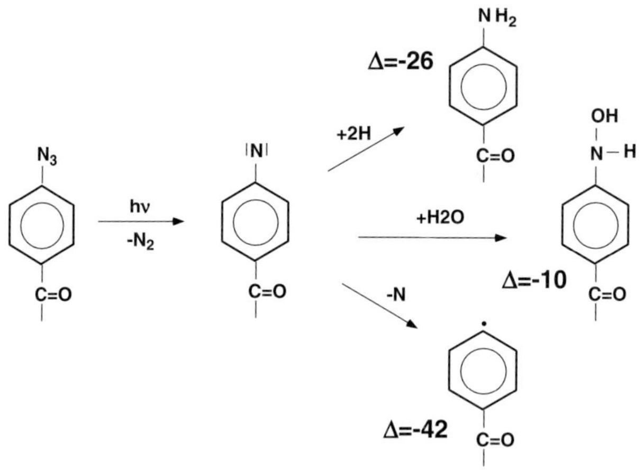

Fig. 17.5. Photodegradation products of the benzoylazido group. The mass differences to the starting material and possible end products are shown

N_2 elimination and addition of oxygen. The species at [M-H-10]$^-$, which is not observed in the analysis of azido-ATP is proposed to be the result of N_2 elimination and addition of H_2O. Based on the study by Chen et al. [9] and our observations, we propose the following interpretation of the masses observed for the photodegradation products (Fig. 17.5). The [M+H-10]$^+$ species results from elimination of N_2 and addition of H_2O, possibly leading to te formation of a hydroxylamine. The species at [M+H-26]$^+$ is most likely the result of N_2 elimination and the addition of two hydrogen atoms which could lead to the formation of an amine. Loss of the azido group (N_3) should lead to a species at [M+H-42]$^+$, whereas we observe a signal at Δ=–43. We interpret this as a non-protonated species, thus [M-42]$^+$ corresponds with the positively charged molecule ion that has lost the N_3^- anion.

To investigate the differences in the matrices ability to protect photolabile peptides from degradation by laser illumination, we recorded the UV spectra of the matrices employed (Fig. 17.6). Of the three matrices, sinapinic acid exhibits the highest absorbance near the wavelength of the laser employed (337 nm). The absorbance of both ACHC and DHB is approximately 50% lower in this wavelength region. It is, therefore, likely that more of the laser energy is absorbed by sinapinic acid rather than by the other matrices, which leads to a protection of the analyte from photodegradation.

The results presented here show that azidobenzoyl peptides undergo photoactivation during MALDI-MS analysis to a different extent depending on the matrix employed. Analysis by MALDI-MS resulted in the formation of mainly two reaction products during laser irradiation, the [M+H-26]$^+$ and [M+H-10]$^+$ species, when ACHC or DHB are used as a matrix. When sinapinic acid is employed, the authentic azido peptide [M+H]$^+$ becomes the predominant signal. The reaction products are proposed to result from loss of N_2 and the addition of 2H or H_2O, respectively. These differences in signal abundance

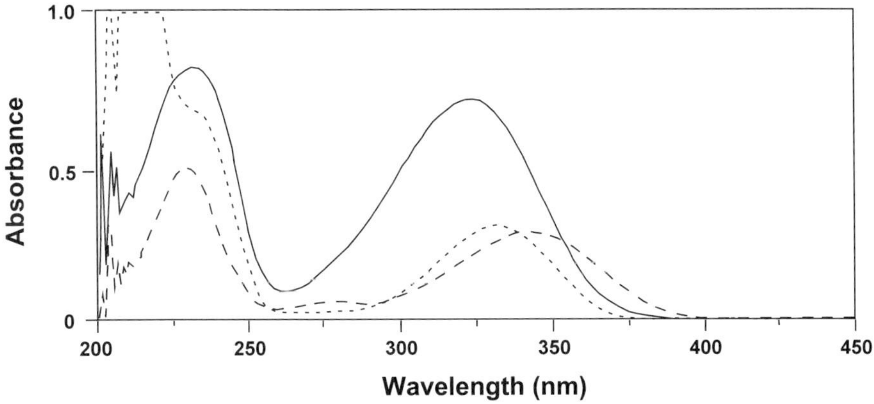

Fig. 17.6. UV spectra of sinapinic acid (*solid line*), ACHC (*dashed line*) and DHB (*dotted line*). The matrices were dissolved at a concentration of 0.01 mg/ml in 0.3 % trifluoroacetic acid, 30 % acetonitrile/water

for the degraded species are likely to be a consequence of the UV absorbance characteristics of the matrix molecule compared to the analyte. The matrix sinapinic acid appears to *protect* the analyte with the highest efficiency from UV irradiation.

17.3 Experimental Procedures

17.3.1 Azidobenzoylation of Astressin

DTyr12-astressin (0.07 mg) was reacted with N-hydroxysulfosuccinimidyl-4-azidobenzoate (1.8 mg) in 20 mM sodium phosphate buffer (pH 7) for 70 min at RT. The reaction mixture was separated by reversed-phase HPLC. The three major components were collected and further analyzed.

17.3.2 V8 Peptidase Digestion of Modified Peptides

The modified peptides collected after HPLC separation were lyophilized and digested with 1 µg V8 protease (Roche Biochemicals, Indianapolis) in 50 µl ammonium bicarbonate buffer (50 mM, pH 8.3). The digests were resolved by reversed phase HPLC. The fractions were analyzed by MALDI-MS to identify the modified proteolytic fragments.

17.3.3 MALDI-MS Analysis

MALDI-MS spectra were measured on an ABI-Perseptive DE-STR instrument. The instrument employs a nitrogen laser (337 nm) at a repetition rate of 20 Hz. The spectra were recorded in the delayed extraction mode. The accelerating voltage was 20 kV. All spectra were recorded in the positive reflector mode. Spectra are sums of 100 laser shots. Matrices (sinapinic acid and a-cyano-4-hydroxycinnamic acid) were prepared as saturated solutions in 0.3 % trifluoroactetic acid and 30 % acetonitrile. The matrix 2,5-dihydroxybenzoic acid was prepared at 1 mg/ml in the same solvent mixture.

17.3.4 UV Spectra

UV spectra were recorded at a concentration of 0.01 mg/ml in 0.3 trifluoroacetic acid and 30 % acetonitrile/water as solvent. This is the same solvent used for MALDI matrices in the analyses described above.

References

1. Chowdhry V, Westheimer FH (1979) Photoaffinity labeling of biological systems. Annu Rev Biochem 48:293
2. Hazum E (1983) Photoaffinity labeling of peptide hormone receptors. Endocr Rev 4:352
3. Vale W, Spiess J, Rivier C, Rivier J (1981) Characterization of a 41 residue ovine hypothalamic peptide that stimulates the secretion of corticotropin and ß-endorphin. Science 213:1394
4. Lewis K, Li C, Perrin MH, Blount A, Kunitake K, Donaldson C, Vaughan J, Reyes TM, Gulyas J, Fischer W, Bilezikjian L, Rivier J, Sawchenko PE, Vale WW (2001) Identification of Urocortin III, an additional member of the corticotropin-releasing factor (CRF) family with high affinity for the CRF2 receptor. Proc Natl Acad Sci USA 98: 7570
5. Chen R, Lewis KA, Perrin MH, Vale WW (1993) Expression cloning of a human corticotropin releasing factor (CRF) receptor. Proc Natl Acad Sci, USA 90:8967
6. Perrin MH, Donaldson C, Chen R, Blount A, Berggren T, Bilezikjian L, Sawchenko P, Vale WW (1995) Identification of a second corticotropin-releasing factor receptor gene and characterization of a cDNA expressed in heart. Proc Natl Acad Sci USA 92:2969
7. Perrin MH, Fischer WH, Kunitake KS, Craig AG, Koerber SC, Cervini LA, Rivier JE, Groppe JC, Greenwald J, Nielsen SM, Vale WW (2001) Expression, purification and characterization of a soluble form of the first extracellular domain of the human type 1 corticotropin releasing factor receptor. J Biol Chem 276:31528
8. Rivier JE, Kirby DA, Lahrichi SL, Corrigan A, Vale WW, Rivier CL (1999) Constrained corticotropin releasing factor (CRF) antagonists (Astressin analogues) with long duration of action in the rat. J Med Chem 42:3175
9. Chen X, Siems WF, Asbury GR, Yount RG (1999) Fingerprint patterns from laser-induced azido photochemistry of spin-labeled photoaffinity ATP analogs in matrix-assisted laser desorption/ionization mass spectrometry. J Am Soc Mass Spectrom 10:1337

18 A New Edman-Type Reagent for High Sensitive Protein Sequencing

Christian Wurzel, Barbara zu Lynar, Christoph Radcke, Ralf Krüger, Michael Karas and Brigitte Wittmann-Liebold

Abstract

A variety of applications in the proteomics field still rely on the Edman sequencing: The method is used to confirm MS data, to identify the amino-terminus, for heterogeneity assignment, domain structure analysis or the determination of proteolysis sites. However, the major benefit of the Edman sequencing method is clearly its de novo sequencing and quantification capability and, hence, it is still a powerful alternative to mass spectrometry.

To overcome the limited sensitivity of present-day sequencers, we would like to introduce in this text two innovations in Edman sequencing:

- The Chip Sequencer, a microreaction system which delivers minute amounts of reagents and solvents adapted to low femtomole protein amounts.
- A new Edman-type coupling reagent which results in comparable coupling and degradation yields as PITC in the typical Edman chemistry. Additionally, this reagent offers a detection limit in the low attomole range of the released thiohydantoin amino acid derivatives.

Nowadays, we still perform automated wet-phase sequencing, with off-line detection by GC-MS. The method, the chemical adaptation to the Chip Sequencer and an on-line interface are being optimized. The combination of the new coupling reagent with the Chip Sequencer and detection in a GC system makes amino acid sequencing in the low femtomole range feasible and will largely support any future proteome research. A sequencer combining these features will be commercially available in spring 2004.

Principles and Practice
Methods in Proteome and Protein Analysis
R.M. Kamp, J. J. Calvete, T. Choli-Papadopoulou (Eds.)

18.1 Introduction

Various methods for the identification of proteins and peptides by mass spectrometry have been developed and refined in the last few years. For example, identification of protein spots from a high resolution 2-DE gel by enzymatic digestion and MALDI peptide fingerprinting is straightforward if the protein is known. With the nanospray, ion-trap and Q-TOF electrospray instruments, it is claimed that de novo sequencing is possible if the peptide is not too large. However, chemical sequencing is still necessary in various applications.

Since Pehr Edman [1] published the identification by stepwise chemical degradation of proteins the sensitivity was enhanced from micromole to high femtomole amounts. Namely, technical innovations like the Protein Sequenator introduced by Edman and Begg [2], dead volume free delivery valves and the automatic conversion by Wittmann Liebold [3], the design of the cartridge by Hood and Hunkapillar [4] or the on-line coupling to liquid chromatography [5] made this possible. In the last few years, further refinement of the valve blocks or the introduction of capillary liquid chromatography improved the performance of protein sequencers. Recently, Tempst and his coworkers have shown sequencing of 100 femtomole β-lactoglobulin with a self-constructed sequencer and a 300-μm capillary LC [6].

We developed a microreaction system [7, 8] as the next step in miniaturizing automates for Edman chemistry. In this system, all essential valves, as well as the reaction chamber, are arranged on one chip. Inner volumes and the liquids for delivery are reduced to the submicroliter scale. The miniaturization of the volumes and the inner surfaces are advantageous for the application to Edman sequencing. The delivery volume of the reagents is adapted to minute protein amounts; washing and drying steps are more effective and fast. The consumption of reagents and solvents is reduced, but miniaturization of the sequencer hardware alone is not sufficient to gain a sensitivity which is competitive to the different mass spectrometric methods for the identification of proteins. In present-day sequencers, the UV detection limits the scale of high sensitive identification of the released amino acids. In the past, many attempts were made to introduce different detection systems for Edman sequencing or alternative isothiocyanate derivatives (see 9 for a review), but none of the methods have progressed into routine usage. The difficulty in this task is to find a reagent, a derivatization or a detection system which fulfills three conditions at the same time: (1) the Edman chemistry must run at high repetitive yields, (2) the reagent should be suitable for gas-phase or wet phase sequencing without the need of covalent attachment of the proteins and (3) the released amino acids should be detectable at very low concentrations, in the low femtomole to subfemtomole range.

Here, we have introduced 1,3 bis-(trifluoromethyl)-phenylisothiocyanate as a new Edman-type coupling reagent which fits perfectly well to these conditions. This reagent shows high yields in the coupling reaction as well as in the cleavage and conversion reaction. The repetitive degradation yields are as high and as obtainable with the standard PITC chemistry. The reagent can be used in the normal pulsed liquid phase, in the gas-phase or in the wet-phase mode without covalent attachment of the proteins. Due to its high content of fluorine, this reagent and its Edman derivatives can be detected by electron capture detection or negative chemical ionization and it offers detection limits in the low attomole range of the thiohydantoin derivatives.

18.2 Materials and Methods

Reagents and solvents for manual and automated sequencing were purchased from Applied Biosystems, Sigma Aldrich and Fluka. No further purification was made. 1,3-bis-(trifluoromethyl)-phenylisothiocyanate was purchased from Sigma. Thiohydantoin derivatives of all amino acids were produced by manual coupling of 400 nmol of each amino acid with 12 μmol 1,3-bis-(trifluoromethyl)-phenylisothiocyanate (FM-PITC) in acetonitrile/water/pyridine/ethanol/TMA (40/20/20/15/5) at 55 °C for 1 h. After drying the FM-PTC amino acids the conversion was done with 20 % TFA in water/acetonitrile (30/70) at 58 °C for 1 h. The reaction rate was measured by adding the same amounts of PITC and FM-PITC to the amino acid as described above. Manual sequencing was performed as described in [10]. Automated sequencing was done in the Chip Sequencer which we developed and constructed in ConSequence GmbH, Teltow or in a Knauer Sequencer Modell 910. As an HPLC system, an isocratic HPLC from Knauer (column YMC ODS AM i.d. 3 mm, length 250 mm, particle size 3 μm, buffer 42 % acetonitrile, 20 mM sodium acetate, 140 mg SDS/lt, or a 130A gradient system from ABI equipped with a Shimadzu SPD 10AV (column Sepserv ODS, gradient 35–70 %) was used. GC measurements were done on an Agilent 6890 N equipped with a PTV injection system and a 5973 N mass selective detector in negative chemical ionization mode (column HP5 MS, methane as reagent gas).

18.3 Results

18.3.1 Chip-Sequencer

We constructed a new innovative sequencer. This Chip Sequencer offers a number of advantages compared to other systems. Delivery of reagents is between 0.2 and 1 μl and is adapted to minute protein amounts. The consumption of reagents and solvents as compared to the ABI Procise Sequencer is reduced to 30 %. Due to the small inner surfaces and volumes – the reaction chamber has a total volume of 3 μl including all delivery lines – washing and

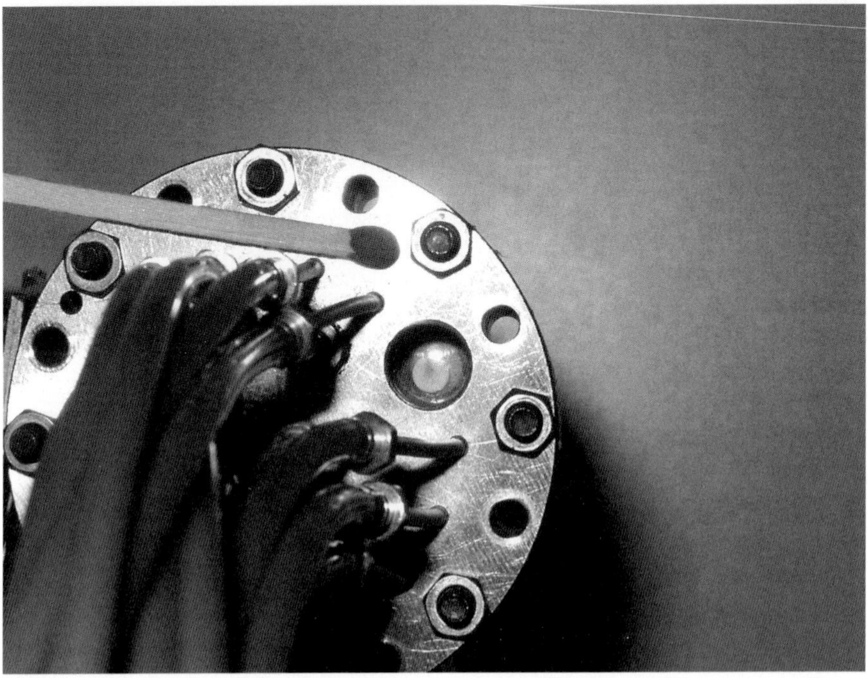

Fig. 18.1. A Coomassie stained blot of 3 mm diameter prior to sequencing in the Chip Sequencer

drying subsequently after the reaction are much more efficient and need less time. The chemical background is reduced, leading to a better signal-to-noise ratio and precise sequence interpretation. Typically, blots of 2–3 mm diameter are applied and fit perfectly into the reaction chamber (Fig. 18.1). The Chip Sequencer can be used in the wet-phase, solid-phase or gas-phase mode. Classical PITC chemistry with HPLC detection or the new FM-PITC chemistry presented here can be applied on the same machine.

18.3.2 Evaluation of 1,3-bis-(Trifluoromethyl)-Phenylisothiocyanate as a New Coupling Reagent in Edman Chemistry

A new coupling reagent which can substitute PITC must have the following properties: high coupling yields, high cleavage rates, a low detection limit and it should fit into automated wet-phase sequencing without covalent attachment of the protein.

We developed, designed and tested a number of isothiocyanate derivatives with different detection capabilities including fluorescent dyes or isothiocyanates suitable for MS detection. We performed three tests with these derivatives to evaluate them for Edman chemistry. First, we performed the coupling reaction with different free amino acids, converted them and identified the derivatives by HPLC or MALDI-MS to judge the reaction. To compare the reaction rate of the new isothiocyanate to PITC, we applied the same concentrations of the new isothiocyanate together with PITC to a solution of free amino acids, performed the coupling reaction, converted the amino acids and measured the amount of both released thiohydantoin derivatives. In the third test, the first step was to apply the new isothiocyanate and then after a delay of 30 min, the next step was to add PITC, to measure if the reaction is complete or if there is any remaining PTH amino acid. A number of isothiocyanates, e.g. DANSA-PITC [11] or 311-PITC [12] already known in the literature, showed high coupling efficiency with a comparable reaction velocity as PITC. We found one derivative, namely 1,3-bis-(trifluoromethyl)-phenylisothiocyanate which passed these tesst and showed a much faster coupling rate compared to PITC. When applied simultaneously, the amino acids reacted >90 % with FM-PITC and were detected as FM-PTH-amino acids.

To check whether the new reagent offered a high cleavage efficiency, we performed manual sequencing of synthetic peptides and standard proteins with and without covalent attachment to solid-phase support. The sequence could easily be detected by HPLC. This showed that FM-PTC derivatives can easily be cleaved off the protein. First, we performed sequencing with the double-coupling method to control the overlap. In case of FM-PITC as a first coupling reagent and PITC as a second, PTH amino acids could not be detected. After these successful tests we applied 5 % 1,3-bis-(trifluoromethyl)-phenylisothiocyanate in heptane as a coupling reagent to the Knauer sequencer. We used the normal wet-phase program with no relevant changes to the step-file, the reagent and solvent deliveries or the other steps. At first, the initial yields were at 10 % and the repetitive yields at 82 %. Subsequently, we optimized the coupling conditions, the extractions and conversions. Figure 18.2 shows a run done in the wet-phase mode in the Knauer sequencer. Lactoglobulin (50 pmol) was applied on a PVDF membrane to the reactor. The amino acids were detected offline with gradient HPLC and UV at 269 nm. An aliquot of 50 % was injected. The initial yield in this case was 42 %. All amino acids were detected and the run showed a repetitive yield of 92 %. With these results it is clearly shown that 1,3-bis-(trifluoromethyl)-phenylisothiocyanate fits perfectly into the desired properties for Edman sequencing.

18.3.3 High Sensitive Detection of Thiohydantoin Derivatives

As shown above, the thiohydantoin amino acid derivatives released from Edman sequencing with 1,3-bis-(trifluoromethyl)-phenylisothiocyanate as a coupling reagent can be identified by HPLC gradient separation and UV detection in a comparable sensitivity to the usual PTH amino acids. The main

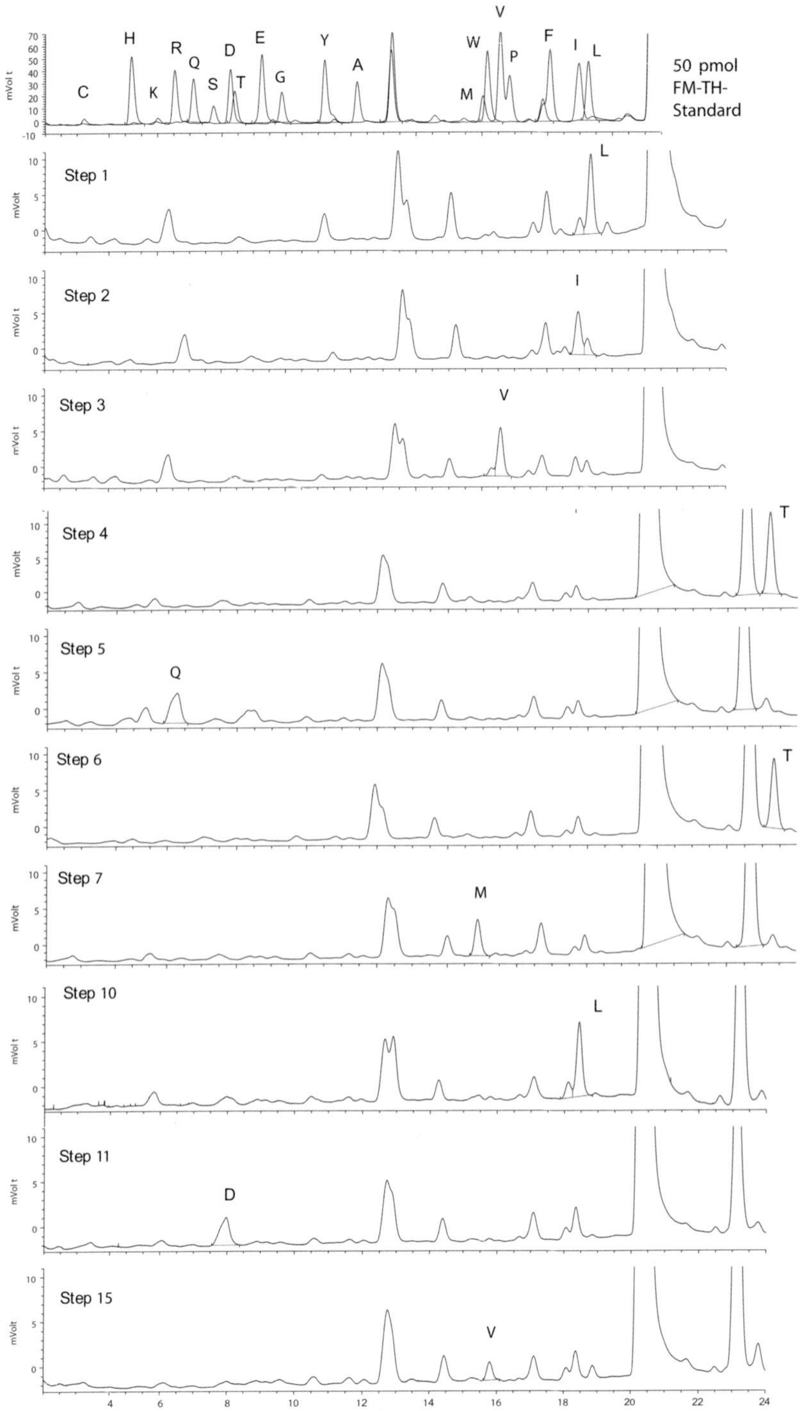

Fig. 18.2. Protein sequencing with our new coupling reagent 1,3-bis-(trifluoromethyl)-phenylisothiocyanate, 50 pmol of lactoglobulin was sequenced. Repetitive yield is 92 %

advantage of the new reagent is that it offers an ultra-high sensitive detection. Due to the high content of fluorine in 1,3-bis-(trifluoromethyl)-phenylisothiocyanate and its thiohydantoin derivatives, this reagent is dedicated to the detection by electron capture detection or negative chemical ionization. Additionally, the boiling point of 1,3-bis-(trifluoromethyl)-phenylisothiocyanate is only 63 °C compared to 218 °C of PITC. This leads to much better properties of the derivatized amino acids for the separation by gas chromatographic methods.

Figure 18.3 shows in an overlay of two total ion chromatograms the separation of 14 amino acids derivatized by 1,3-bis-(trifluoromethyl)-phenylisothiocyanate. Four picomoles of each amino acid derivative were applied in the solvent vent mode to the GC system. Negative chemical ionization with methane as the reagent gas was used. The MS was set to the scan mode from 260 to 460 mass units. The low background of the total ion chromatogram is remarkable. Groupwise selected ion monitoring mode further enhances the signal-to-noise ratio. In this chromatogram the derivatized amino acids of asparagine, glutamine, histidine and serine yields lower responses due to imperfect deactivation of the liner. The GC method is currently being opti-

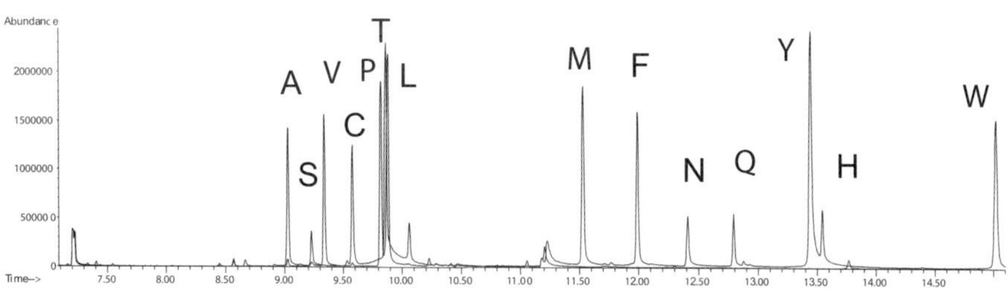

Fig. 18.3. Separation by GC and detection by negative chemical ionization of 14 1,3-bis-(trifluoromethyl)-penylthiohydantoin amino acid derivatives. Four picomoles of each amino acid derivative detected in scan mode with a high signal-to-noise ratio

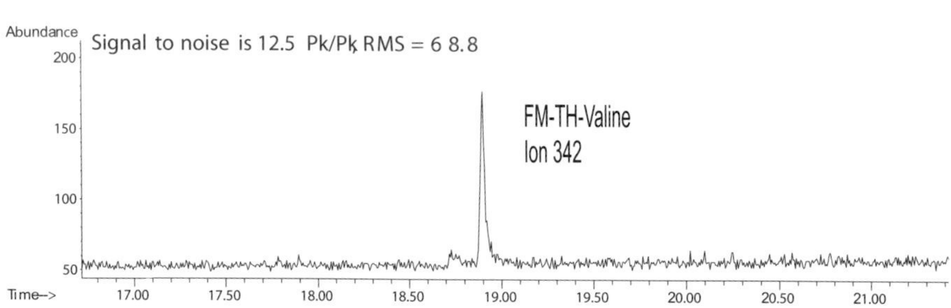

Fig. 18.4. Highly sensitive detection of the valine derivative by GC-NCI: 50 attomole measured at the detector monitored in the selected ion-monitoring mode

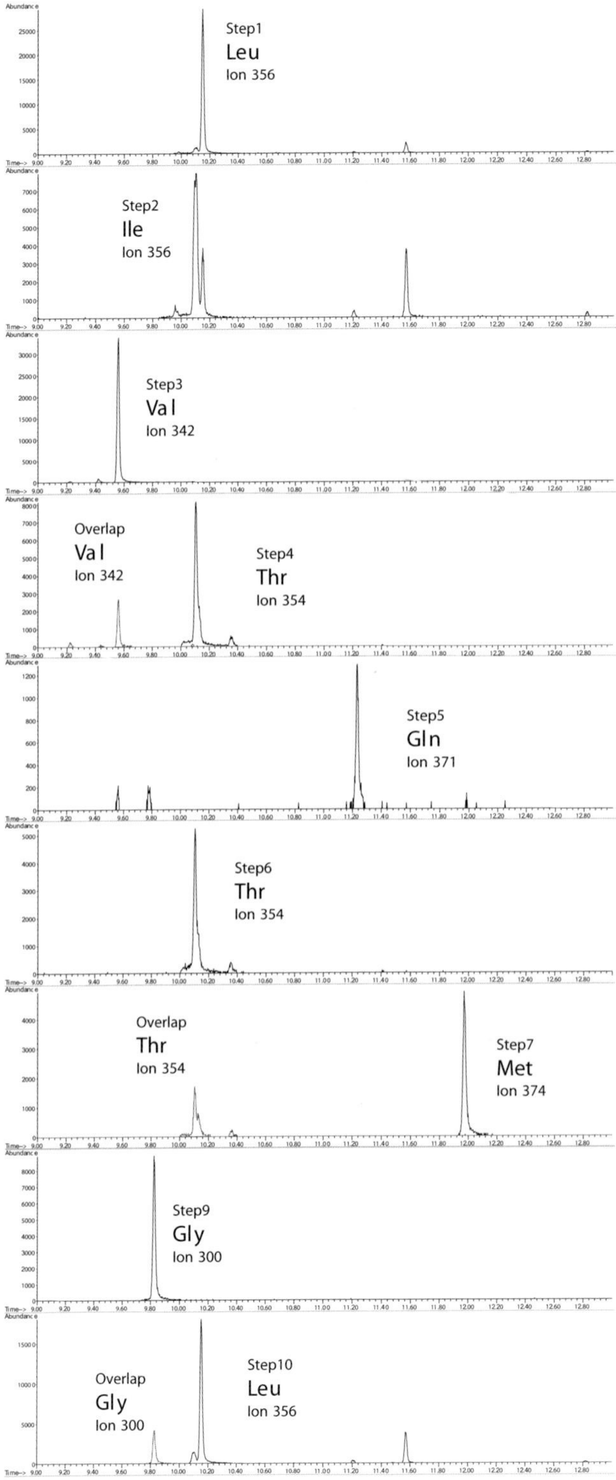

Fig. 18.5. Sequencing with our new coupling reagent 1,3-bis-(trifluoromethyl)-phenyl-isothiocyanate and off-line detection by GC MS: 30 pmol lactoglobulin was applied to the sequencer

mized, e.g. application of fast GC methods can reduce the separation time of the 20 amino acids below 5 min. The electron capture negative ionization offers a very high sensitive detection of the thiohydantoin derivatives. Figure 18.4 shows detection of 50 attomoles of the valine derivative in the selected ion monitoring mode. The RMS noise is 68.8.

We performed Edman sequencing with the new Edman-type reagent and offline measurement of the released PTH-amino acids. Figure 18.5 shows the data obtained from a run with 30 pmol lactoglobulin. The converted amino acids were redissolved in 50 μl acetone, and an aliquot of 1 μl was injected offline into the GC. The MSD was set to scan mode from 290 to 460 mass units. In the figure the extracted ion of the current amino acid is shown. Additionally, the overlap is shown in steps 4, 7 and 10. One can see the high coupling efficiency. The repetitive yield of leucine in steps 1 and 10 calculates to 94.1 %. The amount of the amino acid derivative at the detector is calculated to femtomole amounts. The signal-to-noise ratio can easily be enhanced by applying the total volume in the solvent vent mode. Due to complete chromatographic separation, the selected ion monitoring mode can be used to identify the amino acid derivatives with a high signal-to-noise ratio at their specific retention times which gives a further enhancement of the sensitivity.

18.4 Discussion and Outlook

The new Edman-like chemistry we have developed offers a high sensitivity detection of the amino acids. GC separation is done in 10 min, which leads to a higher throughput in Edman sequencing. We have tested our isothiocyanate

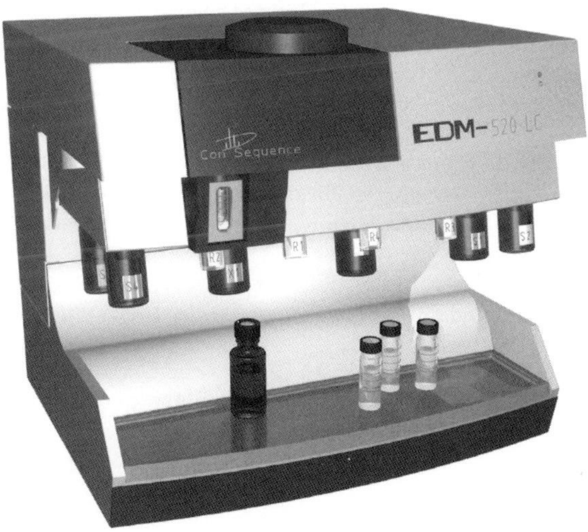

Fig. 18.6. Design study of the new sequencer generation

derivative in normal wet-phase chemistry with electroblotted and dot-blotted proteins. Initial yields of 50 % and mean repetitive yields of 93 % represent the current state of development. Optimization of the chemistry and the GC method is in progress. Currently, the GC-MS system is used for the first investigations. The next step towards sequencing in the low femtomole range is the construction of an on-line interface between the Chip Sequencer and the GC system. Figure 18.6 shows a design study of the new sequencer generation.

Acknowledgements. We would like to acknowledge the financial support through a Futour-Project no. FU 0482 given to ConSequence GmbH and a BMBF project no. FKZ 01GG9842 given to WITA GmbH in the Leitlinienverbund "Proteom-Analyse des Menschen".

References

1. Edman, P. (1950): Method for Determination of the Amino Acid Sequence in Peptides, *Acta Chem Scand* 4:283–293
2. Edman, P. and Begg, G. (1967): A Protein Sequenator, *Eur J Biochem* 1, 80–97
3. Wittmann-Liebold, B.; Graffunder, H. & Kohls, H. (1976): A Device Coupled to a Modified Sequenator for the Automated Conversion of Anilinothiazolinones into PTH Amino Acids, *Analyt. Biochem.* 75, 621–633;
4. Hewick, R.,M., Hunkapiller, M.W., Hood, L.E., Dreyer, W.J. (1981): A Gas-Liquid Solid Phase Peptide and Protein Sequenator, *J. Biol. Chem.* 256, 7990–7997
5. Wittmann-Liebold, B.; Ashman, K. (1985): On-Line Detection of Amino Acid Derivatives Released by Automatic Edman Degradation of Polypeptides. in Tschesche, H. (ed): *Modern Methods in Protein Chemistry*, Walter de Gruyter, Berlin, New York, 303–327
6. Powell, M. and Tempst, P. (2001): Microflow-Based Automated Chemistries: Application to Protein Sequencing, *Anal. Chem.2001*, 73,776–786
7. Wurzel C. and Wittmann-Liebold B., (1998): A New Design of a Wafer Based Micro Reaction System, in: *Microreaction Technology*, W. Ehrfeld (ed.) Springer 219–224
8. Wurzel C. and Wittmann-Liebold B., (2000): New approaches for innovations in sensitive Edman sequence analysis by design of a wafer based chip sequencer, in P. Jolles and H. Jörnvall (eds.): *Proteomics in Functional Genomics*, Birkhäuser Verlag, Basel, 145–157
9. Shively, J.E. (2000): The chemistry of protein sequence analysis, in P. Jolles and H. Jörnvall (eds.): *Proteomics in Functional Genomics*, Birkhäuser Verlag, Basel, 99–117
10. Kamp, R.M., Choli-Papdopoulou, T., Wittmann-Liebold, B. (1997): Protein Structure Analysis, Springer Berlin Heidelberg,137–151
11. Jin, S.-W., Chen, G-X., Palacz, Z., Wittmann-Liebold, B. (1986): A New Sensitive Edman-Type Reagent: 4–(N-1-Dimethylaminonaphthalene-5–sulfonylamino) phenyl-isothiocyanate, FEBS Lett 198, 150
12. Hess, D., Nika, H., Chow, D. T., Bures, E. J., Morrison, H. D. and Aebersold, R. (1995): Liquid chromatography-electrospray ionization mass spectrometry of 4-(3-pyridinylmethylaminocarboxypropyl) phenylthiohydantoins, Anal. Biochem. 224, 373–381

19 Amino Acid Sequencing of Sulfonic Acid-Labeled Tryptic Peptides Using Post-Source Decay and Quadratic Field MALDI-ToF Mass Spectrometry

RAMA BHIKHABHAI, MATTIAS ALGOTSSON, ULRIKA CARLSSON, JOHN FLENSBURG, LENA HÖRNSTEN, CAMILLA LARSSON, JEAN-LUC MALOISEL, RONNIE PALMGREN, MARI-ANN PESULA and MARIA LIMINGA

Abstract

Knowledge of protein and peptide sequences is fundamentally important for understanding many physiological and biochemical processes at the molecular level. Chemically assisted fragmentation by MALDI (matrix-assisted laser desorption/ionization mass spectrometry) is a new approach for amino acid sequencing of tryptic peptides. The technology is based on a new class of water stable sulfonation reagents, which strongly improves post-source decay (PSD) analysis and also simplifies the interpretation of the fragmentation spectra of the peptides. Convenient derivatization methods have been developed and optimized on a solid-phase support, enabling fast, simple and robust sample preparation.

The quadratic field reflectron of Ettan™ MALDI-ToF Pro allows fast PSD analysis, focusing all fragments independent of size in a single run. Together with its software-containing tools for automated protein identification from chemically assisted fragmentation-MALDI data, the technique enables rapid, sensitive and precise peptide sequencing and protein identification. Several examples of using chemically assisted fragmentation-MALDI for protein identification, peptide sequencing and characterization of phosphorylation sites are shown here.

19.1 Introduction

Peptide mass fingerprinting using matrix-assisted laser desorption/ionization (MALDI) time-of-flight (ToF) mass spectrometry has become a major

Principles and Practice
Methods in Proteome and Protein Analysis
R.M. Kamp, J. J. Calvete, T. Choli-Papadopoulou (Eds.)

tool for identifying proteins in proteomic research. The method provides high speed, sensitivity and mass accuracy, but for a substantial fraction of the proteins analyzed, identification is not definitive. Under these circumstances, amino acid sequence information from one or more peptides is required for unambiguous identification. By labeling the N-terminus of tryptic peptides with an acidic group, fragmentation efficiency of the peptides is improved, interpretation of their fragmentation spectra is simplified and sequence data are easily elucidated [6].

Introduction of a negative charge at the N-terminus enhances the fragmentation towards the amide bond, producing mainly b- and y-ion fragments. The negative charge at the amino terminus makes b-ion fragments neutral overall and thus only y-ion fragments are seen on the spectrum. This chemistry is applicable to tryptic peptides having arginine C-termini. For peptides with lysine C-termini, an additional guanidination reaction is required to block the ε-amino group of the lysine side chain [5]. The original reagents were very unstable in water and the derivatization reaction had to be performed in a non-aqueous environment. To overcome the drawbacks of the earlier chemistry, a novel water-stable reagent (sulfopropionic acid NHS-ester) was developed. A simple and robust protocol to derivatize the tryptic peptides (both lysine and arginine-peptides) when bound to a solid phase has also been developed. All the chemicals used in the protocol are commercially available and are included in the Ettan CAF MALDI Sequencing Kit (Amersham Biosciences).

The PSD analysis of derivatized peptides using various MALDI and electrospray instruments have been reported [2, 7, 8]. In this chapter, the Ettan MALDI-ToF Pro mass spectrometer (Amersham Biosciences) was used. The quadratic field reflectron of the Ettan MALDI-ToF Pro allows fast PSD analysis, focusing all fragments in a single run, independent of size. The accompanying software includes tools for automated protein identification from the data generated.

In the present work, examples using the Ettan CAF™ MALDI Sequencing Kit for amino acid sequencing of tryptic peptides, identification of increased numbers of proteins from 2-D gels and characterization of phosphorylation site are demonstrated.

19.2 Material and Methods

19.2.1 Chemicals

The Ettan CAF MALDI Sequencing Kit was used for the derivatization of peptides. The kit contains lysine modifier for guandination (blocking) of lysines, CAF reagent for labeling with the sulfonic acid group, hydroxylamine, buffers and control peptides. Additional reagents included μC_{18} ZipTip™ (μZT, Millipore Corporation, Bedford, MA, USA), a-cyano-4-hydroxy-cinnamic acid matrix (HCCA), trifluoroacetic acid and acetonitrile (Sigma-Aldrich), and ultrapure water (18 MΩ/cm).

One of the synthetic peptides was a kind gift from Dr. Ulf Hellman and Ulla Engström from the Ludwig Institute for Cancer Research, Uppsala, Sweden.

19.2.2 CAF Labeling Protocol

The protocol is described in detail in the instructions accompanying the Ettan CAF MALDI Sequencing Kit and is summarized as follows: The dried sample is dissolved in 0.1 % TFA and adsorbed to the pre-equilibrated matrix in the μZT. To block the ε-amino group of lysine, a fresh solution of ortho-methyl isourea hydrogen sulfate is allowed to react with bound peptides at pH 10 at room temperature overnight. To label the amino termini of peptide with sulfonic acid in the next step, the CAF reagent is dissolved in CAF buffer at pH 9.4 just prior to the reaction. CAF labeling is very fast and is complete within 3 min. Unwanted sulfonation products on hydroxyl group of serine, threonine and tyrosine residues are selectively reversed by washing the μZT with 5 % hydroxylamine. Samples are eluted using 80 % acetonitrile/0.5 % TFA, and loaded onto Ettan MALDI-ToF Pro target using HCCA as matrix.

19.2.3 Analysis of Peptides by MALDI-ToF Mass Spectrometry

All peptides were analyzed using Ettan MALDI-ToF Pro mass spectrometer. Sample and matrix (HCCA) were loaded onto the target as described in the kit instructions. All samples for PSD analysis, labeled or non-labeled were analyzed in reflector mode prior to PSD fragmentation. The software allows selection of peptides for fragmentation in PSD mode and the ion gate for the precursor ion (i.e. parent ion).

19.2.4 Interpretation of Spectra

Ideally, a spectrum contains mainly y-ion fragments, with decreasing intensity from high to low masses. By simple calculation or using the accompanying software, differences between adjacent y-ion fragments can be calculated and the amino acid sequence can be easily interpreted. The labeled peptide loses the CAF label (m/z 136) from the precursor ion and hence the difference between the two highest mass peaks is 136 Da. The y1-ion fragment for derivatized arginine-peptide is m/z 175.3 and for lysine-peptide (guanidinated and sulfonated) it is m/z 189.11 (147.11+42).

19.2.5 Protein Identification

For protein identification, the fragment masses obtained from chemically assisted fragmentation-MALDI are analyzed by the integrated protein database search engine in Ettan MALDI-ToF Pro. The mass of native (non-derivatized) peptide and five fragment masses or more (depending on the protein) are needed for protein identification. When Ettan MALDI-ToF Pro was not used, the fragment masses together with the mass of the native parent ion was submitted to a protein database search engine, e.g. PepFrag: www.expasy.org/proteomics and sequence sequence analysis tools/PROWL/ProteinInfo or MS-Fit:prospector.ucsf.edu.

In cases where chemically assisted fragmentation-MALDI data for a lysine-peptide were used, mass of 42 Da was subtracted before submitting the data.

19.2.6 Analysis of Synthetic Phosphopeptides

Synthetic phosphopeptides were used to demonstrate the applicability of Chemically Assisted Fragmentation-MALDI for sequencing of phosphorylated peptides. As the peptides had arginine C-termini, the guanidination step was omitted and only CAF labeling was performed. The derivatized phosphopeptides were analyzed by Chemically Assisted Fragmentation-MALDI to check the mass of phosphorylated amino acid. The non-derivatized phosphopeptides were also analyzed in reflectron mode and peptides were subjected to PSD analysis to check the presence of phospho-amino acids in the peptide.

19.3 Results and Discussion

Chemically assisted fragmentation-MALDI is defined as a method where the fragmentation in PSD is promoted by introduction of an acidic group to the N-terminus of tryptic peptides. The derivatization reaction of chemically assisted fragmentation is divided into two steps (Fig. 19.1). The first step converts the ε-amino group of each lysine side chain to homo-arginine (generating a mass addition of 42 Da). This step is necessary to protect the lysine groups from being labeled with sulfonic acid in the following step. The second step introduces a sulfonic acid group at the N-terminus (generating a mass addition of 136 Da). Thus, if lysine-peptide is guanidinated and CAF labeled, an increase in total mass of 178 Da (136+42 Da) will be generated.

Chemically assisted fragmentation-MALDI was performed on a number of peptides (tryptic or synthetic) to illustrate the vast application area of this methodology. Using the Ettan MALDI-ToF Pro, the PSD spectra of derivatized

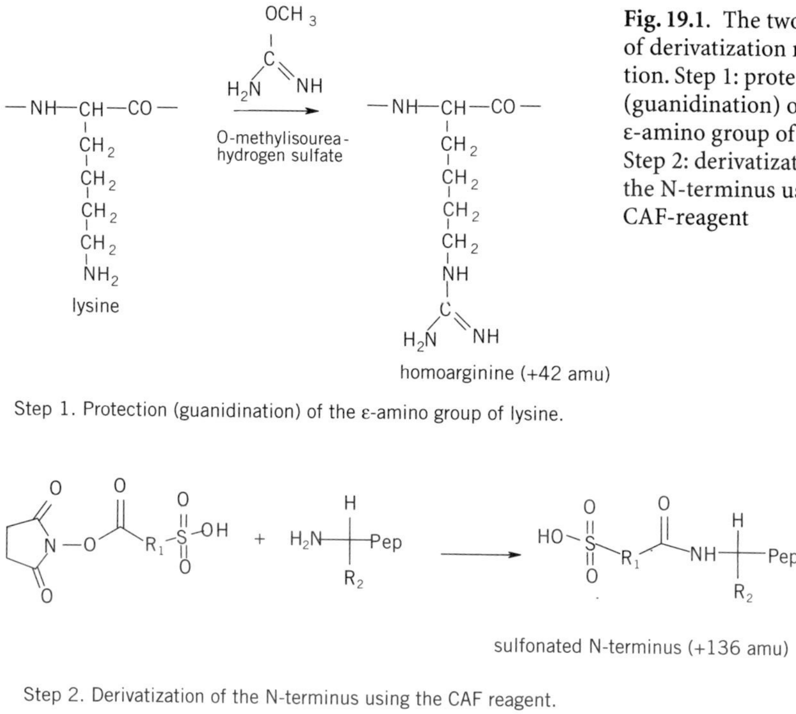

Fig. 19.1. The two steps of derivatization reaction. Step 1: protection (guanidination) of the ε-amino group of lysine. Step 2: derivatization of the N-terminus using the CAF-reagent

peptides were obtained within 30 s and no stitching or pasting of spectra was necessary.

19.3.1 Sequencing of a Synthetic Peptide

An example of complete amino acid sequencing of a model arginine-peptide, 16 amino acids long, is shown in the PSD spectrum in Fig. 19.2. Only y-ions are observed and from the differences between the y-ions, the amino acid sequence is deduced. The CAF label was lost during PSD from the precursor ion and is identified by the differences between the two y-ion fragments of higher mass, while the y1-ion fragment indicates the presence of an arginine C-terminus. The only amino acid residues which cannot be distinguished are leucine and isoleuscine, they have almost the same residual mass.

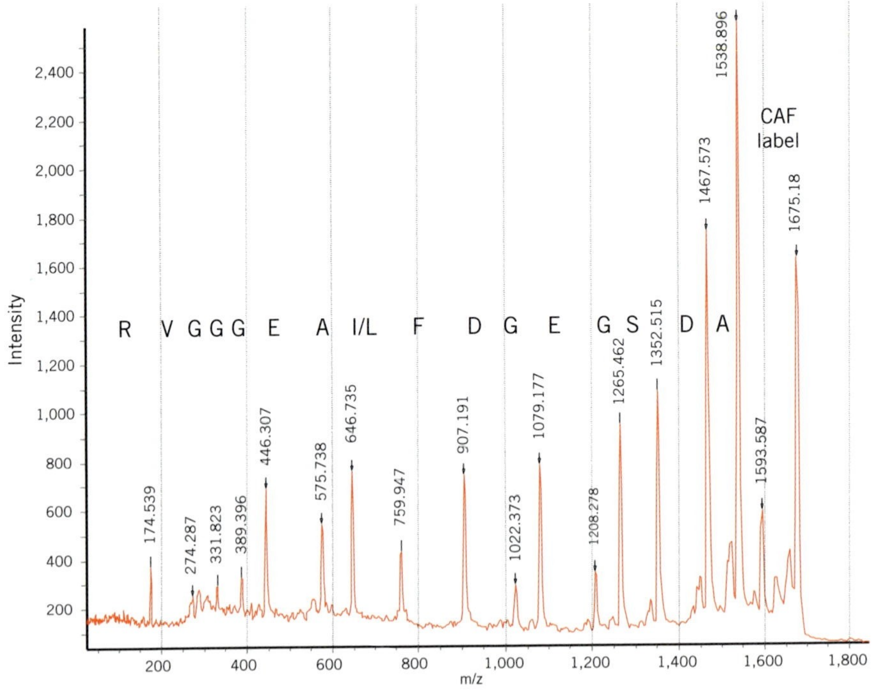

Fig. 19.2. Amino acid sequencing of a model peptide, 16 amino acids long, Sequence of peptides: A D S G E G D F I/L A E G G G V R

19.4 Identification/Confirmation of Recombinant Protein

Recombinant human serum albumin (rHSA), expressed in yeast was characterized by MALDI-ToF mass spectrometry [3]. The tryptic peptides of reduced and alkylated rHSA were CAF derivatized. Figure 19.3 shows the reflectron spectrum of derivatized tryptic peptides of rHSA. The acquired m/z values were submitted to the internal search engine, which resulted in more than 40 % coverage of the protein and the assignment of rHSA as the highest ranking candidate.

The mass of tryptic peptides of derivatized (Fig. 19.3A) and non-derivatized rHSA were compared. Two of the labeled peptides, which increased the mass by 136 and 178 Da, respectively, were selected for PSD analysis. Fig-

Fig. 19.3. **A** Spectrum of trypsin digested and CAF labeled rHSA in the reflectron mode. Peaks denoted with *arrows* were chosen for sequencing by PSD. **B** Chemically assisted fragmentation-MALDI analysis of the derivatized peptide, m/z 1836, 13 amino acids long. The derived peptide sequence (Q N CM-C E L/ IF E Q L/I G E Y K) shows the presence of one carboxymethylated cysteine and y1-ion, m/z 188, indicating that the peptide has a lysine C-terminus

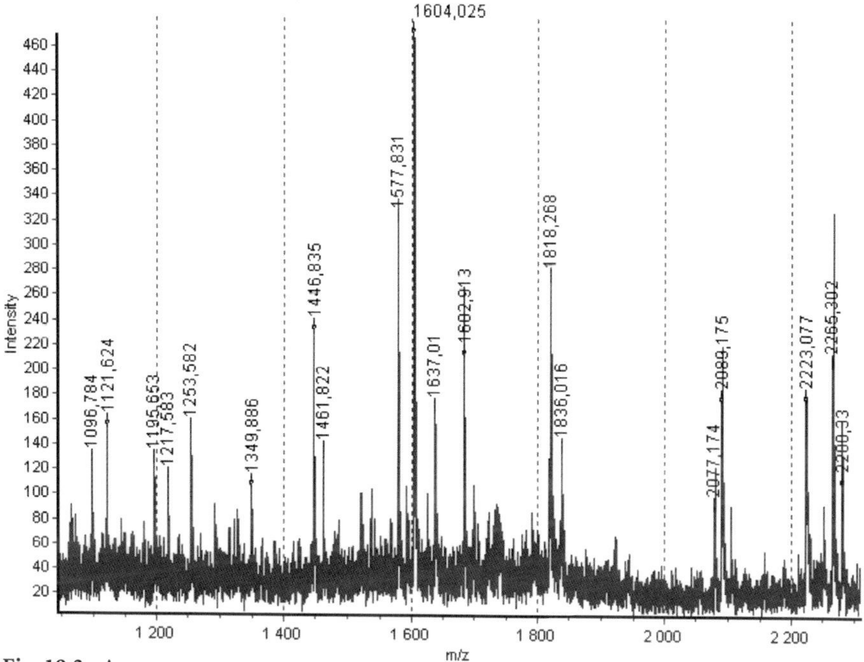

Fig. 19.3. A

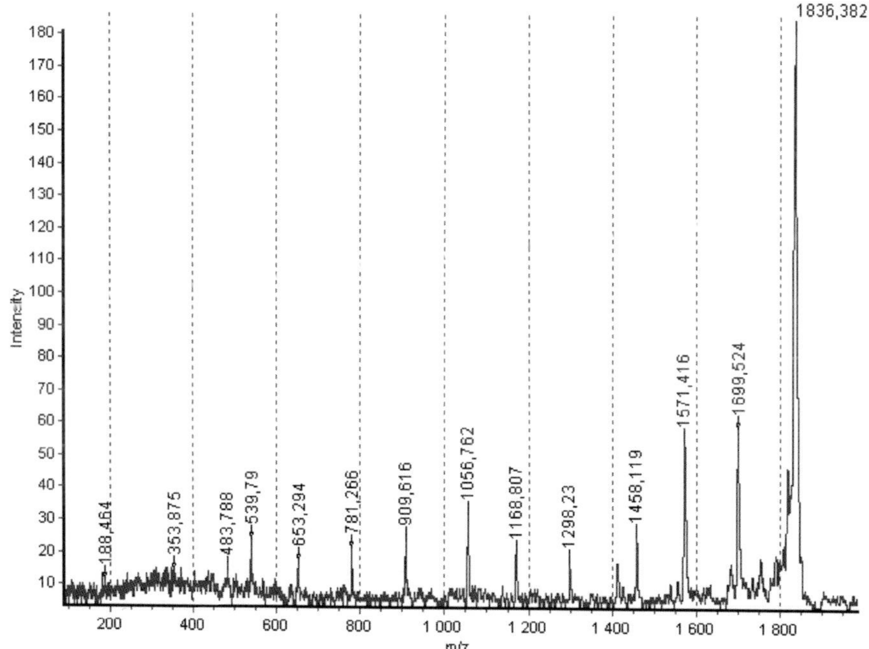

Fig. 19.3. B

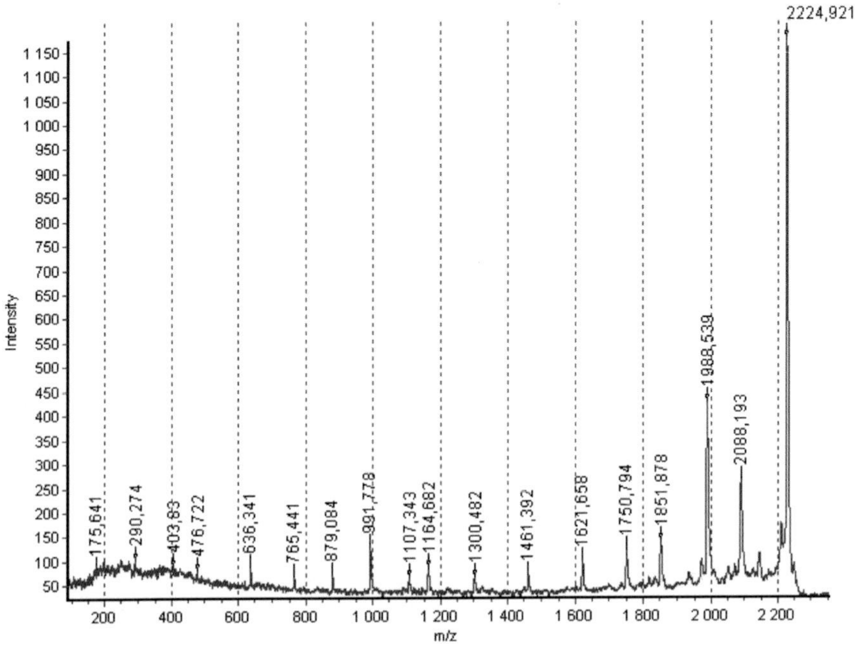

Fig. 19.3. (*Continued*) **C** Chemically assisted fragmentation-MALDI analysis of the derivatized peptide, m/z 2224, 17 amino acids long. The presence of three carboxymethylated cysteines can be observed in the sequence. (**A**, **B** and **C** are from Flensburg and Belew, unpublished results)

ure 19.3B, C shows the PSD spectra of derivatized peptides at m/z 2224 and at m/z 1836. Identification was performed using the Ettan MALDI-ToF Pro software and the protein was identified correctly.

19.4.1 Sensitivity

The sensitivity limit, defined as the lowest amount of sample needed for protein identification from chemically assisted fragmentation-MALDI data, when Ettan MALDI-ToF Pro was employed, was determined for both lysine- and arginine-terminated peptides from different sources. Model peptides (Figs. 19.4 and 19.5), in-solution digests (Fig. 19.6) and in-gel digests (Fig. 19.7), were investigated. In all cases, peptide identification was performed.

For synthetic peptides, 400 fmol of two peptides were derivatized and onetenth (40 fmol) of the CAF labeled sample was applied onto the target slide and analyzed. For lys-terminated peptide, the complete amino acid sequence containing 14 amino acids was read and identified as derived from

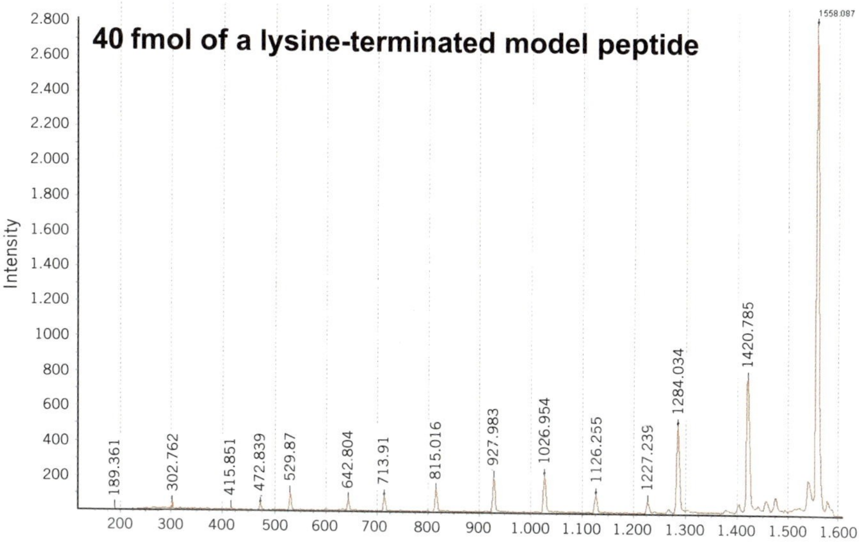

protein results

#	Expect	Result
1.	6.1×10^{-7} ●●●●	(nr-Other-Mammalia) 17.0 kDa—gi|1942750|pdb|1RSE| Myoglobin (Horse Heart) Mutant With Ser 92 Replaced By Asp (S92d) *Redundant [4]*: 1. (nr-Other-Mammalia) 16.9 kDa—gi|2506462 2. (nr-Other-Mammalia) 16.9 kDa—gi|70561 3. (nr-Other-Mammalia) 16.9 kDa—gi|2554649 4. (nr-Other-Mammalia) 17.0 kDa—gi|999870

	a:b:y	$^zm/z^{m-a}$	Sequence
6.1×10^{-7}	0:0:13	$^{1}1378.8^{-0.9}$	^{64}HGTVVLTALGGILK^{77}

Fig. 19.4. Chemically assisted fragmentation-MALDI analysis of synthetic peptide, containing the same sequence as a Lys-terminated tryptic peptide of horse myoglobin. 400 fmol was derivatized, and onetenth of the sample (40 fmol) was analyzed

fibrinopeptide A (Fig. 19.5). For arg-terminated peptide, a partial sequence of eight amino acids was read and identified as derived from myoglobin (Fig. 19.4).

Horse myoglobin (400 fmol) was subjected to tryptic digestion followed by CAF derivatization of the peptides. Onetenth of the sample (40 fmol) was analyzed. From the reflectron spectrum of the derivatized peptides, two peptides were subjected to PSD analysis and peptide identification was performed from the resulting sequence data. The search results identified that the two peptides, arg- and lys-terminating peptide (Fig. 19.6A, B) originated from the protein, myoglobin.

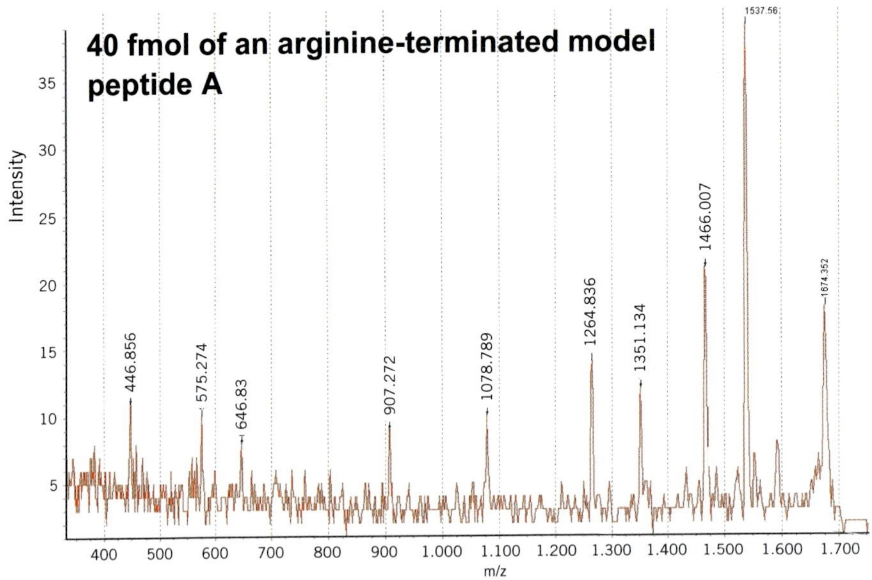

protein results

#	Expect	Result
1.	1.7×10^{-3}	(nr-Other-Primates) 1.5 kDa—gi|229333|prf||701211A fibrinopeptide A [Pan troglodytes] *Redundant [2]:* 1. (nr-Homo-sapiens) 49.4 kDa—gi|223918 2. (nr-Homo-sapiens) 1.5 kDa—gi|229185

	a:b:y	$^{z}m/z^{m-a}$	Sequence
1.7×10^{-3}	0:0:9	$^{1}1537.6^{-0.0}$	1 ADSGEGDFLAEGGGVR 16

Fig. 19.5. Chemically assisted fragmentation-MALDI analysis of synthetic peptide containing the same sequence as an Arg-terminated tryptic peptide of fibrinopeptide A. 400 fmol was derivatized and onetenth of the sample (40 fmol) was analyzed

For the study of in-gel digests, phosphorylase B was used to investigate the sensitivity limit. Seven hundred and fifty fmol of protein was spiked into a gel plug. The protein was digested with trypsin. The peptides were extracted from the gel plug, followed by lysine modification and CAF derivatization. One-tenth of the resulting peptides was loaded onto the MALDI target and analyzed (Fig. 19.7). Positive identification of the protein was provided even when the peptide sequence information was for only five amino acids.

The sensitivity limit varied from 100 to 750 fmol of starting material before derivatization of the samples of different origin. It should be noted, however, that the sample applied to the target was one-tenth of the eluted derivatized sample from the μZT. Hence the sensitivity level can be increased by altering

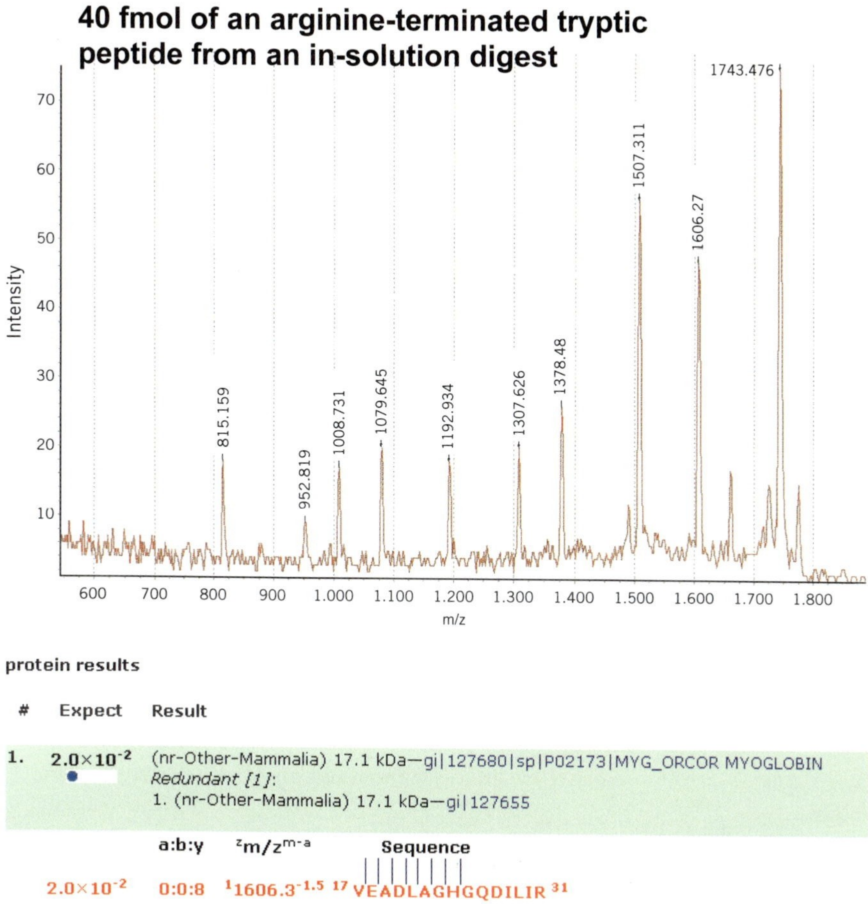

Fig. 19.6. Analysis of 40 fmol of a tryptic peptide of horse myoglobin in solution by mass spectrometry. Tryptic peptides of horse myoglobin (400 fmol) was derivatized and onetenth (40 fmol) was analyzed. From the reflectron spectrum of CAF-derivatized peptides, two peptides were selected for PSD analysis. **A** PSD spectrum of Arg-terminated peptide (m/z 1743) and **B** PSD spectrum Lys-terminated peptide (m/z 1492)

the amount applied to the target, either by concentrating the eluted sample or by eluting the derivatized peptides with matrix solution directly on the target slide.

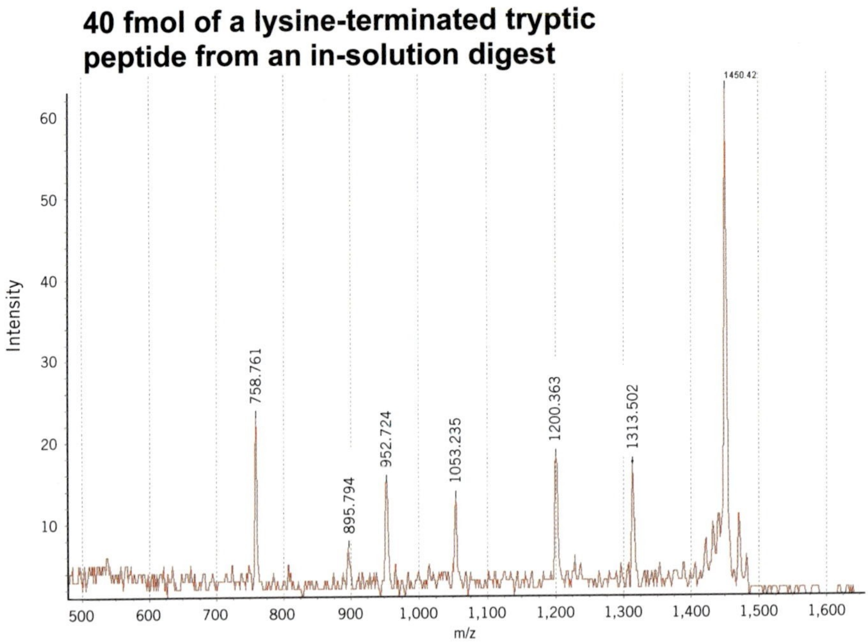

protein results

#	Expect	Result
1.	0.08	(nr-Other-Mammalia) 17.1 kDa—gi|127638|sp|P02192|MYG_BOVIN MYOGLOBIN *Redundant [13]*: 1. (nr-Other-Primates) 16.9 kDa—gi|127667 2. (nr-Other-Mammalia) 17.0 kDa—gi|494711 3. (nr-Other-Mammalia) 16.9 kDa—gi|2506462 4. (nr-Other-Mammalia) 16.9 kDa—gi|127643 5. (nr-Other-Mammalia) 16.9 kDa—gi|127694 6. (nr-Other-Mammalia) 16.9 kDa—gi|70561 7. (nr-Other-Mammalia) 17.0 kDa—gi|999870 8. (nr-Other-Mammalia) 17.0 kDa—gi|127668 9. (nr-Other-Mammalia) 17.0 kDa—gi|70559 10. (nr-Other-Mammalia) 16.9 kDa—gi|2914321 11. (nr-Other-Mammalia) 17.0 kDa—gi|2811090 12. (nr-Other-Mammalia) 16.9 kDa—gi|2554649 13. (nr-Other-Mammalia) 17.0 kDa—gi|1942750

	a:b:y	$^{z}m/z^{m-a}$	Sequence
0.08	0:0:5	$^{1}1271.5^{-0.9}$	^{33}LFTGHPETLEK^{43}

Fig. 19.6. (*Continued*)

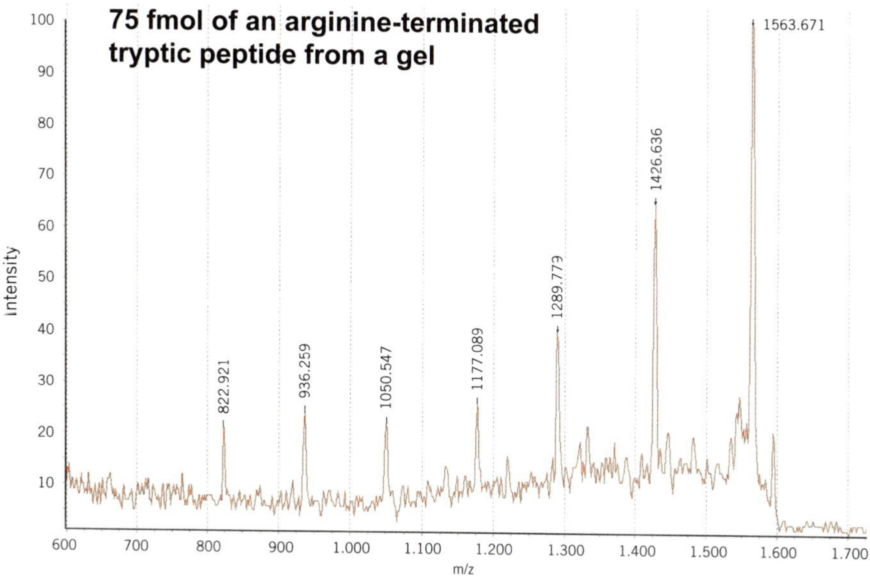

protein results

#	Expect	Result
1.	0.50	(nr-Other-Mammalia) 96.0 kDa—gi|442605|pdb|1ABB|A Chain A, Glycogen Phosphoryla (E.C.2.4.1.1) Complex With Pyridoxal-5'-Diphosphate

Redundant [18]:

1. (nr-Other-Mammalia) 97.7 kDa—gi|8569507
2. (nr-Other-Mammalia) 97.7 kDa—gi|14916625
3. (nr-Other-Mammalia) 97.6 kDa—gi|223003
4. (nr-Other-Mammalia) 97.8 kDa—gi|1664
5. (nr-Homo-sapiens) 97.5 kDa—gi|5032009
6. (nr-Other-Mammalia) 97.7 kDa—gi|8569398
7. (nr-Mus-musculus) 97.7 kDa—gi|6755256
8. (nr-Rattus) 97.8 kDa—gi|1730556
9. (nr-Other-Mammalia) 96.2 kDa—gi|6730143
10. (nr-Other-Mammalia) 97.7 kDa—gi|1633147
11. (nr-Homo-sapiens) 97.6 kDa—gi|87570
12. (nr-Other-Mammalia) 97.7 kDa—gi|14916628
13. (nr-Other-Mammalia) 97.6 kDa—gi|3318708
14. (nr-Homo-sapiens) 97.4 kDa—gi|225897
15. (nr-Other-Mammalia) 96.4 kDa—gi|231300
16. (nr-Rattus) 97.6 kDa—gi|479548
17. (nr-Other-Mammalia) 97.7 kDa—gi|8569323
18. (nr-Other-Mammalia) 97.8 kDa—gi|6093713

	a:b:y	$^{z}m/z^{m-a}$	Sequence
0.50	0:0:5	$^{1}1426.0^{-1.6}$	^{390}HLQIIYEINQR^{400}

Fig. 19.7. Chemically assisted fragmentation-MALDI analysis of an Arg-terminated tryptic peptide of phosphorylase B isolated from a 2-D gel. A total of 750 fmol was derivatized, and onetenth was analyzed

19.4.2 Sequencing of Phosphopeptides

Mass spectrometry currently provides the best general method for identifying and locating phosphorylated sites in peptides. Peptides containing phosphorylated threonine and serine lose a mass of 98 Da upon ionization due to loss of H_3PO_4. When phosphate ester is lost by beta-elimination, phosphoserine is converted to dehydroalanine (m/z 69) and phosphothreonine to dehydroaminobutyric acid (m/z 83) residues. The phosphopeptides containing the phosphate group on tyrosine (m/z 243) residues do not show any beta-elimination [1, 4, 9, 10]

Ettan MALDI-ToF Pro was used to analyze three synthetic phosphopeptides, each containing phosphoamino acid either at Tyr, Ser or Thr. The peptides were CAF labeled and subjected to Chemically Assisted Fragmentation-MALDI (Figs. 19.8, 19.9 and 19.10). The results showed, that for the peptides containing phosphoserine or phosphothreonine residues, beta-elimination occurred and a mass difference of 98 Da (mass of H_3PO_4) between the first two peaks indicated the removal of phosphoric acid from the peptide. From

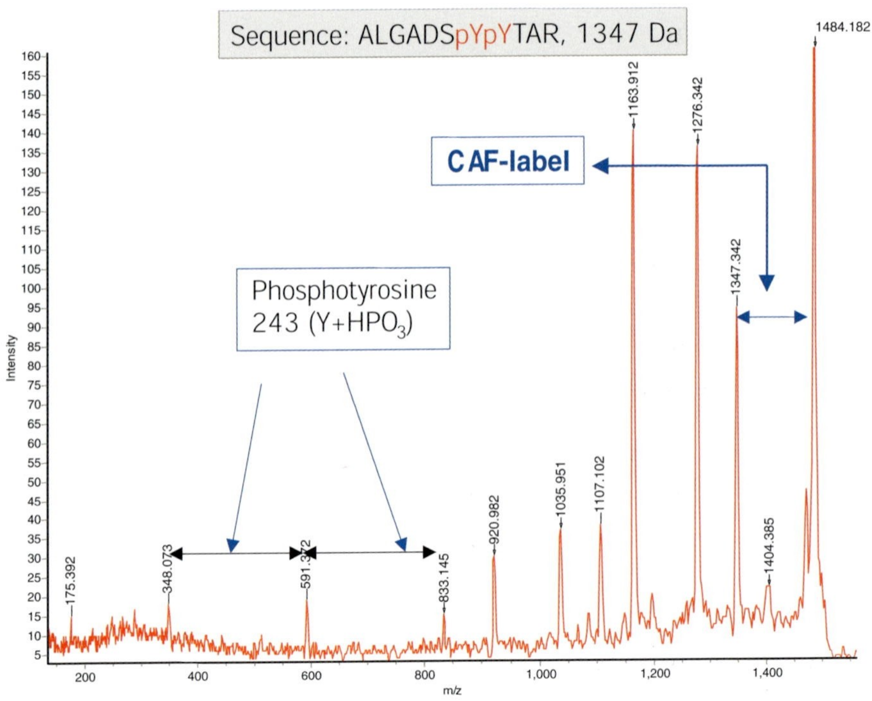

Fig. 19.8. Chemically assisted fragmentation-MALDI analysis of derivatized, diphosphorylated (Tyr/Tyr) synthetic peptide. The results show that the phosphorylated tyrosines at positions 7 and 8 remained intact with a mass difference of 243 Da (163 Da for tyrosine + 80 Da for HPO_3) between y5 and y4 ions, and y6 and y5 ions

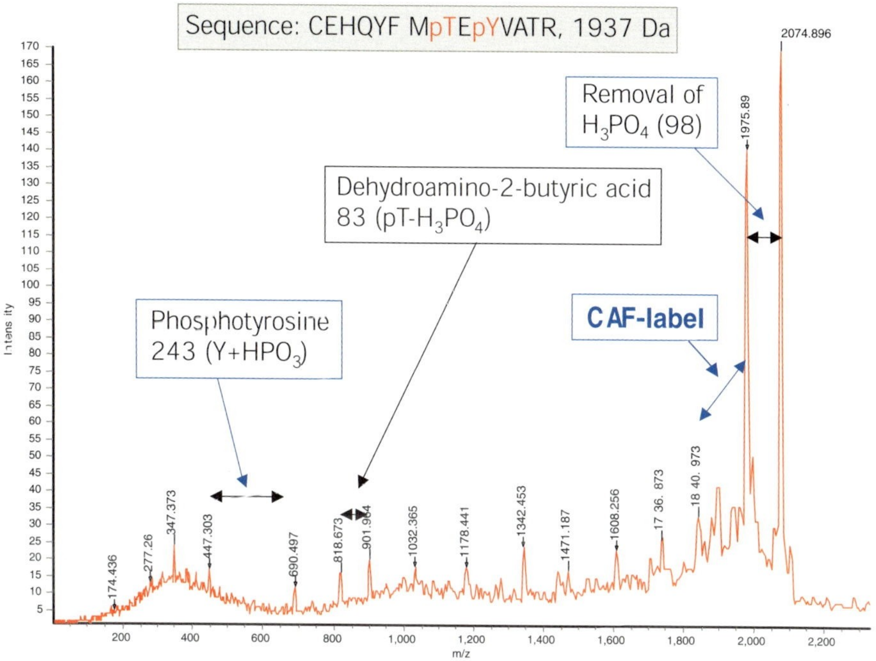

Fig. 19.9. Chemically assisted fragmentation-MALDI analysis of derivatized, diphosphorylated (Tyr/Thr) synthetic peptide. The mass difference of 98 Da between the first two peaks indicates removal of phosphoric acid from the peptide. The mass difference between y7 and y6 ions was 83 Da, which corresponds to the mass of a dehydroamino-2-butyric acid product formed by beta-elimination. The mass difference between y4 and y5 ions was 243 Da, indicating a phosphotyrosine (163 Da for tyrosine+80 Da for HPO_3)

the deduced sequence data, the mass difference of 69 Da indicated dehyroalanine and 83 Da indicated dehydroamino-2-butyric acid, which are products formed by beta-elimination of phosphoserine and phosphothreonine, respectively. For the peptide containing phosphotyrosine, a mass difference of 243 Da between two adjacent ions indicated a phosphotyrosine residue (163 Da for tyrosine; +80 Da for HPO_3).

19.4.2.1 Identifcation of Phosphopeptides

One of the methods for identifying phosphopeptides is to compare the observed peptide masses with those from the known sequences in a reflectron spectrum. A mass of 80 Da greater than predictions indicates the presence of phosphorylated amino acids [1]. Another method to identify the presence of the phospho group, is to perform PSD analysis of the non-derivatized peptides. A typical characteristic pattern was observed when a non-derivatized phosphopeptide was subjected to PSD [4, 9 and 10]. The characteristic pattern was

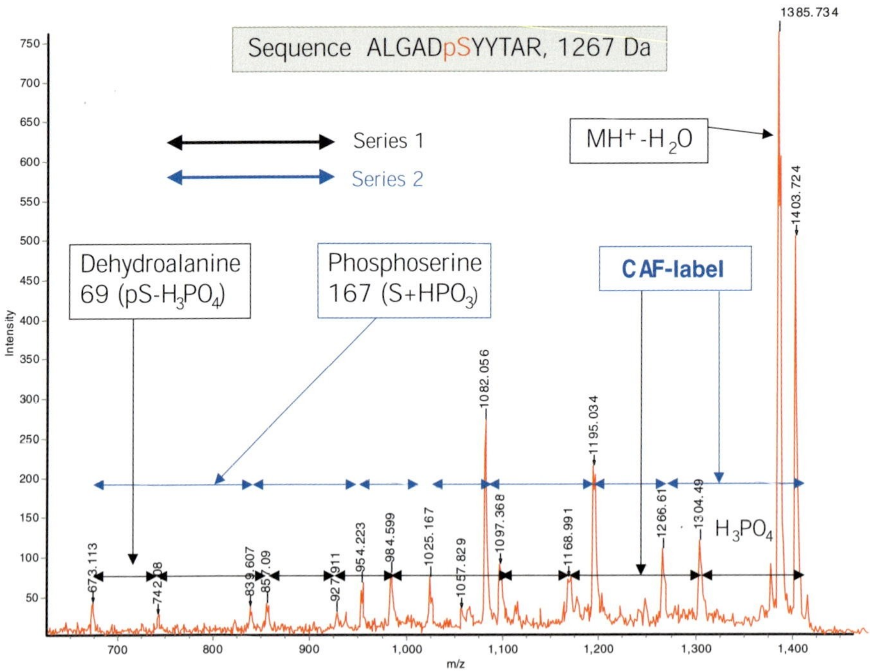

Fig.19.10. Chemically assisted fragmentation-MALDI analysis of derivatized, monophosphorylated (Ser) synthetic peptide. Two series of y-ions were observed. In the first series, a mass of 98 Da (phosphoric acid) was cleaved off the parent ion, followed by cleavage of the CAF label. In the second y-ion series, the CAF label was cleaved off first. The results show that even though there are two series of y-ions, the phosphorylation site was easily identified using PSD

the formation of two fragment ions from the losses of HPO_3 ($[MH\text{-}HPO_3]^+$) and H_3PO_4 ($[MH\text{-}H_3PO_4]^+$), corresponding to mass losses of 80 and 98 Da, respectively.

The three non-derivatized phosphopeptides were subjected to PSD and the results are shown in Figs. 19.11, 19.12 and 19.13. The spectra show average masses of the parent ions and fragment ions. The parent ion produces a strong signal for the Ser phosphopeptide (Fig. 19.11). However, the intensities of the fragment ion $-H_3PO_4$ (-98 ± 1 Da) are only about 50 % of those of the parent ion; and those of the fragment ion $-HPO_3$ (-80 Da) are only 10–25 %.

The two sets of characteristic patterns for Tyr/Thr phosphopeptide (Fig. 19.12) suggest that it is a diphosphopeptide. In the case of the Tyr/Tyr diphosphopeptide (Fig. 19.13), only one set of dephosphorylation patterns with a low intensity of the $-H_3PO_4$ fragment ion was observed. This suggests that there was no beta-elimination of the phosphate group of phosphotyrosine.

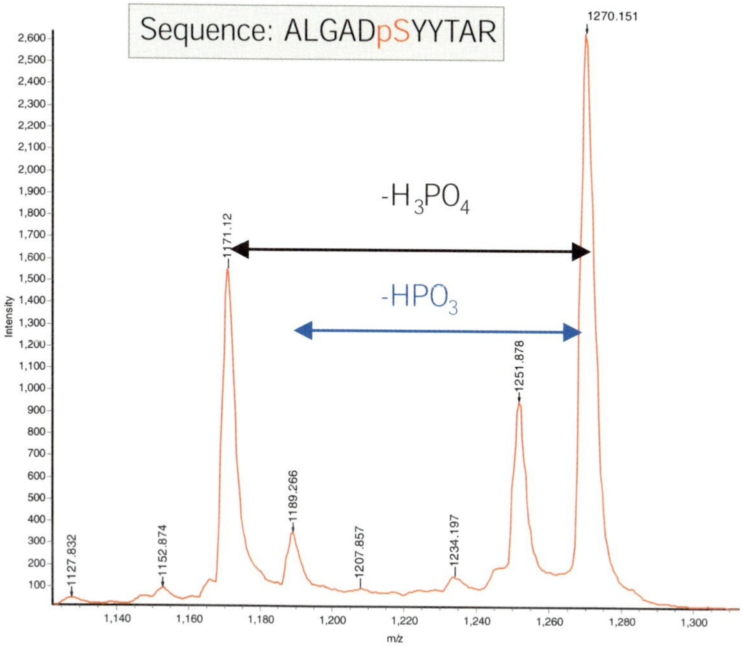

Fig. 19.11. PSD dephosphorylation spectrum of non-derivatized, monophosphorylated (Ser) synthetic peptide

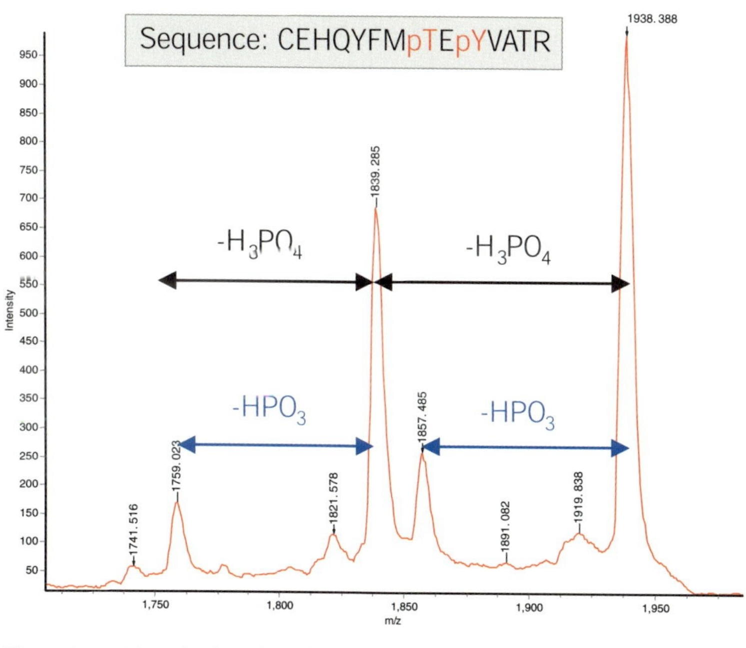

Fig. 19.12. PSD dephosphorylation spectrum of non-derivatized, diphosphorylated (Tyr/Thr) synthetic peptide

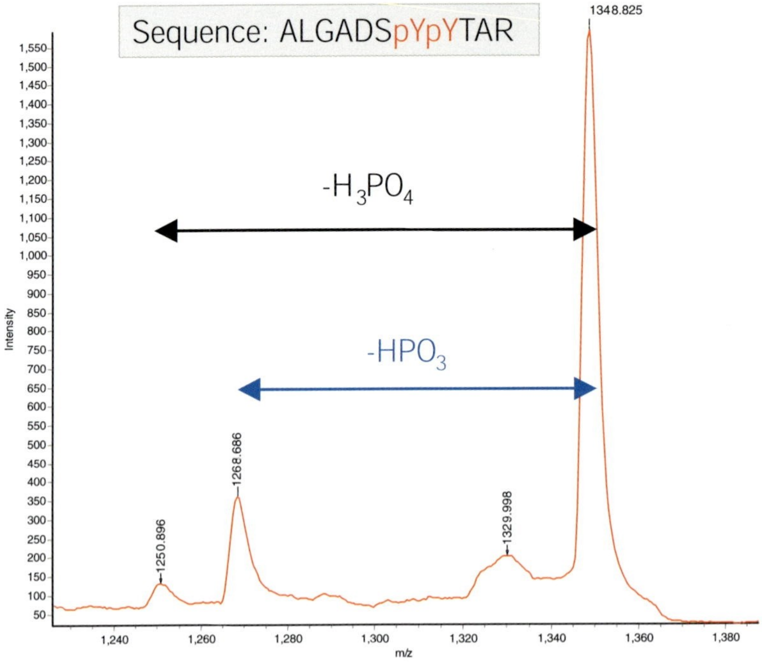

Fig. 19.13. PSD dephosphorylation spectrum of non-derivatized, diphosphorylated (Tyr/Tyr) synthetic peptide

The results show that by performing PSD of non-derivatized phosphopeptide, the presence of phosphate groups will be confirmed from the characteristic pattern, thus identification of phosphopeptides is possible.

19.5 Conclusions

We have shown that chemically assisted fragmentation using the Ettan CAF MALDI Sequencing Kit provides a powerful method for peptide sequencing. Together with Ettan MALDI-ToF Pro, enabling automated single run PSD analysis and automated protein identification, a sensitive and rapid technique for amino acid sequencing of tryptic peptides is offered.

Acknowledgement. We are thankful to Prof. Tom Keough and his colleagues at the Procter and Gamble Company, USA for fruitful collaboration. We would also like to thank Prof. Ulf Hellman for providing one of the phosphopeptides.

References

1. Aebersold R and Goodlett D R (2001) Mass Spectrometry in Proteomics. Chem Rev 101:269–295
2. Hellman U and Bhikhabhai R (2002) Easy amino acid sequencing of sulfonated peptides using post-source decay on a matrix–assisted laser desorption/ionization time-of-flight mass spectrometer equipped with a variable voltage reflector. Rapid Commun Mass Spectrom 16:1851–1859
3. Flensburg J and Belew M (2003) Characterization of Recombinant Human Serum Albumin using MALDI-ToF mass spectrometry. Submitted for J Chrom A ISPPP Symposium volume
4. Hoffmann R, Metzger S, Spengler B and Laszlo O Jr. (1999) Sequencing of peptides Phosphorylated on Serines and Threonines by Post-source Decay in Matrix-assisted Laser Desorption/Ionization Time-of-flight Mass Spectrometry. J Mass Spectrom 34: 1195–1204
5. Keough T, Lacey MP and Youngquist RS (2000) Derivatization procedures to facilitate *de novo* sequencing of lysine-terminated tryptic peptides using post-source decay matrix-assisted laser desorption/ionization mass spectrometry. Rapid Commun Mass Spectrom 14:2348–2356
6. Keough T, Youngquist RS and Lacey MP (1999) A method for high-sensitivity peptide sequencing using postsource decay matrix-assisted laser desorption ionization. Proc Natl Acad Sci USA 96:7131–7136
7. Keough T, Lacey MP and Strife RJ (2001) Atmospheric pressure matrix-assisted laser desorption/ionization ion trap mass spectrometry of sulfonic acid derivatised peptides. Rapid Commun Mass Spectrom 15:2227–2239
8. Keough T, Lacey MP and Youngquist RS. (2002) Solid-phase derivatisation of tryptic peptides for rapid protein identification by matrix-assisted laser desorption/ionization mass spectrometry. Rapid Commun Mass Spectrom 16:1003–1015
9. Metzger S and Hoffmann R (2000) Studies On The Dephosphorylation Of Phosphotyrosine-Containing Peptides During Post-source Decay In Matrix-Assisted Laser Desorption/Ionization. J Mass Spectrom 35:1165–1177
10. Qin J and Chait B T (1997) Identification and Characterization of Post Translational Modification of Proteins by MALDI Ion Trap Mass Spectroscopy. Anal Chem 69:4002–4009

20 Separation of Peptides and Amino Acids using High Performance Capillary Electrophoresis

HONG JIN and ROZA MARIA KAMP

20.1 Introduction

Capillary electrophoresis (CE) has been introduced for analysis of charged molecules. It has been introduced as high performance capillary electrophoresis (HPCE), and is a powerful method for separation of nucleic acids, proteins and peptides. In this chapter, we also describe the application of HPCE for separation of PTH amino acids after protein sequencing or protein hydrolysis. The wide range of different separation conditions, including: buffers pH, temperature, ion-forming reagents, micelle-forming reagents, complex reagents, packed capillaries, makes the separation of different biomolecules possible.

Capillary electrophoresis was introduced by Jorgerson using a 75-μm fused silica capillary and 30 kV voltage. The new generation of CE equipment allows full automatization of the separation and data evaluation.

The advantages of CE are that it has a very short separation time, only nanoliter sample volumes, very high efficiency of the separation and flexibility for detection systems.

HPCE is a comparable method to HPLC, but the method optimization is simpler and expenses for eluents and columns are much lower.

For optimal separation of biomolecules the following parameters can be optimized:

- pH of buffer
- temperature
- salt concentration
- ion strength
- capillary coating or packing
- additives such as SDS, cyclodextrine, ion pairing reagents, complexing reagents

Principles and Practice
Methods in Proteome and Protein Analysis
R.M. Kamp, J. J. Calvete, T. Choli-Papadopoulou (Eds.)

The difference between using capillary electrophoresis as opposed to using the conventional method is the use of glass capillaries instead of polyacrylamide slab gels. The separation is performed using a high capacity buffer, which is responsible for the constant pH during the electrophoresis. The buffers usually applied are:

- phosphate
- citrate
- borate
- TRIS

Depending on buffer additives, different capillary techniques can be used:

- zone electrophoresis
- micellar capillary electrophoresis
- isotachophoresis
- isoelectric focusing

It is important to overcome the adsorption problems on the capillary surface. This is done by using a buffer which has a pH higher than the isoelectric point and a pKa lower than the silanol groups, and which suppresses the ionization of the capillary surface.

The special application of capillary electrophoresis is the use of coated capillaries, e.g. after chemical modification of silanol groups or coating with polymers. For separation of protein and peptides, the techniques usually used are zone electrophoresis, gel electrophoresis and isoelectric focusing electrophoresis.

The simplest technique is capillary zone electrophoresis, which separates molecules depending on their mobility, which varies with size and charge of separated molecules. It is important to reduce protein and wall interaction, because of the unspecific binding of molecules.

Proteins or peptides ions are separated according to different mobility under constant voltage. Proteins and peptides are separated usually at low pH and migrate all as cations. CE can be used for peptide maps after enzymatic cleavage of proteins. CE is also used in the purity control of synthetic peptides. The advantage of CE is coupling it with mass spectrometry, which allows on-line mass determination of separated peptides. The most difficult is separation of amino acids, because of the properties of amino acids. Neutral molecules, which do not migrate using zone electrophoresis system, have to be modified using detergent for micelle formation and the mobility of such molecules increased . For micellar electrokinetic electrophoresis anion detergent SDS can be used for micelle forming.

20.2 Separation of Peptides

All separations of peptides were performed using Applied Biosystem 270 E Capillary Electrophoresis systems.

β-Lactoglobulin and cytochrome c were cleaved using TPCK-trypsin and peptides were separated using capillary electrophoresis apparatus.

20.2.1 Trypsin Cleavage

20.2.1.1 Digestion of β-Lactoglobulin

- 20 μg protein (β-lactoglobulin) was dissolved in 20 μl of 0.25 M Tris buffer containing 8 M urea, pH 7.8
- 5 μl of 50 mM dithiothreitol (DTT) was added and incubated at 50 °C for 15 min.
- 75 μl deionized water was added and diluted with Tris-urea buffer to 0.05 and 2 M, respectively.
- 10 μg protease (trypsin) was dissolved in 10 μl water, then 2 μl added to the protein solution.
- The protein sample was incubated in a water bath at 37 °C for 4 h for enzymatic cleavage.
- After cleavage, the peptides were injected directly into the capillary electrophoresis system or lyophilized and dissolved in deionized water.

20.2.1.2 Trypsin Digestion of Cytochrome C

- 500 μg protein (cytochrome C of horse heart) was dissolved in 50 μl double-deionized water in a small 500-μl Eppendorf tube.
- 50 μl 0.2 M of N-methyl morpholine acetate buffer (pH 8.1) was added.
- 100 μg N-tosyl-l-phenylalanylchloromethylketon (TPCK)-trypsin was dissolved in 10 μl deionized water.
- 10 μl enzyme solution was added to 100 μl protein sample solution and incubated at 37 °C for 4 h.
- The tryptic peptides were injected into the CE instrument for peptide separation.

20.2.2 Separation Conditions for HPCE

20.2.2.1 Separation of β-Lactoglobulin Tryptic Peptides

For separation of β-lactoglobulin tryptic peptides, the following HPCE conditions were used (Fig. 20.1):

- sample was dissolved in double-deionized water
- separation buffer: 50 mM sodium phosphate buffer (pH 2.3)

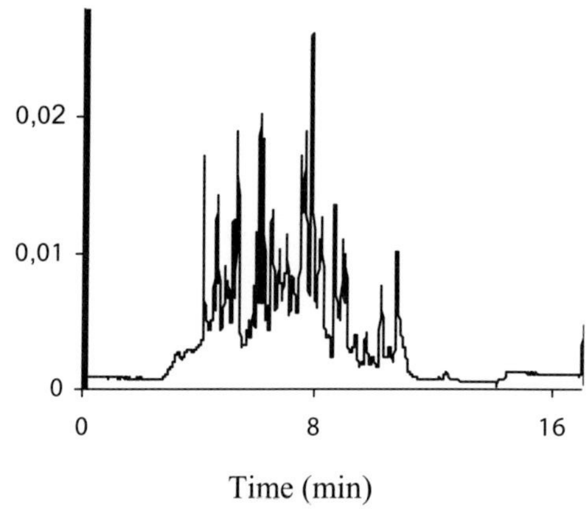

Fig. 20.1. Separation of tryptic peptides of lactoglobulin. Buffer 50 mM sodium phosphate pH 2,3; capillary 75 µm i.d., 78 cm length, fused silica, uncoated; detection at 214 nm; temperature 25 °C; voltage 25 kV; sample injection 10 s at 0.5 psi

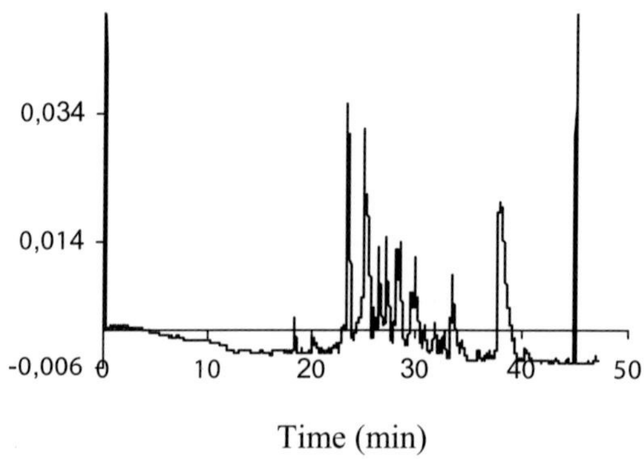

Fig. 20.2. Separation of tryptic peptides of cytochrome C (horse). Separation buffer 0.02 M sodium citrate pH 2,5; capillary 50 µm i.d., 78 cm length, fused silica, uncoated; detection at 214 nm; temperature 30 °C; voltage 15 kV; sample injection 3 s at 0.5 psi

- capillary: 75 µm inner diameter, 78 cm length, fused silica, uncoated
- UV detection: 214 nm
- run temperature: 25 °C
- voltage: 25 kV
- sample injection: 10 s at 0.5 psi (10 pmol)

20.2.2.2 Separation of Cytochrome C After Trypsin Digestion

For separation of cytochrome C after trypsin digestion, the following conditions were used (Fig. 20.2):
- sample was dissolved in double-deionized water
- separation buffer: 0.02 M sodium citrate buffer (pH 2,5)
- capillary: 50 µm inner diameter, 78 cm length, fused silica, uncoated
- UV detection: 214 nm
- run temperature: 30 °C
- voltage: 15 kV
- sample injection: 3 s at 0.5 psi (3 pmol)

20.3. Sequencing of Proteins and PTH Amino Acid Analysis

Manual sequencing of bradykinin was performed by Edman degradation. The PTH amino acids obtained after each cycle were analyzed using capillary electrophoresis (Fig. 20.3).

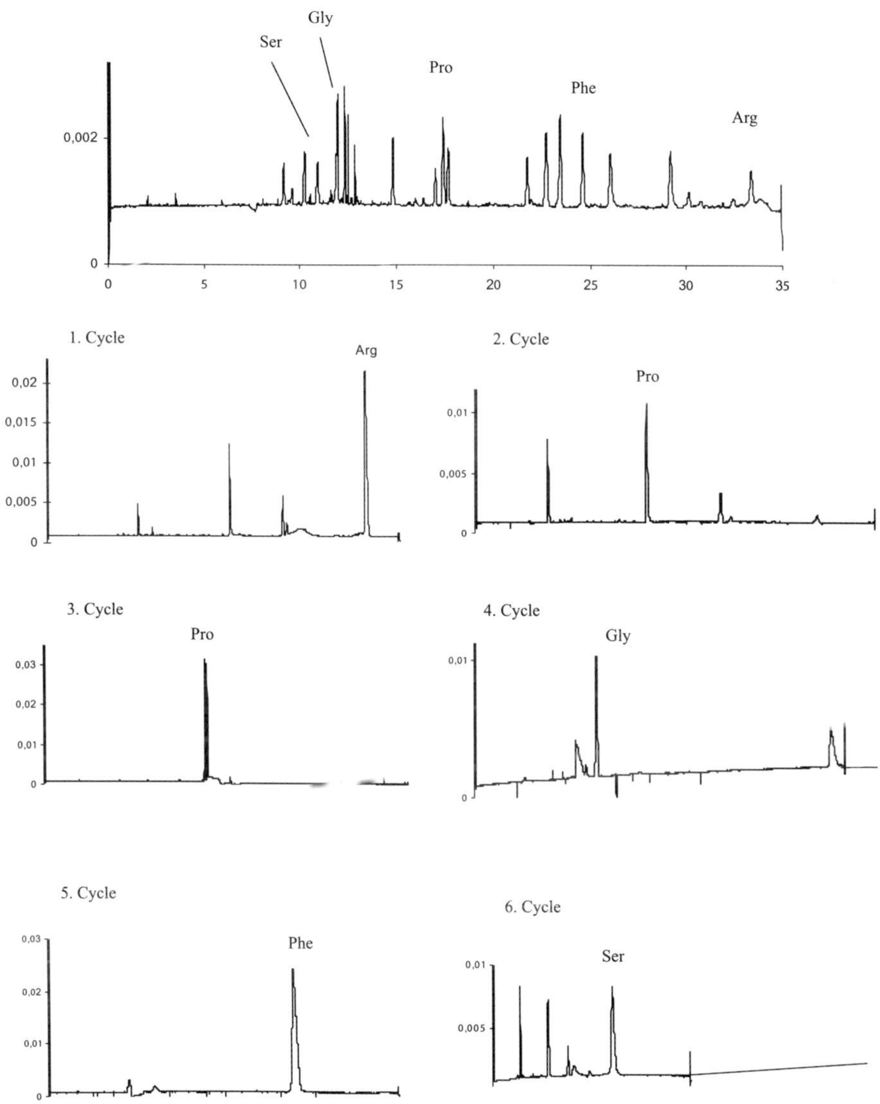

Fig. 20.3. Electropherograms of PTH standard and PTH amino acids after sequencing of bradykinin. Separation buffer 25 mM phosphate pH 6,5 containing 50 mM SDS; capillary 50 µm i.d, 95 cm length, fused silica, uncoated; detection at 260 nm; temperature 25 °C; voltage: 27 kV; sample injection: 5 s at 0.5 psi

20.3.1 Chemicals

Phenylthiohydantoine (PTH) amino acids standards and bradykinin were purchased from Sigma (St. Louis, MO, USA), phenylisothiocyanate (PITC) was purchased from Pierce (Rochford, IL,USA), trifluoracetic acid (TFA), pyridine (distilled three times) and n-butylacetate were purchased from Merck (Darmstadt, Germany). Sodium dodecyl sulfate (SDS) was purchased from BioMol (Hamburg, Germany). Buffering salts were obtained from Merck (Darmstadt, Germany).

HPLC-grade acetonitrile was purchased from J.T. Baker (Deventen, Holland).

20.3.2 Amino Acid Standard Preparation

The mixture of 20 standard PTH-amino acids was purchased from Sigma, dissolved in acetonitrile and stored at –20 °C. Twenty single PTH-amino acids were weighted and dissolved in double-deionized water. 100 µl amino acid standard (10 nmol) dissolved in acetonitrile was dried in speed-vac, then dissolved in double-deionized water for CE analysis.

20.3.3 Sequencing of Bradykinin

- 5 nmol bradykinin was dissolved in 50 µl double-deionized water
- 50 µl 5 % PITC in pyridine was added under the hood, incubated at 55 °C under N_2 for 60 min
- The sample was dried using lyophilization
- 100 µl 100 % TFA was added and then incubated at 55 °C for 20 min
- The sample was dried using speed-vac
- 50 µl water and 200 µl n-butylacetate were added and mixed thoroughly
- The sample was centrifuged for a short period of time and the upper organic layer (first degraded amino acid) was removed and dried, the water phase was used for the next cycle
- 200 µl n-butylacetate was added again, and mixed thoroughly
- After it was centrifuged for a short period of time, the upper organic layer was collected
- The water layer was stored at –20 °C for the next cycle
- The organic layer was dried using speed-vac
- n-butylacetate extract was dissolved in 20 % TFA and incubated at 55 °C for 30 min
- The sample was dried and directly analyzed with capillary electrophoresis
- The water layer was used for the next cycle
- The procedure was repeated for the six cycles

20.3.4 HPCE Separation Conditions for PTH Amino Acid

- PTH standard amino acids were dissolved in double-deionized water
- separation buffer: 25 mM phosphate buffer (pH 6.5) with 50 mM SDS
- capillary: 50 μm inner diameter, 95 cm length, fused silica, uncoated
- UV detection: 260 nm
- run temperature: 25 °C
- voltage: 27 kV
- sample injection: 5 s at 0.5 psi (5 pmol)

20.3.5 Optimization of the PTH Amino Acid Separation

The conditions for the separation of PTH amino acids were optimized. Different buffers as citrate, phosphate or acetate were investigated for their use for separation of amino acids. The best results were achieved using 25 mM phosphate buffer. The separation using buffers containing SDS showed a clear, better resolution as detergent-free buffers, because the neutral molecules were charged using SDS and formed micelles with higher electrophoretic mobility.In contrast, neutral molecules migrate in the electric field. The influence of the pH of 25 mM phosphate buffer was investigated, e.g. pH 6,0, 6.5, 7.0, 7.2 and 8.0. The optimal resolution was achieved at pH 6.5.

20.4 Conclusion

Separation of peptides for peptide mapping using capillary electrophoresis is a very fast and sensitive method, which requires only nanoliter amounts of the sample. Usually, low pH buffers are applied for separation of peptides and all peptides migrate as cations. In the case of high pH, the separation will be performed under conditions of electro-osmotic flow. All ions moved in the direction of the cathode, because electro-osmotic velocity is higher than electrophoretic velocity. The neutral peptides are transported by the electro-osmotic flow and can also be detected at the cathode. The electrophoretic mobility depends on the molecular mass and the charge.

The capillary electrophoresis of peptides in combination with mass spectrometry can be used as an on-line, fully automated method for structural studies of proteins. In off-line mode, separated peptides can be used for determination of amino acid sequences after Edman degradation.

The separation of PTH amino acids, resulting from Edman degradation, was optimized for capillary electrophoresis. After each degradation step, the PTH amino acids were separated by capillary electrophoresis and identified by comparison of retention time with standard amino acids. Amino acid analysis by capillary electrophoresis requires only nanoliter amounts, has a higher separation efficiency and a shorter running time. The uncoated silica capillary is much cheaper than the microbore HPLC column does not requires organic solvents. Because some of the amino acids are not charged, micellar electrokinetic capillary electrophoresis (MECC) has to be used. SDS can be applied as a useful detergent for the formation of micelles. The hydrophilic amino acids elute very fast, because of very weak interaction with SDS. All other amino acids elute according to their hydrophobicity. Stronger basic amino acids are retained in the capillary longer because of ionic interaction.

The results show that capillary electrophoresis can be applied not only for protein separation, but also for protein sequencing, instead of conventional HPLC systems and it is an additional option for protein sequencing.

References

1. Kim J.K., Kim J. H., and Lee K.-J.(1995), Application of capillary electrophoresis to amino acid sequencing of peptide. Electrophoresis 16, 510–515
2. Smith A. J. (2003), Analytical and Micropreparative Capillary Electrophoresis of Peptides, in *Methods in Molecular Biology, vol. 211: Protein Sequencing Protocols* (Smith, B. J. ed.), Humana Press Inc
3. F. Lottschpeich, H. Zorbas (1998), Bioanalytik, Spektrum Akademischer Verlag GmbH Heidelberg-Berlin
4. R. Kellner, F. Lottspeich, H.E. Meyer (1999), Wiley-VCH, "Microcharacterization of Proteins"

21 InterPro and Proteome Analysis – *In silico* Analysis of Proteins and Proteomes

Nicola Jane Mulder, Manuela Pruess and Rolf Apweiler

21.1 Introduction

In the current silicon era of science and technology, scientists at large are no longer concerned with single gene analysis. The rapid appearance of completely sequenced genomes has shifted the focus of research in different ways. For the bench scientist concerned with single or small sets of genes, the focus has widened to include context information. For example, it is now possible to study complete pathways and biological processes rather than single reactions or interactions, since the sequences of all players in the processes have more than likely been elucidated. Comparisons of the genes or processes of interest can also be made between a range of different species or organisms. For scientists in the field of genomics and proteomics, the profusion of new data has increased the scale on which it is feasible to work. Large-scale proteomics experiments like 2-D gel analysis and mass spectrometry as well as the availability of complete proteomes (all proteins encoded by a genome) has facilitated global analysis of protein function in or between different proteomes.

Proteomics data generated by genome sequencing projects or high-throughput laboratory experiments can no longer be analysed manually, *in silico* analysis is a necessity. With the increase in proteomics data in the public and private domains, so follows the need for reliable proteomics databases and tools. A common interest between all proteomics scientists is in the elucidation of protein function. The first step in this process is finding a reliable data source for protein sequences generated from the nucleotide sequence databases or proteomics experiments. The second step is applying a number of different tools to classify the protein(s) into its(their) functional group, whether it is a protein family or a functional domain the protein contains. The first problem is partly solved in the form of the Swiss-Prot and TrEMBL databases (Bairoch and Apweiler 2000). Swiss-Prot is a high quality manually annotated database of protein sequences, while TrEMBL is largely unanno-

Principles and Practice
Methods in Proteome and Protein Analysis
R.M. Kamp, J. J. Calvete, T. Choli-Papadopoulou (Eds.)

tated. TrEMBL evolved with the need to supplement Swiss-Prot with all available protein sequence data that have not yet been hand-curated due to the increasing load of the database from genome sequencing projects. Some automatic annotation procedures are applied to TrEMBL to provide the public with some information on the protein where possible. The second issue is that of the proteomics tools for protein functional analysis. The use of similarity searches against protein sequence databases is widespread and has proven very useful, particularly if a hit to a well-annotated Swiss-Prot entry results. However, this is not always the case, and some distant relationships between proteins are not detected in similarity searches. Alternative methods use protein signatures diagnostic for protein families or domains and associated software for searching query sequences against these signatures. These have proven to be invaluable for protein functional classification.

21.2 Protein Analysis Tools

The use of protein signatures for identifying related proteins or elucidating protein functions has become increasingly popular and many databases providing these signatures have emerged and developed. These databases use different methods for creating signatures, most of which initiate from multiple sequence alignments of proteins known to belong to the family of interest. Diagnostic signatures can be created from a hand-curated protein sequence alignment over short, highly conserved regions using regular expressions e.g. PROSITE patterns (Falquet et al. 2002), or extended to the full length of a domain or the protein sequence using profiles, e.g. PROSITE profiles (Falquet et al. 2002), fingerprints, e.g. Prints (Attwood et al. 2002), or hidden Markov models (HMMs), e.g. Pfam (Bateman et al. 2002), SMART (Letunic et al. 2002), and TIGRFAMs (Haft et al. 2001). Additional methods used by databases like ProDom (Corpet et al. 2000) are based on sequence clustering algorithms, which generally have a higher coverage than the former methods, but the biological relevance of the clusters may be questionable. In combination however, all the methods and databases described above are efficient, reliable tools for protein sequence classification.

While application of protein signature analysis provides a means of classifying proteins into their families or functional classes, nomenclature of protein functions may vary considerably among database users and specialised genome databases. This makes proteome comparisons tedious and time consuming. A solution is found in the Gene Ontology (GO) project (The Gene Ontology Consortium 2001) which strives to provide a universal ontology for describing genes and gene products. The consortium conceived three ontologies under which terms are added in a direct acyclic graph with each hierarchy reflecting biological reality. The ontologies are: molecular function (action characteristic of a gene product), biological process (a phenomenon marked by changes that lead to a particular result, mediated by one or more gene products) and cellular component (the part of a cell of which a gene product is a component; GO includes the extracellular environment of cells). Genes or gene products can be mapped to GO terms in one or all of the ontologies at any level in the hierarchy. The aim of the project is to provide ontology terms for use across all taxonomic ranges so that a user may rationalise gene products and their functions between unrelated organisms.

Consolidating all the tools described above may sound a daunting task, but it has been made simple by a database and a resource, which collectively do the consolidating for all scientific users. These are the InterPro database (Apweiler et al. 2001a) and the Proteome Analysis resource (Apweiler et al. 2001b) respectively.

21.2.1 InterPro

InterPro is an integrated documentation resource of protein families, domains and functional sites that includes the member protein signature databases PROSITE, Prints, Pfam, ProDom, SMART and TIGRFAMs. It is an attempt to rationalise the different protein signatures into a comprehensive resource where those signatures designed to find the same protein family or domain are integrated into single InterPro entries. This enables researchers to use all the major protein signature databases at once, receiving the results in a single format, thus drawing on the strengths of all the databases and at the same time compensating for any downfalls they may have.

21.2.1.1 Content and Features

InterPro consists of over 6000 entries with unique accession numbers and names, which describe different protein families, domains, repeats and post-translational modifications. Each entry contains one or more signatures from the individual member databases which all describe the same group of proteins. For example, all Prints fingerprints, Prosite patterns and profiles, ProDom domains, and Pfam, SMART and TIGRFAMs HMMs that provide diagnostic signatures for identifying the same domain within protein sequences are grouped together in a single InterPro entry. Some signatures may, however, describe a subset of proteins described by another entry, which led to two types of relationships that can exist between InterPro entries: the parent/child and contains/found in relationship. Parent/child relationships are used to describe a common ancestry between entries whereas the contains/found in relationship generally refers to the presence of genetically mobile domains. Signatures from the member databases are integrated manually at regular intervals by a team of biologists, whose role is also to annotate the new or existing entries with an abstract, providing information about the protein family or domain. Additional annotation is provided by the mapping of InterPro entries to GO terms. This is done manually using information from the abstract of the entries and annotation of proteins in the match lists, and mapping of the appropriate GO terms of any level which apply to the whole protein. The associated GO terms should also apply to all proteins with true hits to all signatures in the InterPro entry. For each associated term the name of the term and GO accession number is given, and these are visible in InterPro entries with links to the EBI QuickGo browser. The mappings provide an automatic means of large-scale assignment of GO terms to the corresponding proteins in InterPro entries.

An important part of each InterPro entry is the list of precomputed matches against a composite of the Swiss-Prot and TrEMBL databases. The match lists may be viewed in a tabular form, which lists the protein accession numbers and the positions in the amino acid sequence where each signature from that InterPro entry hits. The match list can also be viewed graphically in an overview or detailed view. In the former only the InterPro entries matched are displayed on the protein sequence, while in the latter the sequence is split into several lines, one for each hit by unique signatures from different entries. The proteins can also be viewed graphically in a condensed view, which computes the consensus domain boundaries from all signatures within each entry, and splits the protein sequence into different lines for each InterPro entry matched. From this view, all proteins sharing a common domain architecture can be grouped, and the sequences aligned using Jalview (http://www.ebi.ac.uk/~michele/jalview/) or DisplayFam (Corpet et al. 1999). The protein matches are computed using the InterProScan software described below.

21.2.1.2 Searching InterPro

The InterPro database is implemented in an Oracle relational database, and is available via text or sequence searches. The databases and protein matches are also available in XML (eXtended Markup Language) format via the FTP (File Transfer Protocol) site or in SRS (Sequence Retrieval System) (Etzold et al. 1996). The database is available for text searching via two different options, searching the database directly, or using SRS. In the former it is possible to search for keywords, GO terms and protein signature, sequence and InterPro accession numbers. Via SRS it is possible to search a combination of different fields in both InterPro and protein sequence entries.

The sequence search package, InterProScan (Zdobnov and Apweiler 2001) combines the search methods from each of the databases into a single package and provides an output with all results in a single format, which may be HTML(HyperText Markup Language) (Fig. 21.1), text or XML. Independent researchers may submit their sequences using a web interface and obtain results of hits in InterPro in both a graphical and tabular view. Groups requir-

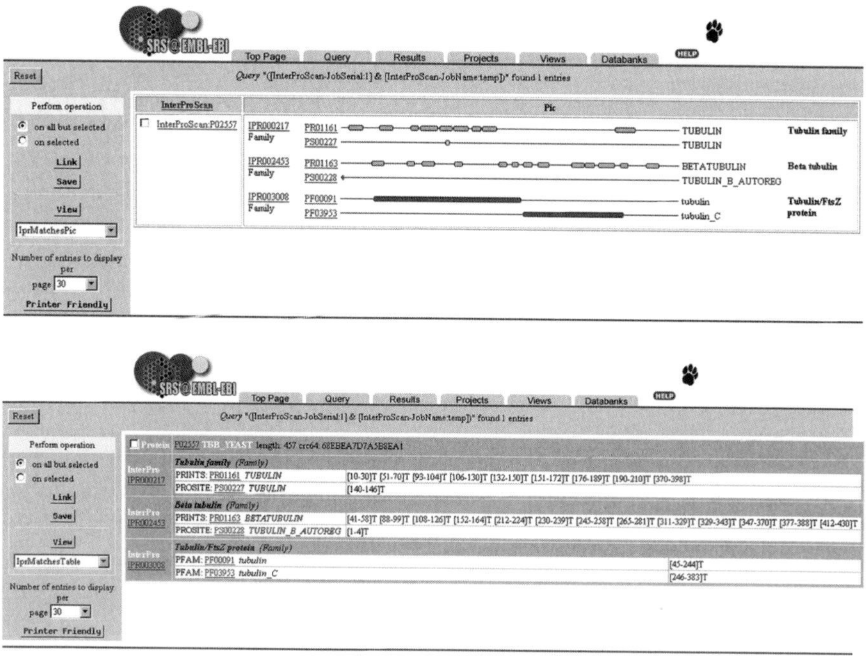

Fig. 21.1. Example of an HTML output of the InterProScan sequence search result. The results can be viewed in **a** graphic or **b** tabular view showing which protein signatures are matched and their corresponding InterPro entries. In this example, a beta tubulin protein was scanned

ing confidentiality or bulk sequence searches may download a Perl standalone InterProScan package, which can be run locally. The results will give an indication of what family the unknown protein belongs to, and its domain composition. For more information the user can read about related proteins or the protein family in the annotation of the InterPro entries it hits.

21.2.1.3 Applications

The primary application of InterPro's family, domain and functional site definitions is in the computational functional classification of newly determined sequences that lack biochemical characterisation. InterPro has also been used internally for enhancing the automated annotation of TrEMBL (Fleischmann et al. 1999). This process is more efficient and reliable than using each of the pattern databases separately, because InterPro provides internal consistency checks and deeper coverage. InterPro has become a major resource for the annotation of newly sequenced genomes. The database and InterProScan software package have thus far been used for: the comparative genome analysis of *Drosophila melanogaster, Caenorhabditis elegans*, and *Saccharomyces cerevisiae* (Rubin et al. 2000), comparative analysis of malaria genomes (Carlton et al. 2001), the study of fish genomes (Biswas et al. 2001), initial annotation of the human genome (The International Human Genome Consortium 2001), and analysis of the mouse cDNAs (Kawaji et al. 2002) and the rice (*Oryza sativa*) genome (Yu et al. 2002, Goff et al. 2002), exemplifying the utility of the resource in analysis and comparison of complete genomes. InterPro is also used extensively in the Proteome Analysis Resource, which provides an analysis of complete proteomes, and is discussed in more detail below.

21.2.2 Proteome Analysis

The Proteome Analysis database is a resource for the *in silico* analysis of proteins and of whole proteomes. It provides comprehensive statistical analyses of the predicted proteomes of fully sequenced organisms, thus aiming at overcoming the lack of in vivo gathered knowledge about the functions of predicted proteins. Information from a variety of sources is integrated, which facilitates the classification of proteins in complete proteome sets. The Proteome Analysis database provides a broad view of the proteome data classified according to signatures describing particular sequence motifs or sequence similarities and at the same time affords the option of examining various specific details like structure or functional classification.

21.2.2.1 Content and Features

In November 2002, the Proteome Analysis database contained proteome sets for 100 complete genomes; a complete statistical analysis is available for 92 complete genomes.

The proteome sets are built from the Swiss-Prot and TrEMBL protein sequence databases that provide reliable, well-annotated data as the basis for the analysis. Proteome analysis data are available for all the completely sequenced organisms present in Swiss-Prot and TrEMBL, spanning archaea, bacteria and eukaryotes. For the statistical analysis of the proteomes the InterPro and CluSTr (Kriventseva et al. 2001) resources are used to classify and group proteins according to families, domains and functional sites. Links to structural information databases like the homology derived secondary structure of proteins (HSSP) database (Dodge et al. 1998), the protein data bank (PDB; Berman et al. 2000), and the structural classification of proteins (SCOP) database (Lo Conte et al. 2002) are provided, for individual proteins from each of the proteomes. A functional classification of proteomes is performed according to the assignment of proteins to a selection of high level terms from each of the three Gene Ontology sections, molecular function, biological process and cellular component, called GO Slim. A functional classification of

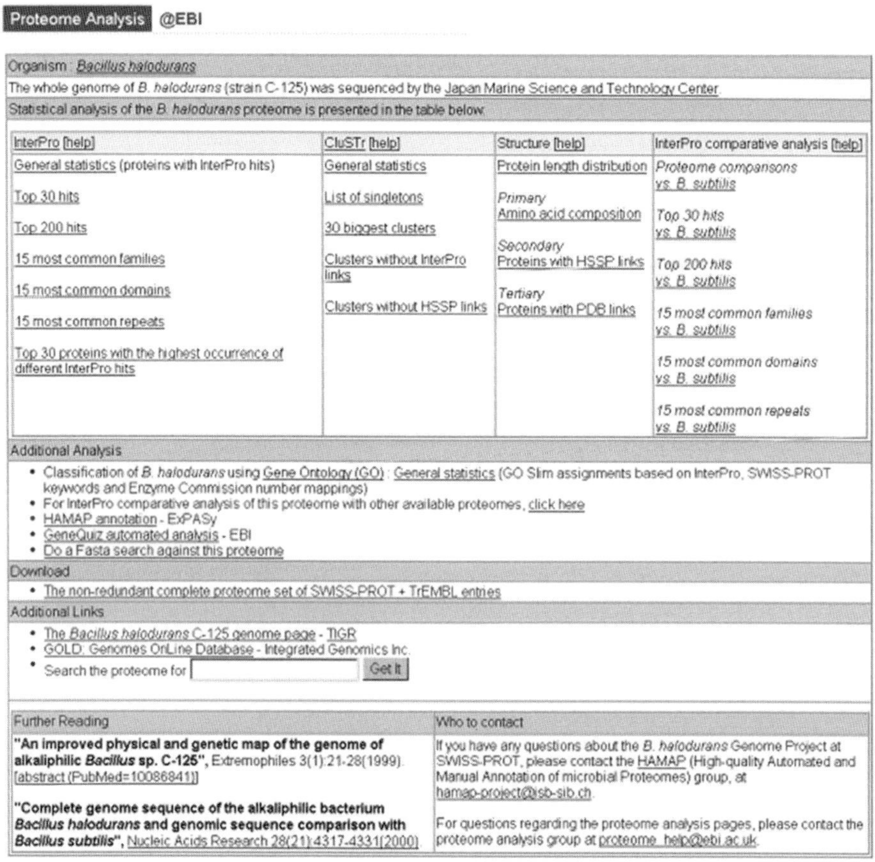

Fig. 21.2. Precomputed Proteome Analysis page for *Bacillus halodurans,* which shows the available statistical analyses and additional information

the proteins within each proteome set has been generated to show the percentage of proteins involved in certain functions. All organisms are linked to the NEWT taxonomy browser (http://www.ebi.ac.uk/newt/index.html), and each entry contains links to related resources (Fig. 21.2).

The International Protein Index (IPI) (http://www.ebi.ac.uk/IPI/IPIhelp.html), which is closely linked to the Proteome Analysis pages, provides a top-level guide to the main databases that describes the human and mouse proteome, namely Swiss-Prot, TrEMBL, RefSeq (Pruitt and Maglott 2001) and Ensembl (Hubbard et al. 2002). IPI maintains a database of cross-references between the primary data sources with the aim of providing a minimally redundant yet maximally complete set of human proteins (one sequence per transcript).

21.2.2.2 Statistical Analysis

Several different forms of analyses of proteins and proteomes are performed in the Proteome Analysis database, some are precomputed, others are carried out dynamically.

The Proteome Analysis pages make available InterPro-based statistical analysis of each of the proteomes, for example, general statistics (matches per genome and the number of proteins matched for each InterPro entry), Top 30 and Top 200 InterPro entries, 15 most common families, and 15 most common domains. The CluSTr-based analysis comprises data on general statistics about the protein clusters, a list of singletons, and 30 biggest clusters. It also suggests a list of candidates for novel protein domains as well as targets for structural genomics. Structural information in the Proteome Analysis pages includes protein length distribution and amino acid composition for each of the proteomes, which are also represented graphically. The structure of proteins that are linked to PDB can be displayed with special visualisation software.

Comparative analysis data are presented in two different versions; static and dynamic HTML pages. The static HTML pages contain the most obvious proteome comparisons, based on InterPro statistics. The dynamic HTML pages allow the user to compare a reference proteome with any other (one or more) proteomes (Kanapin et al. 2002).

Chromosome tables are available for most of the proteomes in the database, providing an ordered list of genes, together with their chromosome location, information about the protein they encode and useful links to other databases.

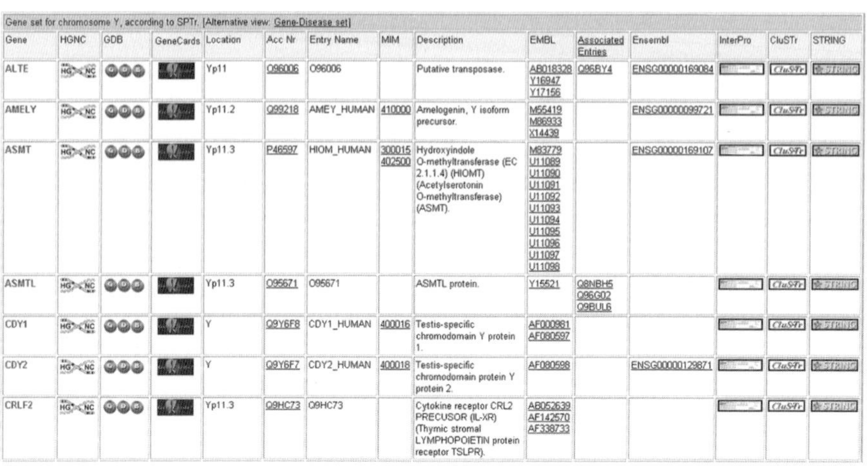

Gene set for chromosome Y, according to SPTr. [Alternative view: Gene-Disease set]

Gene	HGNC	GDB	GeneCards	Location	Acc Nr	Entry Name	MIM	Description	EMBL	Associated Entries	Ensembl	InterPro	CluSTr	STRING
ALTE				Yp11	O96006	O96006		Putative transposase.	AB018328 Y16947 Y17156	Q96BY4	ENSG00000169084			
AMELY				Yp11.2	Q99218	AMEY_HUMAN	410000	Amelogenin, Y isoform precursor.	M55419 M86933 X14439		ENSG00000099721			
ASMT				Yp11.3	P46597	HIOM_HUMAN	300015 402500	Hydroxyindole O-methyltransferase (EC 2.1.1.4) (HIOMT) (Acetylserotonin O-methyltransferase) (ASMT).	M83779 U11089 U11090 U11091 U11092 U11093 U11094 U11095 U11096 U11097 U11098		ENSG00000169107			
ASMTL				Yp11.3	O95671	O95671		ASMTL protein.	Y15521	Q8NBH5 Q96G02 Q9BUL6				
CDY1				Y	Q9Y6F8	CDY1_HUMAN	400016	Testis-specific chromodomain Y protein 1.	AF000981 AF080597					
CDY2				Y	Q9Y6F7	CDY2_HUMAN	400018	Testis-specific chromodomain protein Y protein 2.	AF080598		ENSG00000129871			
CRLF2				Yp11.3	Q9HC73	Q9HC73		Cytokine receptor CRL2 PRECUSOR (IL-XR) (Thymic stromal LYMPHOPOIETIN protein receptor TSLPR).	AB052639 AF142570 AF338733					

Fig. 21.3. Chromosome table for the *Homo sapiens* Y chromosome, full view listing all genes, their location on the chromosomes, and information about the corresponding proteins

Tables are provided for each chromosome, and also for organelle genomes and plasmids, where these are considered to comprise part of the normal genome of a completely-sequenced organism. Two views for each chromosome are available; a full view, listing all genes in the specified chromosome together with additional links to specialised databases (Fig. 21.3), and a gene-disease view which displays only genes that are annotated in Swiss-Prot and TrEMBL as linked to one or more diseases. Further to the chromosome tables for the human proteome, a gene search facility allows the user to search using the gene name, chromosome location, keyword and/or text.

21.2.2.3 Applications

The Proteome Analysis database provides information on domain structure and function, gene duplication and protein families in different genomes. A variety of ways to query and compare the data, depending on the objectives of the analysis, is offered. The tools to interrogate and compare entire proteomes of organisms by domain and/or protein family distributions and combinations provide the means that make it possible to identify systematically conserved proteins, conserved families that are missing in a given genome, or proteins unique to a particular species. Although complete coding sequence predictions for *Homo sapiens* and *Mus musculus* are not yet available in the EMBL Nucleotide Sequence database (Stoesser et al. 2002), Swiss-Prot, TrEMBL and Ensembl jointly offer a draft complete proteome for these species. This data is available as part of the Proteome Analysis Database.

The Proteome Analysis home page (http://www.ebi.ac.uk/proteome/) provides a hyperlinked list of the proteomes analysed, arranged under the classification of archaea, bacteria and eukaryotes. The top level Proteome Analysis page of each organism provides hyperlinks to the data generated by the types of analyses already mentioned, which are all organised in the form of a table. In addition, the index page of each organism contains further information such as a brief description of the organism, where the complete genome sequencing was carried out, hyperlinks to the first publication of the complete genome, additional relevant sites and contact information. Links are provided to the EBI genome and proteome Fasta server (http://www.ebi.ac.uk/fasta33/genomes.html). This server allows users to perform FASTA searches with their own query sequences against one or many proteomes or genomes.

21.3 Discussion

InterPro and Proteome Analysis have several uses for the scientific community, where they capitalise on the individual strengths of the different methods and sources and provide a comprehensive view of proteins and proteomes. The Proteome Analysis resource provides a useful summary of available data on complete genomes with the option to delve deeper for specific details. InterPro provides a useful tool in the resource for analysis and comparison of protein function in and between the proteomes.

Using protein signatures as a means of elucidating protein function has numerous advantages over the traditional methods of searching the query sequence against an existing protein database. The signatures facilitate identification of distantly related and multi-domain proteins, and are not reliant on the integrity and potential redundancy of the protein sequence databases. InterPro also has uses for the member databases themselves by reducing duplication of effort in the labour-intensive process of manual annotation, and facilitating communication between the disparate resources. The integrated resource serves as a quality control mechanism for assessing individual methods, and also highlights the areas where all the member databases are lacking in representation. The increasing availability of complete genome sequences also helps to identify uncharacterised protein families that may be unique to single or groups of related organisms. Future plans of InterPro include extension into the field of protein secondary and tertiary structure by integration data from SCOP and CATH (Pearl et al. 2002) as well as known 3D structures.

It is evident that there are currently a number of high quality signature databases, integrated databases and proteome data resources available for the analysis of proteins and proteomes via automatic and large-scale protein classification. However, the challenge is still in the transference of useful biological knowledge to protein sequences. Automatic methods may provide some useful suggestions of protein architecture or function, but only a biologist can truly assign the function to a protein based on these results, and the ultimate confirmation of these assignments is experimental evidence. *In silico* and laboratory analysis therefore go hand in hand and both are required for efficient analysis of proteins and proteomes.

References

Apweiler R, Attwood TK, Bairoch A, Bateman A, Birney E, Biswas M, Bucher P, Cerutti L, Corpet F, Croning MD, Durbin R, Falquet L, Fleischmann W, Gouzy J, Hermjakob H, Hulo N, Jonassen I, Kahn D, Kanapin A, Karavidopoulou Y, Lopez R, Marx B, Mulder NJ, Oinn TM, Pagni M, Servant F (2001a) The InterPro database, an integrated documentation resource for protein families, domains and functional sites. Nucleic Acids Res 29(1):37–40

Apweiler R, Biswas M, Fleischmann W, Kanapin A, Karavidopoulou Y, Kersey P, Kriventseva EV, Mittard V, Mulder N, Phan I, Zdobnov E. (2001b) Proteome Analysis Database: online application of InterPro and CluSTr for the functional classification of proteins in whole genomes. Nucleic Acids Res 29:44–48

Attwood TK, Blythe MJ, Flower DR, Gaulton A, Mabey JE, Maudling N, McGregor L, Mitchell AL, Moulton G, Paine K, Scordis P (2002) PRINTS and PRINTS-S shed light on protein ancestry. Nucleic Acids Res 30:239–241

Bateman A, Birney E, Cerruti L, Durbin R, Etwiller L, Eddy SR, Griffiths-Jones S, Howe KL, Marshall M, Sonnhammer ELL (2002) The Pfam Protein Families Database. Nucleic Acids Res 30:276–280

Bairoch A, Apweiler R (2000) The SWISS-PROT protein sequence database and its supplement TrEMBL in 2000. Nucleic Acids Res 28:45–48

Berman HM, Westbrook J, Feng Z, Gilliland G, Bhat TN, Weissig H, Shindyalov IN, Bourne PE (2000) The Protein Data Bank. Nucleic Acids Res 28:235–242

Biswas M, Kanapin A, Apweiler R (2001) Application of InterPro for the functional classification of the proteins of fish origin in SWISS-PROT and TrEMBL. J Biosci 26(2): 277–284

Carlton JM, Muller R, Yowell CA, Fluegge MR, Sturrock KA, Pritt JR, Vargas-Serrato E, Galinski MR, Barnwell JW, Mulder N, Kanapin A, Cawley SE, Hide WA, Dame JB (2001) Profiling the malaria genome: a gene survey of three species of malaria parasite with comparison to other apicomplexan species. Mol Biochem Parasitol 118(2):201–220

Corpet F, Gouzy J, Kahn D (1999) Browsing protein families via the "Rich Family Description" format. Bioinformatics 15:1020–1027

Corpet F, Servant F, Gouzy J, Kahn D (2000) ProDom and ProDom-CG: tools for protein domain analysis and whole genome comparisons. Nucleic Acids Res 28:267–269

Dodge C, Schneider R, Sander C (1998) The HSSP database of protein structure-sequence alignments and family profiles. Nucleic Acids Res 26(1):313–315

Etzold T, Ulyanov A, Argos P (1996) SRS: information retrieval system for molecular biology data banks. Methods Enzymol 266:114–128

Falquet L, Pagni M, Bucher P, Hulo N, Sigrist CJA, Hofmann K, Bairoch A. (2002) The PROSITE database, its status in 2002. Nucleic Acids Res 30:235–238

Fleischmann W, Möller S, Gateau A, Apweiler R (1999) A novel method for automatic functional annotation of proteins. Bioinformatics 15:228–233

The Gene Ontology Consortium (2001) Creating the gene ontology resource: design and implementation. Genome Res 11:1425–1433

Goff SA, Ricke D, Lan TH, Presting G, Wang R, Dunn M, Glazebrook J, Sessions A, Oeller P, Varma H, Hadley D, Hutchison D, Martin C, Katagiri F, Lange BM, Moughamer T, Xia Y, Budworth P, Zhong J, Miguel T, Paszkowski U, Zhang S, Colbert M, Sun WL, Chen L, Cooper B, Park S, Wood TC, Mao L, Quail P, Wing R, Dean R, Yu Y, Zharkikh A, Shen R, Sahasrabudhe S, Thomas A, Cannings R, Gutin A, Pruss D, Reid J, Tavtigian S, Mitchell J, Eldredge G, Scholl T, Miller RM, Bhatnagar S, Adey N, Rubano T, Tusneem N, Robinson R, Feldhaus J, Macalma T, Oliphant A, Briggs S (2002) A draft sequence of the rice genome (*Oryza sativa* L. ssp. japonica). Science 296(5565):92–100

Haft D H, Loftus BJ, Richardson DL, Yang F, Eisen JA, Paulsen IT, White O (2001) TIGRFAMs: a protein family resource for the functional identification of proteins. Nucleic Acids Res 29(1):41–43

Hubbard T, Barker D, Birney E, Cameron G, Chen Y, Clark L, Cox T, Cuff J, Curwen V, Down T, Durbin R, Eyras E, Gilbert J, Hammond M, Huminiecki L, Kasprzyk A, Lehvaslaiho H, Lijnzaad P, Melsopp C, Mongin E, Pettett R, Pocock M, Potter S, Rust A, Schmidt E, Searle S, Slater G, Smith J, Spooner W, Stabenau A, Stalker J, Stupka E, Ureta-Vidal A, Vastrik I, Clamp M (2002) The Ensembl genome database project. Nucleic Acids Res 30:38–41

The International Human Genome Consortium. (2001) Initial sequencing and analysis of the human genome. Nature 409(6822):860–921

Kanapin A, Apweiler R, Biswas M, Fleischmann W, Karavidopoulou Y, Kersey P, Kriventseva EV, Mittard V, Mulder N, Oinn T, Phan I, Servant F, Zdobnov E (2002) Interactive InterPro-based comparisons of proteins in whole genomes. Bioinformatics 18:374–375

Kawaji H, Schonbach C, Matsuo Y, Kawai J, Okazaki Y, Hayashizaki Y, Matsuda H (2002) Exploration of novel motifs derived from mouse cDNA sequences. Genome Res 12(3):367–378

Kriventseva EV, Fleischmann W, Zdobnov EM, Apweiler R. (2001) CluSTr: a database of clusters of SWISS-PROT+TrEMBL proteins. Nucleic Acids Res. 29:33–36

Letunic I, Goodstadt L, Dickens NJ, Doerks T, Schultz J, Mott R, Ciccarelli F, Copley RR, Ponting CP, Bork P (2002) Recent improvements to the SMART domain-based sequence annotation resource. Nucleic Acids Res 30:242–244

Lo Conte L, Brenner SE, Hubbard TJ, Chothia C, Murzin AG (2002) SCOP database in 2002: refinements accommodate structural genomics. Nucleic Acids Res 30(1): 264–267

Pearl FM, Lee D, Bray JE, Buchan DW, Shepherd AJ, Orengo CA (2002) The CATH extended protein-family database: providing structural annotations for genome sequences. Protein Sci 11(2):233–244

Pruitt KD, Maglott DR (2001) RefSeq and LocusLink: NCBI gene-centered resources. Nucleic Acids Res 29:137–140

Rubin GM, Yandell MD, Wortman JR, Gabor Miklos GL, Nelson CR, Hariharan IK, Fortini ME, Li PW, Apweiler R, Fleischmann W, Cherry JM, Henikoff S, Skupski MP, Misra S, Ashburner M, Birney E, Boguski MS, Brody T, Brokstein P, Celniker SE, Chervitz SA, Coates D, Cravchik A, Gabrielian A, Galle RF, Gelbart WM, George RA, Goldstein LS, Gong F, Guan P, Harris NL, Hay BA, Hoskins RA, Li J, Li Z, Hynes RO, Jones SJ, Kuehl PM, Lemaitre B, Littleton JT, Morrison DK, Mungall C, O'Farrell PH, Pickeral OK, Shue C, Vosshall LB, Zhang J, Zhao Q, Zheng XH, Zhong F, Zhong W, Gibbs R, Venter JC, Adams MD, Lewis S (2000) Comparative genomics of the eukaryotes. Science 287:2204–2215

Stoesser G, Baker W, van den Broek A, Camon E, Garcia-Pastor M, Kanz C, Kulikova T, Leinonen R, Lin Q, Lombard V, Lopez R, Redaschi N, Stoehr P, Tuli MA, Tzouvara K, Vaughan R. (2202) The EMBL Nucleotide Sequence Database. Nucleic Acids Res 30(1):21–26

Yu J, Hu S, Wang J, Wong GK, Li S, Liu B, Deng Y, Dai L, Zhou Y, Zhang X, Cao M, Liu J, Sun J, Tang J, Chen Y, Huang X, Lin W, Ye C, Tong W, Cong L, Geng J, Han Y, Li L, Li W, Hu G, Huang X, Li W, Li J, Liu Z, Li L, Liu J, Qi Q, Liu J, Li L, Li T, Wang X, Lu H, Wu T, Zhu M, Ni P, Han H, Dong W, Ren X, Feng X, Cui P, Li X, Wang H, Xu X, Zhai W, Xu Z, Zhang J, He S, Zhang J, Xu J, Zhang K, Zheng X, Dong J, Zeng W, Tao L, Ye J, Tan J, Ren X, Chen X, He J, Liu D, Tian W, Tian C, Xia H, Bao Q, Li G, Gao H, Cao T, Wang J, Zhao W, Li P, Chen W, Wang X, Zhang Y, Hu J, Wang J, Liu S, Yang J, Zhang G, Xiong Y, Li Z, Mao L, Zhou C, Zhu Z, Chen R, Hao B, Zheng W, Chen S, Guo W, Li G, Liu S, Tao M, Wang J, Zhu L, Yuan L, Yang H (2002) A draft sequence of the rice genome (Oryza sativa L. ssp. indica). Science 296(5565):79–92

Zdobnov EM, Apweiler R (2001) InterProScan–an integration platform for the signature-recognition methods in InterPro. Bioinformatics 17(9):847–848

22 Prediction of Functional Sites in Proteins by Evolutionary Methods

Pedro López-Romero, Manuel J. Gómez, Paulino Gómez-Puertas and Alfonso Valencia

Abstract

Functional sites are well-defined regions that are relevant for protein function, and that include characteristic groups of amino acids. These regions may be involved in the interaction between proteins and other molecules, such as other proteins, nucleic acids, small ligands and substrates. Interaction sites have been studied in great detail in representative protein families, and their relationship with natural substrates and drugs has been characterized, as well as their mediation in protein complex formation. In many cases they have been studied in relation to their potential for engineering protein activity. Protein binding sites have also been studied at a more general level by characterizing the typical structure of binding sites, and their general residue preferences. However, it is the relationship between the conservation of sequence features and protein active sites and binding sites that constitutes the basis of the development of prediction methods. The conservation of the chemical characteristics of the amino acids in specific groups of sequences, in the context of large protein families, is a particular method used in a growing collection of methods aimed at predicting protein binding sites at a genomic scale. In this review we analyze these methods, discuss their similarities, and describe a number of key unsolved problems.

22.1 Protein Function and Amino Acids Involved

Protein function typically depends on a subset of the amino acid residues of a protein. However, it is also true that the full protein sequence and structure are the result of a process of evolution driven by the necessity of a certain function. In this sense all (or at least most) residues contribute to the functionality of a protein. Protein function is commonly mediated by regions on

Principles and Practice
Methods in Proteome and Protein Analysis
R.M. Kamp, J. J. Calvete, T. Choli-Papadopoulou (Eds.)

the surface that interact with external factors. For example, transcriptional regulators exert their function by binding to specific DNA sequences, and enzymes bind their specific ligands in active sites commonly accessible to the solvent. Therefore, both interaction surfaces and binding sites are sites of molecular recognition, which are important for protein function, that are constituted by clusters of amino acids in surface patches. They have been referred to as hot spots to stress their intimate relation with the key function of the proteins (Clackson and Wells, 1995; Bogan and Thorn, 1998).

22.2 Interaction Sites and Their Structural and Chemical Properties

Interaction sites can be small ligand binding sites and also surfaces of interaction between proteins and macromolecules such as other proteins, DNA, RNA or carbohydrates. With the availability, albeit a limited number, of published structures of protein-molecule complexes, several studies have attempted to identify features of protein interaction interfaces. The general strategy is to obtain a collection of structures of complexes from the Protein Data Bank database, identify the amino acid residues that are part of the interface, calculate their frequencies relative to some reference, evaluate their chemical properties and, according to their coordinates, determine certain structural properties of the interaction sites (Wodak and Janin, 2003).

In the case of complexes between proteins and small ligands, the amino acid residues that are part of the interaction site can be defined as those located within a specific distance from any of the atoms of the ligand. This strategy was followed by Villar and Kauvar (1994), who used a collection of protein-ligand complexes, most of which corresponded to enzymes, to identify which amino acid residues were located close to the ligands. They concluded that some residues, particularly Arg, His, Trp and Tyr, are more frequent in binding sites than in the rest of the sequences. In addition, the geometric features of small ligand binding sites are relatively well defined, since most of them consist of cavities or areas of increased surface roughness (Kuntz et al., 1982; Pettit and Bowie, 1999).

22.3 Functional Role of Conserved Residues in Multiple Sequence Alignments

Since natural selection works at the level of molecular function, amino acid residues that are involved in the function of a given protein family are more likely to remain unchanged during evolution and might be identified by their characteristic conservation in multiple sequence alignments. In this sense, the most obvious observations were about completely conserved residues: as early as 1965, Zuckerkandl and Pauling established the relationship between sequence conservation and function, and discussed the use of sequence conservation information for the prediction of protein binding sites and protein function, for the two protein families for which a few sequences were known at that time.

Surprisingly, this commonly accepted relationship between sequence conservation and protein function has been studied systematically in a relatively small number of cases. Grishin and Philips (1995) studied the degree of conservation at protein-protein interfaces of oligomeric enzymes and concluded that conservation was only slightly better in interfaces than in other protein regions. Valdar and Thornton (2000), and Luscombe and Thornton (2002), analyzed residue conservation at homodimer protein-protein interfaces and protein-DNA interfaces, respectively, and found that interface conservation was higher than expected by chance. An explanation of the discrepancy with the previous observation is that their definition of conservation took into account amino acid similarity, while Grishin and Philips considered only amino acid identity to identify conserved positions. Ouzounis et al. (1998), established the quantitative relationship between the type of amino acid, its conservation, and its participation in biding sites. Interesting trends were discovered between the chemical type of the residues, their degree of conservation, and the possibility of participating in binding sites. For example, Pro, Glu, Gly and Trp (among others) show tendency to be part of binding sites when they are conserved. Although conservation of functional sites implies that evolutionary information can be used for their prediction, it is important to note that a high degree of variability may exist. For example, Devos and Valencia (2000) observed that binding sites were the least conserved feature in a set of related proteins, and Todd et al (2001) observed catalytic residue migrations, and variation of enzyme functions, in one-third and one-fourth, respectively, of the superfamilies defined in the CATH database. Similarly, Rost (2002) compared the proteins in an unbiased dataset, and pointed out that sequence identities of 50 % and over were associated with identical enzyme function in only 30 % of the cases he studied.

22.4 Why Predicting Functional Sites?

The first obvious application of functional site prediction is to find clues about protein function. The availability of the complete sequence of hundreds of genomes opens new possibilities for understanding the biology of organisms, although it requires predicting the function of the majority of the genes in a particular genome. This is usually done by automatic methods that rely on the identification of general similarity to known genes, which are able to assign functions to 40 –80 % of the genes in a genome (Iliopoulos 2000). For the rest, the prediction of functional sites may serve to propose functions that can be tested experimentally. In this sense, functional site predictions should be facilitated by the large number of structures that will be produced by the structural genomics initiatives, and the associated computational efforts in protein modeling (Rost et al. 2002).

The reliable prediction of these key residues is also important in applications related to the identification of targets for pharmaceutical design, elucidation of molecular pathways through site-directed mutagenesis, and the analysis of the function of specific protein families. Indeed, for the characterization of functional interfaces, structural information alone is insufficient, since it does not directly provide specific information on the functionally significant protein surfaces. Therefore, mutational analysis is still the main method to identify functional sites. The lack of effective experimental approaches, together with the increasing number of known protein sequences, have stimulated the development of a variety of computational methods that, based on the analysis of multiple sequence alignments, can predict binding regions and protein surfaces that play a role in functional specificity. The analysis of these methods is the main focus of this review.

22.5 The Use of Sequence Information for the Prediction of Functional Sites

Biological evolution can be expressed in terms of mutation, selection and fixation of the variants in the population. According to the neutral theory of molecular evolution, most mutations are neutral for the selective process, which implies that changes in the nucleotide sequence of genes accumulate with time. This is the basis for the concept of evolutionary divergence. Since selective pressure is exerted at the level of function, which is, in general, more related with structure than with primary sequence, structure tends to evolve at lower rates than sequence. A given biological species can split into two new species if the accumulated changes impose a reproductive barrier. Even in that case, the comparison of biological sequences from different species, may reveal those that come from the same ancestral sequence. These genes, or proteins, are classified as orthologs. Although sequence similarity is taken as the main indicator of orthology, other evidences are considered sometimes, such as the position in the chromosome, or conservation of the gene context. In practical terms, the identification of orthology relationships between proteins is used for function prediction, under the general trend of orthologous sequences sharing the same function. Some evolutionary mechanisms interfere with the identification of orthologs, for example, events of gene duplication and gene transfer. In general, once a gene is duplicated, one of the copies evolves independently of the selective constraints that affect the other, and therefore, it may acquire a different functionality. In the context of function prediction it is important, therefore, to differentiate between orthologs and duplicated genes, or paralogs. Both orthology and paralogy are grouped under the term homology, which denotes common ancestry.

The identification of paralogs and orthologs may be accomplished by the comparison of sequences in multiple sequence alignments and the construction of a specific type of phylogenetic trees, called gene trees. Standard phylogenetic trees are used to reconstruct the evolutionary history of a group of biological species. In molecular phylogeny they are built from the multiple sequence alignment of a collection of orthologous genes or proteins, selected because they are supposed to evolve at a similar rate in all species considered. On the other hand, gene trees are built from the multiple sequence alignment of collections of orthologs and paralogs, and they are used to visualizing the history and relationships between the sequences considered, rather than those of the species involved. These studies are usually possible only for completely sequenced genomes, since all paralogs of any given gene, in a given species, must be considered. The analysis of gene trees helps to identify groups of sequences that have more in common with each other than with other groups. These groups of related sequences are classified as families and subfamilies. The distinction between these two terms is subjective and denotes, simply, that a hierarchy of relationships exists. Proteins belonging to the same family have a common evolutionary origin, related function and detectable sequence similarity. For proteins belonging to the same subfamily, the degree of sequence similarity is higher and they are considered to have the same function.

Once multiple sequence alignments for the members of a given protein family have been generated, it is possible to identify regions with different degrees of variability. Conserved regions or positions point to residues that are supposed to be under stronger evolutionary constraints and that, therefore, may be important for the functionality of the protein. Moreover, given a protein family, residues that are specifically conserved in subfamilies with a certain functional specialization, point to sequence changes that occurred along divergence from the common ancestor and implied functional changes that were selected during evolution. Since these amino acids are identified as conserved for specific subfamilies, they have been named tree-determinant residues.

Tree-determinant residues are, therefore, conserved residues that are specific and determinant of protein subfamilies. From their evolutionary importance it is inferred that they are relevant for the functionality of proteins. Tree-determinant residues could be related to specific sites of certain functions (active sites or substrate binding sites) or, in many cases involved in protein-protein interaction or binding of proteins to other molecules, but also with key structural constraints.

22.6 Methods for Predicting Tree-Determinant Residues

The analysis of multiple sequence alignments for the detection of family specific residues has traditionally been performed by visual inspection, which makes it difficult to detect complex conservation patterns, especially in large multiple sequence alignments. These approaches are also unable to split the multiple sequence alignments into functional subfamilies, or to identify the residues responsible for such division. In this section, we review, chronologically, computational methods to identify positions in multiple sequence alignments of families and subfamilies of proteins, that may correspond to functional residues (for a review of the earlier methods see Livingstone and Barton, 1993).

Livingstone and Barton (1993) proposed an extension of a previous study by Zvelebil et al. (1987) to quantify and compare protein residue conservation, taking into account the physicochemical properties of the amino acids (Taylor, 1986). The extension, in comparison to the work of Zvelebil, is based on the identification of conservation within subfamilies of proteins in the multiple sequence alignment. One disadvantage of this method is that protein subfamilies are not automatically identified. The subsets have to be defined a priori, and the algorithm computes the conservation within the given subfamilies, assigning a degree of conservation to each one. These authors consider a tree-determinant residue as any position that is conserved in any of the subfamilies in which the multiple sequence alignment has been divided, even if the position is not conserved in the remaining subfamilies.

Casari et al. (1995) developed SequenceSpace, which aims to classify the members of a protein family in different subfamilies, identifying at the same time which tree determinants are responsible for such separation. This method identifies positions that are shared between subfamilies, as well as the residues that are conserved across the protein families. Protein sequences from a multiple sequence alignment are codified as high dimensional vectors. Principal component analysis is used to project protein sequences and residue-position point variables, onto two orthogonal spaces that maximize the originally observed variability. Finally, these projections allow clustering of the proteins and the residue-positions, responsible for grouping of proteins into clusters. The method was tested in representative families, Ras-like proteins, SH2 domains and cyclins, and later it was applied to biological systems. In a number of cases, the predictions were followed by experiments that demonstrate the ability of the method to detect residues specific to functions associated with different subfamilies (see last section of the chapter).

Lichtarge et al. (1996) implemented the Evolutionary Trace method (ET) in order to highlight residues comprising functional sites. With the ET method, gene trees are partitioned at different cut off levels to generate groups of branches, that may correspond with protein subfamilies. Then, possible sets of functional residues are elicited from the identification of conservation patterns in the putative subfamilies. Conserved functional clusters whose conservation correlate with the division of the multiple sequence alignment into subfamilies, are required to cover all sequences in the alignment in an attempt to identify class-specific residues. The method tries to find an equilibrium between the size of the sequence groups, their divergence and their capacity to contain representative sequence variation. In the first publication, the selection of the partition cut off was left to the user. Hence, ET did not provide an automatic classification of the protein family into subfamilies, and the definition of clusters of functionally important residues was subjective. One particular disadvantage of the ET method is that the definition of the conserved position is based on an "all-or-none" criterion, by which complete conservation in terms of residue identity, for all sequences belonging to a given subfamily, is required. This requirement produces a dramatic reduction in the number of predicted functional residues, and biased results when divergent sequences are present (Pazos et al., 1997).

Andrade et al. (1997) introduced the self-organized-map clustering algorithm (SOM, Kohonen, 1982) to divide sequences of protein families into subfamilies. Then, they used the vectors associated with the neurons in the SOM map, to identify the key residues responsible for the classification. The input for the algorithm is a multiple sequence alignment, coded in a similar way to that of SequenceSpace. The main disadvantage of SOM is that the presence of overriding groups of proteins in the multiple sequence alignment, may occur as a result of an uneven sampling. This would lead to domination over the others, thus causing a loss of detail in the classification of the sequences.

Landgraf et al. (1999) introduced the concept of weighted evolutionary tracing (WET) by using a quantitative description of the variability at every position in the multiple sequence alignment, weighted by the level of sequence similarity according to Sibbald and Argos (1990). This implementation aims to decrease the influence of very similar sequences on the data.

Hannenhalli and Russell (2000) stated that different functional subfamilies, in which a family of homologous proteins have evolved, cannot always be determined from a gene tree, as for example in the case of convergent evolution, or when the differences between the protein subfamilies is too great to render a reliable tree. Consequently, these authors, instead of grouping proteins in a gene tree, identify regions that fit with groups defined a priori. The conserved regions are identified comparisons of hidden Markov profiles and the evaluation of the relative entropy between positions and sequence subfamilies. This method uses similarity matrices to detect positions with similar physicochemical properties.

Armon et al. (2001) introduced a variation, ConSurf, of the basic scheme of the original ET method. Instead of the poor UPGMA method used by ET for building the gene tree, which considers equal rates of evolution along all branches of the tree, they use a more rigorous maximum parsimony (MP) method. ConSurf also takes into account the physicochemical properties of the amino acids using a residue similarity matrix (Miyata et al., 1979). They also introduce a weighting scheme in an attempt to reduce the effect of bias in the sequence sampling. Another interesting property of ConSurf is its capacity to identify the branch where an amino acid replacement takes place, circumventing the problem of establishing cut offs for the definition of sequence groups. As in the case of ET, ConSurf also bases the analysis on visual inspection of the location of conserved residues on the structures of the corresponding proteins. Despite the basic similarities of both methods, the fine-tuning introduced in ConSurf enabled the detection of an additional specific contact area in the surface of SH2 domains, in comparison to the predictions made by ET.

Pupko et al. (2002) computed a maximum likelihood estimate (ML) of the rate of evolution, or replacement probabilities among residues, at each position of the multiple sequence alignment, as a measure of position conservation. Their Rate4Site ML estimation of the rate of evolution is based on an underlying gene tree, where independence between sites is assumed. It is worth noting that the use of the tree in this particular case is not to determine the functional clusters in the sequence, like in previous methods, e.g. ET and ConSurf. The improvement of ConSurf over ET by weighting the residue replacements based on their physicochemical properties was still insufficient to account for the differences in branch lengths and to weigh the amino acid replacements accurately. This problem was overcome in Rate4Site by the use of branch lengths as a parameter in the ML estimation of the evolution rate, so amino acid changes are weighted in terms of their branch length. That difference is particularly important for the detection of specific residue patches when very similar or very distant homologues are used as input.

Landgraf et al. (2001) presented 3D cluster analysis (3DC) as another extension of ET, but independent of the gene tree, assuming that for some protein regions a gene tree does not reflect relationships between sequence similarity and function. 3DC detects functionally important residues with a low degree of conservation.

Madabushi et al. (2002) proposed an improvement of the ET method introducing first the treatment of gaps as additional amino acids, and by statistically determining the number of significant clusters of residues, using a non parametric test based on a simulated random distribution. The results were shown to be comparable to the manual selection of a threshold, by inspection of the corresponding protein structures.

A different approach for the identification of specific residues in proteins, that does not use gene trees, was proposed by Mirny and Gelfand (2002). These authors divided groups of orthologs and paralogs of the same family in order to identify specificity determinant residues, under the assumption of a larger conservation of functional specificity in orthologous sequences rather than in paralogous sequences. The method for the detection of conserved residues is based on a mutual information formulation. The obvious limitation of the method is the detection of orthologs, a well-known, very difficult problem.

It is interesting to consider in this category other approaches that were initially formulated, not for the problem of detecting tree-determinant residues, but for addressing a variety of other goals in the study of protein structures and functions. For instance, the detection of unusually conserved or unusually variable peptides in protein sequences (Dopazo, 1997), or the detection of atypically evolving positions in multiple sequence alignment using gene trees (Dorit and Ayala, 1995). Other examples include a set of methods to identify positions with correlated patterns of variation. This co-variability between different positions in the alignment is estimated, in most of the cases, either implementing concepts derived from information theory (Shannon and Weaver, 1949), such as the estimate of the mutual information content between different positions in the alignment, or estimating the correlation between matrices that represent every position in the multiple sequence alignment. The basic principle is that all positions that have a distinctive pattern of variation corresponding to their specific conservation in the main protein subfamilies will, by definition, have a correlated pattern of replacement (Altschuh et al., 1987; Taylor and Harrick, 1994; Singer et al., 1995; Lockless and Ranganathan, 1999; Clarke, 1995; Giraud et al., 1998; Atchley et al., 1999; Atchley et al., 2000; Süel et al. 2003).

22.7 Methods for Predicting Functional Sites Based on Structural Information

Fetrow and Skolnick (1998) used fuzzy functional forms (FFFs) to describe the geometry, amino acid distribution and conformation of protein active sites. The descriptors were generalized for the identification of remotely similar binding sites when searching not only in structure databases, but also in low-resolution protein models. De Rinaldis et al (1998) used compilations of the location of surface residues in a reference three-dimensional grid to define three-dimensional profiles. These grid-based three-dimensional profiles, containing sequence and structure information, were used to screen structure databases for similar arrangements of residues in other proteins. This conceptually attractive approach has been tested in examples such as SH2, SH3 and p-loop containing proteins. It is important to realize that these methods rely more on the importance of protein structure for protein function, than on the chemistry of the residues in the binding pockets. Other approaches have been developed to identify structural similarities between unrelated binding sites (Gribskov et al 1998; Aloy et al 2001).

22.8 Comparisons Between Methods

The first comparison of the results of different methods was carried out by Pazos et al., (1997). They analyzed two protein families representing extreme cases of sequence composition: the alpha subunit of trimeric G-proteins, which represents a set of paralogs having different cellular functions in the same organism, and the *Drosophila* short-chain alcohol dehydrogenase, which is composed of all the orthologs of this protein from different species. The main difference between methods was related with the treatment of small subfamilies and poorly represented classes. Methods based on the analysis of gene trees (for example, ET, Lichtarge et al., 1996) were particularly sensitive to the variations in the composition of protein families. In these cases, SequenceSpace was able to predict an additional set of correct predictions based on the detection of residues not-strictly conserved in the corresponding subfamilies.

A more systematic analysis has recently been carried out by Del Sol et al. (2003). They developed the following three basic approaches designed to represent all the available basic methods (Fig. 22.1):

- First, an implementation to identify protein families automatically and the corresponding tree-determinant residues from the output of SequenceSpace. The automation was achieved by clustering the sequences and residues using the information of their coordinates projected in their respective n-dimensional orthogonal spaces.
- A second method based on the quantification of the correlation between matrices of replacements in positions of the multiple sequence alignment, and the matrix representing the distances between all the sequences in the multiple sequence alignment. This implementation is a generalization of the methods related with the identification of statistical dependencies or associations between different positions.
- The third method is based on automatically partitioning the structure of a gene tree, to define subfamilies and to optimize the number of differentially conserved residues. In particular, they maximize the difference between subfamilies and, at the same time, they minimize the differences within the subfamilies. This method represents all those based on the use of gene trees to separate subfamilies, and the subsequent comparison of the subfamilies at the level of individual positions in the multiple sequence alignment.

The performance of those methods was evaluated by systematically comparing the distance of the detected tree-determinant residues to bound heteroatoms (compounds such as GTP, Mg, NADH, etc) used as indicators of specificity sites in protein structures. The authors concluded that the three different types of methods could all be used to find tree-determinant residues, with variations in their coverage and reliability. The detected tree-

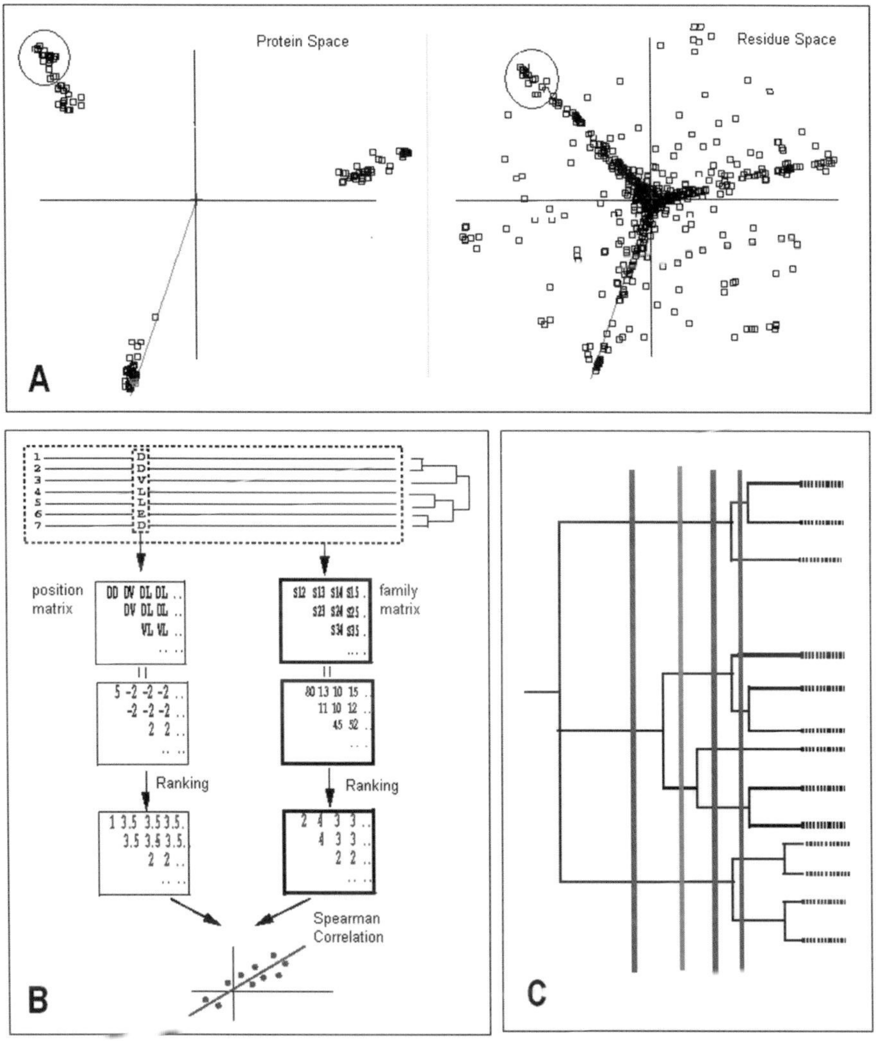

Fig. 22.1. Schematic representation of three main approaches for the detection of tree-determinant residues. **A** SequenceSpace method, representing the two orthogonal spaces in which protein-sequence and residue-position coordinates are projected. **B** Methods based on positional co-variability, showing the one based on estimating the correlation between positions in the multiple sequence alignment and the distance distribution between all the sequences. **C** Approaches related to the analysis of gene trees to divide families into subfamilies. Gene trees are cut at different levels to define branch groups that could correspond to subfamilies. Then conservation and variation of the positions in the multiple sequence alignment is computed, within and between the different subfamilies

determinant residues are statistically closely bound to heteroatoms, even if it is clear that the results represent an under-estimation of the potential of the methods, since the substrate binding sites are not a perfect representation of binding or interaction sites. Results of the comparison can be summarized as follows:

- SequenceSpace returned slightly worse results than the other two methods, probably because the implementation for the automatic analysis was not able to capture all the details of the protein families. A new version of SequenceSpace is being implemented which includes a different automation algorithm for the classification of proteins and tree-determinant residues. This new implementation includes the computation of the confidence in the organization of the family in subfamilies, and the corresponding detection of tree-determinant residues (López-Romero and Valencia, in prep.).
- The main drawback of the method based on detecting dependencies among separate positions was the requirement of a larger number of sequences per subfamily to satisfactorily identify enough tree-determinant residues.
- The implementation based on the optimal cut of gene trees faces the problem of determining the optimal division of a family, a difficult decision that in many methods is left to the user (Lichtarge et al., 1996; Armon et al., 2001). This problem has been addressed later by other methods (Madabushi et al., 2002, Yao et al., 2003) which introduce a non-parametric test to evaluate the significance of the number of subfamilies. Del Sol et al. used mutual information based formulation to evaluate the optimal segregation of different residues between subfamilies, as the optimal point to split the gene trees. The method was developed as an attempt of providing a protein family division, where it is still possible to assign functional residues that are responsible for the observed divergence. Functional or structural conserved residues can be distinguished from those that are conserved just by chance, but the method still cannot discriminate from those residues that are conserved because of a short time of evolution.

The positive message from these comparative studies is that the various methodologies available have the capacity for detecting residues that are close to specificity sites, using exclusively sequenced information, and that combining their predictions could be useful, since they tend to have different drawbacks and capabilities.

22.9 Main Problems in the Characterization of Tree-Determinant Residues

Sampling of the Sequence Space. A major problem in dividing a family into functional subfamilies, based on the identification of tree-determinants, is having heterogeneous distributions of sequences in which over-represented groups of them dominate the multiple sequence alignment, while other groups are represented by fewer sequences. In other cases, protein subfamilies are composed of sequences that are very similar or very dissimilar. All these common deviations from homogenous sequence distributions can cause practical problems in differentiating residues that are conserved for functional reasons, from others that are only apparently conserved, according to the different artifacts of the distribution of sequences within protein families.

Quality of the Alignments. The computation of family specific residues is, in general, strictly dependent on the quality of the corresponding alignments, and the same applies to the construction of gene trees. In many cases, large protein families are difficult to align and contain many sequences that populate the space between families. Common multiple sequence alignment programs tend to misalign residues of different chemical types that are conserved in subfamilies, and to introduce gaps instead. This may create an artificial increase in the number of tree-determinant residues. This problem is especially relevant for the less conserved regions.

Division of a Protein Family into Subfamilies. The issue of dividing a protein family into subfamilies, in an attempt to identify subfamily-specific residues that would be associated with functional specificity, has yet to be resolved. Some methods (like ET and ConSurf) divide the protein families at different levels, according to an evolutionary tree specified a priori. The drawback of this approach is the difficulty in distinguishing tree-determinants from merely conserved positions. Indeed, the consideration of high levels of partition involves a greater number of subfamilies of decreasing size. In this situation, with a lot of clusters of proteins which are hardly populated, residue conservation becomes meaningless, and the tree-determinant identification is shadowed by the great amount of positions that become conserved. This increase in the number of conserved positions may be due to two facts: first, there is a higher probability of a position appearing as conserved as the cluster size decreases, and, secondly, there is more similarity among the proteins contained in clusters obtained at high levels of partition, so it is not possible to decide whether conservation is due to functionality or to a short divergence time .

Introduction of Additional Functional Information. It is quite common to have additional information about the functions of the protein families and subfamilies. Ideally, this information should be combined with that derived from the alignments (phylogenetic distributions). If no functional knowledge is available about the protein to be analyzed, the only resource is the identification of conservation patterns within the groups in the partitions of the gene trees, in which we have referred to as the typical tree-determinant detection application. For example, groups of orthologous sequences can be formed, assuming that specificity will be more conserved than in groups of paralogs (Mirny and Gelfand, 2002). In biological systems, it is common to have additional information about functional characteristics of some of the proteins. In this sense, the proposals by several authors (Livingstone and Barton, 1993; Kuipers et al., 1997; Hannenhalli and Russell, 2000) include external functional groups integrated in the analysis of the families. The correct way to incorporate additional information, about functional classes, with relationships derived from the analysis of protein sequences has yet to be established.

Structural versus Functional Conservation. One of the key biological problems in the interpretation of tree-determinant residues is the differentiation between structural and functional constraints. It is obvious that the contribution of specific residues to protein function and structure cannot always be assigned, and in many cases both of them are mixed. At the same time, the structural constraints can be a mixture of protein stability and protein folding, where key residues can be key contributors to both of them simultaneously. It is quite possible to imagine that tree-determinant residues can confer specific properties related with folding and stability to protein subfamilies, if this confers some selective advantage. The work of Dokholyan et al (2002) and Reva et al (2000), has indeed revealed interesting theoretical possibilities in this direction using lattice models, pair potentials, and simplified folding strategies. One interesting system in which tree-determinants seems to have a key functional role that requires a direct connection with protein structure is the analysis of PDZ domains (Lockless and Ranganathan, 1999) and the G protein-coupled receptors, the chymotrypsin-family of serine proteases and the hemoglobin families (Süel et al., 2003). In these cases, tree-determinant residues seem to form part of a functional network of interacting residues necessary for the allosteric communication between distant zones in the protein structure, as an essential function in the transmission of cellular signals that requires a direct coupling with the protein architecture. Despite these interesting approaches, one must acknowledge that it is very difficult in practice to distinguish between structural and functional conservation, an issue that is still also essentially unsolved for the completely conserved residues, although they have been studied for much longer than tree-determinants.

22.10 The Use of Information on Tree-Determinant Residues in Molecular Biology

A possible way of assessing the influence of the theoretical bioinformatics developments in the study of binding and specific sites is to follow the usage of SequenceSpace, the first program addressing directly the problem of protein specificity prediction, in experimental molecular biology. The original SequenceSpace publication (Casari et al., 1995) has been quoted in 85 articles, including a wide variety of biological applications related to the functional characteristics of various protein families (Table 22.1) such as: chaperonins, ADP-ribosyltransferases, complement proteases, allergens, Ras super-family, bacterial transcription factors, carnitine acyltransferases and others. The program has contributed to deciphering a wide range of problems related to protein function, including: differential binding of substrates and inhibitors, molecular basis of functional divergence, architecture of mimotopes, epitopes and the corresponding recognition regions, distance constraints for protein structure prediction, comparison of families for the detection of remote homologous, and detection of protein domains and protein core boundaries.

An illustrative example of the use of SequenceSpace to define family-specific residues responsible for functions was related to the selection of key residues for the differentiation by Ras and Ral of their specific regulating proteins (Bauer et al., 1999). Ras and Ral are members of two families of the large ras superfamily: they share a considerable level of sequence similarity (around 70 %), a common fold and the same basic biochemical functions. Together with their basic similarities they interact with different effectors: Rlip in the case of Ral and a number of proteins with a common Ras-binding-domain. Mediated by these effectors they trigger different pathways of cellular signaling. SequenceSpace analysis of the multiple alignment of the ras-like family of proteins and the comparison among the sequences of different families such as Ras, Rho, Rab and Ral subfamilies indicated that only a small number of positions could be considered responsible for the differences between families. In this case, the basic information provided by SequenceSpace was additionally validated by assessing the structure of the known complexes of ras with ras-binding domains. In these complexes the predicted tree-determinant residues are localized in the accessible region of the complex interfaces. The combined bioinformatic analysis revealed that two positions were potentially enough for controlling the binding specificity (Ral residue K47 and 48A). Ral wild-type proteins and the variants including mutations in positions 47 and 48 were screened for binding to their natural Rlip effectors and for the binding to the non-natural effectors (Ras-binding domains). The symmetric experiment was carried out in ras wild-type proteins their double mutants, and the corresponding ras-binding and Rlip effectors. The experimental approach included yeast two-hybrid assays and

Table 22.1. Use of SequenceSpace in different biochemical approaches

Binding of substrates and inhibitors to carnitine acyltransferases	Morillas et al. (2001, 2002, 2003)
Specificity-determining residues in bacterial transcription factors	Mirny et al. (2002)
Functional divergence of Jak protein kinase domains	Gu et al. (2002)
Genetic engineering of allergens	Ferreira et al. (2002); Kraft et al. (1999); Ferreira et al. (1998)
Antibody response to mimotopes of the hepatitis C virus hypervariable region 1	Zucchelli et al. (2001); Roccasecca et al. (2001); Puntoriero et al. (1998)
Functional divergence in the caspase gene family	Wang and Gu. (2001)
Analysis of basic phospholipase A(2) myotoxin isoforms from *Bothrops asper*	Lizano et al. (2001)
Structure prediction of alpha-glucosidase-like gene from *Penicillium minioluteum*	Garcia et al. (2001)
Prediction of ligand-binding function in families of bacterial receptors	Johnson et al. (2000)
Analysis of prevalence of specific T-cells in HCV-infected individuals	Del Porto et al. (2000)
Analysis of heregulin symmetry	Landgraf et al. (1999)
Effector recognition by the small GTP-binding proteins Ras and Ral	Bauer et al. (1999)
Model of the Ran-RCC1 interaction	Azuma et al. (1999)
SequenceSpace analysis of Lys49 phospholipases A(2)	Ward et al. (1998)
Use of sequence comparison to detect identities in tRNA genes	Sagara et al. (1998)
Structural model for family 32 of glycosyl-hydrolase enzymes	Pons et al. (1998)
Prediction of the papain prosegment folding pattern	Padilla-Zuniga and Rojo-Dominguez (1998)
Complement and blood coagulation proteases: domain interactions detected at the sequence level	Gaboriaud (1998)
Proposed architecture for the central domain of the bacterial enhancer-binding proteins	Osuna et al. (1997)
Functional diversity of PH domains	Blomberg et al. (1997)
Sequence and structural links between distant ADP-ribosyltransferase families	Bazan and KochNolte (1997)
Helical fold prediction for the cyclin box	Bazan (1996)
Prediction of the structure of GroES and its interaction with GroEL	Valencia et al. (1995)

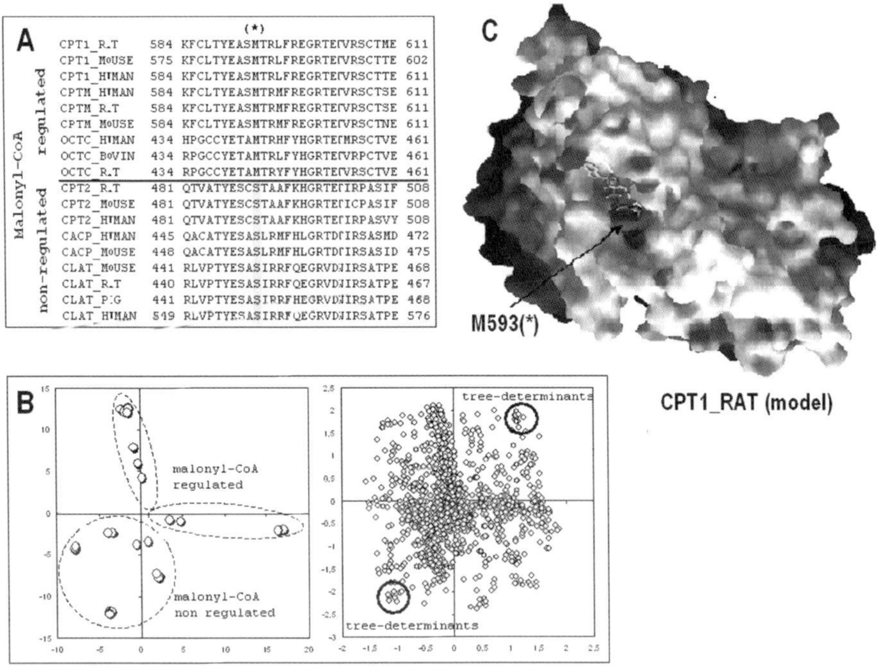

Fig. 22.2. An example of SequenceSpace applied to the analysis of malonyl-CoA regulation of enzymes belonging to the carnitine-choline acyltransferase family. **A** Fragment of the multiple sequence alignment of the protein family. **B** Protein-sequences and residue-position coordinates projected onto two-dimensional space defined by the SequenceSpace algorithm. Proteins are clustered according to their regulation properties and the tree-determinant residues are located in the corresponding spatial position. The tree determinants are mapped on the multiple sequence alignment (**A**): malonyl-CoA regulated enzymes contains a methionine (*) whereas non-regulated ones contain a serine. **C** Three-dimensional model for the malonyl-CoA regulated enzyme CPT1_RAT. The position of the methionine 593, close to the substrate binding site, is indicated. Mutation of this methionine by serine completely abolished the malonyl-CoA regulation effect. (Modified from Morillas et al., 2003)

detailed quantitative calorimetric measurements, and showed that the replacement of these precise two residues completely abrogates the binding of the proteins to their natural effectors, and instead promotes the binding to the effectors of the other proteins, producing an effective swapping of binding specificities.

A second demonstrative application of the capacity of SequenceSpace to explore molecular systems is described in Morillas et al. (2003). Using the multiple sequence alignment of the complete family of the carnitine-choline acyltransferases as input to the SequenceSpace analysis, the authors identified residue positions responsible for the inhibition of the catalytic activity of the enzymes of the carnitine palmitoyltransferase I (L- and M- isoforms) and carnitine octanoyltransferase subfamilies (Fig. 22.2). Despite their sequence similarity, the enzymatic activities of the members of the other subfamilies, carnitine palmitoyltransferase II, carnitine acetyltransferase and choline acetyltransferase are not regulated by malonyl-CoA. The bio-computing study showed that five amino acids are specifically conserved in all inhibitor-regulated enzymes from various organisms and they were not present in non malonyl-CoA-inhibitable acyltransferases. The subsequent mutational analysis confirmed that the L-carnitine palmitoyltransferase mutants in position M593, one of the five predicted tree-determinant residues, completely lacked malonyl-CoA sensitivity. Other examples of the application of the Sequence-Space method to the same group of enzymes regarding, not the inhibitor sensitivity, but the substrate specificity are described in Morillas et al., (2001) and Morillas et al., (2002).

References

Aloy P, Querol E, Aviles FX, Sternberg MJ. Automated structure-based prediction of functional sites in proteins: applications to assessing the validity of inheriting protein function from homology in genome annotation and to protein docking. *J Mol Biol.* 2001, 311(2):395–408

Altschuh D, Lesk AM, Bloomer AC, Klug A. Correlation of coordinated amino acid substitutions with function in virus related to tobacco mosaic virus. *J. Mol. Biol.* 1987, 193:693–707

Andrade MA, Casari G, Sander C, Valencia A. Classification of protein families and detection of the determinant residues with an improved self-organizing map. *Biol Cybern.* 1997, 76:441–450

Armon A, Graur D, Ben-Tal N. ConSurf: an algorithmic tool for the identification of functional regions in proteins by surface mapping of phylogenetic information. *J Mol Biol.* 2001, 307:447–463

Atchley, W. R., Terhalle, W., Dress, A. Positional dependence, cliques and predictive motifs in the bHLH protein domain. *J. Mol. Evol.* 1999, 48:501–516

Atchley, W. R., Wollenberg, K. R., Fitch, W. M., Terhalle, W. Correlations among amino acid sites in bHLH protein domains: an information theoretic analysis. *Mol. Biol. Evol.* 2000, 17:164–178

Azuma Y, Renault L, Garcia-Ranea JA, Valencia A, Nishimoto T, Wittinghofer A. Model of the Ran-RCC1 interaction using biochemical and docking experiments. *Journal of Molecular Biology.* 1999, 289:1119–1130

Bauer B, Mirey G, Vetter IR, Garcia-Ranea JA, Valencia A, Wittinghofer A, Camonis JH, Cool RH. Effector recognition by the small GTP-binding proteins Ras and Ral. *Journal of Biological Chemistry.* 1999, 274:17763–17770

Bazan JF, KochNolte F. Sequence and structural links between distant ADP- ribosyltransferase families. In *Adp-Ribosylation in Animal Tissues.* Edited by; 1997:99–107. Advances in Experimental Medicine and Biology, vol 419.]

Bazan JF. Helical fold prediction for the cyclin box. *Proteins-Structure Function and Genetics.* 1996, 24:1–17

Blomberg N, Nilges M. Functional diversity of PH domains: an exhaustive modelling study. *Folding & Design.* 1997, 2:343–355

Bogan AA, Thorn KS. Anatomy of hot spots in protein interfaces. *J Mol Biol.* 1998, 280(1): 1–9

Casari G, Sander, C., Valencia, A. A method to predict functional residues in proteins. *Nature Struct Biol.* 1995, 2:171–178

Clackson T, Wells JA. A hot spot of binding energy in a hormone-receptor interface. *Science.* 1995, 267(5196):383–6

Clarke, N. D. Covariation of residues in the homeodomain sequence family. *Protein Sci.* 1995, 4:2269–2278

del Porto P, Puntoriero G, Scotta C, Nicosia A, Piccolella E. High prevalence of hypervariable region 1-specific and cross-reactive CD4(+) T cells in HCV-infected individuals responsive to IFN-alpha treatment. *Virology.* 2000, 269:313–324

del Sol, A., Pazos, F., Valencia, A. Automatic methods for predicting functionally important residues. *J. Mol. Biol.* 2003, 326:1289–1302

de Rinaldis M, Ausiello G, Cesareni G, Helmer-Citterich M. Three-dimensional profiles: a new tool to identify protein surface similarities. *J Mol Biol.* 1998, 284:1211–1221

Devos D, Valencia A. Practical limits of function prediction. *Proteins.* 2000, 41:98–107

Dokholyan NV, Li L, Ding F, Shakhnovich EI.. Topological determinants of protein folding. *Proc Natl Acad Sci U S A.* 2002, 99(13):8637–41

Dopazo J. A new index to find regions showing an unexpected variability or conservation in sequence alignments. *Comput Appl Biosci.* 1997, 13(3):313–7

Dorit RL, Ayala FJ. ADH evolution and the phylogenetic footprint. *J Mol Evol.* 1995, 40(6):658–62

Ferreira F, Ebner C, Kramer B, Casari G, Briza P, Kungl AJ, Grimm R, Jahn-Schmid B, Breiteneder H, Kraft D, et al. Modulation of IgE reactivity of allergens by site-directed mutagenesis: potential use of hypoallergenic variants for immunotherapy. *Faseb Journal.* 1998, 12:231–242

Ferreira F, Wallner M, Breiteneder H, Hartl A, Thalhamer J, Ebner C. Genetic engineering of allergens: Future therapeutic products. *International Archives of Allergy and Immunology.* 2002, 128:171–178

Fetrow JS, Skolnick J. Method for prediction of protein function from sequence using the sequence-to-structure-to-function paradigm with application to glutaredoxins/thioredoxins and T1 ribonucleases. *J Mol Biol.* 1998, 281(5):949–68

Gaboriaud C, Rossi V, Fontecilla-Camps JC, Arlaud GJ. Evolutionary conserved rigid module-domain interactions can be detected at the sequence level: The examples of complement and blood coagulation proteases. *Journal of Molecular Biology.* 1998, 282:459–470

Garcia B, Castellanos A, Menendez J, Pons T. Molecular cloning of an alpha-glucosidase-like gene from *Penicillium minioluteum* and structure prediction of its gene product. *Biochemical and Biophysical Research Communications.* 2001, 281:151–158

Giraud, BG, Lapedes A, Liu LC. Analysis of correlation between sites in models of protein sequences. *Physical Rev E.* 1998, 58(5):6312–6322

Gribskov M, Homyak M, Edenfield J, Eisenberg D. Profile scanning for three-dimensional structural patterns in protein sequences. *Comput Appl Biosci.* 1988, 4(1):61–6

Grishin NV, Phillips MA. The subunit interfaces of oligomeric enzymes are conserved to a similar extent to the overall protein sequences. *Protein Sci.* 1994, 3(12):2455–8

Gu JY, Wang YF, Gu X. Evolutionary analysis for functional divergence of Jak protein kinase domains and tissue-specific genes. *Journal of Molecular Evolution.* 2002, 54:725–733

Hannenhalli SS, Russell RB. Analysis and Prediction of Functional Sub-types from Protein Sequence Alignments. *J Mol Biol.* 2000, 303:61–76

Iliopoulos I, Tsoka S, Andrade MA, Janssen P, Audit B, Tramontano A, Valencia A, Leroy C, Sander C, Ouzounis C. A. Genome sequences and great expectations. *Genome Biol.* 2000, 2(1):INTERACTIONS0001

Johnson JM, Church GM. Predicting ligand-binding function in families of bacterial receptors. *Proceedings of the National Academy of Sciences of the United States of America.* 2000, 97:3965–3970

Kohonen, T. Self-organized formation of topologically correct feature maps. *Biol. Cybern.* 1982, 43:59–69

Kraft D, Ferreira F, Vrtala S, Breiteneder H, Ebner C, Valenta R, Susani M, Breitenbach M, Scheiner O. The importance of recombinant allergens for diagnosis and therapy of IgE-mediated allergies. *International Archives of Allergy and Immunology* 1999, 118:171–176

Kuipers W, Oliveira L, Vriend G, Ijzerman AP. Identification of class-determining residues in G protein-coupled receptors by sequence analysis. *Receptors Channels.* 1997, 5(3–4):159–74

Kuntz ID, Blaney JM, Oatley SJ, Langridge R, Ferrin TE. A geometric approach to macromolecule-ligand interactions. *J Mol Biol.* 1982, 161(2):269–88

Landgraf R, Fischer D, Eisenberg D. Analysis of heregulin symmetry by weighted evolutionary tracing. *Protein Engineering.* 1999, 12:943–951

Landgraf R, Xenarios I, Eisenberg D. Three-dimensional cluster analysis identifies interfaces and functional residue clusters in proteins. *J Mol Biol.* 2001, 307:1487–1502

Lichtarge O, Bourne HR, Cohen FE. An Evolutionary Trace Method Defines Binding Surfaces Common to Protein Families. *J Mol Biol.* 1996, 257:342–358

Livingstone CD, Barton GJ. Protein sequence alignments: a strategy for the hierarchical analysis of residue conservation. *Comput Appl Biosci.* 1993, 6:645–756

Lizano S, Lambeau G, Lazdunski M. Cloning and cDNA sequence analysis of Lys(49) and Asp(49) basic phospholipase A(2) myotoxin isoforms from Bothrops asper. *International Journal of Biochemistry & Cell Biology.* 2001, 33:127–132

Lockless, S. W., Ranganathan, R. Evolutionary conserved pathways of energetic connectivity in protein families. *Science.* 1999, 286:295–299

Luscombe NM, Thornton JM. Protein-DNA interactions: amino acid conservation and the effects of mutations on binding specificity. *J Mol Biol.* 2002, 320(5):991–1009

Madabushi S, Yao H, Marsh M, Kristensen DM, Philippi A, Sowa ME, Lichtarge O. Structural clusters of evolutionary trace residues are statistically significant and common in proteins. *J Mol Biol.* 2002, 316:139–154

Mirny LA, Gelfand MS. Using orthologous and paralogous proteins to identify specificity-determining residues in bacterial transcription factors. *Journal of Molecular Biology.* 2002, 321:7–20

Miyata, T., Miyazawa, S., Yashunaga, T. Two types of amino acid substitutions in protein evolution. *J. Mol. Evol.* 1979, 12:219–236

Morillas M, Gomez-Puertas P, Bentebibel A, Selles E, Casals N, Valencia A, Hegardt FG, Serra D Identification of conserved amino acid residues in rat liver Carnitine palmitoyltransferase I critical for malonyl-CoA inhibition. *Journal of Biological Chemistry.* 2003, 278:9058–9063

Morillas M, Gomez-Puertas P, Roca R, Serra D, Asins G, Valencia A, Hegardt FG. Structural model of the catalytic core of carnitine palmitoyltransferase I and carnitine octanoyltransferase (COT) - Mutation of CPT I histidine 473 and alanine 381 and COT alanine 238 impairs the catalytic activity. *Journal of Biological Chemistry.* 2001, 276:45001–45008

Morillas M, Gomez-Puertas P, Rubi B, Clotet J, Arino J, Valencia A, Hegardt FG, Serra D, Asins G. Structural model of a malonyl-CoA-binding site of carnitine octanoyltransferase and carnitine palmitoyltransferase I– Mutational analysis of a malonyl-CoA affinity domain. *Journal of Biological Chemistry.* 2002, 277:11473–11480

Osuna J, Soberon X, Morett E. A proposed architecture for the Central domain of the bacterial enhancer-binding proteins based on secondary structure prediction and fold recognition. *Protein Science.* 1997, 6:543–555

Ouzounis C, Perez-Irratxeta C, Sander C, Valencia A. Are binding residues conserved? *Pacific Symposium on Biocomputing.* 1998, 3:399–410

Padilla-Zuniga AJ, Rojo-Dominguez A. Non-homology knowledge-based prediction of the papain prosegment folding pattern: a description of plausible folding and activation mechanisms. *Folding & Design.* 1998, 3:271–284

Pazos F, Sanchez-Pulido L, García-Ranea JA, Andrade MA, Atrian S, Valencia A. Comparative analysis of different methods for the detection of specificity regions in protein families. In: Lundh D, Olsson, B., Narayanan A. (ed) *Biocomputing and Emergent Computation.* 1997, World Scientific, Singapore, New Jersey, London, Hong Kong, p 132–145

Pettit FK, Bowie JU. Protein surface roughness and small molecular binding sites. *J Mol Biol.* 1999, 285(4):1377–82

Pons T, Olmea O, Chinea G, Beldarrain A, Marquez G, Acosta N, Rodriguez L, Valencia A. Structural model for family 32 of glycosyl-hydrolase enzymes. *Proteins-Structure Function and Genetics.* 1998, 33:383–395

Puntoriero G, Meola A, Lahm A, Zucchelli S, Ercole BB, Tafi R, Pezzanera M, Mondelli MU, Cortese R, Tramontano A, et al. Towards a solution for hepatitis C virus hypervariability: mimotopes of the hypervariable region 1 can induce antibodies cross-reacting with a large number of viral variants. *Embo Journal.* 1998, 17:3521–3533

Pupko T, Bell RE, Mayrose I, Glaser F, Ben-Tal N. Rate4Site: an algorithmic tool for the identification of functional regions in proteins by surface mapping of evolutionary determinants within their homologues. *Bioinformatics.* 2002, 18:S71-S77

Reva BA, Finkelstein AV, Skolnick J. Derivation and testing residue-residue mean-force potentials for use in protein structure recognition. *Methods Mol Biol.* 2000, 143:155–74

Roccasecca R, Folgori A, Ercole BB, Puntoriero G, Lahra A, Zucchelli S, Tafi R, Pezzanera M, Galfre G, Tramontano A, et al. Mimotopes of the hyper variable region I of the hepatitis C virus induce cross-reactive antibodies directed against discontinuous epitopes. *Molecular Immunology.* 2001, 38:485–492

Rost B. Enzyme function less conserved than anticipated. *J Mol Biol.* 2002, 318:595–608

Rost B, Honig B, Valencia A. Bioinformatics in structural genomics. *Bioinformatics.* 2002, 18(7):897–8

Sagara JI, Shimizu S, Kawabata T, Nakamura S, Ikeguchi M, Shimizu K. The use of sequence comparison to detect 'identities' in tRNA genes. *Nucleic Acids Research.* 1998, 26:1974–1979

Shannon CE, and Weaver W. The Mathematical Theory of Communication. The University of Illinois Press, Urbana, 1949

Sibbald PR, Argos P. Weighting aligned protein or nucleic acid sequences to correct for unequal representation. *J Mol Biol.* 1990, 216(4):813–8

Singer, M. S., Oliveira, L. Vriend, G., Shepherd, G. M. Potential ligand-binding residues in rat olfactory receptors identified by correlated mutation analysis. *Receptor and Channels.* 1995, 3:89–95

Süel, G. M., Lockless, S. W., Ranganathan, R. Evolutionary conserved networks of residues mediate allosteric communication in proteins. *Nat. Struct. Biology.* 2003, 10(1): 59–68

Taylor, W. R., Harricks, K. Compensating changes in protein multiple sequence alignments. *Prot. Eng.* 1994, 7:342–348

Taylor, W. R. Classification of amino acid conservation. *J. Theor. Biol.* 1986, 119:205–218

Todd AE, Orengo CA, Thornton JM. Evolution of function in protein superfamilies, from a structural perspective. *J Mol Biol.* 2001, 307:1113–1143

Valdar WS, Thornton JM. Protein-protein interfaces: analysis of amino acid conservation in homodimers. *Proteins.* 2001, 42:108–124

Valencia A, Hubbard TJ, Muga A, Banuelos S, Llorca O, Carrascosa JL, Valpuesta JM. Prediction of the Structure of Groes and Its Interaction with Groel. *Proteins-Structure Function and Genetics.* 1995, 22:199–209

Villar HO, Kauvar LM. Amino-acid preferences at protein binding sites. *FEBS Lett.* 1994, 349:125–130

Wang YF, Gu X. Functional divergence in the caspase gene family and altered functional constraints: Statistical analysis and prediction. *Genetics.* 2001, 158:1311–1320

Ward RJ, Alves AR, Neto JR, Arni RK, Casari G. A SequenceSpace analysis of Lys49 phospholipases A(2): clues towards identification of residues involved in a novel mechanism of membrane damage and in myotoxicity. *Protein Engineering.* 1998, 11:285–294

Wodak SJ, Janin J. Structural basis of macromolecular recognition. *Advances in Protein Chemistry.* 2003, 61:9

Yao, H., Kristensen, D. M., Mihalek, I., Sowa, M. E., Shaw, C., Kimmer, M., Kavraki, L., Lichtarge, O. An accurate, sensitive, and scalable method to identify functional sites in protein structures. *J. Mol. Biol.* 2003, 326:255–261

Zucchelli S, Roccasecca R, Meola A, Ercole BB, Tafi R, Dubuisson J, Galfre G, Cortese R, Nicosia A. Mimotopes of the hepatitis C virus hypervariable region 1, but not the natural sequences, induce cross-reactive antibody response by genetic immunization. *Hepatology.* 2001, 33:692–703

Zuckerkandl E, Pauling L. Evolutionary Divergence and Convergence in Proteins. In: Bryson V, Vogel HJ (eds) *Evolving Genes And Proteins.* Academic Press, 1965, New York, p 97–166

Zvelebil, M. J. J. M., Barton, G. J., Taylor, W. R., Stenberg, M. J. E. Prediction of protein secondary structure and active sites using the alignment of homologous sequences. *J. Mol. Biol.* 1987, 195:957–961

23 Extracting and Searching for Structural Information: A Multiresolution Approach

NATALIA JIMÉNEZ-LOZANO, MÓNICA CHAGOYEN,
PEDRO ANTONIO DE-ALARCÓN and JOSÉ MARÍA CARAZO

23.1 From Protein to Function

Nowadays, the scientific community is aware of the importance of the structure of proteins in order to understand the functional events in which they are involved. Indeed, a wide range of diseases are induced by modifications in their structural properties, leading to a loss of protein function (e.g. muscular dystrophy). Thus, protein structure elucidation can provide crucial information about its biochemical function. The two most widely spread methods of protein structure determination, at high resolution, are X-ray diffraction and nuclear magnetic resonance spectroscopy (NMR).

X-ray crystallography allows the study of diverse biological specimens ranging from peptides to viruses. The success of the X-ray analysis lies in the availability of large and well-ordered 3D crystals of the specimen and on obtaining isomorphous heavy atom derivatives for the initial phasing of the diffraction data. Integral membrane proteins and filamentous proteins, such as actin and tubulin, often form aggregates rather than well-ordered 3D crystals which hinders their study using this method.

NMR allows the study of small molecules and individual domains of proteins in solution. To obtain an interpretable signal by this method, it is necessary to have large amounts of protein with high solubility in water. The main disadvantage of NMR is the theoretical size limit for the proteins to be studied (Auer M 2000).

In addition to these well-known structural techniques, there is an increasing use of cryo-electron microscopy with image processing tools (three-dimensional electron microscopy or 3D-EM) for the solution of 3D structures. 3D-EM medium resolution data provides slightly different information compared to atomic co-ordinates. The main structural result obtained in a 3D-EM study is a three-dimensional image, or volume, in which each voxel is related to the Coulomb potential of the biological sample at that position

Principles and Practice
Methods in Proteome and Protein Analysis
R.M. Kamp, J. J. Calvete, T. Choli-Papadopoulou (Eds.)

(Hawkes and Kasper 1996). Images taken with an electron microscope can be considered as two-dimensional projections of the specimen being studied. After combining projections at different angles, a complete three-dimensional reconstruction of the sample can be obtained. The exact steps taken for the reconstruction from projection images vary according to the nature and symmetry of the specimen.

Although, by means of 3D-EM, atomic resolution has been obtained in some cases like bacteriorhodopsin (Subramaniam and Henderson 1999), the light-harvesting complex (Wang and Kulhbrandt 1991), tubulin (Lowe et al. 2001) and aquaporin-1 (Ren et al. 2000), these are more the exceptions than the rule. 3D-EM is a technique that provides enough quantitative measurements of the conformation of macromolecules in the range of 8 to 30 Å and it can be used to elucidate structures of about 50 Å (protein subunits) to more than 1000 Å (macromolecular complexes) in diameter. Even though a less detailed description of the structure under study is obtained because of the limitation set by the resolution, it provides clues about conformational changes and on the overall organization of subunits and domains within the machines. The advantage of this experimental approach is not only that it is not necessary for specimens to be arranged in crystal lattices (as is the case of X-ray diffraction) or to be below a given molecular weight (as with NMR) but that it can be studied in its physiological conformation.

It is now clear that most functions in the cell are not carried out by single protein enzymes, colliding randomly within the cellular jungle, but by macromolecular complexes containing multiple subunits with specific functions (Alberts 1998). 3D-EM has been used to study these macromolecular complexes as microtubules (Li et al. 2002), icosahedral viruses (Baker et al. 1999; Bernal et al. 2003), the apoptosome (Acehan et al. 2002), the CCT chaperonin (Llorca et al. 2001), the splicesome (Stark et al. 2001), the ribosome (Stark 2002), as well as whole subcellular elements using tomographic approaches (McEwen and Marko 2001; Kozubek et al. 2002; Baumeister 2002).

These and other macromolecular assemblages have also been studied by a combination of high (X-ray diffraction and NMR spectroscopy) and medium resolution (3D-EM) techniques (*fitting studies*). The motivation of these experiments is clear, as nowadays, it is still not possible to obtain the structure at atomic resolution of very large macromolecular machines or complexes. By means of these combined techniques, the quaternary structural information coming from 3D-EM together with atomic level structures of individual components, provide a detailed picture of the overall architecture of the complexes and on the interactions between the components. This is the case of large viruses (Zhang et al. 2002), viral complexes (Thouvenin and Hewat 2000), muscle related filaments (Hatch et al. 2001), membrane joined molecules (Orlova EV et al. 2003), the ribosome (Spahn et al. 2001; Agrawal et al. 2001; Klaholz et al. 2003), and chaperons (Carrascosa et al. 2001).

23.2 Structural Feature Relevance in Macromolecular Complexes

The storage of atomic co-ordinates within the framework of the Protein Data Bank since its establishment in 1971, has provided us with the possibility of studying large amounts of structural data. Many studies have been geared towards the analysis and further comparison of structures in PDB with two final aims: to classify its domains into meaningful categories (Pearl et al. 2003; Lo Conte et al. 2002) or to extrapolate a common function from structural similarity (Eidhammer et al. 2000). Thus, tools to analyze and extract structural features from data at a high resolution have been tested, validated and sometimes discarded over the years. The application of these tools has resulted in the characterization of structural features from high-resolution data. That is the case of small cavities (Liang et al. 1998a, b; Hubbard et al. 1994) or protein-protein interaction surfaces (Jones and Thornton 1996, 1997; Larsen et al. 1998). Moreover, as the result of the use of the previously mentioned tools, some general rules about structural features have been established.

In contrast to high-resolution techniques, 3D-EM is a more recently developed technique. Due to the fact that a public archive for 3D-EM data did not exist until very recently (EMDB database: http://www.ebi.ac.uk/msd/), little attention has been paid to large structural features like, for example, those observed in macromolecular complexes that are formed by the joining of subunits. Some important examples of these kinds of structural features are found in literature. As is with hexameric helicases, whose central channel encircles ssDNA during the unwinding process, aquaporins, a family of transmembrane channel proteins that selectively allow water to cross the cell plasma membrane in various human tissues, ion pumps, like Na^+ and K^+-ATPases which are responsible for the establishment and maintenance of ion gradients in kidney and nerve tissue, or the Ca^{2+}-ATPase, which affects the contraction and relaxation of heart muscle (Auer M 2000). Further examples are found in the cavity of chaperonins, like GroEL, CCT or thermosomes that lodge the protein to be folded or the connector of some bacteriophages which allow the passage of DNA to be packed.

Nowadays, each author describes their volume's structural features in a rather personal manner because there is no way of comparing their results with others. They are mainly centred on symmetry and topology. Comparative studies have concentrated on differences among few data under very controlled conditions; that is the case of differences between wild and mutant states of macromolecules (Bottcher et al. 2000; Ferlenghi et al. 1998) or among the binding of different substrates to the same complex (Llorca et al. 2001). No systematic studies have been performed on the analysis of 3D-EM structural features in order to give an explanation of their geometric properties. The absence of structural conclusions obtained by the comparison of large and diverse data sets solved by 3D-EM, has encouraged us to develop and apply some tools to fill this gap. In this chapter, the basis that will allow making an exhaustive structural analysis of 3D-EM volumes will be established, regardless of the resolution achieved. This will enable one to carry out similarity searches among data and consequently assign common structures. At best, structure similarity will reveal functional similarity.

23.3 Extraction and Characterisation of Structural Features

In the context of this work, the general term *structural feature* will include cavities, channels, protrusions, surface and topology of the macromolecules, regardless of the technique used to solve them. Although this chapter will be focused on cavities, channels, surface and topology, we are aware of the functional relevance of protrusions, and therefore we will leave the door open to explore them in the future.

The tools used to extract the structural features are based on two principles (see below in this section) and their application depends on geometric properties. In order to clarify concepts, a structural feature classification from the medium resolution point of view has been established, attending to geometric properties:

- *Surface* is the two-dimensional boundary of a three-dimensional macromolecular structure.
- *Cavity* is a groove or hollow observed on the macromolecular surface (*external cavity or pocket*) or an interior empty space inside the macromolecule (*internal cavity or void*) and whose relationship with the outside world is described by its number of mouths or openings. Thus, a void will have no mouth and a pocket will have one mouth. A *shallow cavity* is a special kind of cavity in which none of its infinite number of possible cross sections is larger than the mouth opening (shallow principle).
- *Channel* is a passage in the macromolecular structure that connects with the outside world by means of two or more openings. Like in the case of cavities, we can find macromolecules with *shallow channels* in which at least one mouth in the channel complies with the shallow principle.
- *Protrusion* is a protuberance on the macromolecular surface.
- *Topology* refers to the structural properties of macromolecules, which are preserved through deformations, twistings, and stretchings. For example, in the topological sense, a circle will be equivalent to an ellipse into which it can be deformed by stretching, and a sphere will be equivalent to an ellipsoid. Topology can be used to abstract the inherent connectivity of objects while ignoring their detailed form.

One of the most successful methods employed to extract structural features in high-resolution data is based on the alpha-shape theory (Edelsbrunner H and Mücke EP 1994). We will briefly describe the alpha-shape theory as a generalization of the *convex hull*[1] of a point set in the three-dimensional space. Each alpha shape is a well-defined three-dimensional polyhedron derived from the *Delaunay triangulation*[2] of the point set, with a parameter $\alpha \in \Re$ controlling the desired level of detail. When $\alpha=0$ (zero shape), the actual topological structure of the molecule is obtained (as alpha decreases, the shape shrinks and gradually develops cavities); however, when $\alpha=\infty$, the alpha complex is the convex hull of the initial set (Figs. 23.1 and 23.2).

The *simplicial complex*[3], associated with an α value, is a subcomplex of the Delaunay complex and is called alpha complex. The alpha shape is a part of the space occupied by *simplices*[3] in the alpha complex.

(1) Convex hull: The convex hull of a set S of points, denoted hull(S), is the smallest polyhedron P for which each point of S is either on the boundary or in the interior of P.

(2) Delaunay triangulation: The Delaunay triangulation is the geometric dual of the Voronoi diagram. The Voronoi diagram of a set S of points divides the space into Voronoi regions, one per point. A Voronoi region is generated by a point, and consists of the part of the space closest to this point.

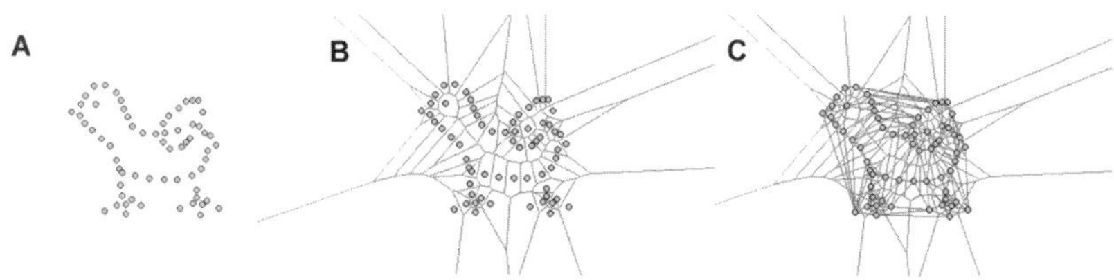

Fig. 23.1. Illustration of the Voronoi diagram and Delaunay triangulation duality. **A** Set of points representing a chicken shape; **B** Voronoi diagram generated from this point set; **C** Delaunay triangulation and Voronoi diagram of the same point set. Thanks are extended to Francois Bélair for the utilization of his applet (http://www.cgrl.cs.mcgill.ca/~godfried/teaching/projects97/belair/alpha.html)

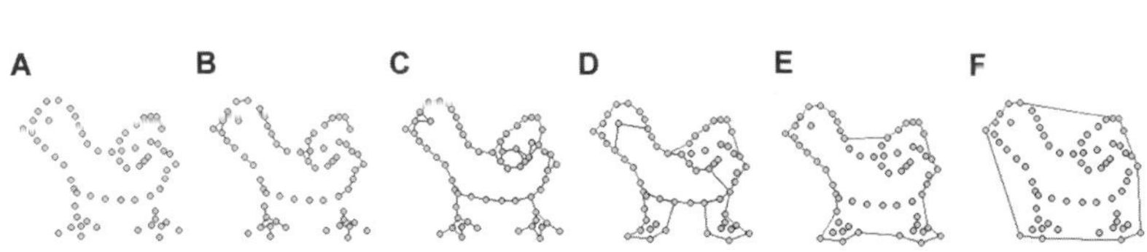

Fig. 23.2. Relevance of the α parameter to obtain a valid representation of the object under study. **A** Point set representing a chicken shape; **B–F** alpha shape for increasing α values. Observe that when α increases, the points start joining until reaching the real shape of the chicken (**C**). At higher α values, far away points start joining until the maximum α value where the alpha shape is equivalent to the convex hull (**F**). Thanks are extended to Francois Bélair for the utilization of his applet (http://www-cgrl.cs.mcgill.ca/~godfried/teaching/projects97/belair/alpha.html)

The Delaunay triangulation is the set of triangles obtained by drawing a line between any two points whose Voronoi domains touch. Generally, this triangulation is unique.

(3) *Simplex and simplicial complexes*: topology creates a unified language and notation by calling a vertex a 0-simplex, an edge a 1-simplex, a triangle a 2-simplex, and a solid tetrahedron a 3-simplex. Complicated geometric objects can be built from a collection of simplices. This is achieved by adhering to the following rules: (1) for every simplex used, its faces are also part of the construction, and (2) the common intersection of any two simplices is either empty or a face of both simplices. If the above rules are followed, the resulting object is called a simplicial complex (Liang et al. 1998)

(4) *Geometry*: is a branch of mathematics concerning shapes and their relationship to one another. The central notion in geometry is that of congruence. In Euclidean geometry, two figures are said to be congruent if they are related by a series of reflections, rotations and translations.

Alpha shape allows the representation of the geometric and topologic information presented by atomic co-ordinates in such a way that enables the detection and, consequently, the extraction and characterization of some interesting structural features such as channels, cavities, topology and surface. Intuitively, this way of representation is based on growing the initial point set (at high resolution it is composed of atomic co-ordinates) according to α parameter until a definite value (given by the maximum resemblance) at which these atoms will be connected forming edges, triangles or tetrahedra (Fig. 23.3).

In order to extract and study structural features from 3D-EM volumes, it is necessary to obtain a first discrete point set. For this purpose a new quantita-

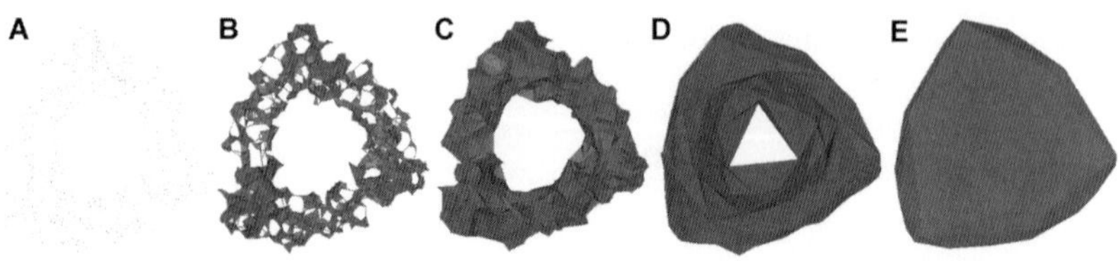

Fig. 23.3. By means of Alvis 4.1 3D Alpha-Shape Visualizer, successive alpha-shape representations for sliding clamp gp45 from bacteriophage RB69 are obtained with the α parameter increasing from *left* to *right*. Note the channel crossing through the structure. **A** Representation of macromolecular atomic co-ordinates downloaded from PDB that constitute the vertices (0-simplices) of the alpha shape; **B** at higher α values (1.17), vertices join to form edges (or 1-simplices), triangles (or 2-simplices) and tetrahedra (or 3-simplices); **C** alpha-shape representation that best resembles the original object and corresponding to an α value of 2.32; **D** from this point the increase in the α value (7.94) means the loss of the original object shape (the channel disappears); **E** convex hull

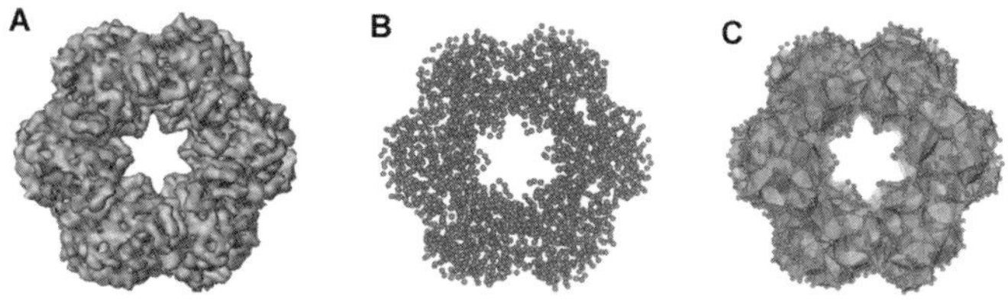

Fig. 23.4. This picture summarizes the steps taken towards obtaining the alpha shape with the glutamine synthase example. **A** 3D-EM map representation at 5-Å resolution; **B** output of kernel C-means application: 2500 pseudo-atoms are represented as *balls*. The pseudo-atoms connectivity provided by the application of alpha-shape theory results in the alpha shape shown in **C**. As can be seen, the similarity between the original 3D-EM volume and the final alpha representation is clear

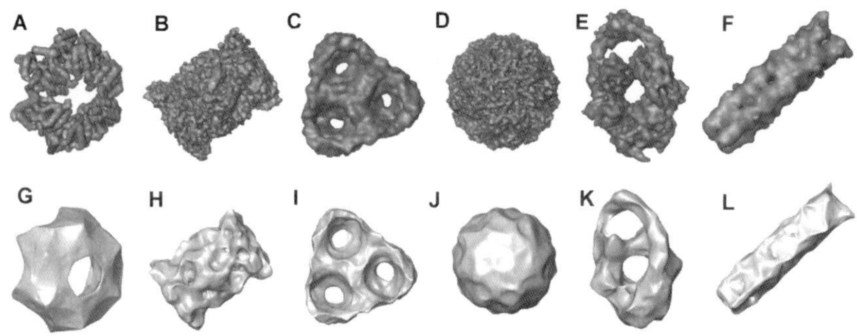

Fig. 23.5. 3D-EM representations (*first row*) and surface representations of alpha-shapes (*second row*) for recA (**A** and **G**); proteosome (**B** and **H**); porin (**C** and **I**); thermosome (**D** and **J**); mutS (**E** and **K**) and tetrabrachion (**F** and **L**). The aim of this picture is to show the preservation of the geometry and the topology of the macromolecular structures between 3D-EM data and alpha-shape representation

tive technique called kernel c-means (Pascual-Marqui et al. 2001), was employed. Kernel c-means provides the subset of 3D points that optimally approximates the probability density function (pdf) of the voxel values in a maximum likelihood sense. These points are called pseudo-atoms (Fig. 23.4). Further details on the exact pseudo-atoms extraction procedure can be found in de-Alarcón et al. (2002.)

The alpha shape is a compact representation of the 3D-EM maps with the specific property of preserving all the geometric and topologic information of the original 3D-EM volumes (Fig. 23.5). Thus, in a whole collection of 3D-EM data, any question to be asked about geometry and topology can be done to *light* alpha-shape representation instead of the *bulky* volume. Moreover, it automatically allows the segmentation and measurement of a number of interesting structural features such as channels and cavities, which can be used to solve interesting questions about sub-structures of the whole macromolecule.

One of the advantages of the representation employed, apart from the compression in terms of size, is that it can be applied to medium and high-resolution data indistinctly. As described above, a novel method for obtaining pseudo-atoms (initial point set) allows the building of alpha shapes from medium-resolution data while, in the case of high- resolution data, the initial point set is constituted by the atomic positions.

As already mentioned, the principles used to extract structural features depend on the nature of the structural feature. Surface and topology are inherent to the alpha shape, but cavities, channels and protrusions need a further step of analysis in order to be detected, isolated and quantified starting from the alpha-shape representation. For detection and isolation of external cavities and channels, we used discrete-flow methods (Liang et al. 1998). A model-based extension of the above-mentioned method, which incorporates a form of a priori knowledge (de-Alarcón PA, 2002), was used for the specific case of shallow features. Analysis of the complement space has been applied in the case of internal cavities (Liang et al. 1998).

Once cavities/channels have been detected and segmented, it is necessary to derive some metric properties such as volume and area. The analytical method called inclusion-exclusion (Edelsbrunner et al. 1995; Liang et al. 1998) has been applied to compute the volume and area of macromolecules, as well as of their cavities/channels from the alpha-shape definition.

Another interesting characteristic that can be extracted from the alpha-shape representation of channels and cavities is the skeleton. The skeleton is a global descriptor of the structure and it allows the obtaining of objective information about ramifications (number of branches, length of each branch).

23.4 FEMME Database: Feature Extraction in a Multi-Resolution Macromolecular Environment

Traditionally, a great effort has been concentrated on high-resolution data archiving. As a result, atomic resolution data have been organized in structural databases that are now accessible to the scientific community. One of the most well-known is the Protein Data Bank (PDB). The number of structures stored in PDB has increased exponentially over the years. As a consequence, several derived databases have been built from information stored in PDB. This is the case with structural domains classification databases such as CATH (Pearl et al. 2003) or SCOP (Murzin et al. 1995), the database of homology derived secondary structure of proteins (Sander and Schneider 1991), the protein quaternary structure database (Henrick and Thornton 1998), PDBsum database (Laskowski 2001), MSD ligand sevice (http://www.ebi.ac.uk/ msd/) or the nucleic acid database (Berman et al. 1992).

FEMME (Feature Extraction in a Multiresolution Macromolecular Environment: http://biocomp.cnb.uam.es/FEMME/) database is presented in this context. FEMME emerges from previous efforts toward the development of structural databases containing information at several resolution levels like the Bioimage (Carazo and Stelzer 1999) or the EMDB database (Tagari et al. 2002). Therefore, FEMME database collects information supplied by structures solved by any technique and it is boosted by data in the EMDB database (http://www.ebi.ac.uk/msd/MSDProjects/IIMShome.html). The overall goal of EMDB is to develop a system to integrate the results of three-dimensional electron microscopy with models from X-ray and NMR methods.

The present state of development of FEMME database in the analysis of 3D-EM experimental data follows the data population rate of the EMDB database (publicly available since February 2003). For this reason and, in order to demonstrate the FEMME contents and utilities, as well as to test the methodology with medium resolution volumes, data from the low pass filtering of atomic co-ordinates have been used. This 3D-EM-like data test was developed using the following steps: firstly, a bibliographic search was carried out to find a suitable set of macromolecular complexes with structural features susceptible to being detected by the methodology described (channels/cavities with a given size); secondly, 3D-EM maps simulations of each macromolecule at 5- and 15-Å resolution were generated using the ccp4 package (http://www.ccp4. ac.uk/); thirdly, the methods explained in the previous section were applied in order to build the alpha shape representation and to extract and characterize the structural features; and finally, all the extracted information was stored in a structural and ordered way by means of XML files. All those XML files coming from simulated 3D-EM maps constituted the first FEMME entries. Then FEMME was extended with data now available in the EMD database. Therefore, users can now find structural information from, not only simulated, but also real 3D-EM data in FEMME (Fig. 23.6). Future plans to increase the number of entries in FEMME include the automatic generation, processing and storage of alpha-shape representations from data in the still growing EMDB and from PDB/PQS databases (http://www.ebi.ac.uk/msd/ and http://pqs.ebi.ac.uk/).

The FEMME database has arisen as a way of storing alpha shapes of macromolecular complexes and macromolecular structural features, together with all the information that can be derived from this kind of representation. Thus, each database entry is an XML file in which a description of the macromolecular surface and topology, in terms of alpha shape representation, can be found. Moreover, a detailed description about the number and kind of structural features contained in the macromolecule, together with their characteristics is supplied (such as volume, skeleton, number of mouths) (Figs. 23.7 and 23.8). With the aim of putting all this in depth information in a functional

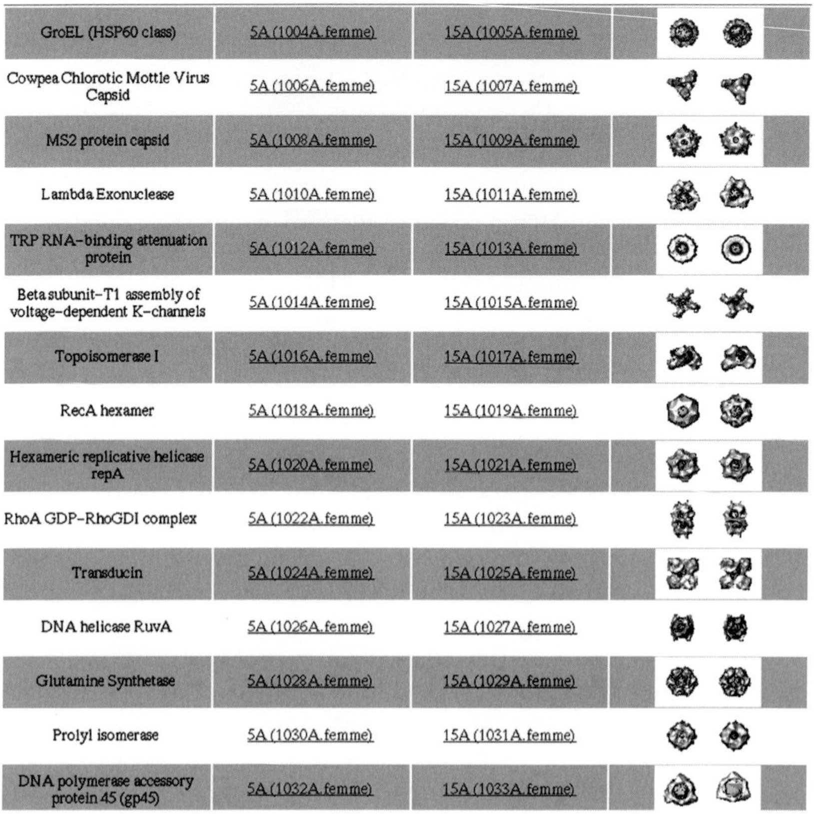

GroEL (HSP60 class)	5A (1004A.femme)	15A (1005A.femme)	
Cowpea Chlorotic Mottle Virus Capsid	5A (1006A.femme)	15A (1007A.femme)	
MS2 protein capsid	5A (1008A.femme)	15A (1009A.femme)	
Lambda Exonuclease	5A (1010A.femme)	15A (1011A.femme)	
TRP RNA-binding attenuation protein	5A (1012A.femme)	15A (1013A.femme)	
Beta subunit-T1 assembly of voltage-dependent K-channels	5A (1014A.femme)	15A (1015A.femme)	
Topoisomerase I	5A (1016A.femme)	15A (1017A.femme)	
RecA hexamer	5A (1018A.femme)	15A (1019A.femme)	
Hexameric replicative helicase repA	5A (1020A.femme)	15A (1021A.femme)	
RhoA GDP-RhoGDI complex	5A (1022A.femme)	15A (1023A.femme)	
Transducin	5A (1024A.femme)	15A (1025A.femme)	
DNA helicase RuvA	5A (1026A.femme)	15A (1027A.femme)	
Glutamine Synthetase	5A (1028A.femme)	15A (1029A.femme)	
Prolyl isomerase	5A (1030A.femme)	15A (1031A.femme)	
DNA polymerase accessory protein 45 (gp45)	5A (1032A.femme)	15A (1033A.femme)	

Fig. 23.6. An example of a browser page with a subset of FEMME entries. The *first column* refers to the names of the macromolecules. In the *second* and *third columns*, there are links to the entries at 5 and 15 Å, respectively. In the *fourth column*, images of the alpha-shape representation of macromolecules are presented in the two resolutions mentioned above

context, the entries are provided with an explanation of the biochemical and, in some cases, the biological role of the macromolecule with links to the best annotated protein function databases.

This database uses a methodology that can be applied to any kind of macromolecular structure representation like atomic co-ordinates or volumes. Thus, it is a powerful way to store the previously extracted and characterized structural features shown in high or medium resolution data. One of the strong points of the FEMME database lies in the possibility of making comparative studies among structural features from all macromolecules solved by any technique. Thus, the FEMME database opens a way for users not only to analyze and compare their data, but also to share this structural information with the scientific community.

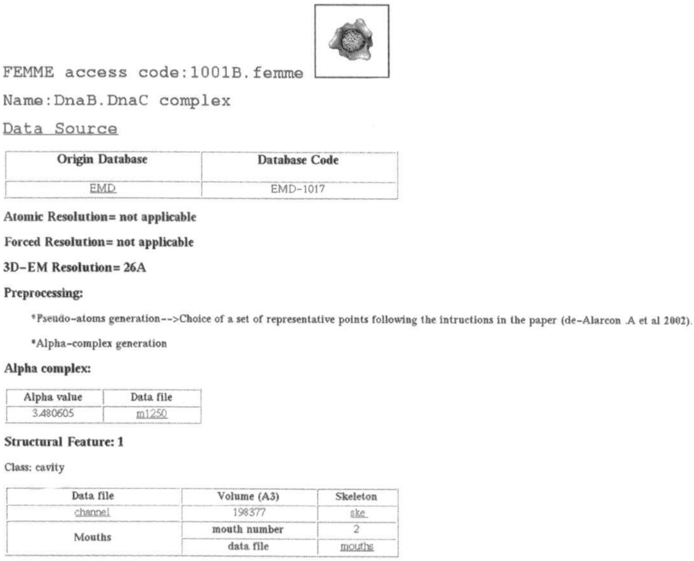

FEMME access code:1001B.femme

Name:DnaB.DnaC complex

Data Source

Origin Database	Database Code
EMD	EMD-1017

Atomic Resolution= not applicable

Forced Resolution= not applicable

3D-EM Resolution= 26A

Preprocessing:

*Pseudo-atoms generation-->Choice of a set of representative points following the intructions in the paper (de-Alarcon .A et al 2002).

*Alpha-complex generation

Alpha complex:

Alpha value	Data file
3.480605	m1250

Structural Feature: 1

Class: cavity

Data file	Volume (A3)	Skeleton
channel	198377	ske
Mouths	mouth number	2
	data file	mouths

Fig. 23.7. The result of a query made to the FEMME database is shown. The specific entry (1001B.femme) corresponds to the DnaB·C complex involved in DNA replication and solved by 3D-EM at 26-Å resolution (Bárcena et al. 2001). FEMME contains information about the generation of the alpha shape (preprocessing), the alpha shape itself and the structural features are identified. Files containing the previously described characteristics can be downloaded

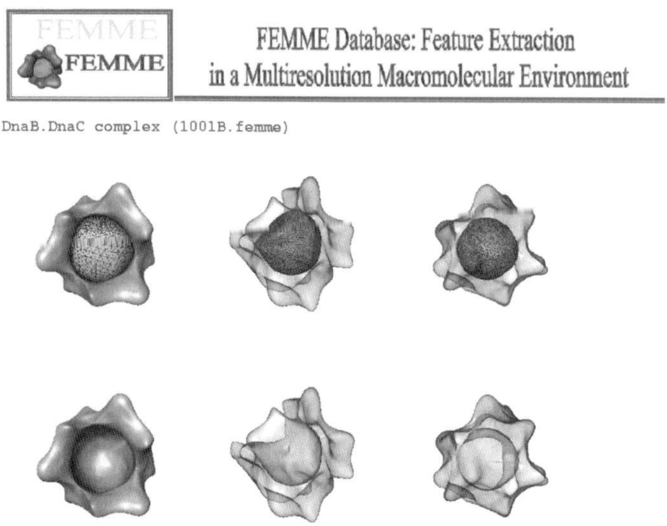

Fig. 23.8. Gallery of images of DnaB·C complex found in FEMME. The *first row* of images represents the alpha shape with the highlighted extracted channel (front, side and rear view). *Second row* corresponds to the alpha-shape representation of the DnaB·C complex. The alpha shape in both cases is *shaded* in the front view and is *transparent* in the other views in order to observe the cavity

23.5 One of the FEMME Utilities: Query by Content

The storage of alpha-shape representations of about 60 different molecular machines in the FEMME database facilitates the study of methods to retrieve objects by their content. One of these methods has been developed in our group and combines the spin-image representation and neural networks (de-Alarcón et al. 2002). Therefore, spin-image representation and neural networks have been tested as a way to find shape similarities with several subsets of the FEMME database. The final goal of this methodology is to be able to extract some functional information from a large set of related structures.

Considering the shape as a geometric concept derived from the object's surface, spin images are a rotational and translational invariant representation of a 3D object in 2D representative images that collect different views of the object surface. Then, a self-organized map (SOM) is built from the stack of spin images of a given object to "summarize" the original set of images into a set of representative spin images. In order to group representative views in the SOM map into a reduced set of clusters, a clustering algorithm (*kcmeans*) is applied. A final database, which contains the ten most representative spin-images for every object is created. At query time, spin images from the query object will be compared with the spin images of the database. The similarity between the query object and the target is measured as the percentage of query spin images that were closest to that target.

By means of the methodology explained above, data from the FEMME database have been compared at two different levels. Firstly, all channels and cavities existing in FEMME (a total of 26 channels and 38 cavities forming the channel database) and, secondly, the surfaces of all FEMME entries (macromolecular surface database), were compared. In the first case, some interesting similarities among channels coming from macromolecules that interact with DNA were observed. In all the searches using such channels as query objects, an average of six channels out of ten corresponds to other macromolecules that interact with DNA as well. More detailed similarities have also been detected. This is the case of PCNA and gp45, two proteins that interact with DNA and perform the same biochemical function: acting as sliding clamps during the DNA replication process. When we queried the database looking for the most similar structures to PCNA, the first match retrieved was gp45, higher than other DNA-interacting proteins acting in other processes (Fig. 23.9).

Another important similarity found was the one between the K^+-channel and the F1-ATPase channel. Both macromolecules are primarily active transporters belonging to the diphosphate bond hydrolysis-driven transporters subclass (Fig. 23.10).

In the second case, the surfaces of all FEMME macromolecules have been compared with very consistent results. When the database was queried with

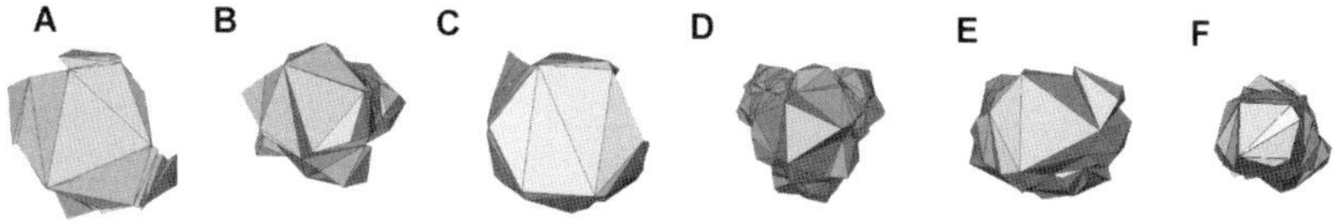

Fig. 23.9. Result of a query in the channel database: **A** gp45; **B** PCNA; **C** repA; **D** exonuclease; **E** topoisomerase and **F** tetrabrachion channel. The first channel corresponds to the query object (gp45 channel)

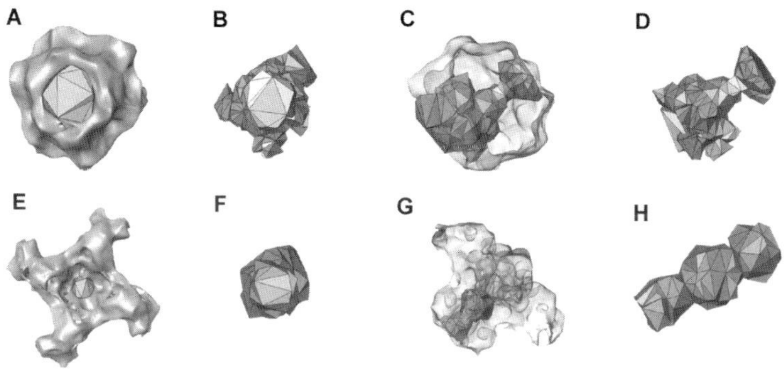

Fig. 23.10. **A** Front and **C** side views of surface alpha-shape representations for F1-ATPase; **B** front and **D** side views of the extracted channel; **E** front and **G** side views of surface alpha-shape representations for K^+-channel; **F** front and **H** side views of the extracted channel

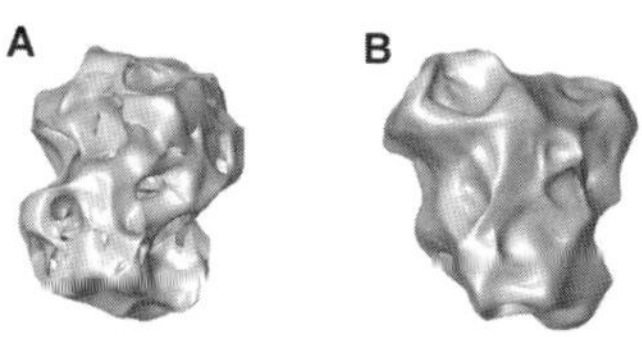

Fig. 23.11. Surface alpha-shape representations of **A** myoglobin; **B** hemoglobin

myoglobin, the hemoglobin monomer was obtained as the most similar structure (Fig. 23.11). These two proteins have the same biochemical function: oxygen binding. Hemoglobin transports oxygen and myoglobin stores it until it is used during metabolism. When using gp45 as the query object, the result was that the most similar structure to gp45 is PCNA, concluding that they are not only similar in their channels, but in their shape too (Fig. 23.12).

When the database was queried with MVM capsomer, four virus capsomers were found among the five most similar structures (Fig. 23.13).

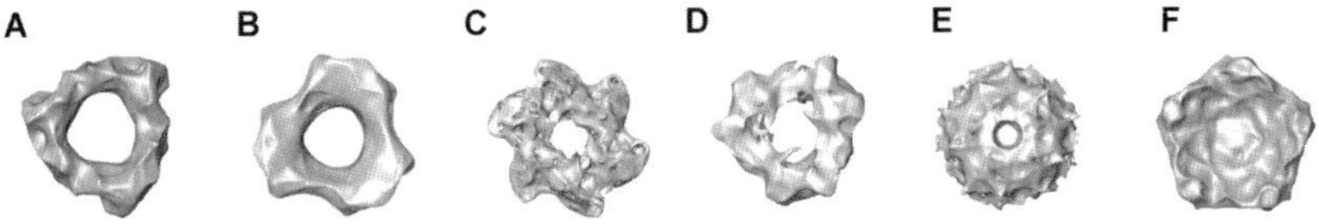

Fig. 23.12. Result of a search in the database using Gp45 as query object (**A**). From *left* to *right* the alpha-shape representations of the most similar structures to the query object are shown: **B** PCNA; **C** Gp4; **D** exonuclease; **E** MS2; **F** Southern bean mosaic virus

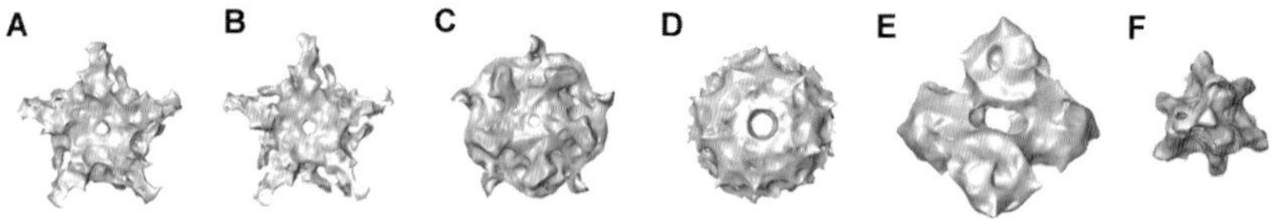

Fig. 23.13. Result of a search using the virus MVM capsomer as query object (**A**). From *left* to *right* the surface alpha-shape representation of the most similar structures to the query object: **B** Panleukopenia capsomer; **C** tomato bunshy stunt virus capsomer; **D** MS2 capsomer; **E** transducin; **F** P22 catalytic subunit. The first macromolecule corresponds to the query object

23.6 Conclusions

In this chapter a new and powerful structural database, FEMME database, has been presented. FEMME stores an important number of entries corresponding to alpha-shape representations of macromolecules and macromolecular structural features obtained from experimental data regardless of the resolution achieved. Each entry is a consistent and detailed description of the macromolecular structural features together with all the parameters chosen to extract them. The storage of this kind of information together with the use of an efficient comparison algorithm based on spin-image representation and neural networks, allows the detection of shape similarities among complexes and features. Thus, the preliminary results obtained applying this methodology with the current set of FEMME database have shown its power for finding structural similarities. As the number of deposited entries increases, the possibility to explore similarities among a larger set of structures in a multi-resolution context will hopefully result in novel functional insights.

Acknowledgements. N.J.L. is supported by a BEFI Grant 00/9395 (Instituto de Salud Carlos III, Ministerio de Sanidad y Consumo). This work was supported partially by the European Commission through contract QLRI-CT-2001-00015 for TEMBLOR under the specific RTD programme "Quality of Life and Management of Living Resources" and by CICYT (Comisión interministerial de ciencia y tecnología) through contract BIO2001-1237.

The EMDB database was developed during the IIMS project funded by the European Commission through contract QLRI-CT-2000-31237 under the RTD programme "Quality of Life and Management of Living Resources".

We would like to thank the developers of the CGAL software library (Computational Geometry Algorithms Library (www.cgal.org) as well as Ernst Mücke, Herbert Edelsbrunner and collaborators for their implementation of the alpha-shape algorithms.

References

Acehan D, Jiang X, Morgan DG, Heuser JE, Wang X, Akey CW (2002) Three-dimensional structure of the apoptosome: implications for assembly, procaspase-9 binding, and activation. Mol Cell 9:423–32

Agrawal RK, Linde J, Sengupta J, Nierhaus KH, Frank J (2001) Localization of L11 protein on the ribosome and elucidation of its involvement in EF-G-dependent translocation. J Mol Biol 311:777–87

Alberts B (1998) Three-dimensional fold of the human AQP1 water channel determined at 4 A resolution by electron crystallography of two-dimensional crystals embedded in ice. Cell 92:291–4

Auer M (2000) Three-dimensional electron cryo-microscopy as a powerful structural tool in molecular medicine. J Mol Med 78:191–202

Baker TS, Olson NH, Fuller SD (1999) Adding the third dimension to virus life cycles: three-dimensional reconstruction of icosahedral viruses from cryo-electron micrographs. Microbiol Mol Biol Rev 63:862–922

Bárcena M, Ruiz T, Donate LE, Brown SE, Dixon NE, Radermacher M, Carazo JM (2001) The DnaB.DnaC complex: a structure based on dimers assembled around an occluded channel. EMBO J 20:1462–8

Baumeister W (2002) Electron tomography: towards visualizing the molecular organization of the cytoplasm. Curr Opin Struct Biol 12:679–84

Berman HM, Olson WK, Beveridge DL, Westbrook J, Gelbin A, Demeny T, Hsieh SH, Srinivasan AR, Schneider B (1992) The Nucleic Acid Database: A Comprehensive Relational Database of Three-Dimensional Structures of Nucleic Acids. Biophys. J 63:751–759. http://ndbserver.rutgers.edu/

Bernal RA, Hafenstein S, Olson NH, Bowman VD, Chipman PR, Baker TS, Fane BA, Rossmann MG (2003) Structural studies of bacteriophage alpha3 assembly. J Mol Biol 325:11–24

Bottcher B, Bertsche I, Reuter R, Graber P (2000) Direct visualisation of conformational changes in EF(0)F(1) by electron microscopy. J Mol Biol 296:449–57

Carazo JM, Stelzer EH (1999) The BioImage Database Project: organizing multidimensional biological images in an object-relational database. J Struct Biol 125:97–102

Carrascosa JL, Llorca O, Valpuesta JM (2001) Structural comparison of prokaryotic and eukaryotic chaperonins. Micron 32:43–50

de-Alarcón PA, Pascual-Montano A, Carazo JM (2002) Spin images and neural networks for efficient content-based retrieval in 3D object databases. Lecture notes on computer science 2383:225–234

de-Alarcón PA, Pascual-Montano A, Gupta A, Carazo JM (2002) Modeling shape and topology of low-resolution density maps of biological macromolecules. Biophysical Journal 83:619–632

Edelsbrunner H, Liang J, Woodward C (1998) Anatomy of protein pockets and cavities: measurement of binding site geometry and implications for ligand design. Protein Sci 7:1884–1897

Edelsbrunner H, Mücke EP (1994) Three-dimensional alpha shapes. ACM Tr. On Graphics 13:43–72

Eidhammer I, Jonassen I, Taylor W (2000) Structure Comparison and Structure Patterns. Journal of Computational Biology 7:685–716

Ferlenghi I, Gowen B, de Haas F, Mancini EJ, Garoff H, Sjoberg M, Fuller SD (1998) The first step: activation of the Semliki Forest virus spike protein precursor causes a localized conformational change in the trimeric spike. J Mol Biol 283:71–81

Hatch V, Zhi G, Smith L, Stull JT, Craig R, Lehman W (2001) Myosin light chain kinase binding to a unique site on F-actin revealed by three-dimensional image reconstruction. J Cell Biol 154:611–7

Hawkes P.W and Kasper E (1996) Principles of electron optics. Academic Press, vol 3, London

Henrick and Thornton (1998) PQS: a protein quaternary structure file server. Trends Biochem Sci 23:358–61. PQS:http://pqs.ebi.ac.uk/

Hubbard SJ, Gross KH, Argos P (1994) Intramolecular cavities in globular proteins. Protein Engineering 7:613–626

Hubbard SJ, Argos P (1995) Detection of internal cavities in globular proteins. Protein Eng 8:1011–5

Jones S, Thornton JM (1996) Principles of protein-protein interaction. Proc. Natl. Acad. Sci. USA 93:13–20

Jones S, Thornton JM (1997) Analysis of protein-protein interaction sites using surface patches. J Mol Biol 272(1):121–32

Rawat UB, Zavialov AV, Sengupta J, Valle M, Grassuci RA, Linde J, Vestergaard B, Ehrenberg M, Frank J (2003). A cryo-electron microscopic study of ribosome-bound termination factor RF2. Nature 421:87–90

Kozubek M, Skalnikova M, Matula P, Bartova E, Rauch J, Neuhaus F, Eipel H, Hausmann M (2002) Automated microaxial tomography of cell nuclei after specific labelling by fluorescence in situ hybridization. Micron 33:655–65

Larsen TA, Olson AJ, Goodsell DS (1998) Morphology of protein-protein interfaces. Structure 6:421–7

Laskowski R A (2001) PDBsum: summaries and analyses of PDB structures. Nucleic Acids Res 29:221–222. http://www.biochem.ucl.ac.uk/bsm/pdbsum/

Li H, DeRosier DJ, Nicholson WV, Nogales E, Downing KH (2002) Microtubule structure at 8A resolution. Structure (Camb) 10:1317–28

Liang J, Edelsbrunner H, Fu P, Sudharkar PV, Subramaniam S (1998) Analytic shape computation of macromolecules I: molecular area and volume through alpha shape. Proteins: Structure, Function, and Genetics 33:1–17

Liang J, Edelsbrunner H, Fu P, Sudhakar PV, Subramanian S (1998a) Analytical Shape Computation of Macromolecules: II. Inaccessible cavities in proteins. Proteins 33:18–29

Liang J, Edelsbrunner H, Woodward C (1998b) Anatomy of protein pockets and cavities: Measurement of binding site geometry and implications for ligand design. Protein Sci 7:1884–97

Llorca O, Martin-Benito J, Gómez-Puertas P, Ritco-Vonsovici M, Willison KR, Carrascosa JL, Valpuesta JM (2001) Analysis of the interaction between the eukaryotic chaperonin CCT and its substrates actin and tubulin. J Struct Biol 135:205–18

Llorca O, Martin-Benito J, Gómez-Puertas P, Ritco-Vonsovici M, Willison KR, Carrascosa JL, Valpuesta JM (2001) Analysis of the interaction between the eukaryotic chaperonin CCT and its substrates actin and tubulin. J Struct Biol 135:205–18

Lo Conte L, Brenner SE, Hubbard TJ, Chothia C, Murzin AG (2002) SCOP database in 2002: refinements accommodate structural genomics. Nucleic Acids Res 30:264–7

Lowe J, Li H, Downing KH, Nogales E (2001) Refined structure of alpha beta-tubulin at 3.5 Å resolution. J Mol Biol 313:1045–57

McEwen BF, Marko M (2001) The emergence of electron tomography as an important tool for investigating cellular ultrastructure. J Histochem Cytochem 49:553–64

Murzin AG, Brenner SE, Hubbard T, Chothia C (1995) SCOP: a structural classification of proteins database for the investigation of sequences and structures. J. Mol. Biol. 247:536–540. http://scop.mrc-lmb.cam.ac.uk/scop

Ochoa WF, Kalko SG, Mateu MG, Gomes P, Andreu D, Domingo E, Fita I, Verdaguer N (2000) A multiply substituted G-H loop from foot-and-mouth disease virus in complex with a neutralizing antibody: a role for water molecules. J Gen Virol 81:1495–505

Orlova EV, Papakosta M, Booy FP, van Heel M, Dolly JO (2003) Voltage-gated K(+) Channel from Mammalian Brain: 3D Structure at 18Å of the Complete (alpha)(4)(beta)(4) Complex. J Mol Biol 326:1005–12

Pascual-Montano A, Donate LE, Valle M, Bárcena M, Pascual-Marqui RD, Carazo JM (2001) A novel neural network technique for analysis and classification of EM single-particle images. J Struct Biol 133:233–45

Pearl FM, Bennett CF, Bray JE, Harrison AP, Martin N, Shepherd A, Sillitoe I, Thornton J, Orengo CA (2003) The CATH database: an extended protein family resource for structural and functional genomics. Nucleic Acids Res 31:452–5

Ren G, Cheng A, Reddy V, Melnyk P, Mitra AK (2000) Three-dimensional fold of the human AQP1 water channel determined at 4 A resolution by electron crystallography of two-dimensional crystals embedded in ice. J Mol Biol 301:369–87

Sander C and Schneider R (1991) Database of homology derived protein structures and the structural meaning of sequence alignment. Proteins 9:56–68. http://www.cmbi.kun.nl/gv/hssp/

Spahn CM, Beckmann R, Eswar N, Penczek PA, Sali A, Blobel G, Frank (2001) Structure of the 80S ribosome from Saccharomyces cerevisiae-tRNA-ribosome and subunit-subunit interactions. J Cell 107:373–86

Stark H (2002) Three-dimensional electron cryomicroscopy of ribosomes. Curr Protein Pept Sci 3:79–91

Stark H, Dube P, Luhrmann R, Kastner B (2001) Arrangement of RNA and proteins in the spliceosomal U1 small nuclear ribonucleoprotein particle. Nature 409:539–42

Subramaniam S, Henderson R (1999) Electron crystallography of bacteriorhodopsin with millisecond time resolution. J Struct Biol 128:19–25

Tagari M, Newman R, Chagoyen M, Carazo JM, Henrick K (2002) New electron microscopy database and deposition system. Trends Biochem Sci 27:589

Thouvenin E, Hewat E (2000) When two into one won't go: fitting in the presence of steric hindrance and partial occupancy. Acta Crystallogr D Biol Crystallogr 56:1350– 7

Wang DN, Kuhlbrandt W (1991) High-resolution electron crystallography of light-harvesting chlorophyll a/b-protein complex in three different media. J Mol Biol 217: 691–699

Zhang W, Mukhopadhyay S, Pletnev SV, Baker TS, Kuhn RJ, Rossmann MG (2002) Placement of the structural proteins in Sindbis virus. J Virol 76:11645–58

24 Peak Erazor: A Windows-Based Program for Improving Peptide Mass Searches

KARIN HJERNØ and PETER HØJRUP

24.1 Introduction

Since its introduction in 1993, peptide mass fingerprinting (PMF) based on matrix-assisted laser desorption/ionisation mass spectrometry (MALDI-MS) has become a standard method in proteomics for the identification of proteins (Mann, 1993; Yates, 1993; Henzel, 1993; James, 1993; Pappin, 1993). Although a number of parameters like protein size and pI are also used by PMF search programs, the main parameter for obtaining a positive identification is to locate a sufficient number of peptides that match the theoretical peptides in a given protein database.

The positive identification of a protein is strongly dependent on the number of input mass values, the size of the database searched and the precision with which the mass values have been obtained. Some of these parameters can be optimised quite easily, i.e. by using a smaller database, non-redundant or species-specific, or removing known contaminant peaks prior to database searching. The most important parameter, the precision of the input mass values (Fig. 24.1), can be improved by a careful calibration of the mass spectra using either external or internal calibration. The internal calibration is usually performed using a two-point calibration on tryptic autolytic peptides. These peaks can be introduced into the spectrum with careful adjustment of the enzyme/substrate ratio when performing the in-gel digest.

Here, we have put forward a report on a program utility for the pre-processing of peptide mass lists destined for peptide mass fingerprinting. The program performs a multipoint internal calibration of the spectrum using all known contaminants in the spectrum. In addition, the program will eliminate mass values related to known contaminants, thus reducing the number of false positives. Furthermore, it is able to 'learn' the system-' contaminants. The program runs under all 32-bit Windows versions, and can interface to most acquisition and analysis systems. It is freely available to download from http://welcome.to/gpmaw/.

Principles and Practice
Methods in Proteome and Protein Analysis
R.M. Kamp, J. J. Calvete, T. Choli-Papadopoulou (Eds.)

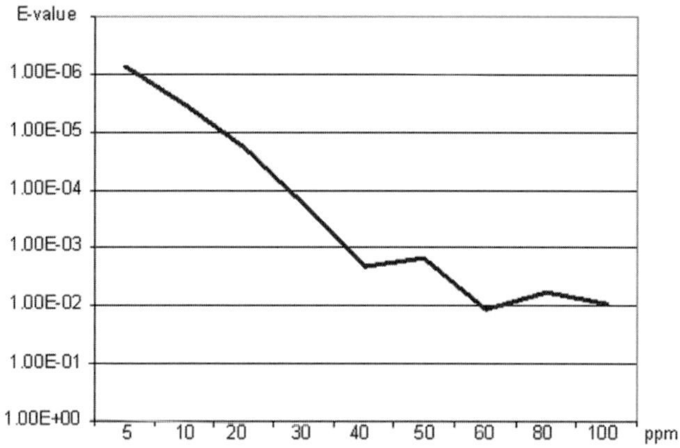

Fig. 24.1. Expectation value as a function of search accuracy for the ProFound peptide mass fingerprinting program (http://www.genomicsolutions.com). The database was searched with eight synthetic peptide masses from human calnexin mixed with eight 'contaminant' masses from keratins and porcine trypsin. For each measuring point random variations within the given accuracy was generated and ten independent mass searches performed and the results averaged. The whole NCBI nr database (2002/07/17) was searched and default search parameters chosen

24.2 Program Layout

The program features a minimalist layout with tabbed pages in the left-hand part and various controls in the right-hand panel (Fig. 24.2). As it is intended for use with other programs, it has been designed to use the smallest amount of screen area possible. Most secondary commands are available in the pop-up menu (right-hand mouse click).

24.2.1 Erazor List

The basis for the program is a list of typical contaminant mass values, called the *Erazor list*, found in tryptic in-gel peptide digests. Selecting the Erazor list tab in the top of the program window gives access to this list. Values can be added to or deleted from the list by using the buttons at the bottom of the list, the pop-up menu or edited by a double-click. Different lists can be used for different projects/instruments/enzymes. When analysing a mass list, the program can enable/disable up to six groups of contaminants. These groups are selected based on the first six characters of the contaminant name. By only using the first six characters, you can differentiate different contaminants in the last part of the name while still enable/disable the group as a single unit (e.g. different keratins like keratin 1 and keratin 10). The contaminant name

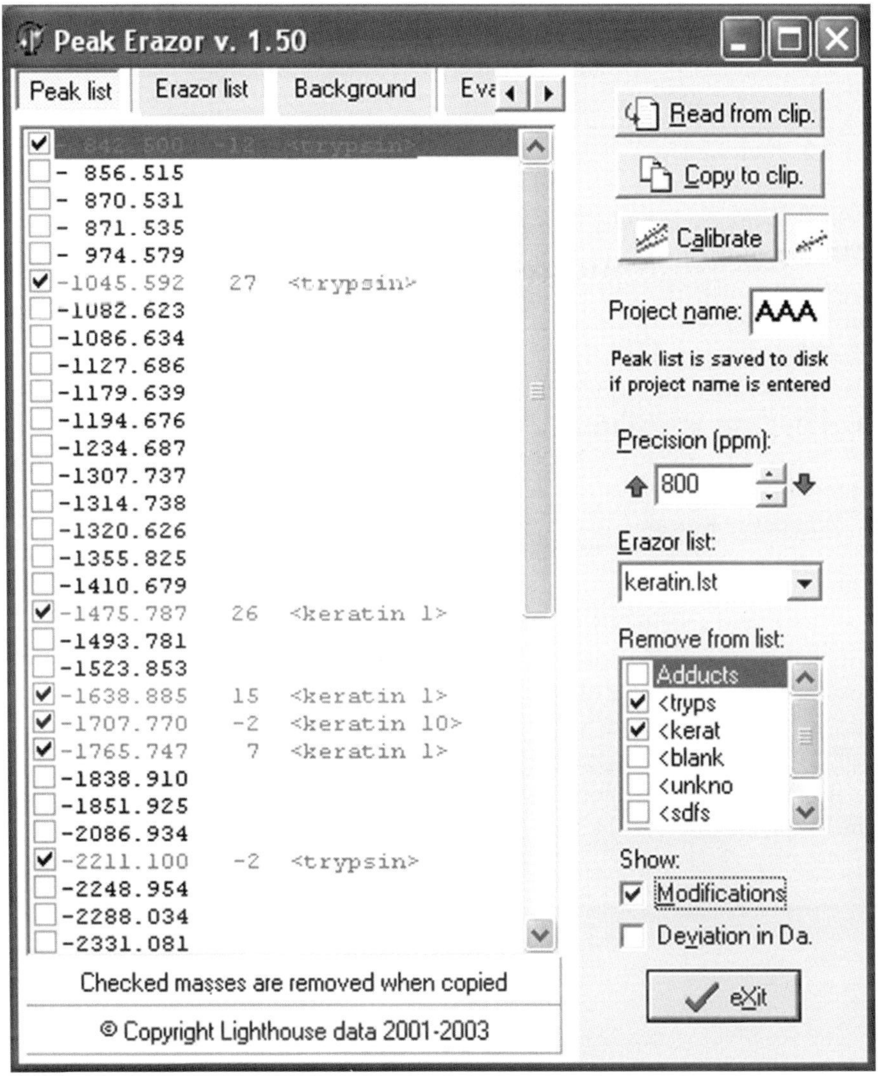

Fig. 24.2. A peak list was pasted into the PeakErazor. The columns show from *left* to *right*: check box, mass value, deviation from known mass value in the Erazor list (in ppm), item found in the Erazor list. The *right-hand panel* shows controls for copying a mass list from the clipboard, pasting the edited list back onto the clipboard, switch to calibration mode, toggle the precision graph on and off (see Fig. 24.2), set the precision for matching against the Erazor list, select an Erazor list, select contaminants to remove, toggle modifications and display of mass differences in daltons. Items in the mass list that are checked will be removed from the list when copied onto the clipboard. Mass values that match values from the Erazor list within the given precision are automatically checked, displayed in *red* and the matching mass value name is displayed

<blank>is reserved for background subtraction for a limited number of mass lists or for a given mass list (e.g. a protein digest), see Section 24.2.3.

24.2.2 Peak List

Peptide peak lists from most mass spectrometry acquisition programs can be pasted into the program by selecting the 'Read from clip' button. The mass values are automatically checked against the currently loaded peak list, and any matches within the given precision are marked (a 'v' in the left-hand check box), colored red, the deviation is listed in ppm (and Dalton if selected), and the name of the contaminant (Fig. 24.2). Adducts and posttranslational modifications, when they can be viewed in the mass spectrum as mass differences, can be checked in the mass list when the appropriate mass values are entered in the initialisation file and checked on the form.

Whenever a peak list is copied to the clipboard (i.e. after calibration), a copy of the mass values is saved to a file called *allmass.mas*. This results in an accumulation of all analysed spectra, from which can then be extracted new contaminants specific for a given system, see Section 24.2.4. If a three-letter project name is not present, the mass values will not be saved nor will the mass list be saved if it is identical to the previously saved list (i.e. the mass values will not accumulate in the disk file if the copy button is pressed repeatedly).

A graph of the precision relative to the mass can be displayed by depressing the graph button (Fig. 24.3).

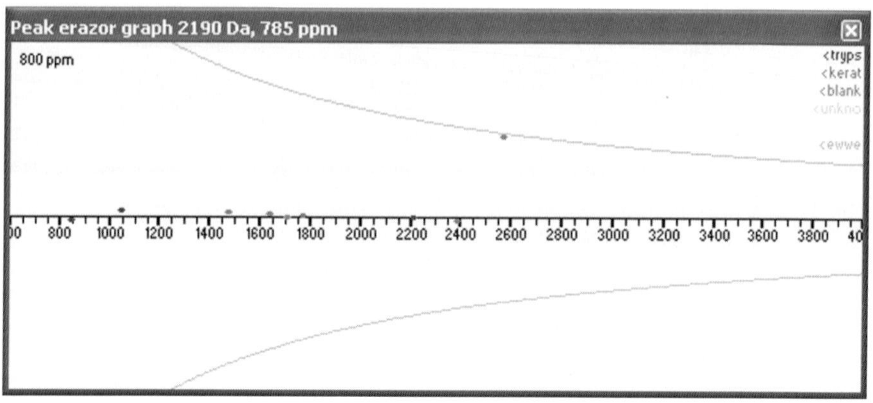

Fig. 24.3. The graph shows the precision of the identified and checked mass values in the peak list. The *vertical scale* shows the deviation in ppm and the *horizontal axis* the m/z value. The vertical scale is controlled from the main application window. The different groups of contaminants are shown in different colors (legend in the *top right corner*). The top and bottom curved *gray lines* indicates ±1 Da. One *dot* in the figure (m/z 2566) shows a +1 Da deviation, indicative of a wrong assignment of the monoisotopic peak

24.2.3 Background

The 'Background' page enables you to extract common mass values from a small series of peak lists.

24.2.4 Evaluate

As the program is being used, the file of combined analysed mass values will increase. After 100–150 mass spectra, it is possible to identify new contaminants to include in the Erazor list, as well as to identify components in the Erazor list that are seldom encountered and may thus be erased from the contaminant list.

24.3 Calibrating for Peptide Mass Fingerprinting

In order to make a successful peptide mass fingerprint you need as many mass values as possible, they have to be as accurate and with as few contaminating mass values as possible. PeakErazor cannot increase the number of mass values, however, the quality of the data can be increased considerably by re-calibration and subtraction of known contaminants.

The operation is carried out in a few steps:

1. Paste the peak list into the table. The list should not be calibrated, as some mass spectrum evaluation systems introduce a non-linear calibration in an effort to increase precision. In our experience these data can be difficult to recalibrate. Make sure that the graph is displayed, the appropriate contaminants are turned on (e.g. trypsin, keratin peptides etc.) and that the search precision is set at a high value (e.g. 800 ppm)
2. Note the presence of a calibration line made up of contaminant peaks (Fig. 24.4A). By adjusting the search precision and/or turning off points by un-checking the corresponding check boxes you can narrow the points on the calibration line.
3. Press the 'Calibrate' button to switch to calibration mode. You may fine-tune the calibration and preview the calibration line by pressing the 'view calib' button. Perform the linear calibration by pressing the 'calibrate' button (Fig. 24.4B). As new data points may move into and out of the displayed precision zone, you can redo the calibration several times.
4. Switch back to normal viewing mode ('calib done' button) and copy the calibrated mass list back onto the clipboard, ready for pasting into a peptide mass fingerprinting program (e.g. Mascot, ProFound etc.).

When performing the calibration you should be careful about which data points are used for the actual calibration. The tryptic autolytic peptides and

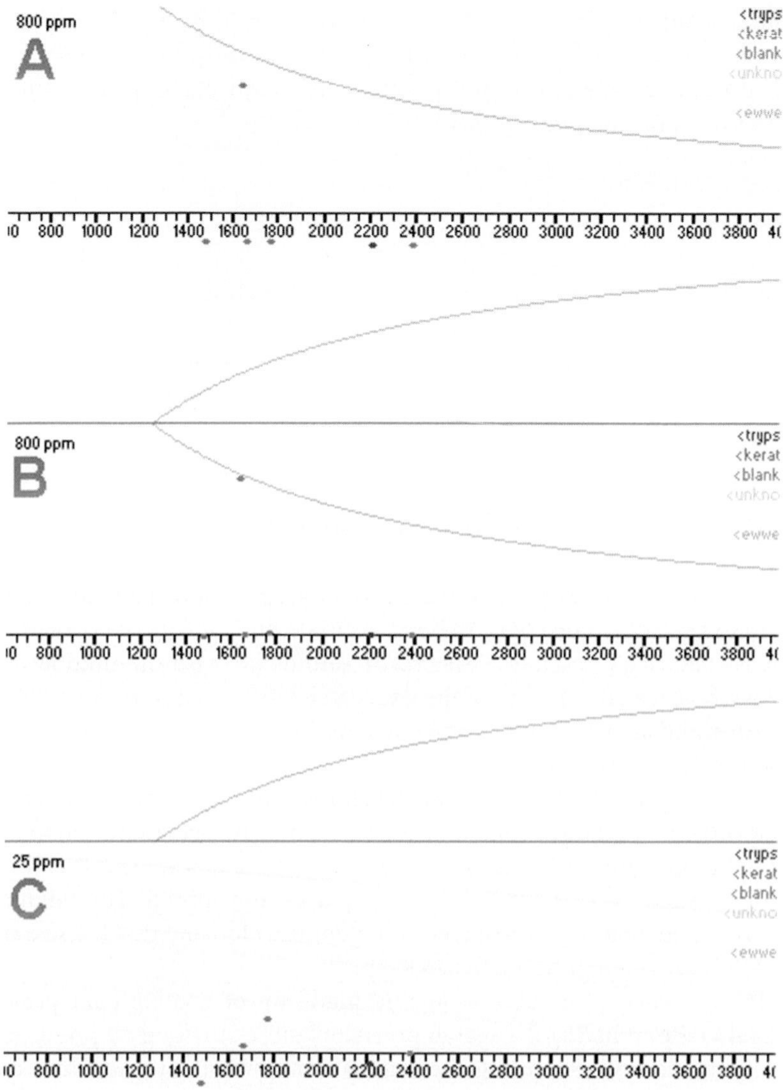

Fig. 24.4. Three stages in the calibration of a peak list as seen in the deviation graph. **A** The uncalibrated mass list viewed in a range of 800 ppm. Deviations are in the order of −120 ppm, and the calibration should correct an off-set and slight slope. **B** After calibration the masses can be seen to be lying closely around the horizontal axis, with only a single point lying on the +1 mass unit line, indicating an incorrectly assigned mass (wrong isotope). **C** Changing the vertical scale to 25 ppm shows the points used for calibration (four keratin and one tryptic peptide) to deviate with less than 5 ppm

the keratin related peptides are known with an exact mass while 'unknowns' are only determined with an approximate mass (that may improve over time, see below). The unknowns are thus well suited for identifying the 'calibration line' and need to be subtracted before the mass list is used for PMS, but the unknowns should be turned off when the actual calibration is performed.

When a sufficient number of contaminants are identified in a mass list, the precision of the identified contaminants will also give an accurate picture of the actual precision of the mass list. Only if the composition of the different components are widely different or analysing different classes of components (e.g. carbohydrates) is it possible that the observed spread in precision should be different in the analyte.

24.4 Mapping Peptide Masses in Known Proteins

In many circumstances it is of great importance to have as high sequence coverage as possible. This can be the case when you want to verify the structure of your protein, or when you are searching for posttranslational modifications, either chemical changes or truncation of terminals.

Peptides that match the cleavage specification of the chosen enzyme are rarely difficult to identify, however, peptides that arise from unusual cleavages or posttranslational modifications can be problematic. In these cases mass precision is essential for a positive identification.

The PeakErazor program can improve on the assignment of mass values to given peptides by first calibrating on the known mass values in order to identify modifications and unusual cleavages (see the following steps):

1. Generate a mass list of peptides from the protein in question. Several programs are available for this purpose, e.g. GPMAW (welcome.to/gpmaw), PeptideMass (www.expasy.org) or PAWS (prowl.rockefeller.edu/software/paws.htm).
2. Paste the list into the 'Peak list' table of PeakErazor.
3. Right-click and select, 'Copy list to <blank>',this will put the entire mass list into the current Erazor list, each mass with the name of <blank>.
4. Generate the mass list to analyse from your favourite mass spectrometric software and paste it into the Peak list table replacing the existing mass list.
5. Make certain that the <blank> subtraction option is turned on, this will highlight all matches to the protein under investigation. Other subtraction options may be turned on to check for contaminant peptides.
6. On the graph a clear calibration line should be visible, the different colouring distinguishes between contaminants and target peptides. Switch to the calibration page and perform the linear calibration.
7. Make certain to turn off the <blank> subtraction before copying the mass list back onto the clipboard. From here you can copy the mass list into a

search program (e.g. GPMAW) to check for unusual cleavages and/or secondary modifications.

In addition to making a linear calibration, the graph gives, even with only a few matches, a clear indication of whether the linear calibration is optimal or a polynomial calibration is optimal. In any case you can determine from the calibrated mass list what the actual deviation is in your mass spectrum. If you need a polynomial calibration, you get an approximation of the deviation in the different mass regions where you have determined peptide hits.

24.5 Identifying Background Peaks

Contaminant peaks may arise from a number of different sources. Some of these sources are common proteins that accidentally enter the sample early enough in the analysis process to be included in the enzymatic digest and thus give rise to contaminant peptides. A typical example is keratin, which can either be included in the gel matrix when casting the gel or can contaminate the surface of the gel when it is handled prior to enzymatic digestion. After digestion protein contamination is usually not a problem, as the contaminants are too large to be recorded in the peptide mass region (500–4000 Da). Another source of contaminants are the proteins added to the sample, e.g. the enzyme trypsin which gives rise to auto digest products. If carefully adjusted, the number of tryptic auto digest products can be minimized and the autolytic peptides can then be used for internal calibration. For porcine trypsin these peaks typically comprise m/z 842.51 and 2211.1.

However, you may also encounter contaminants that are specific to a limited number of experiments (e.g. contamination by a given protein). In order to remove these peaks, PeakErazor have included a table on the Background page, into which up to 10 mass lists can be pasted. From these lists common peaks can then be extracted, based on a given precision and a minimum number of matches. The extracted list of peptides can then be transferred to the current Erazor list under the name <blank>replacing any current instances of this name.

24.6 Evaluation: Extracting Information on Common Contaminants

A feature of the PeakErazor program is that every time a peak list is copied to the clipboard, a copy of the list with information on accepted and rejected mass values is copied to a local file on disk. This makes it possible to evaluate the combined collection of mass values collected when the program has treated a number of peak lists

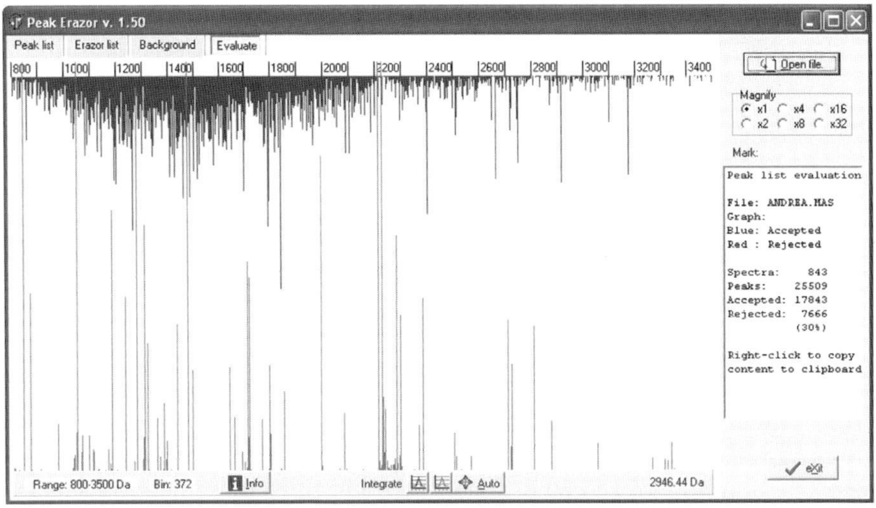

Fig. 24.5. Evaluation of a combined mass file. The m/z scale is shown along the *top* with accepted mass values growing down and rejected mass values growing up. The rejected mass values are typically identified by comparison with the Erazor list. Zooming in on the horizontal scale is done by left-click to the left and right of the zoomed area. The vertical scale is zoomed in on by selecting magnification in the right-hand control box. Statistical and integration information are displayed in the right-hand list box. Integration of the various peaks can be done through the buttons in the central bottom panel

To perform the evaluation select the last tab, Evaluation, upon which the window expands to its wide format (Fig. 24.5). By default the **program** loads the local file and displays it in graphics format with bars for every 0.01 mass unit. Mass values that have been accepted for PMS are shown as blue lines growing down, while rejected mass values are shown as red lines growing up from the bottom of the display. Statistics are shown in the right-hand list box.

The graph can be zoomed in on by clicking on the left and right of the area, and it can be expanded in six steps by selecting the 'magnify' buttons to the right. When you have zoomed in on a 'peak', it can be integrated by selecting the appropriate blue or red 'integrate' button at the bottom centre of the display, followed by defining the integration range by left- and right-click on either side. The result of the integration is displayed in the right-hand list box with the mass range analysed, the average mass and the mass for each point of the integration.

When a sufficient number of mass spectra have been analysed, you can optimise the peak list used for peak extraction by doing the following:

1. Remove mass values that are seldom encountered (i.e. show a low read peak in the evaluation graph;

2. Increase the accuracy of unknown contaminant mass values by integrating the peak. As the number of analysed samples grow, the precision in the determination of the mass of unknowns increase – remember that mass values are saved to disk when they are copied back to the clipboard, i.e. after calibration.
3. Add additional contaminant. When looking at the distribution of accepted peaks (top part of Fig. 24.5) a general 'background' of accepted peaks can be seen with a number of peaks growing out. The most frequently occurring peaks are likely to be contaminants.

Selecting the 'auto' button at the bottom of the display will open a dialog box with an integration of all accepted (blue) peaks in the evaluation spectrum. The right-hand panel contains a few controls for adjusting the integration parameters, the most important of which is the *Min. accepted area*. By selecting a minimum accepted value of 10 % of the total number of spectra analysed, you define that if a peak occurs in more then 1 out of 10 spectra it must be a contamination.

From the pop-up menu (left-mouse click) you can save the resulting list either to clipboard or to a file on disk. As the list is also compared to the Erazor list, it is easy to add new mass values to the existing list or to create a new list.

24.7 Discussion

Optimisation of the peak list prior to peptide mass fingerprinting will in most cases substantially improve protein identification by MALDI-TOF mass spectrometry. The scoring value when using the ProFound search engine shows a general increase in scoring value when increasing the search accuracy from 100 ppm to 40 ppm (Fig. 24.1). Below 40 ppm there is a sharp increase in the scores, most likely resulting from the fact that peptide masses are not uniformly distributed but cluster into narrow mass regions (Wool and Smilansky, 2002). Figure 24.1 shows that an improvement in accuracy from 80 to 50 ppm does not yield a much improved search score, however, an improvement from 40 to 20 ppm results in more than an order of magnitude higher score.

The approach described here works in two ways: one is to use identified contaminants already present in the sample to perform a linear calibration, the other is to remove these contaminants from the mass list, thus giving a smaller dataset for searching which results in better statistics (i.e. a higher score). The calibration is similar to the peptide mass fingerprinting program, MSA, put forward by Egelhofer et al. (2000), but with the important difference that the present program uses contaminants for calibration followed by the removal of the same contaminants before mass searching. The MSA search program works on the assumption that linear calibration yields a much higher improvement on a correct hit than on a random hit. The current approach does not work on a potential hit, but on the contaminants present in the sample and is thus independent of any results. Furthermore, the search is improved as random hits on contaminant peaks are reduced.

Many of the elegant solutions for improving peptide mass searches presented over time, suffer from the fact that only algorithms are put forward while the implementation is left to the reader, who seldom has access to the programming expertise necessary for a practical implementation. Alternatively, the routings may be implemented as programs on the web, with limited capability for integration into existing routines (e.g. Egelhofer, 2000; Gattiker, 2002).

The present program is freely available and can be applied to most mass spectrometric systems and PMS search programs that are not directly linked. A disadvantage is that the program demands human intervention to evaluate which data points to include in the calibration. It will be quite straightforward to include the present approach to an automatic data analysis system, however, in this case the implementation will be specific for each system.

Another feature of the program is the ability to 'improve' over time as all analysed data are saved, which allows for later reanalysis and extraction of system specific contaminants. These 'unknown' contaminants can then be further improved over time as the precision will improve as more samples are being analysed.

A couple of features are missing from the program but are likely to be included in the near future: The possibility to perform a higher order calibration, automated optimised calibration (to enable high throughput), divide the saved spectra according to project name and/or time in order to monitor time dependency of the mass spectrometric data acquisition, and a direct interface to web-based peptide mass fingerprinting search engines. Another aspect is to include the intensity of each peak. As shown by S. Gay (Gay, 2002) although it is not possible to perform an exact match it is possible to improve peptide mass fingerprinting by the inclusion of intensity prediction.

References

Egelhofer V, Bussow K, Luebbert C, Lehrach H, Nordhoff E (2000) Improvements in protein identification by MALDI-TOF-MS peptide mapping. Anal. Chem. 72:2741–2750

Gattiker A, Bienvenut WV, Bairoch A, Gasteiger E (2002) FindPept, a tool to identify unmatched masses in peptide mass fingerprinting protein identification. Proteomics 2:1435–1444.

Gay S, Binz P-A, Hochstrasser, DF, Appel, RD (2002) Peptide mass fingerprinting peak intensity prediction: Extracting knowledge from spectra. Proteomics 2:1374–1391

Henzel WJ, Billeci TM, Stults JT, Wong SC, Grimley C, Watanabe C (1993) Identifying proteins from two-dimensional gels by molecular mass searching of peptide fragments in protein sequence databases. Proc. Natl. Acac. Sci. USA 90:5011–5015

James P, Quadroni M, Carafoli E, Gonnet G (1993) Protein identification by mass profile fingerprinting. Biochem. Biophys. Res. Commun. 195:58–64
Mann M, Højrup P, Roepstorff P (1993) Use of mass spectrometric information to identify protei ns in sequence databases. Biomed. Environ. Mass Spectrom. 22:338–345
Pappin DJ, Højrup P, Bleasby AJ (1993) Rapid identification of proteins by peptide-mass fingerprinting. Curr. Biol. 3:327–332.
Yates JR, Speicher S, Griffin PR, Hunkapiller T, (1993) Peptide mass maps: a highly informative approach to protein identification. Anal. Biochem. 214:397–408
Wool A, Smilansky Z (2002) precalibration of matrix-assisted laser desorption/ionisation-time of flight spectra for peptide mass fingerprinting. Proteomics 2:1365–1373

25 Increasing Throughput and Data Quality for Proteomics

ALFRED L. GAERTNER, NICOLE L. CHOW, BETH G. FRYKSDALE, PAUL JEDRZEJEWSKI, BRIAN S. MILLER, SIGRID PAECH and DAVID L. WONG

Abstract

With the availability of microbial and mammalian genomes combined with dramatic improvements in bioanalytical methods, high-throughput analysis of transcriptomes and proteomes has become a reality for academic and industrial laboratories alike. New technologies have resulted in the discovery of a multitude of novel cellular pathways and interconnective regulatory mechanisms. For instance, the number of drug targets has grown from approximately 500 to the thousands in a short amount of time. While the information from post-genomics techniques is useful in its own right, it does not necessarily accelerate discovery, this is partly because of data management constraints. Bottlenecks also include quality of sample preparation, identification of low abundance compounds, uninterrupted unattended operations, and successful matches of the results with information available in databases.

In this study, we addressed these issues in order to develop robust methods for the discovery of compounds relevant to a product oriented biotechnology environment. Specifically, we developed pre-fractionation methods, deglycosylation protocols, non-radioactive isotopic labeling methods, and improvements in matrix assisted laser desorption (MALDI) matrix techniques to visualize and identify low abundance proteins from complex samples and applied them to a semi-automatic two-dimensional electrophoresis (2-DE) line followed by MALDI time of flight mass spectrometry (MALDI-TOF/MS). The complex exoproteome of a filamentous fungus, *Trichoderma reesei*, an important production organism for a number of biomass related applications, served as a model system to develop and fine tune most of the methods.

Principles and Practice
Methods in Proteome and Protein Analysis
R.M. Kamp, J. J. Calvete, T. Choli-Papadopoulou (Eds.)

25.1 Introduction

Two-dimensional electrophoresis (2-DE) combined with MALDI-TOF (matrix assisted laser desorption–time of flight) mass spectrometry (MS) are widely used tools in proteomic analyses (Quadroni and James 1999, Chapman 2000). With the availability of genomic databases, industrialization of research tools and the development of data analysis algorithms, the quantity of data generated has increased exponentially in the past few years. Unfortunately, data quality has not always kept pace. This is in part due to the fact that many current methodologies for protein identification have not been optimized for high-throughput proteome analysis.

We addressed these issues by developing robust, sensitive methods to enhance sample data quality and throughput. First, we used a membrane-based, preparative electrophoresis system that employs a combination of charge- and size-based separation strategies with the use of mild buffer conditions. This technique greatly enhanced the visualization and identification of unknown gel spots. Secondly, we developed effective deglycosylation methods for the heavily glycosylated fungal exoproteome. We chose to implement acid deglycosylation due to its more complete glycan removal and an enzymatic treatment that gave improved peptide database matching using MALDI-TOF mass spectrometry. Thirdly, we improved automated proteolytic digestion protocols. Concomitantly, we improved MALDI matrix protocols to allow for more automated operation of the mass spectrometer and increased sensitivity. Finally, we explored a non-radioactive internal isotopic labeling method, which allowed quantification of proteins without use of isotopic coded labels (ICAT). Combined, these techniques enabled identification of known and unknown proteins at a rapid pace, thus de-bottlenecking both, discovery and product development.

25.1.1 Prefractionation by Membrane Devices

Two-DE suffers from several limitations such as a bias towards visualizing abundant proteins in a complex proteome. Prefractionation of samples prior to 2-DE holds the potential to resolve many of the dynamic range issues of the technique, thus allowing detection of proteins at lower copy numbers. A strategy most applicable to 2-DE involves prefractionation of proteins according to isoelectric points (p*I*) of proteins. Prefractionation of samples by p*I* can allow the application to a narrow range immobilized pH gradient (nrIPG) strips of more protein within a strips pH range, and for the removal of high abundance proteins, thus increasing the concentration of less abundant proteins within a given p*I* range.

We investigated the enhancement of protein identification and visualization by prefractionation using an electrokinetic membrane-based technology

(Gradiflow BF400). The system uses thin polyacrylamide membranes and operates on different principles compared to the devices based on isoelectric membranes (Horvath et al. 1994, Rylatt et al. 1999). Membranes are either restrictive by pore size (size-based separations) or non-restrictive (charged-based separations). Charged molecules flowing in two co-current streams tangential and on opposite sides of the membrane, partition to either the anode or cathode stream. Protein charge is controlled by setting and monitoring the pH of the buffer streams to create the desired populations of positively and negatively charged proteins. An advantage of the system is its use of mild buffer conditions that can be used to maintain protein function.

25.1.2 Fractionation of a Fungal Exoproteome

We used the Gradiflow technology to prefractionate by charge an enzyme preparation secreted by the filamentous fungus *Trichoderma reesei* (*T. reesei* exoproteome) prior to separation by 2-DE. Following 2-DE, we harvested a subgroup of spots for in-gel trypsin digestion followed by analysis using MALDI-TOF/MS for identification of proteins by peptide mass matching. The results showed that the gels of the fractionated sample yielded a higher number of more confident identifications and greater sequence coverage compared to the unfractionated starting material (Pang et al. 2003).

Since fractionating by charge is more useful in matching the p*I* range of fractions to nrIPG strips for 2-DE, we concentrated our work on this separation mode. Using estimations of the p*I* of the sample proteins based on our previous work (Fryksdale et al. 2002), we chose a fractionation strategy to divide our sample into four fractions. The scheme shown in Fig. 25.1 produced good results with little optimization necessary. To design fractionations according to the scheme, the chosen pH of the buffer was matched to the protein p*I* of the planned dividing point of the two desired fractions. Our main goal was to remove the higher abundance acidic proteins and enhance the more basic lower abundance proteins. After the final separation, it was necessary to concentrate and remove salt from the samples before application to 2-DE. By using sequential fractionation as shown, three separation runs were performed to obtain four theoretical fractions, p*I* <4.5, p*I* 4.5 to 5.5, p*I* 5.5 to 6.7, and p*I* >6.7. These fractions were evaluated by 2-DE using 3–10 NL carrier ampholyte tube gels for the first dimension with a 100 μg protein load.

Figure 25.2 shows the 2-DE gel image of sample neat and the image of the patterns after fractionation. Also shown are the predicted theoretical p*I* cut patterns (black lines) and the actual ones obtained (colored lines). The most obvious result is the increase in spot intensity for the fractions, especially for the more basic low abundance proteins in fractions p*I* 5.5 to 6.7 and p*I* >6.7. Fraction p*I* >6.7 also shows complete removal of all the high abundance, large

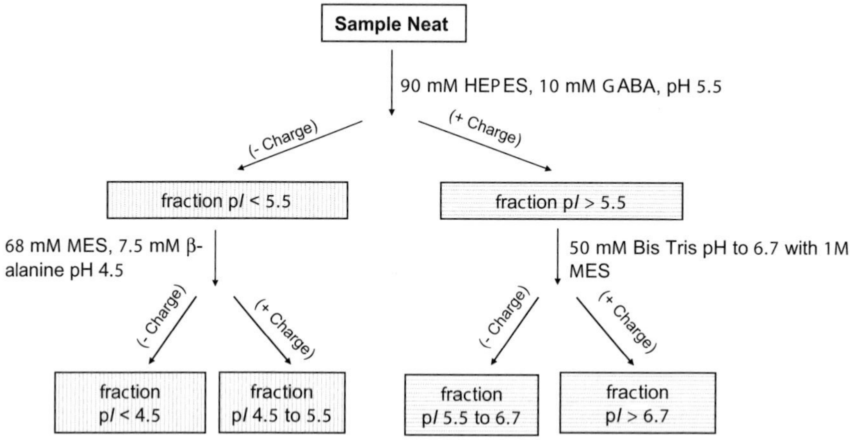

Fig. 25.1. Charge-based fractionation strategy. Sample neat was first fractionated into two fractions: p*I* >5.5 and p*I* l<5.5. Each of these subfractions was then further fractioned into subfractions with p*I*'s >6.7 and <4.5, respectively

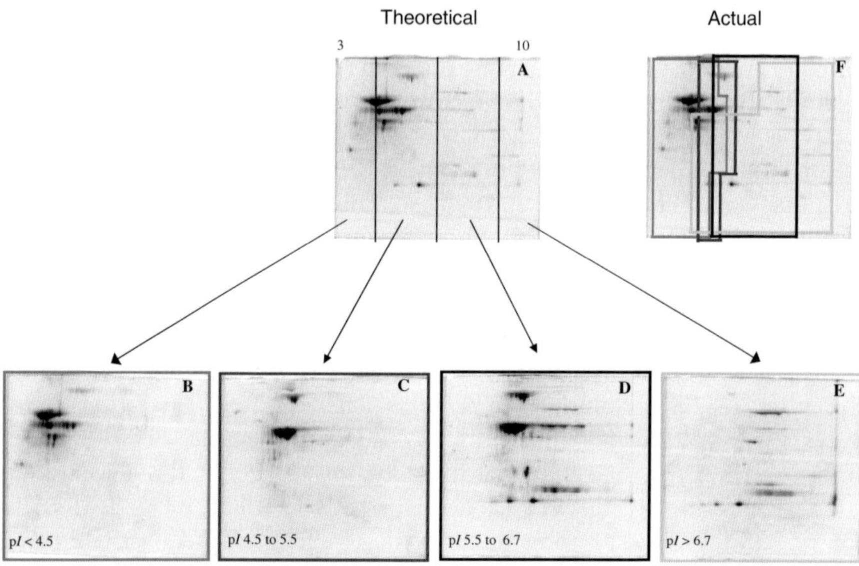

Fig. 25.2. Sample neat (**A, F**) was fractionated into four subfractions: p*I* <4.5 (**B**), p*I* 4.5 to 5.5 (**C**), p*I* 5.5 to 6.7 (**D**), and p*I* >6.7 (**E**). Samples were visualized by 2-DE. 100 µg protein load, 3–10 NL 2-D carrier ampholyte tube gels, SYPRO Ruby stain. Theoretical p*I* cuts are indicated in **A** and actual p*I* cuts are indicated in **F**. The proteins enclosed within the *colored regions* of **F** match those seen in the corresponding *colored gels* shown in **B, C, D,** and **E**

molecular weight, acidic proteins present in the starting material. These spots are known to contain cellobiohydrolase I (CBH I) and cellobiohydrolase II (CBH II), the most abundant proteins in our sample. It should also be noted that in some cases, the same protein spot was present in two different fractions, which indicated that some species of proteins partitioned across two fractions.

We examined the p*I* <5.5 and p*I* >5.5 fractions more carefully and determined the spot volumes for a number of protein spots. In general, a gradient of partitioning was seen; proteins with estimated p*I*'s near the pH 5.5 cut point were split between the fractions. Proteins in the more extreme acid or basic regions were present in a single fraction at levels near 100 %. For separations in general, this implies that some empirical optimization may be necessary to ensure the desired separation of protein species with similar p*I*'s. Proteins with p*I*'s at the pH cut points would be under-represented in any one of these two fractions compared to the theoretical maximum. However, for our objective of increasing the amount of low abundance proteins, the fractions showed large improvements. In addition, it has been noted that some fraction overlap can serve to align a series of fraction gels to create a total gel map of the sample under study (Locke et al. 2002).

In order to improve the characterization of the fractionated sample, we subjected the p*I* >6.7 fraction and the starting control sample to separation on 3–10 NL IPG strips (Fig. 25.3). Using the IPG strips allowed to load a greater amount of protein than using carrier ampholyte tube gels. By increasing the sample load from 100 to 500 μg, we were able to increase the visualiza-

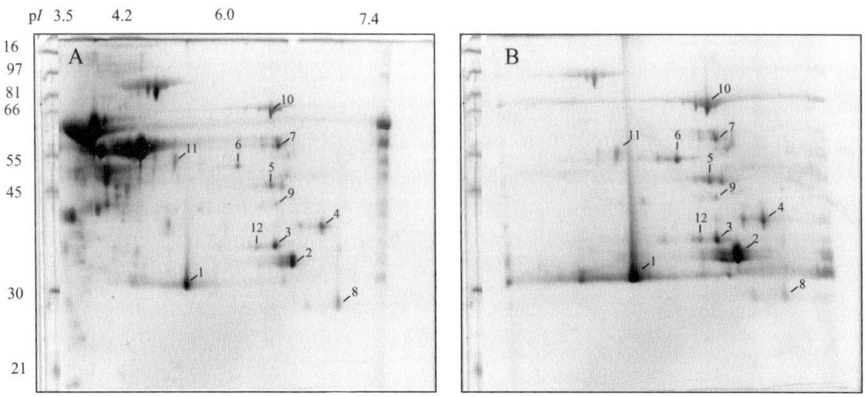

Fig. 25.3. Sample neat (**A**) and fraction p*I* >6.7 (**B**) separated by 2-DE. 500 μg protein load, 3–10 NL IPG strips, SYPRO Ruby stain. Molecular weight standards are indicated on the *y*-axis of the gel, approximate p*I* is indicated on the *x*-axis. *Spots 1–12* indicate spots that were chosen for identification by MALDI-TOF-MS peptide mass matching comparison study

tion of proteins for both the control and p*I* >6.7 fraction (Fig. 25.3) compared to 100-μg samples run on tube gels (Fig. 25.2). The subset of spots seen on the p*I* >6.7 gel shows greater intensity than the control, as expected.

25.1.3 Mass Spectrometry Identification After Prefractionation

With the increased visualization of low abundance proteins seen using the fractionated sample, we investigated whether this would lead to improvements in our ability to identify the proteins by peptide mass matching. To test for improvements, we selected 12 spots matched in both the control and the p*I* >6.7 fraction (500 μg load IPG 2-D gels) (Fig. 25.3). The 12 spots were excised and subjected to in-gel trypsin digestion followed by analysis using MALDI-TOF/MS and peptide mass matching. To compare the differences between the fraction and control, we used two criteria. These were the ability of the spot to be identified with high confidence by the peptide mass-matching program, and if identification was made, the percent sequence coverage of the peptides matched to the entire protein.

The database used was the fungal database within the NCBI public database. This database does not contain the complete genome of *Trichoderma reesei*. Using this limited database may leave some spots unidentified due to their absence from the database. However, this should still allow a valid comparison for those proteins present in the database. We were able to obtain identifications for a total of 6 of the 12 spots from two independent 2-D gel replicates using the same sample neat and fraction p*I* >6.7. In the first set of gels, we obtained improved matched peptide coverage from the fractionated sample for three out of four proteins identified in both gels, including xylanse III, arabinofuranosidase, and endoglucanase III (spots 4, 6 and 12, respectively).

Identification of β-glucosidase I (spot 10) was slightly better from the starting sample gel. We also saw unique weak identifications, xylanase III (multiple spots present) from the original sample and CBH II from the fraction (spots 9 and 11). For the second set of gels, we obtained three good identifications from the fractionated sample of arabinofuranosidase, xylanase III, and β-glucosidase I (spots 6, 9, and 10, respectively), while no identifications were made using spots from the original sample gel. These results not only show the variable nature of protein identification using in-gel digestion, but also show the fractionated sample yields improved results over the unfractionated control sample. As an example of the improvement, Fig. 25.4 shows the MALDI-TOF/MS spectrum from the spot containing arabinofuranosidase from the control and fraction p*I* >6.7. A greater number of peptides and higher signal intensity was seen which correlates with the greater spot volume of the fraction sample. In addition, our results show the improvement made when using higher loads with IPG strips versus the carrier ampholyte tube gels with lower

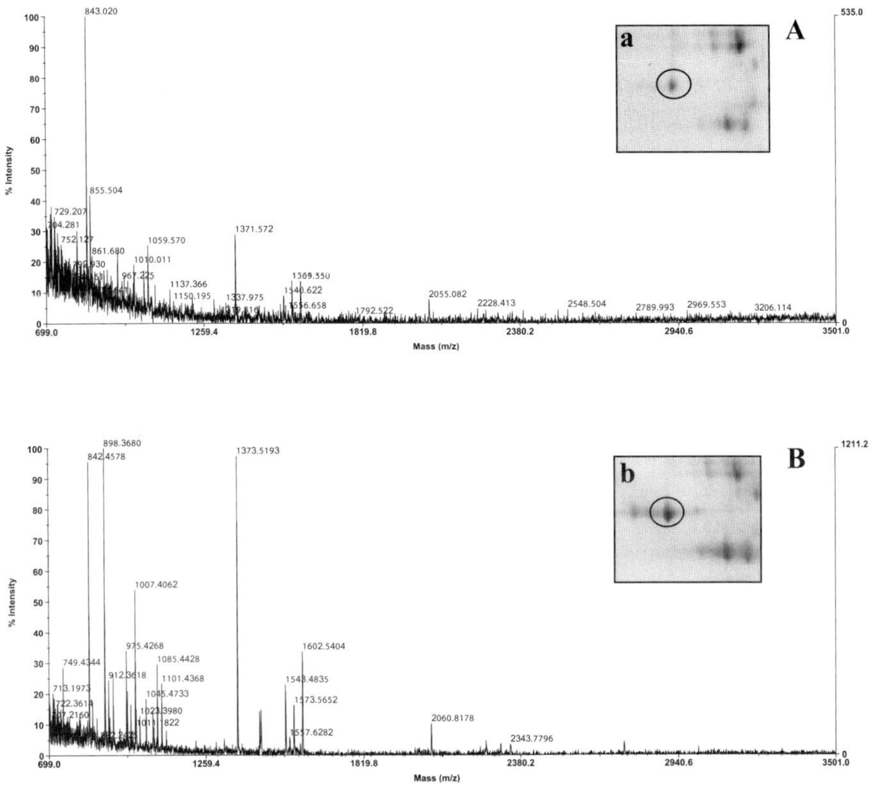

Fig. 25.4. Arabinofuranosidase (named spot 6 in Fig. 3) of the sample neat (**A**) and of fraction pI >6.7 (**B**) separated by 2-DE. Gel spot intensity is greatly enhanced in the pI >6.7 fraction. MALDI-TOF-MS spectra of arabinofuranosidase from sample neat (A) and subfraction pI >6.7 (B)

protein loads. We identified two new protein species, xylanase III and arabinofuranosidase, which were not identified in our previous report from 2-DE using tube gels (Fryksdale et al. 2002). In all cases, these improved results were correlated with increased spot volumes for the spots in the higher load gels and the fractionated sample. This leads to the conclusion that the higher quality identifications made were due to an increase in protein concentration in the gel plugs.

25.2 Deglycosylation as a Means for Improved Protein Identification

Glycosylation is widespread in eukaryotic cells and can add substantial complexity to samples due to the fact that isoforms can vary in pI and by molecular weight, differences that are visible on 2-DE gels. Glycosylation can also interfere with the ability of a protein to be digested by proteases in proteomic analyses. In order to reduce the complexity of the proteome due to glycosylation, deglycosylation of glycoproteins can be accomplished by enzymatic and chemical methods (Dwek et al. 1993). The enzyme peptide N-glycosidase F (PNGase F) cleaves the bond between the asparagine residue and the innermost N-acetylglucosamine or N-acetylgalactosamine of most N-linked glycan groups with the conversion of the asparagine to an aspartic acid (Maley et al. 1989). Peptide N-glycosidase A (PNGase A) can remove the N-linked glycan groups resistant to PNGase F (Dwek et al. 1993). Endoglycosidase H (Endo H) cleaves the chitobiose core of N-linked high mannose and some N-linked hybrid oligosaccharides leaving the innermost N-acetylglucosamine intact (Maley et al. 1989).

The removal of O-linked glycan groups by enzymatic methods is not as straightforward and requires usually a combination of enzymes to remove all possible O-linked glycans (Dwek et al. 1993). All the enzymatic methods may respond to different levels of sample denaturation that can open up glycan groups to enzymatic attack. In addition, chemical methods have been introduced for removing glycan groups from proteins. Treatment with trifluoromethanesulfonic (TFMS) acid has been reported to be very effective at removing all O- and N-linked glycan groups except the linkage N-acetylglucosamine or N-acetylgalactosamine (Sojar and Bahl 1987). Here, we will describe the proteomic characterization of a commercial cellulase enzyme preparation from the filamentous fungus *Trichoderma reesei*. Many of the enzymes found in the cellulase complex mix are known to contain N- and O-linked glycan groups. We can show that the added effort of prior deglycosylation can dramtically reduce the complexity of 1-D SDS-PAGE (sodium dodecyl sulfate-polyacrylamide gel electrophoresis) gel and 2-DE gel patterns, as well as, improve the success of spot identification with MALDI-TOF/MS peptide map matching.

25.2.1 Deglycosylation of a Fungal Proteome

For this study, commercially available *Tricoderma reesei* cellulase mix (LAMINEX) obtained from Genencor International, Inc. was subjected to digestion with enzymes and acid as described (Fryksdale et al. 2002). PNGase F and Endo H were used for removal of N-lined glycan groups. For O-linked glycans, we applied O-Glycanase, sialidase II, β-galactosidase and β-N-acetylhexosaminidase. For acid deglycosylation, TFMS was used.

To allow direct comparisons of the effects of the various deglycosylation methods on a whole cellulase mix we used 1-D SDS-PAGE. Analysis of the enzyme treatments and controls revealed subtle effects (Fig. 25.5A). Both PNGase F and Endo H showed a small reduction in molecular weight for the

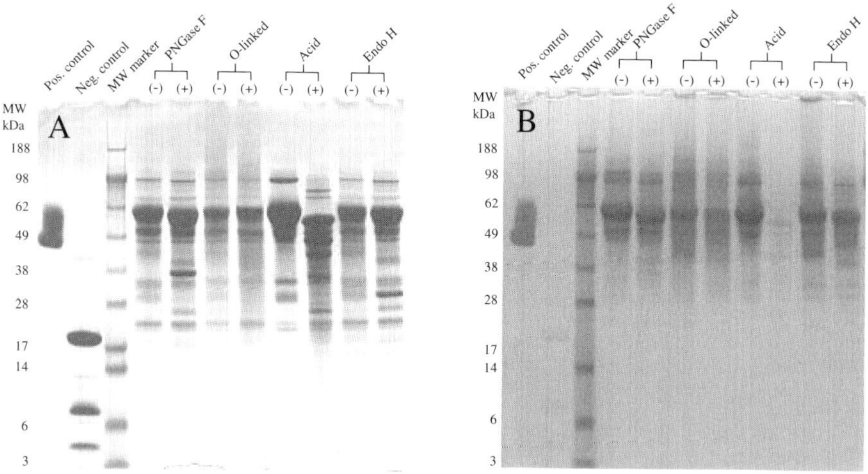

Fig. 25.5. One-dimensional Bis-Tris SDS-PAGE of deglycosylated cellulase samples and controls. For each treatment pair, (+) indicates treatment, (-) indicates control. **A** Stained with CBB R-250. **B** Stained with glycoprotein stain. *Lanes 1* and *2* are glycoprotein stain positive and negative controls, respectively

large band running just below 62 kDa. Other bands were observed to shift at the lower molecular weight ranges. PNGase F was visible at 36 kDa and Endo H was visible at 29 kDa. For the sample digested with the O-linked enzymes, no band shifts were evident. The darker staining area at 62 kDa in the enzyme digest lane was one of the enzymes used for the O-linked glycan removal.

The most dramatic changes were seen with the acid treatment sample. Compared to the control, four major bands appeared to flank the 49-kDa standard. All the material previously located in the major 62-kDa band was shifted to lower molecular weights while forming sharp bands. The large band in the control at 98 kDa was also shifted down forming two lower bands. These results appeared to show limited effects of enzymatic treatments on glycan removal compared to the acid treatment. This hypothesis was confirmed by the staining of a duplicate gel with a stain specific for the glycan groups of glycoproteins (Fig. 25.5B). The glycoprotein stain showed that the most reactive material to the stain was found in three major bands in the control samples. The two major top bands of the PNGase F and Endo H digested sample had a similar band shift as in the Coomassie Brilliant Blue (CBB) stained gel with some loss of stain intensity. This may be due to the partial removal of glycan by the enzymes. While the O-linked enzyme treated sample showed no obvious band shifting, it also stained lighter with the glycoprotein stain. The acid treated sample showed the greatest change with nearly all the glycan detectable by the stain removed. While we did not

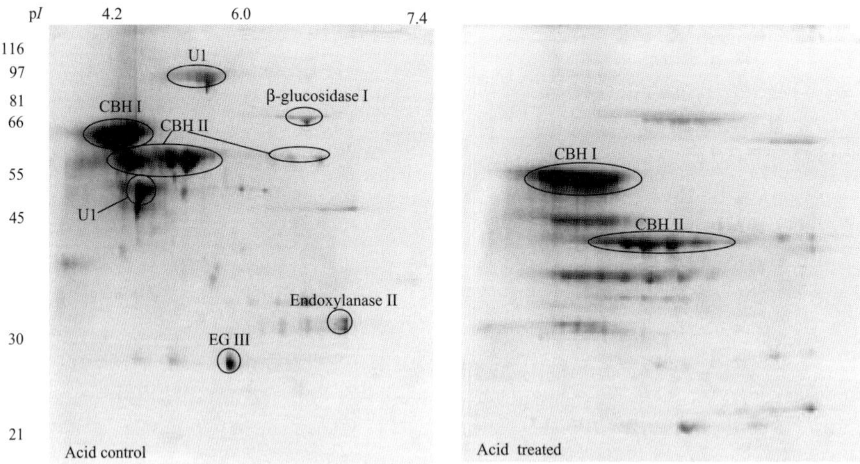

Fig. 25.6. 2-DE maps of acid deglycosylated cellulase sample and control. 100 μg protein loaded/gel. SYPRO Ruby stained. *Circled* and *labeled* proteins on the acid control gel refer to proteins studied and identified in this work on the enzymatic deglycosylation, and corresponding control gels. *Circled* and *labeled* proteins on the acid treated gel refer to proteins identified on this gel only. Molecular weight standards are indicated on the *y-axis* of the gel, approximate p*I* is indicated on the *x*-axis

use a combination of N-linked and O-linked removing enzymes, the theoretical composite of both did not produce the change in molecular weight to equal the acid treatment.

To further reveal the effects of the acid treatment, we separated the acid treated sample and control on 2-DE gels (Fig. 25.6). The acid treatment caused visible decreases in molecular weight with the appearance of tight charge trains. At least eight such trains were visible. These eight charge trains correlated very well with the bands seen on the 1-D SDS-PAGE gel stained with CBB. This treatment also decreased the complexity of the 2-DE gel pattern. Spots that appear as part of these isoform trains most likely consist of the same polypeptide chain or simple mixtures. Lower molecular weight spots appeared as single spots or small trains. These results show that the enzyme treatments were much less effective than the acid treatment in removing glycan from proteins in this sample. We then compared the ability to identify proteins of the deglycosylated samples to the controls. To do this, 2-DE gels were also run for all the enzyme treated samples with their appropriate controls. The spots chosen for identification are indicated in Fig. 25.6 acid control panel. Robotic spot picking and trypsin digestion and extraction were used in a general survey mode. A total of seven proteins from each 2-DE gel were chosen as points of comparison between the deglycosylated samples and their controls. The 2-DE pattern of these proteins for all the enzyme control and enzyme-treated gels was similar (data not shown), which allowed targeting of these spots. These seven spots or charge trains included two previously unidentified proteins, named here as U1 and U2, and five known proteins previously identified on 2-DE gels of similar samples (data not shown); β-glucosidase I, CBH I, CBH II, endoxylanase II, and EG III. Combined, these seven proteins represent approximately 90 % of the total protein in the fungal exoproteome examined.

In general, all three of the enzymatic treatments yielded mass spectra with an increased number of peptides. This effect was mostly evident with specific proteins. Proteins known not to be heavily glycosylated such as endoxylanase II (Lappalainen et al. 2000) showed little or no enhancement of total peptides compared to their controls. While all three enzymatic treatments could increase the total number of peptides seen by MALDI-TOF-MS the sample treated with PNGase F was clearly superior in terms of yielding peptides that resulted in actual protein identification when searched against the database. Four of the seven proteins evaluated, β-glucosidase I, CBH I, CBH II, and endoglucanase III (EG III) were easily identified from the PNGase F treated sample while identification was not possible with gel plugs from the control.

Figure 25.7 shows a comparison of MALDI-TOF/MS spectra from PNGase F treated samples and control for β-glucosidase I. It was evident that peak intensity and peptide coverage had much improved over the control. The enhancement of protein identification by the enzyme treatments is interesting due to the less apparent removal of glycan by these treatments compared to the acid treatment. However, as was noted above, the 1-D SDS-PAGE gel samples clearly showed some effect of the enzymes in decreasing the molecular weight of the major band(s). These treatments appear to cause an enhancement of identification by mass spectrometry without the full effect of deglycosylation. It is possible that even minor deglycosylation renders the protein more susceptible to digestion by trypsin. We were also successful in the identification of CBH I and CBH II and other proteins from acid treated 2-D gels (Fig. 25.6) and 1-D gels (not shown). This demonstrates that acid treatment does not chemically alter the peptides or preclude their identification by mass fingerprinting.

25.2.2 Deglycosylation Summary

We have shown that pretreatment of samples before 1-D SDS-PAGE and 2-D electrophoresis with various deglycosylation techniques can lead to a number of enhancements. These include improved identification of some proteins by peptide mass matching, reduction of complexity, increased resolution of proteins, and identification of proteins that are glycosylated. When dealing with a sample containing glycosylated proteins, the use of deglycosylating pretreat-

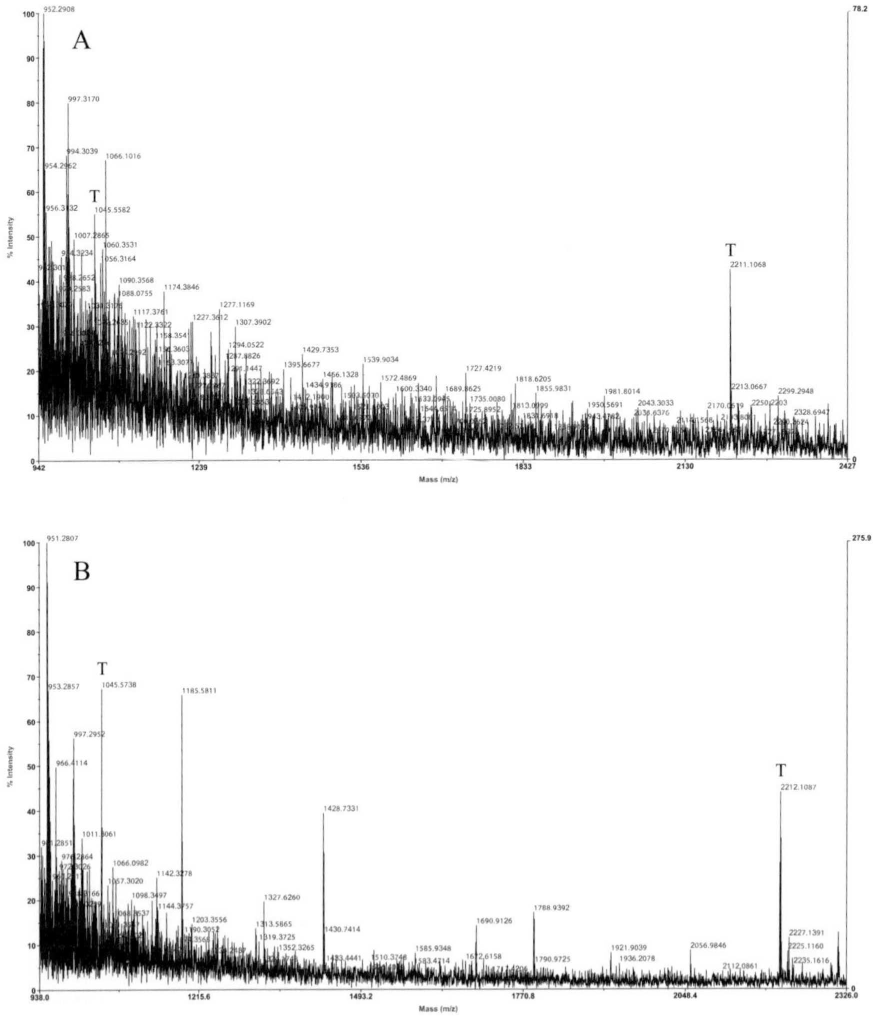

Fig. 25.7. MALDI-TOF mass spectra of representative **A** PNGase F control β-glucosidase I trypsin digest and **B** PNGase F digested β-glucosidase I trypsin digest. Peaks labeled with *T* indicate trypsin autolysis peptides for **A** and **B**

ments could be a beneficial initial approach for whole proteome studies. The acid treatment was shown to remove nearly all glycan from a sample, thus allowing the identification of proteins that are glycosylated, as well as determining accurate protein molecular weights from the gels. We have also determined that these proteins can be identified successfully using the acid treatment, especially from a 1-D SDS-PAGE gel (Fryksdale et al. 2002)). The use of enzyme treatments can be used to follow up and possibly enhance the identification of some proteins that proved difficult to identify in an untreated sample.

25.3 High-Throughput Proteomics Method Optimization

Inadequate sample handling and preparation can lead to reduced ability to correctly identify proteins using mass spectrometry (Speicher et al. 2000, Bornsen 2000). We have evaluated and made improvements in a number of sample preparation methods, including in-gel protease digestion, peptide extraction and sample spotting onto MALDI targets to address this problem. A set of proteins was selected to represent typical challenges faced in today's proteomics research. The proteins included: bovine serum albumin (BSA), the highly hydrophobic bacteriorhodopsin and the highly glycosylated *Trichoderma reesei* cellobiohydrolase I (CBH-I). Our optimization scheme tested the effect of:

- various in-gel digestion methods.
- solvents for peptide extraction.
- various MALDI matrices.
- different sample spotting techniques.

Mass spectra of the samples were evaluated for the total number of peptides detected, average peptide intensity corresponding to its hydrophobicity value, percent protein sequence coverage, and the final protein identification score was obtained from NCBI protein database searching. Although our goal was to optimize in-gel digestion at femtomole levels, we reported here the optimized conditions for both manual preparation and via the use of an automated digestion robot for proteins at picomole levels.

25.3.1 Method Development to Increase Sample Consistency

About 0.5 μg (7.5 pmol) BSA, 4.0 μg CBH-I (75 pmol) and 0.85 μg (30 pmol) bacteriorhodopsin were loaded separately onto NuPAGE 10 % Bis-Tris mini gels (Invitrogen, Carlsbad, CA). The gels were stained with 0.1 % CBB R-250. The bands containing the proteins of interest were excised, destained with 50 % acetonitrile in 25 mM NH_4HCO_3 buffer and dried after dehydration with acetonitrile. Four tryptic digestion buffers were selected with the ratio of trypsin to substrates being 1:20. After tryptic digestion the peptides were extracted using seven different methods and then spotted onto a MALDI target utilizing five commonly used methods (quick and dirty, dried droplet, sandwich, seed-film layer and two-layer).

For high-throughput proteomics, protein gel spots were excised robotically using the Investigator ProPic Robotic Workstation (Genomic Solutions, Ann Arbor, MI). Excised protein spots were destained, reduced, alkylated and proteolytically digested using the Investigator ProGest robot. Peptides were extracted and spotted onto a MALDI target along with alpha-cyano-4-hydroxycinnamic acid matrix (CHCA). The peptide MS data were collected using the Voyager DE STR MALDI-TOF/MS (Applied Biosystems, Foster City, CA) and submitted for protein ID using software programs such as: Data Explorer, Protein Prospector (University of California, San Francisco) and *m/z* searching program – ProFound (Rockefeller University).

All MS and MS/MS data were acquired using the Surveyor LC system coupled to the LCQ Advantage or LCQ Deca XP (ThermoFinnigan, San Jose, CA). The HPLC gradient was programmed from 0 % solvent A: 0.1 % TFA in water to 70 % solvent B: 0.08 % TFA in acetonitrile. Data processing was performed using TurboSEQUEST and Xcalibur (ThermoFinnigan, San Jose, CA).

25.3.2 Method Optimization and Results

Current protocols used for manual or automatic in-gel digestion of proteins often result in less than 50 % of the expected peptides and produces low MS ion signals. For unambiguous protein identification using peptide mass fingerprinting, it is critical to improve the protein total recovery yields as well as the peptide ion intensities. Recent reports suggest that the optimized in-gel protein digestion depends on factors such as digestion buffer, peptide extraction and MALDI matrix spotting methods (Speicher et al. 2000, Bornsen 2000). In order to increase our in-gel digestion efficiency for high-throughput proteomics, we have tested several conditions. We found that the methods used showed variability in improving the peptide recovery and were dependent on the type of protein investigated and of the peptides generated. Overall, all methods tested generated more useful data than the blank.

25.3.2.1 Digestion Buffers

It has been shown that high salt concentrations are problematic in mass spectrometry analysis (Fountoulakis and Langen 1997). For in-gel digestion using a high salt buffer, samples often require further desalting steps (Rosenfeld et al. 1992 and Shaw 1993). Desalting by chromatography is a labor and time-consuming step and the desalting process decreases the overall peptide recovery, and thus reduces the chance of identifying low abundant protein spots. In this study, four digestion buffers have been tested and compared using BSA, CBH-I and bacteriorhodopsin proteins. The digestion buffers were as follows:

1. 25 mM NH_4HCO_3 buffer
2. 25 mM NH_4HCO_3 buffer+20 % methanol
3. 1 mM Tris buffer
4. Reduction and alkylation (R & A), using 25 mM NH_4HCO_3 as a buffer

As an example, Fig. 25.8 shows typical peptide mass fingerprint of in-gel digested BSA using condition No. 1 (left) or No. 3 (right) as the digestion buffer. Note that peak intensity increased by more than twice using condition

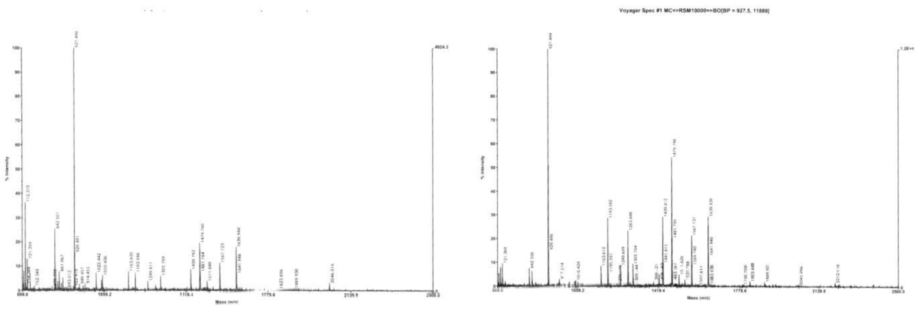

Fig. 25.8. Comparison of MALDI-TOF/MS spectra for various BSA tryptic digestion buffers. The spectrum on the *left* represents the digest using condition No. 1, while the spectra on the *right* represents condition No. 3. Peak intensity increased by more than twice using condition No. 3 versus condition No. 1

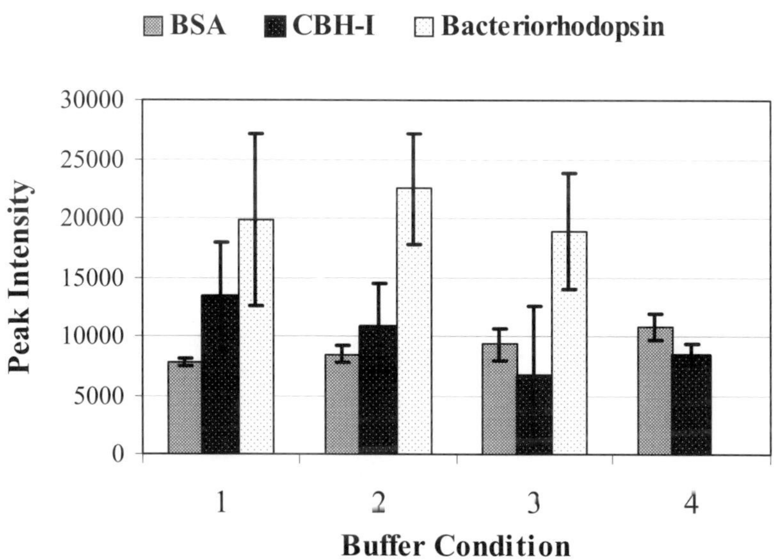

Fig. 25.9. Peak Intensities and percent coverage vs. digestion buffers. Peak intensity was calculated by averaging the intensities of all the peptides matched to the theoretical digest. To continue with extraction, the best buffer was chosen for each protein based on high peak intensities and percent coverage. For BSA, the optimal buffer was 1 mM Tris (condition No. 3). In the case of CBH-I, it was 25 mM NH_4HCO_3 (condition No. 1), and for bacteriorhodopsin, it was 25 mM NH_4HCO_3+20 % methanol (condition No. 2)

#3 versus condition No. 1. We evaluated the MS data with relation to peak intensity and of percent protein coverage (Figs. 25.9 and 25.10). For BSA, Fig. 25.9 shows the highest peak intensities for samples that were reduced and alkylated, yet these samples have a lower percentage coverage than the standard method with 25 mM NH_4HCO_3. This indicates that the sample loss

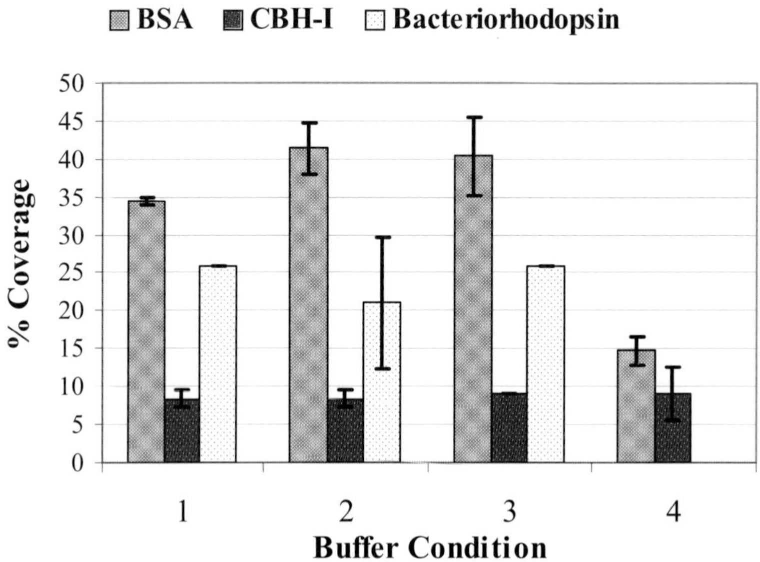

Fig. 25.10. See legend to Fig. 25.9

occurs during the reduction and alkylation procedure. We did not perform reduction and alkylation on the bacteriorhodopsin samples since there is no cysteine in this protein.

In summary, digestion conditions were found to be specific to the protein selected; there was not one condition that universally resulted in significantly greater quality digests. The optimal digestion conditions for each of the proteins are as follows: BSA: 1 mM Tris (condition No. 3), CBH-I: 25 mM NH_4HCO_3 (condition No. 1) and bacteriorhodopsin: 25 mM NH_4HCO_3+20 % methanol (condition No. 2).

It is reasonable to use the standard method (condition No. 1) when performing manual hand digests on unknown proteins. When the protein is known to be hydrophobic, the addition of 20 % methanol to 25 mM NH_4HCO_3 does increase the peak intensity and may increase the percent coverage.

25.3.2.2 Extraction Buffers

The choice of extraction buffer is crucial for the characterization of integral membrane proteins and hydrophobic peptides by 2D-gel and mass spectrometry techniques. Some detergents normally used to solubilize these hydrophobic proteins, and peptides can severely suppress analyte ion formation. This problem has been addressed previously by performing additional detergent removing processes. The most commonly used membrane protein for such studies has been bacteriorhodopsin, since it has a seven transmembrane helical structures and is stable (Grigorieff et al. 1996). In this study, we selected seven different extraction conditions to systematically evaluate peptide recoveries from in-gel tryptic digestion. These extraction buffers were:

1. 50 % acetonitrile with 0.1 % formic acid (FA) (first extraction) + 50 % methanol with 0.1 % FA (second extraction)
2. 50 % acetonitrile with 0.1 % FA (first extraction) + 50 % ethanol/isopropanol (1:1) with 0.1 % FA (second extraction)
3. 50 % acetonitrile with 0.1 % FA, two extractions
4. 70 % acetonitrile with 0.1 % FA, two extractions
5. 70 % acetonitrile with 0.1 % FA, 0.1 % octyl-β-glucopyranoside (OBG), two extractions
6. Two extractions with 0.1 % FA in DI H_2O
7. No extraction (digest buffer was analyzed directly).

The total number of peptides detected have an average peptide intensity corresponding to its hydrophobicity value, and the percent protein sequence coverage was used to evaluate each protein digest under these digestion conditions. Figure 25.11 shows the percent coverage versus various peptide extraction conditions for BSA, CBH-I and bacteriorhodopsin. Extraction conditions were found to be specific for the selected protein. For example, condi-

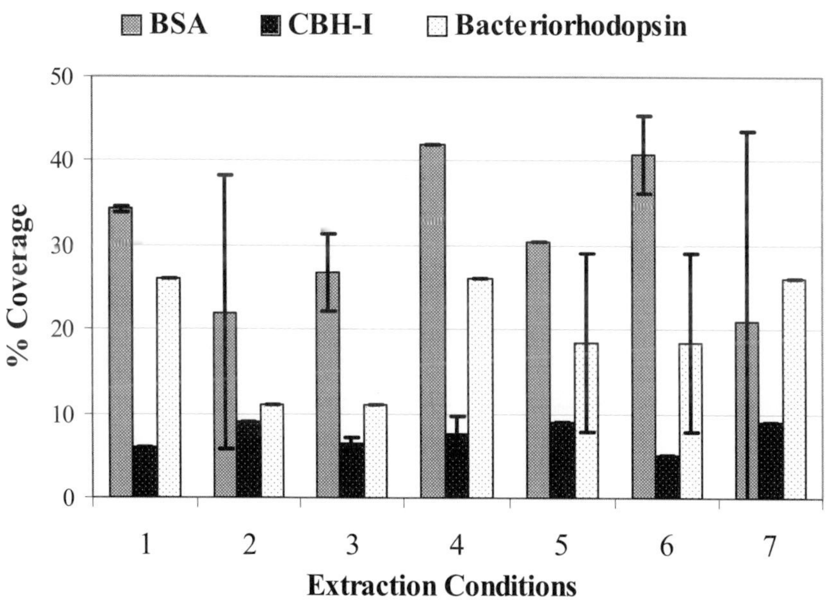

Fig. 25.11. Protein % Coverage vs. Extraction Conditions for all three proteins. Condition #4 was the best overall extraction method. For condition #7, the digest solution was spotted directly on MALDI target without any extraction steps

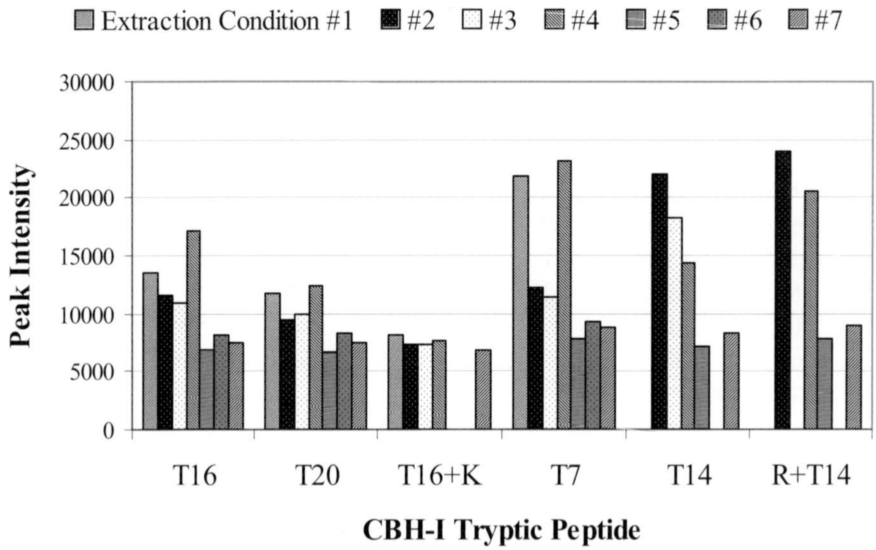

Figs. 25.12. The MS peak intensities of CBH-I peptides that vary in molecular weight and hydrophobicity values, for each of the seven extraction conditions

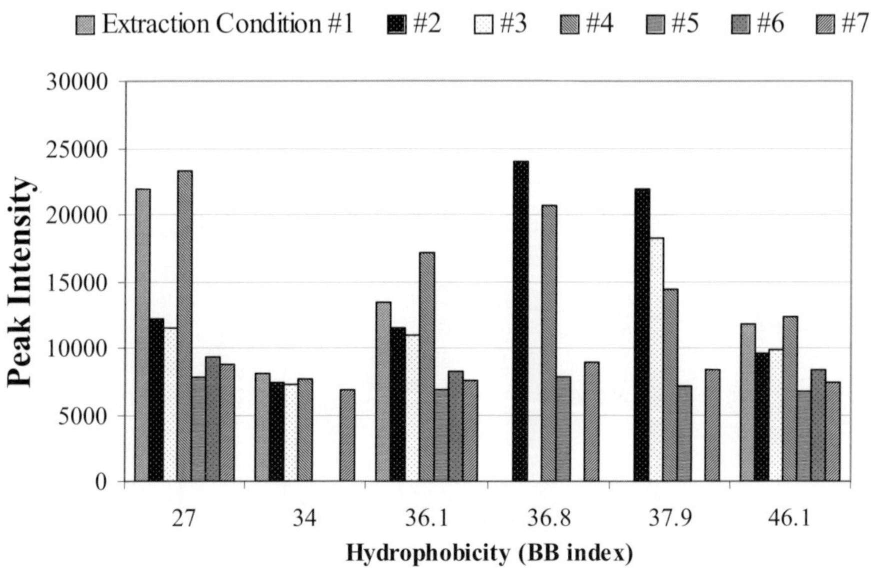

Fig. 25.13. See legend to Fig. 24. 12

tion No. 4 (70% acetonitrile with 0.1% FA, twice extraction) produced the highest peak intensities (data not shown) and percent coverage for BSA. For this protein, the no-organic solvent method proved to be insufficient in producing high quality data.

The effect of extraction buffers on the MS peak intensities of CBH-I peptides that vary in molecular weight and hydrophobicity values (BB Index, Bull and Breese 1974) are shown in Figs. 25.12 and 25.13. The molecular weight of these peptides ranges from 700 to 1800 and the BB index values range from 25 to 50. Larger BB indices indicate higher hydrophobicity of the peptide. Our data demonstrated that the best extraction conditions for CBH-I protein digests based on the peptide data were ranked as follows: Nos. 4, 2 and 3. We also found that condition No. 4 produced greater peak intensities and percent coverage for bacteriorhodopsin digests in comparison to all the other extraction methods. Overall, we concluded that 70% acetonitrile with 0.1% FA (condition No. 4) was the best extraction buffer. The ethanol/isopropanol extraction worked especially well for CBH I peptides with high molecular weights and hydrophobic peptides. Extractions in the absence of organic solvent method worked fairly well, but organics did improve the quality of the MS spectra.

25.3.2.3 Matrix Spotting Methods

Many previous studies have demonstrated that α-Cyano-4-hydroxycinnamic acid (CHCA) is the best MALDI matrix for peptide molecules, since it gives good MS signal intensity (Schuerenberg et al. 2001). It is also clear that the performance of MALDI-TOF/MS for protein digests is very much dependent on the sample/matrix preparation methods (Kussmann and Roepstorff 2002). Therefore, we only used CHCA as matrix in this study to examine various sample preparation methods, such as the Quick and Dirty (Q&D) method, the Dried-Droplet method, the Two-Layer method, the Seed-Film Layer and the Sandwich method (Dai et al. 1999, Karas and Hillenkamp 1988).

In the quick and dirty sample preparation, about 1 μL of matrix solution is added on top of 1 μL of peptide sample. Both solutions are then mixed with the pipette tip before the mixture is dried on the MALDI target. The dried-droplet method allows the pre-mixing of the matrix solution with the analyte solution. This mixture is then deposited onto the MALDI target and dried. For the two-layer sample preparation, the matrix solution is spotted onto the MALDI target and allowed to dry to form the first layer of small crystals. A mixture of matrix and the analyte solution is then added to the top of the matrix layer. The seed-film layer technique involves first laying down a thin layer of small matrix crystals on the MALDI target. A droplet containing the analyte is placed on top of the crystalline deposit. The resulting matrix/analyte solution is then dried, and then the matrix is deposited on the seed sites provided by the undissolved portion of the original layer. The sandwich method is derived from the Q&D method and the two-layer method. In this method, the sample analyte is basically sandwiched between the two matrix layers.

Overall, each of these methods has their own advantage(s) and disadvantage(s). We found that the best matrix method in this study was sample mixed 1:1 with saturated CHCA prepared in 50 % acetonitrile+0.1 % formic acid (the Dried-Droplet method), spotted, then followed by a 5 second wash with 5 % formic acid, and then washed again with distilled water.

25.3.3 High-throughput Proteomics (ProGest) Optimization

In the post-genome era, the demands for the analysis of the proteome, including protein expression, identification and post-translational modifications have increased rapidly. High-throughput can be handled by a semi-automated proteomics analysis system, composed of a 2D-gel system, a protein spot excision robot (i.e. ProPic, Genomic Solutions, Inc.), an in-gel digestion robot (i.e. ProGest), an automated target-spotting system and a mass spectrometer. However, many of these high-throughput systems have only achieved a low success rate in protein identification, even though hundreds of protein spots can be analyzed each day. The problems are mainly due to the limitations of the gel electrophoresis technique for the hydrophobic proteins and the low abundance proteins. Our approach to solve these problems was to apply the above optimized manual in-gel digestion methods, such as the digestion buffer, the peptide extraction buffer and the MALDI matrix spotting method to increase the in-gel digestion efficiency for high-throughput proteomics. In addition, we added an extra gel destain and wash step as well as extending the digestion time from 4 hours to 8 hours. Our results indicate that all of these efforts improved the peptide recovery and protein identification compared to standard methods (data not shown). Both the 'extra destain/ wash' and '70 % acetonitrile extraction' protocols significantly improved the overall MS signal intensity as well as the number of peptides identified. The additional 'reduction and alkylation' step also improved the chance of identifying the cysteine-containing peptides. However, the average MS signal intensity on this 'Red+Alk' sample is lower probably due to the sample loss during the additional wash cycle.

25.3.4 High-throughput Proteomics Summary

In summary, we found that the optimal digestion conditions were specific to the protein selected. For example, 1 mM Tris-HCl buffer for BSA, 25 mM NH_4HCO_3 for CBH-I and 25 mM NH_4HCO_3 + 20 % methanol was the best buffer for bacteriorhodopsin. An additional 20 % methanol added to 25 mM NH_4HCO_3 digest buffer increases the detection of hydrophobic peptides. We also concluded that 70 % acetonitrile with 0.1 % formic acid was the best extraction buffer. However, these results were somewhat protein dependent. The ethanol/isopropanol extraction worked especially well for peptides with high molecular weights and hydrophobic peptides. For the matrix spotting technique, the dried-droplet method with 0.1 % formic acid added to the matrix and a 5 % formic acid wash, followed by a water wash was superior.

For the high-throughput (ProGest) system, our preliminary results show that the standard method with an additional destain/wash step or 70 % acetonitrile extraction gave the best results. We predict that the combinations of extra destain/wash, longer digestion time, reduction/alkylation and 70 % acetonitrile extraction will be the optimized protocol for the high-throughput proteomics (ProGest) system.

25.4 Protein Identification and Quantification Using the N^{14}/N^{15} Isotopic Labeling Technique

In the past few years, many research efforts in proteomics using liquid chromatography/mass spectrometry (LC/MS) methods have focused on protein identification utilizing peptide mass fingerprinting and peptide MS/MS sequencing followed by protein database searching. These techniques provide a rapid, sensitive and robust method for the qualitative analysis of unknown proteins within a complex mixture. For protein quantification, the recently developed ICAT (isotope-coded affinity tags) labeling technique provides a rapid and accurate method for the determination of the protein relative concentrations (Gygi 1999). The most commonly used cleavable ICAT reagents consist of: biotin affinity tags, an acid cleavable linker and an iodine reactive group for cysteine containing peptides. Because the tags are biotinylated, peptides containing cysteine can be enriched by avidin affinity column that also reduces the complexity of the sample. However, there are several limitations to this technique: (1) the label may introduce complications during the peptide MS/MS sequencing, (2) the reagents work exclusively with cysteine-containing peptides, (3) the label slightly changes the peptide profile in chromatographic run and (4) the ICAT reagents are proprietary.

Ideally, an isotopically labeled peptide standard of known quantity used for protein quantification should be made available for each and every peptide generated from a digested protein. Several studies have demonstrated that the use of stably isotope-labeled (N^{15}) peptides for protein quantification in complex samples is feasible (Desiderio 1998, Dewey et al. 1992 and Stocklin 2000).

In this study, the N^{15} isotopically labeled proteins were expressed from cells grown in an isotopically depleted medium (e.g. N^{15} medium). The labeled protein can be combined with proteins expressed from another source prior to digestion and mass spectrometry analysis. Unlabeled peptides have an isotopically labeled internal standard whose mass difference (Δm) will be equivalent to the number of nitrogens on the peptide. The key advances of the technique are: unlike the ICAT method, the Δm is not a fixed value for each peptide and not limited to labeling certain amino acids. In addition, there are no chromatographic changes between labeled and unlabeled peptide since the Δm is relatively small compared to the MW of the peptide. Thus, the N^{14}/N^{15} quantification method is simpler and does not interfere with protein identification.

Sample Preparation was at follows: ^{15}N labeled *Bacillus lentus* subtilisin was produced in a microbial growth medium containing ^{15}N-labeled urea. Tryptic digestion of the N^{14}/N^{15} subtilisin mixture was performed in 500 μL of DI H_2O + 50 μL 1.0 N HCl + 100 μL 50 % TCA (enzyme to substrate ratio was 1:50). The mixture was allowed to react for 15 min at 37 °C.

25.4.1 Identification and Quantification Technique

Subtilisins have no cysteine residues and are not amenable to the sulfide based ICAT method. Using the N^{15} quantification approach, we have analyzed the subtilisin protease, chosen for its commercial importance. The predicted tryptic digested peptide fragments of this subtilisin protease are shown in Table 25.1. Under normal tryptic digestion conditions, 15 peptide fragments (including two incompletely digested fragments) with molecular masses ranging from 275 to 4924 are expected. The tryptic digest of the mutant (N^{15}) and wild-type (N^{14}) subtilisin mixture were separated by reverse-phase HPLC, and the mixture was analyzed by electrospray mass spectrometry. Figure 25.14 shows the tryptic map of the mutant (N^{15}) and wild-type (N^{14}) subtilisin. Each chromatographic peak corresponded to either a peptide doublet (co-eluted N^{14}/N^{15} peptides) or a single peptide (N^{14} wild-type peptide or N^{15}-labeled peptide). In the co-eluted N^{14}/N^{15} peptide doublets, the mass of the mutant (N^{15}) tryptic fragments will be equivalent to the wild type peptides plus one amu for each nitrogen present in the peptide. Figure 25.15 shows the extracted ion traces of the tryptic peptide (T_{12}) (511.30 Da) and its N^{15}-labeled isotope (519.30 Da). The peak areas of corresponding (co-eluting) wild-type and labeled peptides are a relative measure of the concentration. The peak retention time (RT) and mass differences (Δm) of other corresponding peptides are an indication of mutation(s). The completed analysis of all identified tryptic peptides and their peak area ratios are shown in Table 25.2. The average peak ratio (1.523) represented the differences in pro-

Table 25.1. Predicted tryptic fragments of subtilisin protease

Peptide name	Sequence	Molecular weight (daltons)	Peptide sequence
T_1	1–10	1100.59	AQSVPWGISR
T_2	11–19	963.51	VQAPAAHNR
T_3	20–27	718.41	GLTGSGVK
T_4	28–44	1820.99	VAVLDTGISTHPDLNIR
T_5	45–92	4588.31	GGASFVPGEPSTQDGNGHGTHVAGTIAALNNSIGVLGVAPSAELYAVK
T_6	93–143	4924.41	VLGASGSGAISSIAQGLEWAGNNGMHVANLSLGSPSPSATLEQAVNSATSR
T_7	144–164	1933.01	GVLVVAASGNSGAGSISYPAR
T_8	165–180	1709.77	YANAMAVGATDQNNNR
T_9	181–229	4797.38	ASFSQYGAGLDIVAPGVNVQSTYPGSTYASLNGTSMATPHVAGAAALVK
T_{10}	230–231	275.17	QK
T_{11}	232–241	1200.61	NPSWSNVQIR
$T10\text{-}_{11}$	236–247	1456.76	QKNPSWSNVQIR
$T1_2$	242–245	511.30	NHLK
$T1_3$	246–269	2368.17	NTATSLGSTNLYGSGLVNAEAATR
$T1_{2-13}$	242–269	2860.45	NHLKNTATSLGSTNLYGSGLVNAEAATR

Table 25.2. Peptide mapping/quantification report for subtilisin protease

Name	RT	Area	ISTD area	Area ratio
12T	6.57	42653939	29578675	1.442
2T	11.12	148687327	100345402	1.482
3T	12.56	103474196	70999406	1.457
8T	14.83	329677348	231961944	1.421
10–11T	19.02	312998081	221254618	1.415
1T	20.36	153173328	97418802	1.572
7T	22.82	721191291	471243261	1.530
13T	24.50	828942896	492908269	1.682
4T	26.25	813508035	577795833	1.408
9T	33.02	1105448271	683521396	1.617
5T	35.14	1274294087	781505436	1.631
6T	37.99	1250886459	773080324	1.618

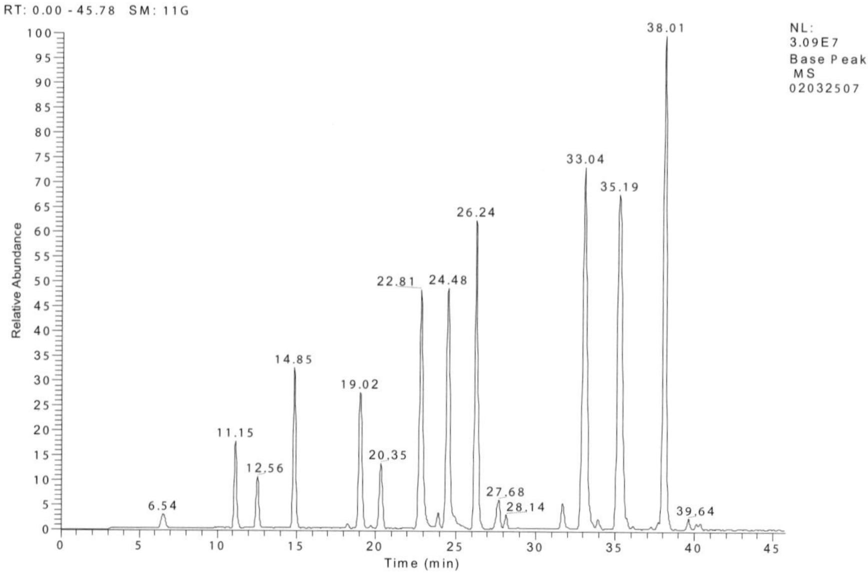

Fig. 25.14. Tryptic map of mutant (N^{15}) combined with wild type (N^{14}) subtilisin protease

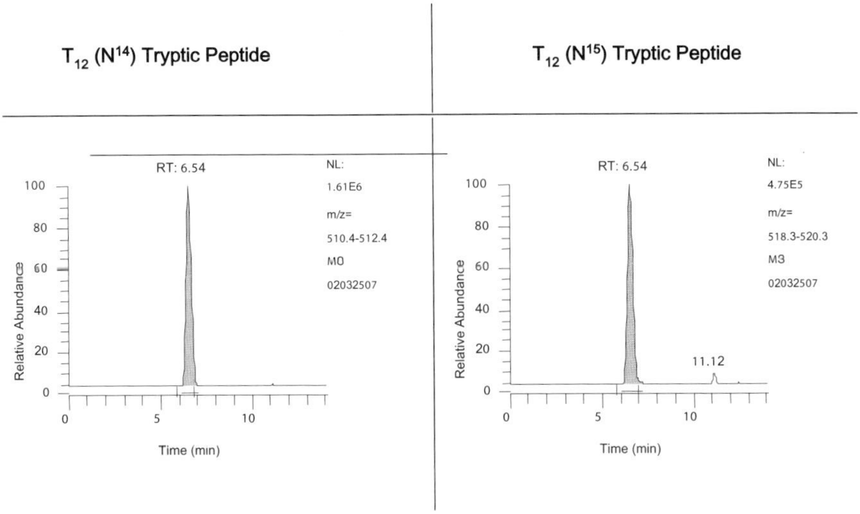

Fig. 25.15. Extracted ion traces for analyte peptide (T_{12} peptide: 511.30 Da) and internal standard (N^{15} isotope labeled peptide: 519.30 Da)

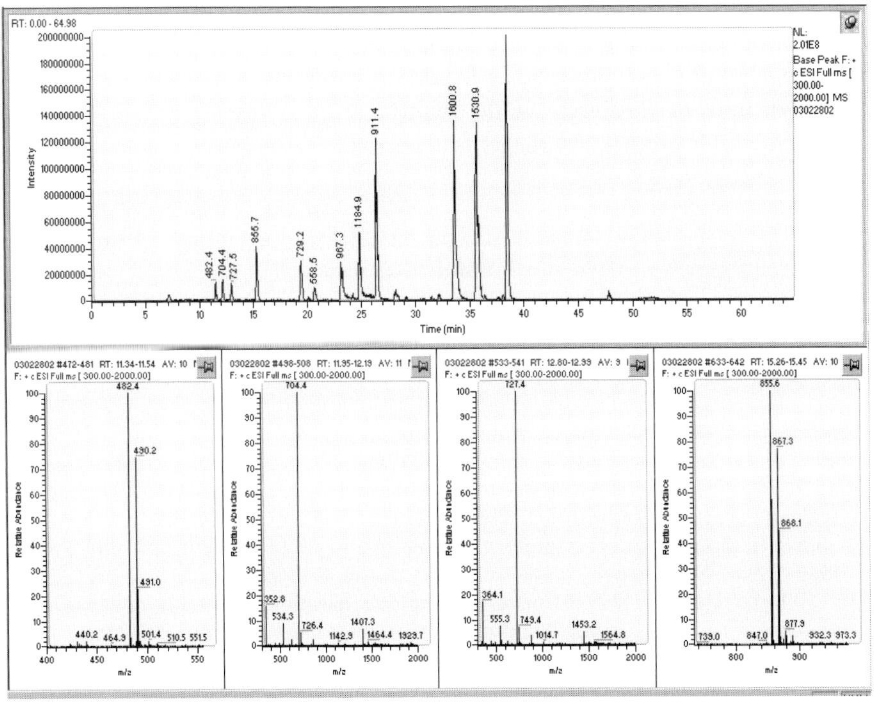

Fig. 25.16. **A** Total ion chromatogram (TIC) of a mixture of N^{15}-subtilisin and a variant subtilisin. **C** Mutant peptide; **D** parent peptide; **B, E** peptide peaks form doublets; All peptides show N^{14} and N^{15} masses unless they are mutated, and the peak area ratio is a measure for quantities

tein concentration between the wild type (N^{14}) and labeled (N^{15}) subtilisin. Moreover, the N^{14}/N^{15} isotopic labeling technique can be used to identify point mutation in the protein. As shown in Fig. 25.15, the identical RT (6.54 min) and its mass difference ($\Delta m=8$) for the tryptic peptide T_{12} doublet indicating there was no mutation occurred in the peptide. Figure 25.16 shows a similar analysis on the N^{15}-subtilisin and a variant subtilisin mixture. Most peptides present their N^{14} and N^{15} doublet masses (B & E). However, Fig. 25.16 (C & D) shows that peptides with significantly different area ratios or no detection of peptide doublet may indicate that a mutation has occurred. Finally, our data-dependent MS/MS can also provide qualitative confirmation and may indicate the identity of such a mutation.

25.4.2 Conclusion

N^{15} stable isotope-labeled protein is an excellent internal standard for the quality control and quantification of proteins. This quantification strategy offers several advantages over other labeling techniques: (1) There are no limitations regarding the presence of specific residues (e.g. cysteine in ICAT),because every amino acid residue contains nitrogen and complete coverage of proteins is possible. (2) The label does not change peptide chromatographic behavior. (3) The label does not complicate MS/MS identification by introducing label fragment ions. (4) Proteome-wide quantification is readily possible. (5) No requirement for proprietary reagents is needed.

Acknowledgments. We are indebted to Shauna Bowden, Grant Ganshaw, Christian Paech for providing samples, analyses and helpful discussions.

References

Börnsen KO (2000) Influence of salts, buffers, detergents, solvents, and matrices on MALDI-MS protein analysis in complex mixtures. Meth. Mol. Biol. 146, 387–404

Bull HB, and Breese K (1974) Surface tension of amino acid solutions: a hydrophobicity scale of the amino acid residues. Arch. Biochem. Biophys 161, 665–670

Chapman JR (2000) Mass spectrometry of proteins and peptides. Meth. Mol. Biol. 146. Hu mana Pres s, New Jersey

Dai Y, Whittal RM, Li L (1999) Two-Layer Sample Preparation: A method for MALDI-MS analysis of complex peptide and protein mixtures. Anal. Chem. 71, 1087–1091

Desiderio DM (1998) Quantitative analysis of methionine enkephalin and beta-endorphin in the pituitary by liquid secondary ion mass spectrometry and tandem mass spectrometry. J. Chromatog. A. 794, 85–96

Dewey RS, Liesch JM, Williams HR, Sugg EE, Dolan CA, Davies P, Mumford RA, Albers-Schonberg G (1992) Purification and characterization by fast-atom-bombardment mass spectrometry of the polymorphonuclear-leucocyte-elastase-generated A alpha

(1–21) fragment of fibrinogen from human blood after incubation with calcium ionophore A23187. Biochem J. 281 (pt 2), 519–24

Dwek RA, Edge CJ, Harvey, DJ, Wormald MR., Parekh RB (1993) Analysis of glycoprotein-associated oligosaccharides. Annu. Rev. Biochem. 62:65–100

Fountoulakis M. and Langen H (1997) Identification of proteins by matrix-assisted laser desorption ionization-mass spectrometry following in-gel digestion in low-salt, nonvolatile buffer and simplified peptide recovery. Anal. Biochem. 250, 153–156

Fryksdale BG, Jedrzejewski PT, Wong DL, Gaertner AL, Miller BS (2002) Impact of deglycosylation methods on two-dimensional gel electrophoresis and matrix assisted laser desorption/ionization-time of flight-mass spectrometry for proteomic analysis. Electrophoresis 23:2184–2193

Gobom J, Schuerenberg M, Mueller M, Theiss D, Lehrach H, Nordhoff E (2001) a-cyano-4-hydroxycinnamic acid affinity sample preparation. A protocol for MALDI-MS peptide analysis in proteomics. Anal. Chem. 73, 434–438

Grigorieff N, Ceska TA, Downing KH, Baldwin JM, Henderson R (1996) Electron-crystallographic refinement of the structure of bacteriorhodopsin. J. Mol. Biol. 259, 393–421

Gygi SP, Rist B, Gerber SA, Turecek F, Gelb MH, Aebersold R (1999) Quantitative analysis of complex protein mixtures using isotope-coded affinity tags. Nat Biotechnol. 17:994–9

Horvath ZS, Corthals GL, Wrigley CW, Margolis J (1994) Multifunctional apparatus for electrokinetic processing of proteins. Electrophoresis 15:968–971

Karas M. and Hillenkamp F (1988) Laser desorption ionization of proteins with molecular masses exceeding 10,000 daltons. Anal. Chem. 60, 2299–2301

Kussmann M. and Roepstorff P (2000) Sample preparation techniques for peptides and proteins analyzed by MALDI-MS. Meth. Mol. Biol. 146, 405–424

Lappalainen A, Siika-Aho M, Kalkkinen N, Fagerstrom R, Tenkanen M (2000) Endoxylanase II from Trichoderma reesei has several isoforms with different isoelectric points. Biotechnol. Appl. Biochem. 31:61–68

Locke VL, Gibson TS, Thomas TM, Corthals GL, Rylatt DB (2002) Gradiflow as a prefractionation tool for two-dimensional electrophoresis. Proteomics 2:1254–1260

Maley F, Trimble RB, Tarentino AL, Plummer TH (1989) Characterization of glycoproteins and their associated oligosaccharides through the use of endoglycosidases. Anal. Biochem. 180:195–204

Pang L, Fryksdale BG, Chow C, Wong DL, Gaertner AL, Miller BS (2003) Impact of prefractionation on two-dimensional gel electrophoresis and protein identification by MALDI-TOF mass spectrometry. Electrophoresis 24:3484–3492

Quadroni M, and James P (1999) Proteomics and automation. Electrophoresis 20, 664–677

Rosenfeld J, Capdevielle J, Guillemot JC, Ferrara P (1992) In-gel digestion of proteins for internal sequence analysis after one- or two-dimensional gel electrophoresis. Anal. Biochem. 203, 173–179

Rylatt DB, Napoli M, Ogle D, Gilbert A, Lim S, Nair C H (1999) Electrophoretic transfer of proteins across polyacrylamide membranes. J. Chromatogr. A 865:145–153

Shaw G (1993) Rapid identification of proteins. Proc. Natl. Acad. Sci. USA 90, 5138–5142

Sojar H.T, Bahl, OP (1987) A chemical method for the deglycosylation of proteins. Arch. Biochem. Biophys. 259:52–57

Speicher K, Kolbas O, Harper S, Speicher DW (2000) Systematic analysis of peptide recoveries from in-gel digestions for protein identifications in proteome studies. J. Biomol. Tech. 11, 74–86

Stocklin R, Arrighi JF, Hoang-Van K, Vu L, Cerini F, Gilles N, Genet R, Markussen J, Offord RE, Rose K (2000) Positive and negative labeling of human proinsulin, insulin, and C-peptide with stable isotopes. New tools for in vivo pharmacokinetic and metabolic studies. Meth. Mol. Biol. 146, 293–315

Subject Index